AN INTRODUCTION TO DIFFERENTIAL EQUATIONS

Deterministic Modeling, Methods and Analysis

Volume

1

AN INTRODUCTION TO DIFFERENTIAL EQUATIONS

Deterministic Modeling, Methods and Analysis

Volume 1

Anil G Ladde
Chesapeake Capital Corporation, USA

G S Ladde
University of South Florida, USA

NEW JERSEY • LONDON • SINGAPORE • BEIJING • SHANGHAI • HONG KONG • TAIPEI • CHENNAI

Published by

World Scientific Publishing Co. Pte. Ltd.
5 Toh Tuck Link, Singapore 596224
USA office: 27 Warren Street, Suite 401-402, Hackensack, NJ 07601
UK office: 57 Shelton Street, Covent Garden, London WC2H 9HE

British Library Cataloguing-in-Publication Data
A catalogue record for this book is available from the British Library.

AN INTRODUCTION TO DIFFERENTIAL EQUATIONS
Deterministic Modeling, Methods and Analysis
(Volume 1)

ISBN-13 978-981-4368-89-6
ISBN-10 981-4368-89-X
ISBN-13 978-981-4368-90-2 (pbk)
ISBN-10 981-4368-90-3 (pbk)

Typeset by Stallion Press
Email: enquiries@stallionpress.com

Printed in Singapore.

Acknowledgement and Dedication

We dedicate this volume to the forces behind the cause and development of the topics and ideals contained in this book. This book also is dedicated to the memory of the loving and supportive parents of the second author. Although they never had a formal education, they dedicated their lives to the growth and development of their children so that each child could make the most out of any opportunities that might come their way.

I also would like to dedicate this work to various people in the village where I lived. This includes my elementary and middle school mathematics teachers. In particular, I would like to highlight my 5th and 6th grade teacher, the late Mr. Madhav P Atre, for intellectually challenging and preparing me to be an independent problem solver. This intellectual challenge coupled with my academic nurturing was furthered by my professors during my undergraduate and graduate studies. These include Dr. C. G. Capse (People's College, Nanded), Dr A. A. Kayande, Dr. S. G. Deo, and Dr. V. Lakshmikantham (Marathwada University, Aurangabad). Moreover, my classmates and friends, Professors J. U. Chamergore, D. Y. Kasture, B. G. Pachpatte and Gopal R. Shendge, have supported me unconditionally in my growth.

I truly appreciate the endless and unconditional love, patience, and sacrifice by my wife, Sushila, and my children, Anil, Jay and Nathan. They had their needs, but they understood the importance of my academic and research efforts as I spent long hours with my students and my mathematical science research efforts. Ultimately, without my family's encouragement, in particular, Anil's comment about my writing an undergraduate level book, this work would not exist. Furthermore, I only chose to act on his suggestion when he agreed to assist me with the effort. Going back to my wife, I want to extend my deepest gratitude to her for constantly asking me on the status of our work. She greeted my request to proof-read our book with a smile. My other sons, Jay and Nathan, also reviewed parts of this work, and their comments proved to be valuable.

Early in my teaching career, I was very fortunate to be associated with a wonderful educator and individual, Professor Clarence F. Stephens at the SUNY-Potsdam. Not only he was a tremendous administrator and educator, but he and his wife, Mrs. Harriet Stephens, were also a great people and cherished friends. Professor

Stephens' insights and leadership at SUNY-Potsdam's math department helped to create a fruitful academic atmosphere for the research and teaching.

I am also grateful for the continuous encouragement and support from the following colleagues: Professors: N. U. Ahmed, Edward Allen, Armando Arciniega, Stephen R. Bernfeld, Mou-Hsiung Chang, Vasily Cateforis, Jagdish Chandra, P.-L. Chow, Kumer Das, D. Kannan, Andrzej Korzeniowski, Negash Medhin, M. Sambandham, S. Sathananthan, A. S. Vatsala, Drs: Mohmoud J. Anabtawi, Janusz Golec, Byron L. Griffin, Roger Dale Kirby, Bonita A. Lawrence, Srinivasa G. Rajalakshmi, Ongard Sirisaengtaksin, Michael S. Smith and Ling Wu. A special thanks goes to Dr. Ling Wu for contributing many of the figures that are noted in this book. In addition, I appreciate the support from my current PhD students: Bong-Jin Choi, Arnut Paothong, Jean-Claude Pedjeu, Olusegun Otunuga, Daniel Siu, Wanduku Divine, and Tadesse Zerihun, In particular, Bong-Jin Choi also helped with many of the figures in this book.

We greatly appreciate the World Scientific Publishers for their kind assistance and confidence in our efforts. In particular, I would like to recognize Ms. He You, Editor-In-House, as she is a very kind, efficient and effective editor.

Over the past several decades (30 years), the US Army Research Office has supported my research work. Many aspects of these research projects went into the development of this book.

Preface

Differential equations are central to the present and future multidisciplinary research on dynamic processes. From personal experience and from speaking to many colleagues, we know that many students frequently ask the following questions:

- Why do I need this course (if required)?
- How can I use it?
- How do we get started?
- Do we need to memorize the formula?
- How can one organize and utilize the given conditions or assumptions or data to solve a given problem?

Unfortunately, the existing books do not provide answers to these questions. The first volume of our book focuses on the deterministic approach, while the second volume focuses on the stochastic approach. Both volumes offer innovations and style in the presentation of topics, methods, and concepts, with adequate preparation for deterministic calculus. In the following, we will outline the basic features of Vol. I: deterministic modeling, methods, and analysis.

1 Deterministic Modeling

This volume begins with an outline about developing mathematical models of dynamic processes under controlled conditions or environments. The mathematical modeling approach is based on theoretical procedures coupled with the basic laws or principles of biology, chemistry, economics, pharmacokinetics, physics, physiology, sociology, etc. No attempt is made to teach all disciplines; however, the aim is to reach out to readers with interdisciplinary background and interests. Modeling of dynamic processes in various disciplines (such as the biological, chemical, engineering, medical, physical, or social sciences) is also based on the conceptual and computational understandings of differential equations. This understanding allows or motivates one to study the advanced topics in differential equations and dynamic modeling. Unfortunately, the importance of this link is not always made clear to the students, and the material learned in the current differential equations' course will not properly prepare a student to take these advanced classes.

2 Methods

This volume incorporates the important cutting-edge research ideas and techniques. There are only three basic methods of solving ordinary differential equations:

(i) Eigenvalue-type method;
(ii) Method of variation of constant parameters;
(iii) Energy/Lyapunov function method.

This unified approach to solving deterministic ordinary differential equations was initially motivated by the development of the methods of solving Itô–Doob-type stochastic differential equations (Vol. 2). In short, this volume serves as a research tool for developing various methods of solving a larger class of ordinary as well as other types of differential equations.

3 Analysis

Unlike most comparable books, this volume emphasizes the importance of using numerous mechanical and conceptual understandings in the problem-solving process. Furthermore, it minimizes the gap between the conceptual and the computational understandings. In fact, the volume systematically exhibits the competitive–cooperative interactions between the conceptual and the computational understanding in the problem-solving process at every level of skill/knowledge. For instance, the presented methods of solving differential equations generate the conceptual ideas as well as provide a basis for raising and addressing several important questions in the modeling of dynamic processes in the biological, chemical, engineering, medical, physical, and social sciences. Examples are:

- Are linear differential equations adequate mathematical models of dynamic processes?
- Prior to finding a solution to a differential equation, how do we know that it has a solution?
- Are there more than one solutions to a given differential equation?
- Are the methods of integration sufficient for finding a closed form solution to each differential equation?
- Are the procedures for solving differential equations applicable to any other types of differential equations?
- Do we really need a closed form solution to a differential equation?
- In the absence of a closed form solution to a differential equation, is the modeling of dynamic processes necessary?
- Without a closed form solution, is there any qualitative knowledge about a solution process that would shed light on the dynamic processes?
- Is it feasible to study the qualitative properties of the solutions to differential equations without their closed-form knowledge?

- Are there any basic universal qualitative properties of solutions to differential equations?

Assuming a basic understanding of the topic, this volume provides answers to questions 1–3 and, particularly, questions 8–10.

4 Presentation Style

From our experience and from speaking with various colleagues, we know that students that take the entry-level differential equations course spend more time on the mechanics than on understanding the concepts. So the question is: How do we teach them to focus on the concepts? They need to be introduced to a different approach to facilitate the learning process, so that they will increase their chances of better understanding advanced differential equations. The approach used by the existing textbooks does not incorporate a process that focuses on a "break down the problem" type of approach. Many tend to be more mechanical or tend to limit the use of citations or fundamental reasons. Also, they do not do a great job of creating a connection to the student's academic or professional background. It has been observed that conceptual skills coupled with algebraic ideas help to create a fundamental conceptual framework. This framework coupled with an analytic argument helps to design solutions to complex problems. Most graduate students lack this type of understanding.

5 Pedagogical Approach

Every section of this volume includes a variety of tools for the reader to better understand the concepts. Such tools or approaches include:

Teaching by example. It is often easier to understand a concept or meaning if we can provide examples of its use in academia as well as the "real world." This better illustrates the global use of many concepts.

Heavy use of step-by-step procedures for problem-solving (critical thinking). This volume focuses on the logical steps to arrive at the solution. This not only allows the reader to get used to the style of the volume but also enables the approach to be taught or enhanced. Once mastered, this approach can be used in other areas outside of this discipline, and this reduces the need to memorize a routine or formula. To encourage its use, the volume notes the challenge, fun, and excitement of this approach.

Learning the language of math. The reader is introduced to various math terms which are used regularly throughout the book. A person not focused on math will be given ample exposure to learning the math language which can be used in situations outside of math. Various examples of such situations are provided throughout the book.

Observations. To nurture curiosity and to minimize the gap between the conceptual and computational understandings, observations are noted at appropriate locations. These observations illustrate: (i) the role and scope of the end product of a procedure or theoretical algorithm (theorems, lemmas, or corollaries) and (ii) its function in the problem-solving process.

Frequent recaps. Our experiences have proven that it is better to have frequent recap than to have one at the end of a chapter. Given this, we offer brief recaps on the lessons learned at several points in a chapter. We also include some examples to allow the reader to better grasp the material.

Flowcharts. To facilitate the problem-solving process, the basic steps used in each procedure are summarized by a flowchart.

Exercises. We have included many exercises to allow students to practice the learned concepts. This also allows them to utilize and master the step-by-step procedure for solving problems. Separately, many of the exercises focus on nonmath disciplines like biology, chemistry, medicine, physics, and the social sciences. This better enables the reader to see their diversity.

6 Organizational Style

The volume is composed of six chapters. Chapter 1 begins by outlining the basic elements of the problem-solving process. The remainder of this chapter introduces the relevant supplementary background material and summarizes learned concepts from linear algebra and deterministic calculus. Except for Section 1.2, the chapter serves as a guide or reference for the instructor and a reference for its readers. The topics covered in Chapters 2–5 can be helpful in an entry-level course on ordinary differential equations or deterministic mathematical modeling, or an applied mathematics course. Chapter 6 offers a brief synopsis of the current trends in the areas of modeling of dynamic processes, and this can be used for undergraduate/graduate research projects.

Chapter 2 lays the groundwork for finding the solution process of linear differential equations. For this purpose, we note that learning and teaching mathematical sciences depends highly on one's past knowledge, skills as well as our understanding of the problem-solving process in Section 1.2. Using this, we develop a procedure termed the "eigenvalue-type method." Fundamentally, this is composed of the following ideas: (a) seeking a suitable form of a solution to a given linear homogeneous scalar differential equation that can be reduced to a linear algebraic equation, and then (b) finding a solution to the linear algebraic equation in (a). This approach is different from the existing textbook approach. We then introduce the fundamental, general, and particular solutions to linear scalar differential equations. The problem of solving linear nonhomogeneous scalar differential equations is analogous to the problem of solving linear nonhomogeneous scalar algebraic equations. Therefore, the conceptual ideas of finding a general solution to linear nonhomogeneous scalar differential equations are described by the following:

(i) Finding a general solution to the first-order linear homogeneous scalar differential equation corresponding to a linear nonhomogeneous differential equation;
(ii) Finding a particular solution to the given linear nonhomogeneous scalar differential equation;
(iii) Setting up a candidate for testing the general solution to the desired differential equation.

Then, this introduces the method of variation of constant parameters, for finding a particular solution to the differential equation. This approach is based on the general solution to linear homogeneous scalar differential equations. By observing the computational procedures and their limitations, the developed procedures (Sections 2.3–2.5) are utilized to develop concepts and theoretical algorithms in Section 2.6 (computational procedures generate concepts and algorithms). In addition, these algorithms (theorems, lemmas, or corollaries) are employed to investigate the fundamental properties of the solution processes of linear scalar differential equations.

Chapter 3 offers a foundation for finding solutions to nonlinear scalar differential equations knowing that they have a solution. Unlike comparable books, this volume introduces an innovative and unified method of solving nonlinear scalar differential equations. This is called the "energy/Lyapunov function method." The four basic ideas are:

(i) Seeking an unknown energy function (nonlinear);
(ii) Associating a simpler conceptual differential equation with an unknown energy function and the original nonlinear differential equation;
(iii) Determining an energy function and rate coefficients of simpler differential equations in the context of a simpler differential equation and the original nonlinear differential equation;
(iv) Finding a representation of a solution to the original differential equation in the context of the energy function and the solution process of a simpler differential equation.

Based on the knowledge gained in Chapter 2, simpler differential equations are:

(i) Directly integrable differential equations;
(ii) First-order linear scalar differential equations;
(iii) Nonlinear solvable scalar differential equations.

Furthermore, the scope of this method is displayed by showing that the variable separable differential equations, the homogeneous differential equations, and essentially time-invariant differential equations can be reduced to integrable differential equations (Section 3.4). Also, the Bernoulli-type differential equations can be reduced to linear scalar differential equations (Section 3.5). In addition, this method solves nonlinear scalar differential equations that cannot be solved by using any of the methods that exist in the current textbooks. In short, the energy/Lyapunov function method opens a new avenue beyond any methods outlined in any current textbook to enlarge the class of finding solutions to differential equations (in either explicit or implicit form). This highlights the role and scope of this method. Due to the limited background of readers, the conceptual aspects of nonlinear differential equations are separately addressed in Chapter 6.

Using linear algebra and matrices, the methods of solving scalar differential equations are directly extended to first-order linear systems of differential equations with constant coefficients in Chapter 4. The problem of solving linear systems of differential equations is not always feasible. This difficulty forces one to consider an alternative way to address this problem. Naturally, the conceptual framework paralleled to the one addressed in Section 2.6 is the most suitable approach. This is addressed in Section 4.7. This further encourages or motivates a student to look beyond the introductory course in ordinary differential equations.

By introducing a change to the dependent variable, higher order linear scalar differential equations reduce to systems of linear differential equations in Chapter 5. Then, techniques developed for solving and analyzing linear systems of differential equations in Chapter 4 are applied to this transformed system of linear differential equations. Furthermore, by knowing the concept and the nature of the Laplace transform, the problem of solving differential equations again reduces to the problem of solving algebraic equations, which leads to a solution to initial value problems.

The aim of modeling is not only to compute a solution process but also to provide qualitative information about a dynamic process and an input to design a dynamic process to meet the specified goals of systems. For this purpose, the specific knowledge of a solution is not very important. Historically, by knowing the fundamental properties of solutions to differential equations, the method of variation of parameters and Lyapunov's second method have played a very significant role in the past century (a brief outline of qualitative properties is presented). Furthermore, dynamic processes operating under more than one timescale, with or without the presence of random perturbations and with or without hereditary effects, are topics of current research. These topics are briefly highlighted in Chapter 6.

This volume's use is intended for the following:

- An interdisciplinary undergraduate/graduate-level course in mathematical modeling or applied mathematics (usually deterministic for biology, chemistry, business, engineering, economics, mathematics, medical sciences, physics, psychology, social sciences, etc.;
- Juniors and seniors majoring in the mathematical, engineering, physical, or material sciences;

- An additional readership including interdisciplinary professionals (in particular, young ones), since this volume would be a good reference book for researchers as a do-it-yourself type of manual;
- A good reference book for undergraduate/graduate students conducting research to develop mathematical models and new methods of solving a larger, a more general class of differential equations.

Anil G. Ladde
G. S. Ladde

Contents

Chapter 1

Elements of Matrices, Determinants, and Calculus

1.1 Introduction

This chapter serves as a review. It begins by highlighting a few ideas about the process of solving mathematics/education/research problems, and then covers relevant mathematical concepts and statements in linear algebra and the calculus of matrix functions. In particular, the algebra of matrices, the properties of determinants, some concepts about vector spaces, sets of independent and dependent vectors, methods for solving systems of linear-algebraic equations, and Wronskians of functions are discussed in Sections 1.3 and 1.4. Moreover, differential calculus of determinant functions, particularly a generalized mean-value theorem and Taylor's formula for determinant functions, are developed in Section 1.5. These results play a very important role in finding solutions to linear differential equations, especially stochastic differential equations. Note that we are not attempting to teach linear algebra and multivariate calculus. This chapter not only serves as a reference guide for the reader, but it also is teaching guide for the instructor.

1.2 Problem-Solving Process

One of the most important goals of education is to gain knowledge about the entire universe, and to apply it for the benefit of living things on the Earth. Its purpose is also to gain knowledge of the problem-solving process to eradicate or understand our ignorance. The presented knowledge about the problem-solving process can easily be applied in any area of science: biological, chemical, engineering, mathematical, medical, physical, political, social, etc.

Currently, it is well recognized that the knowledge and practices employed in the mathematical-problem-solving processes provide a suitable background and the tools for solving problems arising in any discipline. In the 21st century, the question that is important to all of us is how to gain knowledge of the complex system (the whole) with some information about its subsystems (constituent parts). Scientists (biological, chemical, engineering, mathematical, medical, physical, political, social, etc.) have been trying, are still trying, or will be trying to obtain answers to particular problems which have existed since the beginning of the human race.

The goal of this work is to develop and foster the problem-solving process, and then to bring it to the 21st century classroom. By looking at the problem-solving process as a puzzle-solving or game-playing process, we make a serious effort to provide a detailed procedure for understanding, developing, and applying the proposed problem-solving process in a systematic and unified way. In this context, the concept of the decomposition–aggregation problem-solving process is introduced. Furthermore, the decomposition–aggregation problem-solving strategy is outlined in a systematic way.

We begin with a question: *What do we mean by a problem-solving process?* The goal of this work is to provide the answer to this question. For this purpose, let us introduce a few terms that will be used throughout this section:

Definition 1.2.1. A process is a sequence of actions and/or behaviors that constitute a system with a goal-producing function.

Example 1.2.1. Dynamic process: (1) cell growth; (2) substance decay.

Definition 1.2.2. A system is a regularly interacting or interrelated set of elements or items forming a unified whole.

Example 1.2.2. System: (1) a multiple market system; (2) the planetary system.

Definition 1.2.3. A problem-solving process is a sequence of actions, reactions or behaviors that constitute a system with a problem-solving goal(s).

Example 1.2.3. Problem-solving process: (1) traveling to Houston; (2) achieving energy independence.

The problem-solving process is a question-and-answer process until it reaches its goal. Questions are composed (consciously or subconsciously), and are answered by the problem-solver. Roughly speaking, one can call this stage of the problem-solving process the "trial-and-error process." To be an effective problem-solver, it is essential that one understands and masters the mechanisms of the problem-solving process.

Problem analysis and solution strategies are considered to be the two major parts of the problem-solving process. These two components will be discussed in and an example will be given to illustrate this process.

1.2.1 *Problem analysis*

Problem analysis is a very careful study of all aspects of the given problem. The components of this process are (i) reading and understanding, (ii) listing concepts and assumptions, and (iii) determining the goal(s) of the problem. This aspect takes up about 50% of the effort of a problem-solver. The effort of the problem-solver can be distributed among the following three steps:

Step 1. Read and understand the problem. One must make sure that one knows and understands each and every word or term in the statement(s) of the given problem. Simply put, one needs to fully understand the problem.

Step 2. List all terms, definitions, operations, etc.. In this step, one needs to make a list of all the terms, definitions, operations, etc. that appear in the given problem. One must make sure that one understands these definitions, operations, etc. In other words, the problem-solver needs to completely understand the underlining mathematical definitions, operations, the mathematical terms in the language, etc.

Step 3. Determine the goal(s) of the problem. In the process of reading and understanding the problem, one needs to make a note of the goal(s) of the problem. Sometimes, it is helpful for the problem-solver to rewrite the goal(s) of the problem in an equivalent form(s). This will assist in forming a strategy for solving the problem in the most effective and efficient way.

1.2.2 *Solution (proof) strategies*

The solution strategies are the final act of the solution process and is the core of the process. Everyone has equal potential to undertake the task of solving a problem. The only prerequisite is that one channels one's energy into the problem-solving process. Since all of us like music, games, and puzzles, why not 'mathematics'? Of course, the answer to this is left to the individual reader. Now, we are ready to provide a brief outline about potential solution strategies. Note that this summary of strategies is not exhaustive.

Step 4. In the context of steps 2 and 3, one needs to make an exhaustive search for all possible meanings of both the mathematical and linguistic terms, their derivatives, relatives, or opposites, etc. In addition, in the context of the given statement(s) in the problem, one needs to recall all of the mathematical statements or axioms concerning the terms, definitions, operations, etc.

Step 5. From steps 2–4, one formulates the strategies for solving the given problem. As noted at the beginning of this section, the strategy formulation process depends on the individual problem-solver. For a given problem, one may not be able to solve it by using a single-shot strategy. In this case, the problem-solver needs to formulate a complex strategy for solving a complex or difficult problem. This complex strategy consists of the following substrategies: (i) a substrategy(s) to decompose a problem into a finite number of subproblems, (ii) a finite number of solutions substrategies that corresponds to the finite number of subproblems described in (i), and (iii) a substrategy to connect the substrategies in (ii) to the subproblems in (i) to form the solution to the overall problem. We call this problem-solving strategy the decomposition–aggregation problem-solving strategy (DAPSS). This concept is an extension of the concept of the decomposition–aggregation method which has been successfully employed in systems analysis research [126]. This is the most crucial step in the problem-solving process.

Step 6. The strategies formulated in step 5 are executed in this step. This step may consist of a finite series of subproblem-solving processes. Using the strategies formulated in step 5, a given problem can be decomposed into a finite set of simpler subproblems that are easy to solve. Employing the substrategies of these simpler subproblems, the subproblems are solved. Finally, using the strategy developed in step 5, the subsolutions to these subproblems of the given problem are linked to the solution of the original complex problem. Again, this problem-solving process is called the decomposition–aggregation problem-solving process (DAPSP).

A Few General Problem-Solving Tips

Let us shed light on problem analysis (reading and understanding a given problem) and solution strategies (procedures). The presentation of these topics will be in very general and broad-based terms. A particular problem may require very creative ideas in formulating the solution strategy in the context of the following ideas. The presented ideas on the above mentioned topics are based on "LOGIC" [41, 43, 93].

Problem analysis

A written work in a natural language is made up of sentences. A sentence is a finite sequence of words with an appropriate order as per the grammar of the given language. It is obvious that one must understand the meaning of the words in order to understand the given problem. The language provides a number of ways in which sentences are combined or altered to form new compound sentences. The most common connectives between the basic sentences are "AND," "OR," and their derivatives. The word "NOT" and its equivalent derivatives also occur more frequently. The equivalent usage, and its notations or symbols in the mathematical world, are:

(1) "AND" = CONJUNCTION/INTERSECTION = $\wedge/\cap$;
(2) "OR" = DISJUNCTION/UNION = $\vee/\cup$;
(3) "NOT" = NEGATION = $\neg/\neq$.

Example 1.2.4. Let "p" and "q" stand for two assertions, say, "N is a set of natural numbers" and "Z is a set of integers," respectively. Find: (1) p and q, (2) p or q, and (3) not p.

Solution process. (1) "p and q" ($p \cap q$) is N, (2) "p or q" ($p \cup q$) is Z; (3) "not p" ($\neg p$) is "not natural numbers."

In the mathematical world, assertions are made in the form of sentences.

Definition 1.2.4. A statement/proposition is a declarative sentence that is either true or false.

Example 1.2.5. Statement/Proposition: (a) It is raining; (b) The union of any number of open sets is open; (c) 3 is divisible by 2.

The language also provides a number of ways in which statements/propositions are combined or altered by connectives other than "AND," "OR," "NOT" to form new compound statements/propositions. The most common connective of this type in mathematics is:

"IMPLICATION" means
"A CONDITIONAL/IF–THEN STATEMENT EXISTS." (1.1)

It is denoted by "$\Rightarrow$."

Example 1.2.6. Find the mathematical translations of: "If it is raining, then I will not be playing."

Solution process. This example can be described as follows. Let us denote "p" and "q" by "It is raining" and "I will not be playing," respectively. Then the above empirical statement can be translated into any one of the following logically equivalent statements: (1) "If p, then p"; (2) "p implies q"; (3) "p is necessary for q"; (4) "q is sufficient for p"; (5) "$p \Rightarrow q$:"

"EQUIVALENCE" means
"IFF/NECESSARY AND SUFFICIENT." (1.2)

It is denoted by "$\Leftrightarrow/\leftrightarrow$."

Example 1.2.7. Provide the mathematical translations of: "A number is even if and only if (iff) it is divisible by 2."

Solution process. This example can be described as follows. Let us denote "p" and "q" by "A number is even" and "It is divisible by 2," respectively. The logically equivalent forms of the above statement are: (1) "p is necessary and sufficient for q"; (2) "If p, then q" and if q, then p"; (3) "$p \Leftrightarrow q$"; (4) "$p \Rightarrow q$" and "$q \Rightarrow p$."

Definition 1.2.5. A tautology is an expression that is true for all of its possible propositional variables.

Example 1.2.8. A countable union of a countable sets is countable.

Definition 1.2.6. The propositional variables are the variables that are assigned to the propositions.

Example 1.2.9. "It is raining." We define a variable "p" which stands for the proposition "It is raining." In this case, "p" is called the propositional variable.

Definition 1.2.7. A theorem is a mathematical statement that can be shown to be true.

Example 1.2.10. If $A_1, A_2, \ldots$ are countable sets, then $\bigcup_{n=1}^{\infty} A_n$ is countable. (In words, the countable union of countable sets is countable.)

Definition 1.2.8. A lemma is a theorem that is not so interesting in itself, but is used to prove other theorems.

Definition 1.2.9. A corollary is a theorem, and may even be the main result that is an immediate consequence of the result just preceding it.

Example 1.2.11. "The set of rational numbers is countable." This statement is a corollary to the statement in Example 1.2.10 with respect to the definition of the term "theorem."

Remark 1.2.1. We note that the proposition is a mathematical result between lemma and theorem. These terms are subjective. It depends on the author of the mathematical result(s).

Remark 1.2.2. The concepts of proposition, lemma, theorem, or corollary can be called the theoretical problem-solving algorithms (TPSAs).

Definition 1.2.10. A professional function is a function defined on a set X into a set of propositions.

Example 1.2.12. "If x is a prime number, then x is odd."

Solution process. This statement generates a set of propositions, one for each $x \in N$ (a set of natural numbers). Here the value of propositional function $P(x)$ is "If x is a prime number, then x is odd."

Definition 1.2.11. A quantifier is a rule that assigns a single proposition to a propositional function.

There are two types of quantifiers that play a major role in the mathematical sciences.

Definition 1.2.12. An existential quantifier is a rule that assigns to a propositional function "P" the proposition "truth set of "P" not empty".

This rule is denoted by "$\exists$". The result applying the existential quantifier $\exists$ to a P is written as $\exists x\, P(x)$ and it is read as "There exists x such that P(x)".

Example 1.2.13. If P is a propositional function defined on $X = Z$ (a set of integers) by $P(x) =$ "$x > 5$". In this case the truth of P is not empty.

Definition 1.2.13. A universal quantifier is a rule that assigns to a propositional function "P" the proposition "The truth set of P is equal to its domain".

This rule is denoted by "$\forall$" and the result applying the universal quantifier $\forall$ to P is written as $\forall\, x\, P(x)$. This is read as "For all x, $P(x)$" or "for every x, $P(x)$".

Example 1.2.14. If P is a propositional function defined on $X = Z$ (Z is as defined earlier) by $P(x) =$ "$x + 1 = 3$".

Solution process. In this case, $\exists$ is true but $\forall$ is not. This is due to the fact that the statement is valid for $x = 2$, but not valid for all elements of Z. In fact, for $x = 0$, it is false.

Solution strategies

By noting step 6 and employing the collected information in step 4, one is ready to execute step 5 (strategies). These strategies are designed for the finite number of subproblems that are formulated based on the goal(s) of the given problem. In fact, these subproblems are really reactions of a problem-solver in the context of the gathered information about the problem. This is like Newton's third law of motion. Of course, the reactions are also based on an individual's abilities and the background. First, one needs to solve these subproblem(s), and then one has to connect the solutions to these subproblem(s) to form the solution to the overall problem.

In the following, we present a few standard problem-solving strategies, procedures, or methods that have historically been used by the mathematical community. One can use these techniques to solve the subproblems (simpler problems) that are formulated by the problem-solver to solve the given problem.

Definition 1.2.14. A way in which one argues about the validity of the proposition, lemma, theorem, or corollary is called a proof.

Now, we are ready to present the well-known strategies, procedures, or methods that are commonly used to prove the validity of the propositions.

Direct Method 1.2.1. *A direct proof in mathematics is a finite sequence of arguments that leads directly to the conclusion of the solution to the problem. In the following, we present a conceptual framework to illustrate the procedure.*

Illustration 1.2.1 (proposition). We suppose that we are given the following very general mathematical statement: $A \Rightarrow B$, where A is a given mathematical statement (hypothesis(hypotheses), conditions, or assumptions) and B is a goal (conclusion) of the proposition.

Proof strategy (direct method). In this type of proof, one starts with the given statement(s), hypotheses, conditions, or assumptions. One recalls all the possible information and uses ones background to deduce various consequences (subproblems and their solutions) (if necessary) from them or it. Finally, one ends up with the validity of the concluding statement B. This kind of thinking process is also called the deductive reasoning process.

Indirect Method 1.2.2. *An indirect proof in mathematics consists of two stages: (a) subproblem formulation and its solution — at this stage, we solve this subproblem other than the given problem; (b) connection strategy — we use logic to show that the given problem is actually solved. There are two common methods of indirect proof, namely (i) proof by the contrapositive method and (ii) proof by contradiction. We are ready to outline these methods. Again, an illustration is presented and its proof is outlined in a very general framework.*

Illustration 1.2.2 (proposition). We suppose that we are given the following mathematical statement: $A \Rightarrow B$, where A is the given mathematical statement(s)/hypothes(si)es and B is the concluding statement(s).

Proof strategy (contrapositive method). In this type of proof strategy, one formulates the contrapositive statement "$A \Rightarrow B$." It is described by "$\neg B \Rightarrow \neg A$." Now apply the same deductive reasoning strategy that is described in Illustration 1.2.1 to the above-formulated contrapositive mathematical proposition. Finally, one uses the logic to conclude the validity of the original proposition: $A \Rightarrow B$.

Illustration 1.2.3 (proposition). We suppose that we are given the following mathematical statement: $A \Rightarrow B$, where A and B are as defined before.

Proof strategy (contradiction method). One starts with A and $\neg B$ and uses the deductive reasoning. This reasoning leads to a contradiction (*reductio ad absurdum*) to validate the proposition.

Principle of Mathematical Induction Method 1.2.3. This method of proof is used to prove a mathematical proposition that involves the natural (positive) numbers. This kind of mathematical problem arises in all areas of the biological, chemical, engineering, medical, physical, and social sciences. When one conducts an experiment in the laboratory or fieldwork, one collects data. Of course, the amount of data is finite. This is due to the fact that there are limited resources and time. But the experimenter wants to draw a conclusion or inference based on a finite amount of data. These conclusions or inferences are useful for the decision-making processes or for understanding and formulating the laws or principles of nature or mathematical models in particular dynamic processes, such as, dynamic processes in biology, chemistry, engineering, medicine, physics, or the social sciences. The present state of the knowledge about nature justifies the above statement. Frequently, the conclusions are formed from any level of data. This fact is the essence of the mathematical induction method. We are ready to state a principle of mathematical induction.

Principle of Mathematical Induction (PMI). *Suppose that S is a subset of natural numbers such that: (i) S is nonempty, (ii) $1 \in S$, and (iii) if $k \in S$, then $k+1 \in S$. Then $S = N$, where N is a set of natural numbers.*

Proof of PMI. The proof of a proposition involves two steps
Step 1. One needs to show that $1 \in S$.

Step 2. One needs to show that if $k \in S$, then $k+1 \in S$. The statement "If $k \in S$, then $k+1 \in S$" is called the induction hypothesis/assumption.

Remark 1.2.3. In particular, the set S in the PMI is defined as $S = \{n \in N : P(n) \text{ is true}\}$, where P is some propositional function defined of N.

Equivalence Principle 1.2.4. *Let A, B, and C denote arbitrary mathematical propositions.*

(i) *If $A \Leftrightarrow A$ (reflexive statement);*
(ii) *If $A \Rightarrow B$, then $B \Rightarrow A$ (symmetric statement);*
(iii) *If $A \Rightarrow B$ and if $B \Rightarrow C$, then $A \Rightarrow C$ (transitive statement).*

Example 1.2.15. Let x, y, z be real numbers. Then the following statements are valid: (i) $x = x$; (ii) if $x = y$, then $y = x$; (iii) if $x = y$ and if $y = z$, then $x = z$.

Construction of Counterexample 1.2.5. If one feels that a given statement, in particular an implication statement, is false, then one needs to prove this feeling by constructing an example. This type of example is called a counterexample.

Strategy for construction of a counterexample. If we want to disprove an implication statement, we must set up a situation in which "the given hypothesis is true and the conclusion is false." This type of setting up of a situation to meet the above need is indeed a counterexample.

Illustration 1.2.4 (a ninth grade problem). In the following, we present a very simple mathematical problem to illustrate the description and mechanism of the problem-solving process. At this point it is expected (at the beginning level) that interested readers of mathematics, as well as problem-solvers, will try to practice the presented problem-solving skill. Of course, the process seems to be complex and time-consuming. Some mathematical problems are complex, and one needs knowledge and training. In this sense, this process provides the following: how to start and how to solve a given problem in a systematic way. Having said this, let us begin a simple problem whose solution process seems to be complex.

Problem Statement. At the pharmacy, Liz told the clerk, "If you give me double the money I have, I will spend \$4." The clerk agreed to this, if Liz would repeat the process twice. After the third time, Liz had no money left. What did she have to start with?

Solution process. Based on the discussion about the description and mechanism of the problem-solving process, the first step is to read the problem as many times as desired. This depends on the problem-solver. Assuming that the problem has been read, let us start with the second step.

Step 2. List all terms, definitions, operations, etc. Double the money. Spend \$4; repeat twice. After the third time, no money is left. Start with.

Step 3. Determine the goal(s) of the problem. The goal of the problem is "to find how much money Liz had when she entered the pharmacy."

Step 4. Strategy formulation

- Let x be the amount of money Liz has when she enters the pharmacy.
- If the clerk gives her double the money she has, then she will spend \$4. Therefore, at the end of the first purchase, she will have $2x - 4$ dollars.
- Repeat the above process twice.
- If the clerk gives her double the money she has at the end of the first purchase, then she will have $2(2x - 4)$ dollars.
- But she has agreed to spend \$4. Therefore, at the end of the second purchase, she will have $2(2x - 4) - 4$ dollars.
- If the clerk gives her double the money she has at the end of the second purchase, then she will have $2[2(2x - 4) - 4]$ dollars.
- But she has agreed to spend \$4. Therefore, at the end of the third purchase, she will have $2[2x(2x - 4) - 4] - 4$ dollars.
- It is known that at the end of the third purchase, Liz has no money left.

Subproblem formulation. Solve for x:

$$2[2(2x - 4) - 4] - 4 = 0.$$

Step 5. We need to solve the above-formulated subproblem. For this purpose, we repeat the six steps. In this case, steps 1–3 are obvious.

Step 4: Recalling, etc.

- Operations: addition and multiplication of real numbers;
- Distributive law for multiplication over addition;
- Symbols: ( ) and [ ].

Step 5: Subproblem strategy

- Simplify the algebraic numbers in ( );
- Use the distributive law for real numbers;
- Repeat the actions (if necessary).

Step 6: *Execution of strategy*

$$
\begin{aligned}
2[2(2x-4)-4]-4=0& \\
&= 2[4x-8-4]-5=0 \quad \text{(by the distributive law)} \\
&= 2[4x-12]-4=0 \quad \text{(by the simplification process)} \\
&= 8x-24-4=0 \quad \text{(by the distributive law)} \\
&= 8x-28=0 \quad \text{(by the simplification process).}
\end{aligned}
$$

Therefore, we have

$$8x = 28 \quad \text{(by adding 28 to both sides of the equation),}$$

and hence

$$x = \frac{28}{8} \quad \left(\text{by multiplying both sides by } \frac{1}{8}\right).$$

Thus

$$x = 3.50 \quad \left(\text{by approximating the fraction } \frac{28}{8}\right).$$

At the time Liz entered the pharmacy, she had \$3.50 (approximately). This completes the solution process for this problem.

Significance of the Presented Problem-Solving Process

(i) This problem-solving process provides a natural scheme for fostering and strengthening one's (1) style of solving a problem, (2) self-confidence and self-esteem, (3) self-security, and (4) independence.
(ii) It provides a kind of training in attempting to solve a problem.
(iii) It initiates, fosters, and strengthens individuals' natural abilities and freedom: actions and reactions to the problem.
(iv) The questioning process generates curiosity and creativity, and hence it stimulates an interest in learning and in eradiating one's ignorance about the subject matter under consideration.

1.3 Algebra of Matrices

Mathematical models of several dynamic processes in the physical, biological, and social sciences lead, naturally, to both homogeneous and nonhomogeneous types of systems of linear equations. They play an important role in these sciences. They not only simplify the mathematical description but also play the role of setting information out of the model. In this section, we shall present some basic concepts and results that will be utilized throughout this chapter and subsequent chapters.

We start with an algebraic background about the concept of matrices and its usage of various problem-solving processes. Let us first introduce the concept of a matrix.

Definition 1.3.1. Let m and n be any two positive integers. A rectangular array of numbers such as

$$A = \begin{bmatrix} a_{11} & a_{12} & \cdots & a_{1j} & \cdots & a_{1n} \\ a_{21} & a_{22} & \cdots & a_{2j} & \cdots & a_{2n} \\ \cdots & \cdots & \cdots & \cdots & \cdots & \cdots \\ a_{i1} & a_{i2} & \cdots & a_{ij} & \cdots & a_{in} \\ \cdots & \cdots & \cdots & \cdots & \cdots & \cdots \\ a_{m1} & a_{m2} & \cdots & a_{mj} & \cdots & a_{mn} \end{bmatrix} = (a_{ij})_{m\times n} \tag{1.3}$$

is called an $m \times n$ matrix. It has m rows,

$$A_1 = [a_{11}a_{12}\cdots a_{1j}\cdots a_{1n}], \quad A_2 = [a_{21}a_{22}\cdots a_{2j}\cdots a_{2n}], \ldots,$$
$$A_i = [a_{i1}a_{i2}\cdots a_{ij}\cdots a_{in}], \ldots, \quad A_m = [a_{m1}a_{m2}\cdots a_{mj}\cdots a_{mn}],$$

and n columns,

$$A^1 = \begin{bmatrix} a_{11} \\ a_{21} \\ \cdots \\ a_{i1} \\ \cdots \\ a_{m1} \end{bmatrix}, \quad A^2 = \begin{bmatrix} a_{12} \\ a_{22} \\ \cdots \\ a_{i2} \\ \cdots \\ a_{m2} \end{bmatrix}, \ldots, \quad A^j = \begin{bmatrix} a_{1j} \\ a_{2j} \\ \cdots \\ a_{ij} \\ \cdots \\ a_{mj} \end{bmatrix}, \ldots, \quad A^n = \begin{bmatrix} a_{1n} \\ a_{2n} \\ \cdots \\ a_{in} \\ \cdots \\ a_{mn} \end{bmatrix}, \tag{1.4}$$

which are $1 \times n$ and $m \times 1$ matrices, respectively. They are also called $1 \times n$ and $m \times 1$ row and column vectors, respectively. The number (real or complex) a_{ij} is called the element/entry in the ith row and jth column of a matrix. The notation $m \times n$ is called the size of matrix A.

Observation 1.3.1. Matrix A as defined in (1.3) can also be represented by an $m \times 1$ column vector matrix as

$$A = \begin{bmatrix} A_1 \\ A_2 \\ \cdots \\ A_i \\ \cdots \\ A_m \end{bmatrix}, \tag{1.5}$$

and a $1 \times n$ row vector matrix by

$$A = [A^1 \quad A^2 \quad \cdots \quad A^j \quad \cdots \quad A^n]. \tag{1.6}$$

We further note that the entry a_{ij} in the ith row and jth column of the matrix A can also be called the ith and the jth component of the jth column vector A^j and the ith row vector A_i of the matrix A, respectively.

Example 1.3.1.

(a) You are given

$$A = \begin{bmatrix} 1 & 3 \\ 2 & 4 \end{bmatrix}.$$

This is a 2×2 matrix with two rows, $A_1 = [1\ 3]$, $A_2 = [2\ 4]$, and two columns,

$$A^1 = \begin{bmatrix} 1 \\ 2 \end{bmatrix}, \quad A^2 = \begin{bmatrix} 3 \\ 4 \end{bmatrix}.$$

Moreover, the matrix A can be represented as

$$A = \begin{bmatrix} A_1 \\ A_2 \end{bmatrix} = \begin{bmatrix} [1\ 3] \\ [2\ 4] \end{bmatrix}, \quad A = [A^1\ A^2] = \left[\begin{bmatrix} 1 \\ 2 \end{bmatrix} \quad \begin{bmatrix} 3 \\ 4 \end{bmatrix} \right],$$

whose elements are 1×2 row vectors or matrices and 2×1 column vectors or matrices, respectively.

(b) You are given

$$A = \begin{bmatrix} 1 & 0 & 2 & -1 \\ 3 & 4 & 1 & 5 \\ -1 & 5 & 4 & 2 \end{bmatrix}.$$

This is a 3×4 matrix with three rows, $A_1 = [1\ 0\ 2\ -1]$, $A_2 = [3\ 4\ 1\ 5]$, $A_3 = [-1\ 5\ 4\ 2]$, and four columns,

$$A^1 = \begin{bmatrix} 1 \\ 3 \\ -1 \end{bmatrix}, \quad A^2 = \begin{bmatrix} 0 \\ 4 \\ 5 \end{bmatrix}, \quad A^3 = \begin{bmatrix} 2 \\ 1 \\ 4 \end{bmatrix}, \quad A^4 = \begin{bmatrix} -1 \\ 5 \\ 2 \end{bmatrix}.$$

Moreover, the matrix A can be represented by

$$A = \begin{bmatrix} A_1 \\ A_2 \\ A_3 \end{bmatrix} = \begin{bmatrix} [1\ 0\ 2\ -1] \\ [3\ 4\ 1\ 5] \\ [-1\ 5\ 4\ 2] \end{bmatrix},$$

$$A = [A^1\ A^2\ A^3\ A^4] = \left[\begin{bmatrix} 1 \\ 3 \\ -1 \end{bmatrix} \quad \begin{bmatrix} 0 \\ 4 \\ 5 \end{bmatrix} \quad \begin{bmatrix} 2 \\ 1 \\ 4 \end{bmatrix} \quad \begin{bmatrix} -1 \\ 5 \\ 2 \end{bmatrix} \right],$$

whose elements are 1×4 row vectors or matrices and 3×1 column vectors or matrices, respectively.

Definition 1.3.2. An $m \times n$ matrix A is called a square matrix if $m = n$.

Example 1.3.2. The matrix in Example 1.3.1(a) is a square matrix, and the matrix in Example 1.2.1(b) is not.

Illustration 1.3.1. To provide the motivation for the definitions and the development of mathematical formulas, 1×1, 2×2, and 3×3 matrices are frequently used as the basic building blocks. The consideration of these matrices provides a pattern of ideas, and these ideas are used to formulate universal computational algorithms.

(i) For $n = m = 1$. Here, $A = [a_{11}]_{1\times 1}$. In this case, the row and column vector representation described in (1.5) and (1.6) are the same. Moreover, it is a number.

(ii) For $n = m = 2$. In this case,

$$A = \begin{bmatrix} a_{11} & a_{12} \\ a_{21} & a_{22} \end{bmatrix}.$$

Furthermore, the row and column vector representations described in (1.5) and (1.6) are

$$A = \begin{bmatrix} A_1 \\ A_2 \end{bmatrix}, \quad A = [A^1 \; A^2],$$

respectively.

(iii) For $n = m = 3$. Here,

$$A = \begin{bmatrix} a_{11} & a_{12} & a_{13} \\ a_{21} & a_{22} & a_{23} \\ a_{31} & a_{32} & a_{33} \end{bmatrix}.$$

In this case, the row and column vector representations described in (1.5) and (1.6) are

$$A = \begin{bmatrix} A_1 \\ A_2 \\ A_3 \end{bmatrix}, \quad A = [A^1 \; A^2 \; A^3],$$

respectively.

Definition 1.3.3. Let A be an $n \times n$ square matrix. A square matrix A is said to be a diagonal matrix if $a_{ij} = a_{ji} = 0$ for all $i \neq j$, where i, and j satisfy the relations $1 \leq i \leq n$ and $1 \leq j \leq n$, respectively, i.e. its off-diagonal entries are all zero. It is represented by $A = \text{diag}\,(d_1, d_2, \ldots, d_i, \ldots, d_n)$, where $a_{ii} = d_i$ for $1 \leq i \leq n$.

Example 1.3.3.

(a) You are given

$$A = \begin{bmatrix} -2 & 0 \\ 0 & -1 \end{bmatrix}, \quad B = \begin{bmatrix} 0 & 0 & 0 \\ 0 & 3 & 0 \\ 0 & 0 & 2 \end{bmatrix}.$$

The matrices A and B are square matrices of size 2×2 and 3×3, respectively. By the application of Definition 1.3.3, we conclude that they are diagonal matrices.

(b) We are given

$$A = \begin{bmatrix} 1 & 0 \\ 0 & 2 \\ 0 & 5 \end{bmatrix}.$$

The matrix is of size 3×2, but it is not a square matrix. By Definition 1.3.3, it is not a diagonal matrix.

Definition 1.3.4. A diagonal matrix $A = \text{diag}(d_1, d_2, \ldots, d_i, \ldots, d_n)$ is called an identity matrix if $d_i = 1$ for all i satisfying the relation $1 \leq i \leq n$. It is denoted by $A = I(n) \equiv I$.

Example 1.3.4. Let us consider

$$I(2) = \begin{bmatrix} 1 & 0 \\ 0 & 1 \end{bmatrix}, \quad I(3) = \begin{bmatrix} 1 & 0 & 0 \\ 0 & 1 & 0 \\ 0 & 0 & 1 \end{bmatrix}.$$

The matrices $I(2)$ and $I(3)$ are square matrices of size 2×2 and 3×3, respectively. From Definition 1.3.4, we conclude that they are identity matrices.

Definition 1.3.5. Let A be an $n \times n$ square matrix. A square matrix A is said to be an upper (lower) triangular matrix if $a_{ij} = 0$ for all $j < i$ $(i < j)$, and i and j satisfy the relations $1 \leq i \leq n$ and $1 \leq j \leq n$, respectively, i.e. its entries below (above) the main diagonal matrix are zeros.

Example 1.3.5. You are given

$$A = \begin{bmatrix} 1 & 1 \\ 0 & 1 \end{bmatrix}, \quad B = \begin{bmatrix} -1 & 0 & 0 \\ 3 & -2 & 0 \\ 0 & 4 & -2 \end{bmatrix}.$$

The matrices A and B are square matrices of size 2×2 and 3×3, respectively. From the application of Definition 1.3.5, we conclude that A and B are upper and lower triangular matrices, respectively.

Definition 1.3.6. Let $A = (a_{ij})_{m \times n}$ and $B = (b_{ij})_{p \times q}$ be two matrices of size $m \times n$ and $p \times q$, respectively. The two matrices A and B are said to be equal if (i) $m = p$, $n = q$, and (ii) $a_{ij} = b_{ij}$ for all i and j. This is denoted by $A = B$.

Example 1.3.6.

(a) You are given

$$A = \begin{bmatrix} 1 & 1 \\ 1 & 1 \end{bmatrix}, \quad B = \begin{bmatrix} 1 & 1 & 1 \\ 1 & 1 & 1 \\ 1 & 1 & 1 \end{bmatrix}.$$

The matrices A and B are matrices of size 2×2 and 3×3, respectively. From Definition 1.3.6, we conclude that they are not equal.

(b) You are given

$$A = \begin{bmatrix} 1 & 1 \\ 0 & 1 \end{bmatrix}, \quad B = \begin{bmatrix} 1 & 0 \\ 1 & 1 \end{bmatrix}.$$

The matrices are of size 2×2 and 2×2, respectively. They are not equal because the element $a_{12} = 1$ of the matrix A is different from the element $b_{12} = 0$ of the matrix B.

Definition 1.3.7. An $m \times n$ matrix $A = (a_{ij})_{m \times n}$ is called a zero matrix if $a_{ij} = 0$ for all i and j for $1 \le i \le m$, $1 \le j \le n$. It is denoted by $A = 0$.

Example 1.3.7. We are given

$$A = \begin{bmatrix} 0 & 0 \\ 0 & 0 \end{bmatrix}, \quad B = \begin{bmatrix} 0 & 0 & 0 \\ 0 & 0 & 0 \\ 0 & 0 & 0 \end{bmatrix}.$$

From Definition 1.3.7, these matrices are zero matrices of size 2×2 and 3×3, respectively.

Definition 1.3.8 (addition). Let

$$A = (a_{ij})_{m \times n}, \quad B = (b_{ij})_{m \times n}$$

be two matrices of the same size — say, $m \times n$. A sum of two matrices A and B is the $m \times n$ matrix with entries $a_{ij} + b_{ij}$, i.e. the sum of two matrices of the same size is the $m \times n$ matrix obtained by adding the corresponding entries. The sum is represented by

$$A + B = (a_{ij})_{m \times n} + (b_{ij})_{m \times n} = (a_{ij} + b_{ij})_{m \times n} = C.$$

Example 1.3.8.

(a) We are given

$$A = \begin{bmatrix} 2 & 3 \\ 4 & -5 \end{bmatrix}, \quad B = \begin{bmatrix} -1 & 6 \\ 1 & 7 \end{bmatrix}.$$

These matrices have the same size, 2×2. From Definition 1.3.8, the sum of these matrices is

$$A + B = \begin{bmatrix} 2 & 3 \\ 4 & -5 \end{bmatrix} + \begin{bmatrix} -1 & 6 \\ 1 & 7 \end{bmatrix} = \begin{bmatrix} 2 + (-1) & 3 + 6 \\ 4 + 1 & -5 + 7 \end{bmatrix} = \begin{bmatrix} 1 & 9 \\ 5 & 2 \end{bmatrix}.$$

(b) The matrices

$$A = \begin{bmatrix} 2 & 3 \\ 4 & -5 \end{bmatrix}, \quad B = \begin{bmatrix} 2 & 4 \end{bmatrix}$$

have the size of 2×2 and 1×2, respectively. They do not have the same size, so, in this case, $A+B$ is not defined. Thus, the sum of A and B does not make sense.

Definition 1.3.9 (scalar multiplication). Let $A = (a_{ij})_{m\times n}$ be a matrix of size $m \times n$, and let c be a real or complex number. The scalar multiple to a matrix A is the $m \times n$ matrix cA, whose ith row and jth column entry is ca_{ij} for each i and j satisfying the condition, $1 \leq i \leq m$ and $1 \leq j \leq n$, respectively, i.e. the scalar multiplication to matrix A is the $m \times n$ matrix obtained by multiplying each entry of A by c. It is denoted by $cA = (ca_{ij})_{m\times n}$. This type of use of a number c is called a scalar.

Example 1.3.9. Let us consider

$$A = \begin{bmatrix} 1 & 0 & 2 \\ 3 & 1 & 5 \end{bmatrix}$$

and a scalar quantity $c = 2$. The matrix has the size of 2×3. From Definition 1.3.9, the scalar multiple to matrix A is

$$2A = 2\begin{bmatrix} 1 & 0 & 2 \\ 3 & 1 & 5 \end{bmatrix} = \begin{bmatrix} (2)(1) & (2)(0) & (2)(2) \\ (2)(3) & (2)(1) & (2)(5) \end{bmatrix} = \begin{bmatrix} 2 & 0 & 4 \\ 6 & 2 & 10 \end{bmatrix}.$$

Observation 1.3.2. For matrices A and B of the same size, we define the difference of the two matrices as $A+(-1B)$. It is written as $A-B = A+(-B)$. We note that $(-1)B = -B$.

Example 1.3.10.

(a) We are given

$$A = \begin{bmatrix} 2 & 3 \\ 4 & -5 \end{bmatrix}, \quad B = \begin{bmatrix} -1 & 6 \\ 1 & 7 \end{bmatrix}.$$

These matrices have the same size as 2×2. Thus, from Observation 1.3.2, the difference $A - B$ is

$$\begin{aligned} A - B &= \begin{bmatrix} 2 & 3 \\ 4 & -5 \end{bmatrix} + (-1)\begin{bmatrix} -1 & 6 \\ 1 & 7 \end{bmatrix}, \\ &= \begin{bmatrix} 2 & 3 \\ 4 & -5 \end{bmatrix} + \begin{bmatrix} (-1)(-)1 & (-1)6 \\ (-1)1 & (-1)7 \end{bmatrix} \\ &= \begin{bmatrix} 2 & 3 \\ 4 & -5 \end{bmatrix} + \begin{bmatrix} 1 & -6 \\ -1 & -7 \end{bmatrix} \\ &= \begin{bmatrix} 2+1 & 3+(-6) \\ 4+(-1) & -5+(-7) \end{bmatrix} = \begin{bmatrix} 3 & -3 \\ 3 & -12 \end{bmatrix}. \end{aligned}$$

(b) The matrices

$$A = \begin{bmatrix} 2 & 3 \\ 4 & -5 \end{bmatrix}, \quad B = [2 \quad 4]$$

have the size of 2×2 and 1×2, respectively. They do not have the same size, so, in this case, $A - B$ is not defined. Thus, the difference of A and B does not exist.

Definition 1.3.10. Let $A = (a_i)_{1\times n}$ and $B = (b_j)_{n\times 1}$ be $1 \times n$ and $n \times 1$ matrices (row and column vectors), respectively. The product of the two matrices A and B is the 1×1 matrix (the number), whose element is $a_1b_1 + a_2b_2 + \cdots + a_jb_j + \cdots + a_nb_n$. It is denoted by

$$AB = a_1b_1 + a_2b_2 + \cdots + a_jb_j + \cdots + a_nb_n = \sum_{j=1}^{n} a_jb_j. \tag{1.7}$$

We note that this matrix multiplication is like the "DOT PRODUCT" of two directed vectors (in either R^2 or R^3).

Example 1.3.11. You are given

$$A = [1 \ 0 \ 2 \ -1], \quad B = \begin{bmatrix} 2 \\ 4 \\ 0 \\ 6 \end{bmatrix}.$$

The matrices A and B are of size 1×4 and 4×1, respectively. By the application of Definition 1.3.10, the product AB is

$$AB = 1(2) + 0(4) + 2(0) + (-1)(6) = 2 + 0 + 0 - 6 = -4.$$

Definition 1.3.11 (matrix multiplication). Let $A = (a_{ij})_{m\times n}$ and $B = (b_{ij})\, n \times q$ be two matrices of the size $m \times n$ and $n \times q$, respectively. The product of the two matrices A and B is the matrix of size $m \times q$ whose ith row and jth column entry $c_{ij} = A_iB^j$ is the product of the ith row vector A_i of A and the jth column vector B^j of B, where i and j satisfy the conditions $1 \leq i \leq m$ and $1 \leq j \leq q$, respectively. It is denoted by

$$AB = \begin{bmatrix} A_1B^1 & A_1B^2 & \cdots & A_1B^j & \cdots & A_1B^q \\ A_2B^1 & A_2B^2 & \cdots & A_2B^j & \cdots & A_2B^q \\ \cdots & \cdots & \cdots & \cdots & \cdots & \cdots \\ A_iB^1 & A_iB^2 & \cdots & A_iB^j & \cdots & A_iB^q \\ \cdots & \cdots & \cdots & \cdots & \cdots & \cdots \\ A_mB^1 & A_mB^2 & \cdots & A_mB^j & \cdots & A_mB^q \end{bmatrix} = (c_{ij})_{m\times q}, \tag{1.8}$$

where A_i and B^j are as defined in Observation 1.3.1, and A_iB^j is as defined in (1.7), i.e.

$$A_iB^j = a_{i1}b_{1j} + a_{i2}b_{2j} + \cdots + a_{ik}b_{kj} + \cdots + a_{in}b_{nj} = \sum_{k=1}^{n} a_{ik}b_{kj}. \tag{1.9}$$

Example 1.3.12. You are given

$$A = \begin{bmatrix} 1 & 2 \\ -2 & 5 \end{bmatrix}, \quad B = \begin{bmatrix} 1 & 2 & 3 \\ 0 & 2 & 5 \end{bmatrix}.$$

Find AB (if possible).

Solution procedure. A and B are 2×2 and 2×3 matrices, respectively. The product AB is defined, but the product BA is undefined. Thus, by utilizing the formula (1.8) in Definition 1.3.11 and by adopting the following procedure, the product AB is computed as:

Step 1. We write row vectors of the matrix A and the column vectors of the matrix B:

$$A_1 = [1\ 2], \quad A_2 = [-2\ 5], \quad B^1 = \begin{bmatrix} 1 \\ 0 \end{bmatrix}, \quad B^2 = \begin{bmatrix} 2 \\ 2 \end{bmatrix}, \quad B^3 = \begin{bmatrix} 3 \\ 5 \end{bmatrix}.$$

Step 2. For each i, $1 \leq i \leq 2$, and each j, $1 \leq j \leq 3$, by using Definition 1.3.10 we compute A_iB^j:

$$A_1B^1 = 1 + 0 = 1, \quad A_1B^2 = 2 + 4 = 6, \quad A_1B^3 = 3 + 10 = 13,$$
$$A_2B^1 = -2 + 0 = -2, \quad A_2B^2 = -4 + 10 = 6, \quad A_2B^3 = -6 + 25 = 19.$$

Step 3. Now, by substituting for A_iB^j for $i = 1, 2$ and $j = 1, 2, 3$, we have

$$AB = \begin{bmatrix} A_1 \\ A_2 \end{bmatrix} [B^1\ B^2\ B^3] = \begin{bmatrix} A_1B^1 & A_1B^2 & A_1B^3 \\ A_2B^1 & A_2B^2 & A_2B^3 \end{bmatrix} = \begin{bmatrix} 1 & 6 & 13 \\ -2 & 6 & 19 \end{bmatrix}.$$

This completes the solution procedure.

For easy reference, the fundamental properties of matrix algebra are listed in the following result.

Theorem 1.3.1. *Let A, B, C be matrices, and $a, b, c, \ldots$ be numbers. It is assumed that the matrices are appropriately defined to fulfill the following specified operations and relations:*

(A1) $A + B = B + A$ (*commutative law of addition*).
(A2) $(A + B) + C = A + (B + C)$ (*associative law of addition*).

(A3) *For all m, n positive integers, there is an $m \times n$ matrix 0 defined as having all entries 0 with the property*

$$A + 0 = 0 + A$$

for any $m \times n$ matrix A. Furthermore, 0 is the only matrix with this property. It is called the zero matrix or additive identity.

(A4) *Given an $m \times n$ matrix $A = (a_{ij})$, there is a unique matrix denoted by $-A = (-a_{ij})$ with the property*

$$A + (-A) = 0 = (-A) + A.$$

It is called the additive inverse of A.

(S1) $a(A + B) = aA + aB$ *(distributive law for scalar multiplication over matrix addition).*

(S2) $(a + b)A = aA + bA$ *(distributive law for matrix multiplication over scalar addition).*

(S3) $(ab)A = a(bA)$ *(associative law for scalar multiplication to matrix over scalar multiplication).*

(S4) $1A = A$, *where 1 is the number one (scalar multiplicative identity over the matrix multiplication).*

(M1) $(AB)C = A(BC)$ *(associative law for matrix multiplication).*

(M2) $A(B + C) = AB + AC$ *(left distributive law for matrix multiplication over matrix addition) and $(A+B)C = AC+BC$ (right distributive law for matrix multiplication over matrix addition).*

(M3) $a(AB) = (aA)B = A(aB)$ *(associative law for scalar multiplication over matrix multiplication).*

(M4) *For each positive integer n, there is an $n \times n$ matrix I called the multiplicative identity such that*

$$AI = IA = A$$

for any $n \times n$ matrix A, where I is the matrix with 1's on the main diagonal and 0's elsewhere.

Example 1.3.13. Let us consider

$$A = \begin{bmatrix} 1 & 2 & 3 \\ 0 & 2 & 5 \end{bmatrix}, \quad B = \begin{bmatrix} 1 & -2 & 4 \\ 6 & 2 & 1 \end{bmatrix}.$$

These matrices have the same size, 2×3. From Definition 1.3.8, the sums $A + B$ and $B + A$ are

$$A + B = \begin{bmatrix} 1 & 2 & 3 \\ 0 & 2 & 5 \end{bmatrix} + \begin{bmatrix} 1 & -2 & 4 \\ 6 & 2 & 1 \end{bmatrix} = \begin{bmatrix} 2 & 0 & 7 \\ 6 & 4 & 6 \end{bmatrix},$$

$$B + A = \begin{bmatrix} 1 & -2 & 4 \\ 6 & 2 & 1 \end{bmatrix} + \begin{bmatrix} 1 & 2 & 3 \\ 0 & 2 & 5 \end{bmatrix} = \begin{bmatrix} 2 & 0 & 7 \\ 6 & 4 & 6 \end{bmatrix}.$$

Therefore, $A + B = B + A$.

Example 1.3.14. In general, the commutative law for the matrix multiplication is not valid. To justify this, we provide a counterexample to show that $AB \neq BA$. For example, we consider

$$A = \begin{bmatrix} 1 & 2 \\ 3 & 4 \end{bmatrix}, \quad B = \begin{bmatrix} 1 & 2 \\ 2 & 3 \end{bmatrix},$$

and compute AB and BA. These are

$$AB = \begin{bmatrix} 1 & 2 \\ 3 & 4 \end{bmatrix} \begin{bmatrix} 1 & 2 \\ 2 & 3 \end{bmatrix} = \begin{bmatrix} 1+4 & 2+6 \\ 3+8 & 6+12 \end{bmatrix} = \begin{bmatrix} 5 & 8 \\ 11 & 18 \end{bmatrix},$$

$$BA = \begin{bmatrix} 1 & 2 \\ 2 & 3 \end{bmatrix} \begin{bmatrix} 1 & 2 \\ 3 & 4 \end{bmatrix} = \begin{bmatrix} 1+6 & 2+8 \\ 2+9 & 4+12 \end{bmatrix} = \begin{bmatrix} 7 & 10 \\ 11 & 16 \end{bmatrix}.$$

Thus

$$AB = \begin{bmatrix} 5 & 8 \\ 11 & 18 \end{bmatrix} \neq \begin{bmatrix} 7 & 10 \\ 11 & 16 \end{bmatrix} = BA.$$

In the following, we present a nonhomogeneous system of m linear algebraic equations with n unknowns:

$$\begin{aligned} a_{11}x_1 + a_{12}x_2 + \cdots + a_{1j}x_j + \cdots + a_{1n}x_n &= b_1, \\ a_{21}x_1 + a_{22}x_2 + \cdots + a_{2j}x_j + \cdots + a_{2n}x_n &= b_2, \\ \cdots\cdots\cdots\cdots\cdots\cdots\cdots\cdots\cdots\cdots\cdots\cdots\cdots\cdots\cdots & \\ a_{i1}x_1 + a_{i2}x_2 + \cdots + a_{ij}x_j + \cdots + a_{in}x_n &= bi, \\ \cdots\cdots\cdots\cdots\cdots\cdots\cdots\cdots\cdots\cdots\cdots\cdots\cdots\cdots\cdots & \\ a_{m1}x_1 + a_{m2}x_2 + \cdots + a_{mj}x_j + \cdots + a_{1m}x_n &= b_m, \end{aligned} \tag{1.10}$$

where a_{ij}'s are given real numbers and x_j's are unknown real numbers (variables). By using the definitions of matrix multiplication, scalar multiplication to a matrix, addition of matrices, the equality of two matrices, and by using notations in Observation 1.3.1, the system (1.10) can be rewritten in a variety of ways (see below for examples):

$$A^1x_1 + A^2x_2 + \cdots + A^jx_j + \cdots + A^nx_n = b, \tag{1.11}$$

$$Ax = b, \tag{1.12}$$

where A is called the coefficient matrix of the system (1.10); it is defined as in (1.3); $A^1 \ A^2 \ \cdots \ A^j \ \cdots \ A^n$ are $n \times 1$ column vectors or matrices associated with the matrix A defined in (1.4); $b = (b_j)_{m\times 1}$ and $x = (x_j)_{n\times 1}$ are column vectors. This shows that the concept of the matrix economizes the mathematical representation of the linear system of algebraic equations in (1.10). With regard to the system (1.12), we say that $x = (x_j)_{n\times 1}$ is a solution to the matrix equation in (1.12) if and only if (abbreviated to "iff") the **n**-tuple $(x_1, x_2, \ldots, x_j, \ldots, x_n)$ is a solution to the original system of linear algebraic equations (1.10).

Example 1.3.15. We are given

$$\begin{aligned} 2x_1 - x_2 + x_3 &= 1, \\ x_1 + 2x_2 - x_3 &= 3, \\ x_1 + 7x_2 - 4x_3 &= 8. \end{aligned}$$

Find a solution set (if possible).

Solution procedure. The goal is to find a solution set of the given nonhomogeneous system of algebraic equations. We utilize the elementary reduction technique — the Gauss–Jordan method — to solve this algebraic system. The goal is to reduce the coefficient matrix A,

$$A = \begin{bmatrix} 2 & -1 & 1 \\ 1 & 2 & -1 \\ 1 & 7 & -4 \end{bmatrix},$$

into a simpler, upper-triangular form. Then, we use backward substitution to find the solution set for this system.

Step 1. For this purpose, we write the augmented coefficient matrix:

$$[A \mid b] = \left[\begin{array}{ccc|c} 2 & -1 & 1 & 1 \\ 1 & 2 & -1 & 3 \\ 1 & 7 & -4 & 8 \end{array}\right], \quad \text{where } b = \begin{bmatrix} 1 \\ 3 \\ 8 \end{bmatrix}.$$

Step 2. We apply the following admissible elementary row operations to the augmented/partitioned matrix $[A \mid b]$:

(E_1) Multiply any (single) row of the matrix by a nonzero constant.
(E_2) Interchange any two rows of the matrix.
(E_3) Subtract a constant multiple of one row from any other row:

$$[A \mid b] = \left[\begin{array}{ccc|c} 2 & -1 & 1 & 1 \\ 1 & 2 & -1 & 3 \\ 1 & 7 & -4 & 8 \end{array}\right] \to E_2 \to \left[\begin{array}{ccc|c} 1 & 2 & -1 & 3 \\ 2 & -1 & 1 & 1 \\ 1 & 7 & -4 & 8 \end{array}\right] \to E_3$$

$$\to \left[\begin{array}{ccc|c} 1 & 2 & -1 & 3 \\ 0 & -5 & 3 & -5 \\ 0 & 5 & -3 & 5 \end{array}\right] \to E_3 \to \left[\begin{array}{ccc|c} 1 & 2 & -1 & 3 \\ 0 & -5 & 3 & -5 \\ 0 & 0 & 0 & 0 \end{array}\right]$$

$$\to E_1 \to \left[\begin{array}{ccc|c} 1 & 2 & -1 & 3 \\ 0 & 1 & -\frac{3}{5} & 1 \\ 0 & 0 & 0 & 0 \end{array}\right] \to E_3 \to \left[\begin{array}{ccc|c} 1 & 0 & \frac{1}{5} & 1 \\ 0 & 1 & -\frac{3}{5} & 1 \\ 0 & 0 & 0 & 0 \end{array}\right].$$

Note. By using the sequence of elementary operations E_1, E_2, and E_3 (one by one, in turn), the augmented $[A \mid b]$ is transformed into an upper-triangular form that has only zero beneath its principal diagonal. This process is carried out one column at a time, from left to right. This is shown in the above example.

Step 3. The algebraic system corresponding to the last augmented/partitioned matrix,

$$\left[\begin{array}{ccc|c} 1 & 0 & \frac{1}{5} & 1 \\ 0 & 1 & -\frac{3}{5} & 1 \\ 0 & 0 & 0 & 0 \end{array}\right],$$

is

$$x_1 + \frac{1}{5}x_3 = 1,$$
$$x_2 - \frac{3}{5}x_3 = 1.$$

By solving for x_1 and x_2, we have

$$x_1 = 1 - \frac{1}{5}x_3,$$
$$x_2 = 1 + \frac{3}{5}x_3.$$

The solution set for the given linear nonhomogeneous system of algebraic equations is described by

$$\begin{aligned} S &= \left\{(x_1, x_2, x_3) \in R^3 : (x_1, x_2, x_3) = \left(1 - \frac{1}{5}x_3, 1 + \frac{3}{5}x_3, x_3\right), x_3 \in R\right\} \\ &= \left\{(x_1, x_2, x_3) \in R^3 : (x_1, x_2, x_3) = (1, 1, 0) + \left(-\frac{1}{5}, \frac{3}{5}, 1\right) x_3, x_3 \in R\right\}. \end{aligned}$$

This shows that the system has infinitely many nontrivial solutions. This completes the goal of the problem.

Observation 1.3.3. The linear homogeneous system of algebraic equations corresponding to the system (1.10) is

$$\begin{aligned} a_{11}x_1 + a_{12}x_2 + \cdots + a_{1j}x_j + \cdots + a_{1n}x_n &= 0, \\ a_{21}x_1 + a_{22}x_2 + \cdots + a_{2j}x_j + \cdots + a_{2n}x_n &= 0, \\ \cdots\cdots\cdots\cdots\cdots\cdots\cdots\cdots\cdots\cdots\cdots\cdots \\ a_{i1}x_1 + a_{i2}x_2 + \cdots + a_{ij}x_j + \cdots + a_{in}x_n &= 0, \\ \cdots\cdots\cdots\cdots\cdots\cdots\cdots\cdots\cdots\cdots\cdots\cdots \\ a_{m1}x_1 + a_{m2}x_2 + \cdots + a_{mj}x_j + \cdots + a_{1m}x_n &= 0, \end{aligned} \tag{1.13}$$

and its matrix representations are:

$$A^1x_1 + A^2x_2 + \cdots + A^jx_j + \cdots + A^nx_n = 0, \tag{1.14}$$

$$Ax = 0. \tag{1.15}$$

We further note that every linear homogeneous system of algebraic equations always has at least one solution. If $n > m$, the system (1.13) or (1.15) has infinitely many nonzero solutions.

Example 1.3.16. Let us consider

$$\begin{aligned} 6x_1 + 4x_2 &= 0, \\ -6x_1 - 4x_2 &= 0, \\ -6x_1 - 4x_2 &= 0. \end{aligned}$$

Find a solution set of the given homogeneous system of algebraic equations.

Solution procedure. We utilize the elementary reduction technique — the Gauss–Jordan method — to solve this algebraic system. The goal is to reduce the coefficient matrix A,

$$A = \begin{bmatrix} 6 & 4 & 0 \\ -6 & -4 & 0 \\ -6 & -4 & 0 \end{bmatrix},$$

into its upper-triangular form.

In the following, we imitate the procedure outlined in Example 1.3.15.

Step 1. For this purpose, we write the augmented coefficient matrix:

$$[A \mid b] = \left[\begin{array}{ccc|c} 6 & 4 & 0 & 0 \\ -6 & -4 & 0 & 0 \\ -6 & -4 & 0 & 0 \end{array}\right], \quad \text{where } b = \begin{bmatrix} 0 \\ 0 \\ 0 \end{bmatrix}.$$

Step 2. We apply the admissible elementary row operations outlined in Example 1.3.15 to the augmented/partitioned matrix $[A \mid b]$ in step 1:

$$[A \mid b] = \left[\begin{array}{ccc|c} 6 & 4 & 0 & 0 \\ -6 & -4 & 0 & 0 \\ -6 & -4 & 0 & 0 \end{array}\right] \to E_3 \to \left[\begin{array}{ccc|c} 6 & 4 & 0 & 0 \\ 0 & 0 & 0 & 0 \\ 0 & 0 & 0 & 0 \end{array}\right] \to E_1 \to \left[\begin{array}{ccc|c} 1 & \frac{2}{3} & 0 & 0 \\ 0 & 0 & 0 & 0 \\ 0 & 0 & 0 & 0 \end{array}\right].$$

The process is stopped.

Step 3. The algebraic system corresponding to the last augmented/partitioned matrix in step 2,

$$\left[\begin{array}{ccc|c} 1 & \frac{2}{3} & 0 & 0 \\ 0 & 0 & 0 & 0 \\ 0 & 0 & 0 & 0 \end{array}\right],$$

is

$$x_1 + \frac{2}{3}x_2 = 0.$$

By solving for x_1, we have

$$x_1 = -\frac{2}{3}x_2.$$

The solution set for the given linear homogeneous system of algebraic equations is described by

$$S = \left\{(x_1, x_2, x_3) \in R^3 : (x_1, x_2, x_3) = \left(-\frac{2}{3}x_2, x_2, x_3\right) x_2, x_3 \in R\right\}$$
$$= \left\{(x_1, x_2, x_3) \in R^3 : (x_1, x_2, x_3) = \left(-\frac{2}{3}, 1, 0\right) x_2 + (0,0,1)x_3, x_2, x_3 \in R\right\}.$$

This shows that the system has infinitely many nontrivial solutions. This completes the goal of the problem.

In the following, we introduce a concept of transpose operation, which will be used subsequently.

Definition 1.3.12. Let $A = (a_{ij})_{m\times n}$ be an $m \times n$ matrix. A transpose of A is the $n \times m$ matrix with ith row and jth column entry a_{ji} for each i and j satisfying the conditions $1 \leq i \leq m$ and $1 \leq j \leq n$, respectively. It is denoted by $A^T = (a_{ji})_{n\times m}$.

Observation 1.3.4. The rows of an $m \times n$ matrix A are the columns of its transpose matrix A^T and vice versa.

Example 1.3.17. You are given

$$A = \begin{bmatrix} -2 & 3 \\ 0 & -1 \end{bmatrix}, \quad B = \begin{bmatrix} 1 & 2 & 3 \\ 0 & 2 & 5 \end{bmatrix}.$$

The matrices are of size 2×2 and 2×3, respectively. By the application of Definition 1.3.12 and Observation 1.3.4, we have

$$A^T = \begin{bmatrix} -2 & 0 \\ 3 & -1 \end{bmatrix}, \quad B^T = \begin{bmatrix} 1 & 0 \\ 2 & 2 \\ 3 & 5 \end{bmatrix}.$$

The following result exhibits the properties of the transpose matrix in the context of algebraic properties and operations of matrices.

Theorem 1.3.2 (properties of transpose operation). *Let A and B be matrices, and c be a number. It is assumed that the matrices are appropriately defined to fulfill the* *following specified operations*:

(T1) $(A+B)^T = A^T + B^T$, *i.e. the transpose of the sum* $(A+B)$ *of two matrices* A *and* B *is the sum* $(A^T + B^T)$ *of their transpose matrices* A^T *and* B^T.

(T2) $(cA)^T = cA^T$, *i.e. the transpose of the scalar multiple of matrix* cA *is the scalar multiple of its transpose* cA^T.

(T3) $(A^T)^T = A$, *i.e. the transpose of the transpose of a matrix* A *is* A *itself.*

(T4) $(AB)^T = B^T A^T$, *i.e. the transpose of the product* AB *of two matrices* A *and* B *is the product of transpose matrices in reverse order* $B^T A^T$.

(T5) $AB = B^T A^T$, *where* A *(n-dimensional row vector) and* B *(n-dimensional column vector) are* $1 \times n$ *and* $n \times 1$ *matrices, respectively.*

Example 1.3.18. We are given

$$A = \begin{bmatrix} -2 & 3 \\ 0 & -1 \end{bmatrix}, \quad B = \begin{bmatrix} 1 & 2 & 3 \\ 0 & 2 & 5 \end{bmatrix}.$$

The matrices are of size 2×2 and 2×3, respectively. Therefore, the products AB and $B^T A^T$ are defined. Thus, by utilizing the formula (1.8), Definition 1.3.11 and adopting the three-step procedure outlined in Example 1.3.12, the products AB and $B^T A^T$ are computed as follows:

$$AB = \begin{bmatrix} -2 & 2 & 9 \\ 0 & -2 & -5 \end{bmatrix}, \quad A^T = \begin{bmatrix} -2 & 0 \\ 3 & -1 \end{bmatrix},$$

$$B^T = \begin{bmatrix} 1 & 0 \\ 2 & 2 \\ 3 & 5 \end{bmatrix}, \quad (AB)^T = \begin{bmatrix} -2 & 0 \\ 2 & -2 \\ 9 & -5 \end{bmatrix}, \quad B^T A^T = \begin{bmatrix} -2 & 0 \\ 2 & -2 \\ 9 & -5 \end{bmatrix}.$$

From Definition 1.3.6, it follows that $(AB)^T = B^T A^T$. This justifies the validity of the property (T4) in Theorem 1.3.2.

We next present a concept of a nonsingular or invertible matrix. This is very useful for the study of systems of linear algebraic equations, as well as systems of differential equations.

Definition 1.3.13. An $n \times n$ square matrix A is said to be invertible or nonsingular if there is an $n \times n$ matrix P such that $PA = AP = I$, where I is the $n \times n$ identity matrix.

Observation 1.3.5. Let A be an $n \times n$ invertible or nonsingular matrix. This means that one can find an $n \times n$ matrix P such that $PA = AP = I$, where I is the $n \times n$ identity matrix. From the existence of the matrix P, one can conclude that this matrix is unique. Hence, this unique matrix P is called the inverse of the matrix A. It is denoted by A^{-1}. It is completely characterized by $A^{-1}A = AA^{-1} = I$. We further note that any two matrices A and B are said to possess the commutative property with respect to the matrix multiplication iff $AB = BA$.

Example 1.3.19. Let us consider

$$A = \begin{bmatrix} 1 & 2 \\ -1 & 1 \end{bmatrix}.$$

Goal: to find its inverse (if it exists).

Solution procedure. This is a 2×2 square matrix. From Definition 1.3.13 and Observation 1.3.5, we recall that the invertibility of the matrix A is completely characterized by $A^{-1}A = AA^{-1} = I$.

Step 1. Our goal is to find A^{-1} (if it exists). Therefore, we assume that

$$P = A^{-1} = \begin{bmatrix} a & c \\ b & d \end{bmatrix} = [P^1\ P^2], \quad P^1 = \begin{bmatrix} a \\ b \end{bmatrix}, \quad P^2 = \begin{bmatrix} c \\ d \end{bmatrix},$$

where a, b, c, and d are unknown numbers to be determined. From Definition 1.3.11 and (1.8), we observe that

$$AA^{-1} = [AP^1\ AP^2],$$

and hence the problem of finding A^{-1} reduces to the problem of solving two systems of linear nonhomogeneous algebraic equations with the coefficient matrix A:

$$\begin{bmatrix} 1 & 2 \\ -1 & 1 \end{bmatrix}\begin{bmatrix} a \\ b \end{bmatrix} = \begin{bmatrix} 1 \\ 0 \end{bmatrix}, \quad \begin{bmatrix} 1 & 2 \\ -1 & 1 \end{bmatrix}\begin{bmatrix} c \\ d \end{bmatrix} = \begin{bmatrix} 0 \\ 1 \end{bmatrix}.$$

Step 2. Again, we apply the elementary reduction technique — the Gauss–Jordan method outlined in Example 1.3.15 (steps 1–3) — to the above two formulated systems. Its augmented matrix is

$$\begin{aligned}[A \mid I] = \left[\begin{array}{cc|cc} 1 & 2 & 1 & 0 \\ -1 & 1 & 0 & 1 \end{array}\right] &\to E_3 \to \left[\begin{array}{cc|cc} 1 & 2 & 1 & 0 \\ 0 & 3 & 1 & 1 \end{array}\right] \to E_1 \\ \to \left[\begin{array}{cc|cc} 1 & 2 & 1 & 0 \\ 0 & 1 & \frac{1}{3} & \frac{1}{3} \end{array}\right] &\to E_3 \to \left[\begin{array}{cc|cc} 1 & 0 & \frac{1}{3} & -\frac{2}{3} \\ 0 & 1 & \frac{1}{3} & \frac{1}{3} \end{array}\right].\end{aligned}$$

The algebraic systems corresponding to the last augmented/partitioned matrix,

$$\left[\begin{array}{cc|cc} 1 & 0 & \frac{1}{3} & -\frac{2}{3} \\ 0 & 1 & \frac{1}{3} & \frac{1}{3} \end{array}\right],$$

are

$$\begin{aligned} a &= \frac{1}{3}, \quad c = -\frac{2}{3}, \\ b &= \frac{1}{3}, \quad d = \frac{1}{3}. \end{aligned}$$

The solutions to two systems are uniquely determined. Therefore, A^{-1} exists and it is

$$P = A^{-1} = \begin{bmatrix} \frac{1}{3} & -\frac{2}{3} \\ \frac{1}{3} & \frac{1}{3} \end{bmatrix}.$$

Step 3. One can check the correctness of the answer by verifying the correctness of $A^{-1}A = AA^{-1} = I$:

$$\begin{bmatrix} \frac{1}{3} & -\frac{2}{3} \\ \frac{1}{3} & \frac{1}{3} \end{bmatrix} \begin{bmatrix} 1 & 2 \\ -1 & 1 \end{bmatrix} = \begin{bmatrix} \frac{1}{3} + \frac{2}{3} & \frac{2}{3} - \frac{2}{3} \\ \frac{1}{3} - \frac{1}{3} & \frac{2}{3} + \frac{1}{3} \end{bmatrix} = \begin{bmatrix} 1 & 0 \\ 0 & 1 \end{bmatrix},$$

$$\begin{bmatrix} 1 & 2 \\ -1 & 1 \end{bmatrix} \begin{bmatrix} \frac{1}{3} & -\frac{2}{3} \\ \frac{1}{3} & \frac{1}{3} \end{bmatrix} = \begin{bmatrix} \frac{1}{3} + \frac{2}{3} & -\frac{2}{3} + \frac{2}{3} \\ -\frac{1}{3} + \frac{1}{3} & \frac{2}{3} + \frac{1}{3} \end{bmatrix} = \begin{bmatrix} 1 & 0 \\ 0 & 1 \end{bmatrix}.$$

From this and Definition 1.3.6, we have $A^{-1}A = AA^{-1} = I$. Thus, the inverse of the matrix A is

$$A^{-1} = \begin{bmatrix} \frac{1}{3} & -\frac{2}{3} \\ \frac{1}{3} & \frac{1}{3} \end{bmatrix}.$$

This completes the procedure for finding the inverse of the matrix A by the method of Gauss–Jordan reduction.

Theorem 1.3.3 (properties of invertible matrices). *Let A and B be $n \times n$ invertible matrices. Then,*

(I1) $(AB)^{-1} = B^{-1}A^{-1}$, *i.e. the inverse of the product AB of two matrices A and B is the product of inverse matrices in the reverse order $B^{-1}A^{-1}$.*
(I2) $(A^{-1})^{-1} = A$, *i.e. the inverse of the inverse of a matrix A is A itself.*
(I3) $(A^{-1})^{T} = (A^{T})^{-1}$, *i.e. the transpose of the inverse is the inverse of the transpose.*

Theorem 1.3.4. *Assume that $m = n$ in the system of linear inhomogeneous equations in* (1.10) *or* (1.12). *Further assume that the coefficient matrix A is invertible. Then the system of equations in* (1.12) *has a unique solution. Moreover, this unique solution is determined by $x = A^{-1}b$.*

Example 1.3.20. We are given

$$x_1 + x_3 = 1,$$
$$2x_1 + x_2 + x_3 = 0,$$
$$x_1 + x_2 + 2x_3 = 1.$$

Find a solution to the given system (if possible).

Solution procedure. The goal is to find a solution set for the given linear system of nonhomogeneous algebraic equations.

We rewrite the given system in the matrix form as described in (1.12):

$$Ax = b,$$

where A is the coefficient matrix associated with the given system, and b is the given input matrix. They are

$$A = \begin{bmatrix} 1 & 0 & 1 \\ 2 & 1 & 1 \\ 1 & 1 & 2 \end{bmatrix}, \quad b = \begin{bmatrix} 1 \\ 0 \\ 1 \end{bmatrix}.$$

We apply Theorem 1.3.4 to solve this problem. We imitate the procedure described in Example 1.3.19:

$$[A|I] = \left[\begin{array}{ccc|ccc} 1 & 0 & 1 & 1 & 0 & 0 \\ 2 & 1 & 1 & 0 & 1 & 0 \\ 1 & 1 & 2 & 0 & 0 & 1 \end{array}\right] \to E_3 \to \left[\begin{array}{ccc|ccc} 1 & 0 & 1 & 1 & 0 & 0 \\ 0 & 1 & -1 & -2 & 1 & 0 \\ 0 & 1 & 1 & -1 & 0 & 1 \end{array}\right] \to E_3$$

$$\to \left[\begin{array}{ccc|ccc} 1 & 0 & 1 & 1 & 0 & 0 \\ 0 & 1 & -1 & -2 & 1 & 0 \\ 0 & 0 & 2 & 1 & -1 & 1 \end{array}\right] \to E_1$$

$$\to \left[\begin{array}{ccc|ccc} 1 & 0 & 1 & 1 & 0 & 0 \\ 0 & 1 & -1 & -2 & 1 & 0 \\ 0 & 0 & 1 & \frac{1}{2} & -\frac{1}{2} & \frac{1}{2} \end{array}\right] \to E_3$$

$$\to \left[\begin{array}{ccc|ccc} 1 & 0 & 0 & \frac{1}{2} & \frac{1}{2} & -\frac{1}{2} \\ 0 & 1 & 0 & -\frac{3}{2} & \frac{1}{2} & \frac{1}{2} \\ 0 & 0 & 1 & \frac{1}{2} & -\frac{1}{2} & \frac{1}{2} \end{array}\right].$$

Finding the inverse of the coefficient matrix is terminated. The inverse of the coefficient matrix A is

$$A^{-1} = \begin{bmatrix} \frac{1}{2} & \frac{1}{2} & -\frac{1}{2} \\ -\frac{3}{2} & \frac{1}{2} & \frac{1}{2} \\ \frac{1}{2} & -\frac{1}{2} & \frac{1}{2} \end{bmatrix}.$$

Of course, one needs to check the correctness of the inverse. Now, by the application of Theorem 1.3.4, the solution to the given system is

$$x = A^{-1}b = \begin{bmatrix} \frac{1}{2} & \frac{1}{2} & -\frac{1}{2} \\ -\frac{3}{2} & \frac{1}{2} & \frac{1}{2} \\ \frac{1}{2} & -\frac{1}{2} & \frac{1}{2} \end{bmatrix} \begin{bmatrix} 1 \\ 0 \\ 1 \end{bmatrix},$$

$$x = \begin{bmatrix} x_1 \\ x_2 \\ x_3 \end{bmatrix} = \begin{bmatrix} 0 \\ -1 \\ 1 \end{bmatrix}.$$

The solution set for the given system is $S = \{(0, -1, 1)\}$. By the use of Theorem 1.3.4, this concludes the solution process.

Theorem 1.3.5. *Assume that $m = n$ and $b = 0$ in the system of linear inhomogeneous equations in (1.10) or (1.12), i.e. $Ax = 0$. The system of linear homogeneous equations in (1.15) has a unique solution $x = 0$ iff the coefficient matrix A is invertible.*

1.3. Exercises

(1) Let $A = \begin{bmatrix} 1 & 3 \\ 3 & 1 \end{bmatrix}$, $B = \begin{bmatrix} 2 & 0 & -3 \\ 3 & 1 & 4 \end{bmatrix}$, and $C = \begin{bmatrix} -2 & 0 \\ 1 & -2 \\ 1 & -3 \end{bmatrix}$. Compute:

  (a) CB and BC. Are they equal? Justify.
  (b) Is $A + B$ defined? Justify.
  (c) Is $A(BC)$ defined? If "YES," then show that $A(BC) = (AB)C$.
  (d) A^2 and A^3.
  (e) $3A^2 + 5A - 4I$.
  (f) C^T.

(2) Write the following systems in matrix form:

$$\text{(a)}\ \begin{aligned} x &= 0 \\ y &= 0 \\ z &= 0 \end{aligned} \qquad \text{(b)}\ \begin{aligned} x_1 + 2x_2 &= 0 \\ -2x_1 + 3x_3 &= 0 \\ x_1 + x_2 + x_3 &= 0 \end{aligned}$$

$$\text{(c)}\ \begin{aligned} x + y + z &= 1 \\ -y + z &= 0 \\ x + 2y + z &= 1 \end{aligned} \qquad \text{(d)}\ \begin{aligned} 2a + b + 3c + 4d &= 0 \\ 3a - b + 2c &= 3 \\ -2a + b - 4c + 3d &= 2 \end{aligned}$$

(3) Using the Gauss–Jordan reduction technique, solve the systems of linear algebraic equations in Problem 2 (if possible).

(4) Use the method described in Example 1.3.20 to decide whether the following matrices are invertible or not invertible:

$$\text{(a)}\ A = \begin{bmatrix} 1 & 3 \\ 3 & 1 \end{bmatrix} \qquad \text{(b)}\ A = \begin{bmatrix} 2 & 0 & -3 \\ 3 & 1 & 4 \end{bmatrix}$$

$$\text{(c)}\ A = \begin{bmatrix} \cos\theta & \sin\theta \\ -\sin\theta & \cos\theta \end{bmatrix} \qquad \text{(d)}\ A = \begin{bmatrix} 4 & 3 & 2 \\ 3 & 5 & 2 \\ 2 & 2 & 1 \end{bmatrix}$$

(5) Verify that

$$A = \begin{bmatrix} 1 & 2 \\ -1 & 1 \end{bmatrix}, \quad A^{-1} = \begin{bmatrix} \frac{1}{3} & -\frac{2}{3} \\ \frac{1}{3} & \frac{1}{3} \end{bmatrix}$$

are inverses of each other.

(6) Use the method described in Example 1.3.19 to find the unique solutions to the following systems of equations:

(a) $\begin{aligned} x_1 + 3x_2 &= 1 \\ 3x_1 + x_2 &= 0 \end{aligned}$ (b) $\begin{aligned} 4a + 3b + -3c &= 2 \\ 3a + 5b + 2c &= -3 \\ 2a + 2b + c &= 1 \end{aligned}$

1.4 Determinants

The concept of the determinant plays a very important role in verifying the invertibility property of a matrix. Moreover, it is employed in the study of finding the complete set of solutions to a linear system of differential equations.

The concept of the determinant of a square matrix of the size $n \times n$ is based on an inductive process. For this purpose, let us introduce a definition that will be useful for our discussion. For $n > 1$, let us consider an $n \times n$ square matrix A, defined by

$$A = \begin{bmatrix} a_{11} & a_{12} & \cdots & a_{1j} & \cdots & a_{1n} \\ a_{21} & a_{22} & \cdots & a_{2j} & \cdots & a_{2n} \\ \cdots & \cdots & \cdots & \cdots & \cdots & \cdots \\ a_{i1} & a_{i2} & \cdots & a_{ij} & \cdots & a_{in} \\ \cdots & \cdots & \cdots & \cdots & \cdots & \cdots \\ a_{n1} & a_{n2} & \cdots & a_{nj} & \cdots & a_{nn} \end{bmatrix} = (a_{ij})_{n \times n}. \tag{1.16}$$

If we delete the ith row and jth column of the matrix A in (1.16), then the resulting $(n-1) \times (n-1)$ matrix is called (i,j)th submatrix A_{ij}, corresponding to the ith-row and the jth-column entry a_{ij} of the matrix A in (1.16). There are n^2 submatrices of A of the size $(n-1) \times (n-1)$.

Illustration 1.4.1. Let us illustrate the idea of a submatrix for $n = 2, 3$.

(i) First, let us consider the case $n = 1$. From Illustration 1.3.1, it is obvious that the 1×1 matrix $A = (a_{ij})_{1 \times 1}$ has one submatrix of the size 1×1.

(ii) Now, let us consider the case $n = 2$. From Illustration 1.3.1, it is clear that the 2×2 matrix $A = (a_{ij})_{2 \times 2}$ has four submatrices of the size 1×1 namely, $A_{11} = (a_{22})_{1 \times 1}$, $A_{12} = (a_{21})_{1 \times 1}$, $A_{21} = (a_{12})_{1 \times 1}$, and $A_{22} = (a_{11})_{1 \times 1}$. These submatrices correspond to the entries a_{11}, a_{12}, a_{21}, and a_{22}, respectively.

(iii) For the case $n = 3$, the 3×3 matrix $A = (a_{ij})_{3 \times 3}$ in Illustration 1.3.1 has nine submatrices of the size 2×2, namely

$$A_{11} = \begin{bmatrix} a_{22} & a_{23} \\ a_{32} & a_{33} \end{bmatrix}, \quad A_{12} = \begin{bmatrix} a_{21} & a_{23} \\ a_{31} & a_{33} \end{bmatrix}, \quad A_{13} = \begin{bmatrix} a_{21} & a_{22} \\ a_{31} & a_{32} \end{bmatrix},$$

$$A_{21} = \begin{bmatrix} a_{12} & a_{13} \\ a_{32} & a_{33} \end{bmatrix}, \quad A_{22} = \begin{bmatrix} a_{11} & a_{13} \\ a_{31} & a_{33} \end{bmatrix}, \quad A_{23} = \begin{bmatrix} a_{11} & a_{12} \\ a_{31} & a_{32} \end{bmatrix},$$

$$A_{31} = \begin{bmatrix} a_{12} & a_{13} \\ a_{22} & a_{23} \end{bmatrix}, \quad A_{32} = \begin{bmatrix} a_{11} & a_{13} \\ a_{21} & a_{23} \end{bmatrix}, \quad A_{33} = \begin{bmatrix} a_{11} & a_{12} \\ a_{21} & a_{22} \end{bmatrix}.$$

These submatrices correspond to the entries $a_{11}, a_{12}, a_{13}, \ldots$, and a_{33}, respectively.

Example 1.4.1. We are given

$$A = \begin{bmatrix} 4 & 3 & 2 \\ 3 & 5 & 2 \\ 2 & 2 & 1 \end{bmatrix}.$$

Find the submatrices A_{11}, A_{12} and A_{13} which correspond to the entries a_{11}, a_{12}, and a_{13}, respectively. From the definition of the submatrices of a_{11}, a_{12}, and a_{13} we have

$$A_{11} = \begin{bmatrix} 5 & 2 \\ 2 & 1 \end{bmatrix}, \quad A_{12} = \begin{bmatrix} 3 & 2 \\ 2 & 1 \end{bmatrix}, \quad A_{13} = \begin{bmatrix} 3 & 5 \\ 2 & 2 \end{bmatrix}.$$

Now, we present an abstract formulation of the determinant:

Definition 1.4.1. Let $A = (a_{ij})_{n \times n}$ be an $n \times n$ matrix. A determinant of the $n \times n$ square matrix A is the value of a function that is defined on the collection of $n \times n$ square matrices to a set of real/complex numbers. It is denoted by $|A|$ or $\det(A)$.

Illustration 1.4.2. An abstract computational definition of the determinant of an $n \times n$ square matrix A is based on the principle of mathematical induction (PMI). For this purpose, we present the following three special cases:

(i) For $n = 1$, $A = (a_{ij})_{1 \times 1}$. In this case, we define $\det(A)$ as a number a_{11}, and it is denoted by $\det(A) = a_{11} = (-1)^{1+1} a_{11}$. To summarize, for the $n = 1$ determinant of any 1×1 matrix, A is the value of a function defined by $\det(A) = a_{11}$. We note that $\det(A) = a_{11}$ is the first-degree homogeneous polynomial function with one entry, one term and one independent variable. Moreover, the minor of (11) is assumed to be 1. Thus, $\det(A)$ is denoted by $W(a_{11})$.

(ii) For $n = 2$, $A = (a_{ij})_{2 \times 2}$. We define $\det(A)$ as follows:

$$\begin{aligned} \det(A) &= a_{11} \det(A_{11}) - a_{12} \det(A_{12}) = a_{11}a_{22} - a_{12}a_{21} \\ &= -a_{21} \det(A_{21}) + a_{22} \det(A_{22}) = -a_{12}a_{21} + a_{11}a_{22} \\ &= a_{11} \det(A_{11}) - a_{21} \det(A_{21}) = a_{11}a_{22} - a_{12}a_{21} \\ &= -a_{12} \det(A_{12}) + a_{22} \det(A_{22}) = -a_{12}a_{21} + a_{11}a_{22}. \end{aligned}$$

These four computations of $\det(A)$ can be rewritten as

$$\begin{aligned} \det(A) &= a_{11}(-1)^{1+1} \det(A_{11}) + a_{12}(-1)^{1+2} \det(A_{12}) = a_{11}a_{22} - a_{12}a_{21} \\ &= a_{21}(-1)^{2+1} \det(A_{21}) + a_{22}(-1)^{2+2} \det(A_{22}) = -a_{12}a_{21} + a_{11}a_{22} \\ &= a_{11}(-1)^{1+1} \det(A_{11}) + a_{21}(-1)^{2+1} \det(A_{21}) = a_{11}a_{22} - a_{12}a_{21} \\ &= a_{12}(-1)^{1+2} \det(A_{12}) + a_{22}(-1)^{2+2} \det(A_{22}) = -a_{12}a_{21} + a_{11}a_{22}. \end{aligned}$$

The presented discussion suggests a natural pattern for the computation of $\det(A)$. This natural pattern leads to conceptualization of the concept of a determinant. Let us denote $(-1)^{i+j}\det(A_{ij}) = C_{ij}$. The number $(-1)^{i+j}\det(A_{ij})$ is called the (i,j)th cofactor of the matrix A corresponding to the ith-row and the jth-column entry a_{ij} of the matrix A. The number $\det(A_{ij})$ is called the (i,j)th minor of the matrix A corresponding to the ith-row and the jth-column entry a_{ij} of the matrix A. $\det(A_{ij})$ is denoted by M_{ij}. In the context of this notation, $C_{ij} = (-1)^{i+j}M_{ij}$. Moreover, $\det(A_{ij})$'s and hence C_{ij}'s are independent of all the entries of the ith row and jth column of the matrix A. Under this consideration, the $\det(A)$ can be rewritten as

$$\begin{aligned}\det(A) &= \sum_{j=1}^{2} a_{ij}C_{ij}, \quad \text{for any } i = 1,2\\ &= a_{11}a_{22} - a_{12}a_{21};\end{aligned} \tag{1.17}$$

this is called the expansion of "$\det(A)$" by the ith row of the matrix A;

$$\begin{aligned}\det(A) &= \sum_{i=1}^{2} a_{ij}C_{ij}, \quad \text{for any } j = 1,2\\ &= a_{11}a_{22} - a_{12}a_{21};\end{aligned} \tag{1.18}$$

this is called the expansion of "$\det(A)$" by the jth column of the matrix A.

Observation 1.4.1. From the above discussion and the definition of the determinant of any 2×2 matrix A, we further present the following conclusions:

(1) The value of the determinant of any 2×2 matrix A is independent of a row or column expansion of the matrix A. Thus, given an arbitrary 2×2 matrix $A = (a_{ij})_{2\times 2}$, its determinant is uniquely determined by either the row or column expansion. This is due to the fact that for each i or each j, $\det(A)$ in (1.17) or (1.18) is a well-defined linear combination of values of determinants $\det(A_{ij})$ of 1×1 submatrices A_{ij}.
(2) From Conclusion 1, we infer that the determinant of the 2×2 matrix is a function defined on a collection of 2×2 matrices with values in a set of real/complex numbers.
(3) We note that $\det(A)$ of any 2×2 matrix A has $2! = 2.1$ terms. Each term is the product of two distinct entries of the matrix. Moreover, it is the second-degree homogeneous polynomial function of 2^2 entries of the matrix as the independent variables. It is denoted by $\det(A) = W(a_{11}, a_{12}, a_{21}, a_{22})$.
(4) We observe that the size of any (i,j)th submatrix A_{ij} of any 2×2 matrix $A = (a_{ij})_{2\times 2}$ is 1×1. In the light of this, for $k \neq i$ and $l \neq j$, $i,k = 1,2$, the (i,j)th minor is the entry a_{kl} of the matrix A at the kth row and lth column. M_{ij} is denoted by $a_{kl}M_{kl}(ij)$, where $M_{kl}(ij) = 1$. For $n = 2$, we note that $M_{kl}(ij)$

can be considered as the (k,l)th minor of the submatrix A_{ij} corresponding to the kth-row and the lth-column entry a_{kl} of the original matrix in the context of the entry a_{kl} of the submatrix A_{ij} of the matrix A.

(5) Moreover, M_{ij}'s and C_{ij}'s differ by a constant factor, $(-1)^{i+j}$. Therefore, both M_{ij}'s and C_{ij}'s are independent of all the entries of the ith row and jth column of the matrix A. Of course, both $\det(A_{ij})$ and C_{ij} depend on the remaining $(2-1)^2 = 1$ entries. In fact, the ith row expansion of $\det(A)$ depends on all rows of the matrix A, and M_{ij}/C_{ij} depends on all rows of A except for the ith row. The jth column expansion of $\det(A)$ depends on all columns of A, and M_{ij}/C_{ij} depends on all columns except for the jth column. Analogous statements can be made with regard to $M_{kl}(ij)$'s defined in Conclusion 4. We further note that $M_{kl}(ij)$'s/$C_{kl}(ij)$'s defined in Conclusion 4 are independent of the ith row, kth row, jth column, and lth column of the matrix A.

(6) In addition, we note that

$$\det(A) = W(a_{11}, a_{12}, a_{21}, a_{22})$$

is a real/complex-number-valued function of four variables. From Illustration 1.3.1, $\det(A)$ can be considered as the function of two-dimensional row A_1, A_2 or column A^1, A^2 vectors. With this setup, it is represented by

$$\det(A) = W(A_1, A_2) = W(A^1, A^2).$$

Moreover, for any $i = 1, 2$, from (1.4) [(1.5)], $\det(A)$ can be considered as the product of two matrices $A_i(A_i^T)$ and $C_i^T(C_i)$ which are the ith row of the matrix A and the ith column of the matrix $C^T = (C_{ji})_{2\times 2}$, respectively. In other words, $\det(A) = A_i C_i^T = C_i A_i^T$. From the definition of the determinant of any 2×2 matrix A and Conclusion 4, we have another representation of $\det(A)$:

$$\det(A) = \sum_{i=1}^{2} a_{ij}(-1)^{i+j} M_{ij} = \sum_{i=1}^{2} a_{ij}(-1)^{i+j} A_k(ij) C_k^T(ij),$$

with $M_{ij} = A_k(ij)C_k^T(ij)$, where $A_k(ij)$ is the kth row of the matrix A obtained after deleting the ith-row and jth-column vectors of the matrix A for any $i, k = 1, 2$ and $k \neq i$; in this case, it is the entry a_{kl} for $k \neq i$ and $l \neq j$, $i, k = 1, 2$.

(7) In summary, for $n = 2$, $\det(A)$ has the row representations:

$$\begin{aligned} W(A_1, A_2) = \det(A) &= \sum_{j=1}^{2} a_{ij} C_{ij} = A_i C_i^T = \sum_{i=1}^{2} a_{ij}(-1)^{i+j} M_{ij} = C_i A_i^T \\ &= \sum_{i=1}^{2} a_{ij}(-1)^{i+j} A_k(ij) C_k^T(ij), \end{aligned}$$

for any $i, k = 1, 2$ and $k \neq i$. Moreover, C_{ij}'s are independent of the ith row and jth column of the matrix A, and $C_{kl}(ij)$'s are independent of the ith row, kth row, jth column, and lth column of the matrix A. Similar column representations for $\det(A)$ can be formulated.

Example 1.4.2. We consider

$$A = \begin{bmatrix} 5 & 2 \\ 2 & 1 \end{bmatrix}.$$

Find the determinant of A. We use the expansion of the determinant A by the first row expansion (1.4); we get

$$\det(A) = 5(1) - 2(2) = 5 - 4 = 1.$$

The answer is $\det(A) = 1$, which was our objective.

Illustration 1.4.3. For $n = 3$, the 3×3 matrix $A = (a_{ij})_{3\times3}$, by following the argument used in Illustration 1.4.2(ii) and the submatrices defined in Illustration 1.4.1, we define $\det(A)$ as follows:

$$\det(A) = a_{11}\det(A_{11}) - a_{12}\det(A_{12}) + a_{13}\det(A_{13}).$$

Here the submatrices A_{11}, A_{12}, and A_{13} are all 2×2 matrices. Therefore, we use the definition of the determinant of 2×2 matrix in Illustration 1.4.2(ii), and compute them as

$$\begin{aligned} \det(A_{11}) &= a_{22}a_{33} - a_{23}a_{32}, \\ \det(A_{12}) &= a_{21}a_{33} - a_{23}a_{31}, \\ \det(A_{13}) &= a_{21}a_{32} - a_{22}a_{31}. \end{aligned}$$

By substituting these expressions, we get

$$\begin{aligned} \det(A) &= a_{11}(a_{22}a_{33} - a_{23}a_{32}) - a_{12}(a_{21}a_{33} - a_{23}a_{31}) \\ &\quad + a_{13}(a_{21}a_{32} - a_{22}a_{31}) \\ &= a_{11}a_{22}a_{33} - a_{11}a_{23}a_{32} - a_{12}a_{21}a_{33} \\ &\quad + a_{12}a_{23}a_{31} + a_{13}a_{21}a_{32} - a_{13}a_{22}a_{31}. \end{aligned}$$

If we repeat this expansion process with respect to each row or each column of the 3×3 matrix A, then, just like in the case of the 2×2 matrix, we will get the same expression as in the case of the first row expansion. The detailed verification of this statement is left to the reader. The right-hand side expression in the above is the value of the determinant of any 3×3 matrix A. As before, it can be written as:

$$\begin{aligned} \det(A) &= \sum_{j=1}^{3} a_{ij}C_{ij}, \quad \text{for any } i = 1, 2, 3 \\ &= a_{11}a_{22}a_{33} - a_{11}a_{23}a_{32} - a_{12}a_{21}a_{33} \\ &\quad + a_{12}a_{23}a_{31} + a_{13}a_{21}a_{32} - a_{13}a_{22}a_{31}; \end{aligned} \tag{1.19}$$

this is called the expansion of "det (A)" by the ith row of the matrix A;

$$\begin{aligned}\det(A) &= \sum_{i=1}^{3} a_{ij}C_{ij}, \quad \text{for any } j = 1,2,3 \\ &= a_{11}a_{22}a_{33} - a_{11}a_{23}a_{32} - a_{12}a_{21}a_{33} \\ &\quad + a_{13}a_{21}a_{32} + a_{12}a_{23}a_{31} - a_{13}a_{22}a_{31};\end{aligned} \tag{1.20}$$

this is called the expansion of "det (A)" by the jth column of the matrix A.

From the above discussion and definition of the determinant of any 3×3 matrix A, conclusions similar to those regarding 2×2 matrices can be reformulated with respect to 3×3 matrices. To minimize the repetition, details are left to the reader.

Example 1.4.3. We are given

$$A = \begin{bmatrix} 4 & 3 & 2 \\ 3 & 5 & 2 \\ 2 & 2 & 1 \end{bmatrix}.$$

Find $\det(A)$.

Solution procedure.

Step 1. We utilize the expansion of the determinant of A by the first-row expansion (1.19). First, we need to find the cofactors C_{11}, C_{12}, and C_{13} of the matrix A corresponding to the first row and the first column entry $a_{11} = 4$, the first-row and second-column entry $a_{12} = 3$, and the first row and third column entry $a_{13} = 2$, respectively. They are:

$$C_{11} = (-1)^{1+1}\det(A_{11}) = (-1)^{1+1}\det\left(\begin{bmatrix} 5 & 2 \\ 2 & 1 \end{bmatrix}\right) = (-1)^{1+1}(5-4) = 1,$$

$$C_{12} = (-1)^{1+2}\det(A_{12}) = (-1)^{1+2}\det\left(\begin{bmatrix} 3 & 2 \\ 2 & 1 \end{bmatrix}\right) = (-1)^{1+2}(3-4) = 1,$$

$$C_{13} = (-1)^{1+3}\det(A_{14}) = (-1)^{1+3}\det\left(\begin{bmatrix} 3 & 5 \\ 2 & 2 \end{bmatrix}\right) = (-1)^{1+3}(6-10) = -4.$$

Step 2. Now, we are ready to utilize the formula (1.19) to compute the determinant of A:

$$\begin{aligned}\det(A) &= \sum_{j=1}^{3} a_{1j}C_{1j}, \quad \text{for } i = 1 \\ &= a_{11}C_{11} + a_{12}C_{12} + a_{13}C_{13} \\ &= 4(1) + 3(1) + 2(-4) \\ &= -1.\end{aligned}$$

The determinant of A is -1, which was our objective.

Next, we are ready to make the computational definition of the determinant of the $n \times n$ matrix A. This is based on the principle of mathematical induction.

Definition 1.4.2. For $n > 1$, let $A = (a_{ij})_{n\times n}$ be an $n\times n$ matrix. The determinant of the matrix A is defined by:

(1) For any $i = 1, 2, \ldots, n$,

$$\det(A) = \sum_{j=1}^{n} a_{ij}C_{ij} = \sum_{j=1}^{n} a_{ij}(-1)^{i+j}\det(A_{ij});$$

this is called the expansion of "det (A)" by the ith row of the matrix A;

(2) For any $j = 1, 2, \ldots, n$,

$$\det(A) = \sum_{i=1}^{n} a_{ij}C_{ij} = \sum_{i=1}^{n} a_{ij}(-1)^{i+j}\det(A_{ij});$$

this is called the expansion of "det (A)" by the jth column of the matrix A.

Observation 1.4.2. Based on Illustrations 1.4.2 and 1.4.3, Observation 1.4.1, Definition 1.4.2, and Principle of Mathematical Induction 1.2.3, we again summarize the conclusions drawn from Illustration 1.4.2 in the context of $n \times n$ matrices:

(1) The value of the determinant of any $n \times n$ matrix A is independent of a row or column expansion of the matrix A. Thus, given an arbitrary $n \times n$ matrix $A = (a_{ij})_{n\times n}$, its determinant is uniquely determined by either the row or column expansion. This is because of the fact that for each i or each j, $\det(A)$ in Definition 1.4.2(1) or (2) is the finite sum of a well-defined scalar multiple of values of determinants det (A_{ij}) of $(n-1) \times (n-1)$ submatrices A_{ij}.

(2) From Conclusion 1, we infer that the determinant of the $n \times n$ matrix is a function defined on a collection of $n \times n$ matrices with values in a set of real/complex numbers.

(3) We note that $\det(A)$ of any $n \times n$ matrix A has $n! = n(n-1)\ldots 3.2.1$ terms. Each term is the product of n distinct entries of the matrix. Moreover, it is the nth-degree homogeneous polynomial function of n^2 entries of the matrix as the independent variables. It is denoted by

$$\det(A) = W\left(a_{11}, \ldots, a_{1n}, \ldots, a_{i1}, \ldots, a_{in}, \ldots, a_{n1}, \ldots, a_{nn}\right).$$

(4) We note that the size of any (i, j)th submatrix A_{ij} of any $n \times n$ matrix $A = (a_{ij})_{n\times n}$ is $(n-1)\times(n-1)$. In this case, the (i, j)th minor M_{ij} is the determinant of the $(n-1) \times (n-1)$ submatrix A_{ij} of the matrix A corresponding to the ith row and jth column of the matrix A. M_{ij} is independent of all the entries of the ith row and jth column of the matrix A. Moreover, $C_{ij} = (-1)^{i+j}M_{ij}$ can

be considered as a function of either the $n-1$ rows

$$A_1(1j),\ldots,A_{i-1}(i-1j),A_{i+1}(i+1j),\ldots,A_n(nj)$$

or the $n-1$ columns

$$A^1(i1),\ldots,A^{j-1}(ij-1),A^{j+1}(ij+1),\ldots,A^n(in).$$

This is due to the fact that

$$\begin{aligned}&A_1(1j),\ldots,A_{i-1}(i-1j),A_{i+1}(i+1j),\ldots,A_n(nj),\\&A^1(i1),\ldots,A^{j-1}(ij-1),A^{j+1}(ij+1),\ldots,A^n(in)\end{aligned}$$

are row and column vectors of the submatrix $A_{ij}=(a_{kl})_{(n-1)\times(n-1)}$ for $k\neq i$ and $l\neq j$, respectively. For any $k\neq i$, $i,k=1,2,\ldots,n$, $\det(A_{ij})$ can be computed by using the kth-row A_k expansion of the original matrix A, i.e. the kth-row vector $A_k(ij)$ of the submatrix A_{ij} obtained from A_k after deleting ith row the jth column of the matrix A. In short, A_k and $A_k(ij)$ are n- and $(n-1)$-dimensional kth-row vectors of the matrix A and submatrix A_{ij}, respectively. For $k\neq i$ and $l\neq j$, $i,k=1,2,\ldots,n$, a_{kl} is the entry at the kth-row and lth-column of the matrix A, and it is the component of the row vector $A_k(ij)$ of the submatrix A_{ij}. Hence, by using the definition of the determinant of $(n-1)\times(n-1)$, $M_{kl}(ij)$ is determined by this kth-row expansion. For simplicity, for $k\neq i$ and $l\neq j$, $M_{kl}(ij)$ is referred to as the (k,l)th minor of the submatrix A_{ij} corresponding to the kth-row and the lth-column entry a_{kl} of the original matrix in the context of the (k,l)th entry a_{kl} of the submatrix A_{ij} of the matrix A. Of course, its exact representation is not essential to our discussion. We just need to know some information about $M_{kl}(ij)$ and its corresponding cofactor $C_{kl}(ij)$ as the functions of the row and column matrix of the matrix A. From Definition 1.4.2, for any $k=1,2,\ldots,i-1,i+1,\ldots,n$, as per the description in Conclusion 4, we have

$$M_{ij}=\det(A_{ij})=\sum_{l\neq j}^{n}a_{kl}C_{kl}(ij)=A_k(ij)C_k^T(ij)=C_k(ij)A_k^T(ij),$$

where $M_{kl}(ij)$ is the (k,l)th minor of the submatrix A_{ij} corresponding to the kth-row and the lth-column entry a_{kl} of the original matrix in the context of the entry a_{kl} of the submatrix A_{ij} of the matrix A, for any $k\neq i$ and $l\neq j$, $i,k=1,2,3,\ldots,n$.

(5) Moreover, M_{ij}'s and C_{ij}'s differ by a constant factor, $(-1)^{i+j}$. Therefore, both M_{ij}'s and C_{ij}'s are independent of all the entries of the ith row and jth column of the matrix A. Of course, both $\det(A_{ij})$ and C_{ij} depend on the remaining entries $(n-1)^2$. In fact, the ith-row expansion of $\det(A)$ depends on all rows, and M_{ij}/C_{ij} depends on all rows of A except for the ith-row. The jth-column expansion of $\det(A)$ depends on all columns, and M_{ij}/C_{ij} depends on all columns of A except for the jth column. An analogous statement can be made

with regard to $M_{kl}(ij)$'s defined in Conclusion 4. In fact, the kth-row expansion of $\det(A_{ij}) = M_{ij}$, corresponding to the kth row of the original matrix A, depends on all rows of the matrix A except for the ith row, and $M_{kl}(ij)/C_{kl}(ij)$ depends on all rows except for the ith and kth rows. The lth-column expansion of $\det(A_{ij}) = M_{ij}$, corresponding to the lth column of the original matrix A, depends on all columns except for the jth column of A, and $M_{kl}(ij)/C_{kl}(ij)$ depends on all columns of A except for the jth and lth columns of the matrix A. We further note that $M_{kl}(ij)$ and $C_{kl}(ij)$, as defined in Conclusion 4, are independent of the ith row, kth row, jth column, and lth column of the matrix A.

(6) In addition, we note that

$$\det(A) = W(a_{11}, \ldots, a_{1n}, \ldots, a_{i1}, \ldots, a_{in}, \ldots, a_{n1}, \ldots, a_{nn})$$

is a real/complex-number-valued function of n^2 variables. From Illustration 1.3.1, $\det(A)$ can be considered as the function of n n-dimensional row vectors $A_1, A_2, \ldots, A_i, \ldots, A_n$, or column vectors $A^1, A^2, \ldots, A^j, \ldots, A^n$. In this setup, it is represented by

$$\det(A) = W(A_1, \ldots, A_j, \ldots, A_n) = W(A^1, \ldots, A^i, \ldots, A^n).$$

Moreover, for any $i = 1, 2, \ldots, n$, from Definition 1.4.2, $\det(A)$ can be considered as the product of two matrices A_i and C_i^T (A_i^T and C_i) that are the ith row of the matrix A and the jth column of its adjoint matrix C^T, respectively. That is to say, $\det(A) = A_i C_i^T = C_i A_i^T$. From the definition of the determinant of any $n \times n$ matrix A and Conclusion 4, we have another representation of $\det(A)$,

$$\det(A) = \sum_{i=1}^{n} a_{ij}(-1)^{i+j} C_k(ij) A_k^T(ij),$$

where $A_k(ij)$ is the kth row of the matrix A obtained by deleting the ith-row and jth-column vectors of the matrix A for any $i, k = 1, 2, \ldots, n$ and $k \neq i$.

(7) In summary, for any positive integer $n > 1$, $\det(A)$ has the following row representations:

$$W(A_1, A_2, \ldots, A_i, \ldots, A_n) = \det(A) = \sum_{j=1}^{n} a_{ij} C_{ij} = A_i C_i^T = C_i A_i^T$$

$$= \sum_{j=1}^{n} a_{ij}(-1)^{i+j} C_k(ij) A_k^T(ij),$$

for any i and $k, i, k = 1, 2, \ldots, n$ and $k \neq i$. Moreover, C_{ij}'s are independent of the ith row and jth column of the matrix A, and $C_k^T(ij)$'s are independent of the ith row, kth row, and jth, and lth column of the matrix A. A similar column representation for $\det(A)$ can be formulated.

(8) In the case of $n = 1$, $A = (a_{ij})_{1\times 1}$, the cofactor C_{ij} of a_{ij} is defined to be 1. Hence $\det(A) = (-1)^{1+1} a_{11} C_{11} = a_{11}$.

If the size of the matrix is large, the computation of the determinant of the $n \times n$ matrix A by Definition 1.4.2 is not efficient. However, the concept of the determinant possesses several properties that are useful for increasing the efficiency of the computational procedure.

Theorem 1.4.1. *Let $A = (a_{ij})_{n \times n}$ be an $n \times n$ matrix.*

(D_1) $\det(A^T) = \det(A)$, *where A^T is the transpose of A.*
(D_2) $\det(A') = -\det(A)$, *where A' is obtained from A by interchanging two adjacent rows (or columns) of the matrix.*
(D_3) *If A has two identical rows (or columns), then* $\det(A) = 0$.
(D_4) $\det(A') = c\det(A)$, *where A' is obtained from A by multiplying a row (or column) by a scalar c quantity.*
(D_5) $\det(A) = \det(A') + \det(A'')$, *where each entry of the ith row (or jth column) of the matrix A as a sum $(a'_{ij} + a''_{ij})$, A' and A'' have the same entries as A except in their ith row (or jth column), in which they have entries a'_{ij} and a''_{ij}, respectively, $j = 1, 2, \ldots, n$ (or $i = 1, 2, \ldots, n$).*
(D_6) $\det(A') = \det(A)$, *where A' is obtained from A by adding a scalar multiple of a kth row to an ith row (or an lth column to a jth column with $l \neq j$) with $k \neq i$.*
(D_7) *If A is a triangular matrix (upper/lower), then*

$$\det(A) = a_{11}a_{22}\cdots a_{jj}\cdots a_{nn},$$

the product of its diagonal entries.

The following result shows that the determinant of the product is equal to the product of their determinants.

Theorem 1.4.2. *Let $A = (a_{ij})_{n \times n}$ and $B = (b_{ij})_{n \times n}$ be $n \times n$ matrices. Then,* $\det(AB) = \det(A)\det(B)$.

Example 1.4.4. We are given

$$A = \begin{bmatrix} 3 & 2 \\ 2 & 2 \end{bmatrix}, \quad B = \begin{bmatrix} 3 & 2 \\ 2 & 5 \end{bmatrix}.$$

Show that $\det(A)\det(B) = \det(AB)$.

Solution procedure.

Step 1. For this purpose, we need to find $\det(A)$, $\det(B)$, and $\det(AB)$. First, let us find AB, which is

$$AB = \begin{bmatrix} 3 & 2 \\ 2 & 2 \end{bmatrix} \begin{bmatrix} 3 & 2 \\ 2 & 5 \end{bmatrix} = \begin{bmatrix} 13 & 16 \\ 10 & 14 \end{bmatrix}.$$

Step 2. Now we compute $\det(A)$, $\det(B)$, and $\det(AB)$. They are

$$\det(A) = 2, \quad \det(B) = 11, \quad \det(AB) = 182 - 160 = 22.$$

It is clear that $\det(AB) = 22 = 2(11) = \det(A)\det(B)$.

Definition 1.4.3. Let $A = (a_{ij})_{n\times n}$ be an $n \times n$ matrix, and let $C = (C_{ij})_{n\times n}$, where C_{ij} is the (i,j)th cofactor of the matrix A corresponding to a_{ij}. The adjoint matrix of an $n \times n$ matrix A is the transpose of the matrix of cofactors C. It is denoted by $\operatorname{adj} A = C^T$.

Example 1.4.5. Let us consider

$$A = \begin{bmatrix} 4 & 3 & 2 \\ 3 & 5 & 2 \\ 2 & 2 & 1 \end{bmatrix}.$$

Find the adjoint matrix of the given matrix A.

Solution procedure. For this purpose, we first need to find submatrices corresponding to all the entries of the given matrix A.

Step 1. From Illustration 1.4.1(iii), the submatrices corresponding to the entries a_{11}, a_{12}, a_{13}, a_{21}, a_{22}, a_{23}, a_{31}, a_{32}, and a_{33} of the matrix A are

$$A_{11} = \begin{bmatrix} a_{22} & a_{23} \\ a_{32} & a_{33} \end{bmatrix} = \begin{bmatrix} 5 & 2 \\ 2 & 1 \end{bmatrix}, \quad A_{12} = \begin{bmatrix} a_{21} & a_{23} \\ a_{31} & a_{33} \end{bmatrix} = \begin{bmatrix} 3 & 2 \\ 2 & 1 \end{bmatrix},$$

$$A_{13} = \begin{bmatrix} a_{21} & a_{22} \\ a_{31} & a_{32} \end{bmatrix} = \begin{bmatrix} 3 & 5 \\ 2 & 2 \end{bmatrix},$$

$$A_{21} = \begin{bmatrix} a_{12} & a_{13} \\ a_{32} & a_{33} \end{bmatrix} = \begin{bmatrix} 3 & 2 \\ 2 & 1 \end{bmatrix}, \quad A_{22} = \begin{bmatrix} a_{11} & a_{13} \\ a_{31} & a_{33} \end{bmatrix} = \begin{bmatrix} 4 & 2 \\ 2 & 1 \end{bmatrix},$$

$$A_{23} = \begin{bmatrix} a_{11} & a_{12} \\ a_{31} & a_{32} \end{bmatrix} = \begin{bmatrix} 4 & 3 \\ 2 & 2 \end{bmatrix},$$

$$A_{31} = \begin{bmatrix} a_{12} & a_{13} \\ a_{22} & a_{23} \end{bmatrix} = \begin{bmatrix} 3 & 2 \\ 5 & 2 \end{bmatrix}, \quad A_{32} = \begin{bmatrix} a_{11} & a_{13} \\ a_{21} & a_{23} \end{bmatrix} = \begin{bmatrix} 4 & 2 \\ 3 & 2 \end{bmatrix},$$

$$A_{33} = \begin{bmatrix} a_{11} & a_{12} \\ a_{21} & a_{22} \end{bmatrix} = \begin{bmatrix} 4 & 3 \\ 3 & 5 \end{bmatrix}.$$

Step 2. Now we compute the cofactors corresponding to the entries a_{11}, a_{12}, a_{13}, a_{21}, a_{22}, a_{23}, a_{31}, a_{32}, and a_{33} of the matrix A, and they are

$$C_{11} = (-1)^{1+1}\det(A_{11}) = \det\left(\begin{bmatrix} 5 & 2 \\ 2 & 1 \end{bmatrix}\right) = 1,$$

$$C_{12} = (-1)^{1+2}\det(A_{12}) = -\det\left(\begin{bmatrix} 3 & 2 \\ 2 & 1 \end{bmatrix}\right) = 1,$$

$$C_{13} = (-1)^{1+3}\det(A_{13}) = \det\left(\begin{bmatrix} 3 & 5 \\ 2 & 2 \end{bmatrix}\right) = -4,$$

$$C_{21} = (-1)^{2+1}\det(A_{21}) = -\det\left(\begin{bmatrix} 3 & 2 \\ 2 & 1 \end{bmatrix}\right) = 1,$$

$$C_{22} = (-1)^{2+2}\det(A_{22}) = \det\left(\begin{bmatrix} 4 & 2 \\ 2 & 1 \end{bmatrix}\right) = 0,$$

$$C_{23} = (-1)^{2+3}\det(A_{23}) = -\det\left(\begin{bmatrix} 4 & 3 \\ 2 & 2 \end{bmatrix}\right) = -2,$$

$$C_{31} = (-1)^{3+1}\det(A_{31}) = \det\left(\begin{bmatrix} 3 & 2 \\ 5 & 2 \end{bmatrix}\right) = -4,$$

$$C_{32} = (-1)^{3+2}\det(A_{32}) = -\det\left(\begin{bmatrix} 4 & 2 \\ 3 & 2 \end{bmatrix}\right) = -2,$$

$$C_{33} = (-1)^{3+3}\det(A_{33}) = \det\left(\begin{bmatrix} 4 & 3 \\ 3 & 5 \end{bmatrix}\right) = 11.$$

Step 3. The cofactor matrix corresponding to the matrix A is

$$C = \begin{bmatrix} 1 & 1 & -4 \\ 1 & 0 & -2 \\ -4 & -2 & 11 \end{bmatrix}.$$

From Definition 1.4.3, the adjoint matrix of A is

$$\operatorname{adj} A = C^T = \begin{bmatrix} 1 & 1 & -4 \\ 1 & 0 & -2 \\ -4 & -2 & 11 \end{bmatrix},$$

which establishes our goal.

Theorem 1.4.3. *Let $A = (a_{ij})_{n\times n}$ be an $n\times n$ matrix. Then, $A\operatorname{adj}(A) = I\det(A)$.*

Example 1.4.6. We are given

$$A = \begin{bmatrix} 4 & 3 & 2 \\ 3 & 5 & 2 \\ 2 & 2 & 1 \end{bmatrix}.$$

Verify that $A\operatorname{adj}(A) = I\det(A)$.

Solution procedure. For this, we follow the procedure for finding $\operatorname{adj}(A)$, and determine $\operatorname{adj}(A)$.

Step 1. In this case, we have found adj(A) and det(A) in Examples 1.4.5 and 1.4.3, respectively, and they are

$$\operatorname{adj}(A) = \begin{bmatrix} 1 & 1 & -4 \\ 1 & 0 & 2 \\ -4 & -2 & 11 \end{bmatrix}, \quad \det(A) = -1.$$

Step 2. Now, we compute $A\operatorname{adj}(A)$ as follows:

$$\begin{aligned} A\operatorname{adj}(A) &= \begin{bmatrix} 4 & 3 & 2 \\ 3 & 5 & 2 \\ 2 & 2 & 1 \end{bmatrix} \begin{bmatrix} 1 & 1 & -4 \\ 1 & 0 & -2 \\ -4 & -2 & 11 \end{bmatrix} \\ &= \begin{bmatrix} -1 & 0 & 0 \\ 0 & -1 & 0 \\ 0 & 0 & -1 \end{bmatrix} \\ &= (-1) \begin{bmatrix} 1 & 0 & 0 \\ 0 & 1 & 0 \\ 0 & 0 & 1 \end{bmatrix} = I \det(A), \end{aligned}$$

which is our intended result.

The following result provides a test for the invertibility property of matrices.

Theorem 1.4.4. *Let $A = (a_{ij})_{n\times n}$ be an $n \times n$ matrix. The matrix A is invertible iff* $\det(A) \neq 0$.

Observation 1.4.3

(a) Assume that $m = n$ in the system of linear nonhomogeneous Equations (1.10) or (1.12). Further assume that the determinant of the coefficient matrix A is nonzero, i.e. $\det(A) \neq 0$. Then the system of Equations (1.12) has a unique solution. Moreover, this unique solution is determined by $x = A^{-1}b$.

(b) Assume that $m = n$ and $b = 0$ in the system of linear nonhomogeneous Equations (1.10) or (1.12). Then we have $Ax = 0$. Hence, the system of linear homogeneous Equations (1.15) has a unique solution $x = 0$ iff the determinant of the coefficient matrix A is nonzero, i.e. $\det(A) \neq 0$. This statement is equivalent to the statement "A system of linear homogeneous Equations (1.15) has a nonzero solution iff $\det(A) = 0$."

(c) We note that if the determinant of a matrix A is zero, then the matrix A is called a singular matrix.

Example 1.4.7. We consider

$$\begin{aligned} (1-\lambda)\,x_1 + x_2 - x_3 &= 0, \\ x_1 - (1+\lambda)\,x_2 + x_3 &= 0, \\ -x_1 + x_2 + (1-\lambda)\,x_3 &= 0. \end{aligned}$$

Goals:

(1) To find the values of λ so that the given system has nonzero solutions;
(2) To find the values λ so that the given system has a unique solution.

Solution procedure. The coefficient matrix associated with the given system is

$$A = \begin{bmatrix} 1-\lambda & 1 & -1 \\ 1 & -(1+\lambda) & 1 \\ -1 & 1 & 1-\lambda \end{bmatrix}.$$

Step 1. By using the definition of the determinant of a matrix [formula (1.19)], we compute the determinant of A as follows:

$$\begin{aligned} \det(A) &= (1-\lambda)[-(1+\lambda)(1-\lambda)-1] - [(1-\lambda)+1] - [1-(1+\lambda)] \\ &= (1-\lambda)[\lambda^2-2] + \lambda - 2 + \lambda \quad \text{(by simplification)} \\ &= (1-\lambda)(\lambda^2-2) + 2\lambda - 2 \quad \text{(further simplification)} \\ &= (1-\lambda)(\lambda^2-2) - 2(1-\lambda) \quad \text{(by grouping)} \\ &= (1-\lambda)(\lambda^2-4) \quad \text{(by factorization)} \\ &= (1-\lambda)(\lambda-2)(\lambda+2) \quad \text{(further factorization)}. \end{aligned}$$

Step 2. First, we need to find values of λ for which $\det(A)$ is zero. For this purpose, we set $\det(A)$ in step 1 equal to zero, and then solve for λ (if possible). In this case

$$\det(A) = (1-\lambda)(\lambda-2)(\lambda+2) = 0.$$

The solution set for this algebraic equation is $\{-2, 1, 2\}$. For $\lambda : \lambda_1 = -2$, $\lambda_2 = 1$, $\lambda_3 = 2$ are the values of λ for which $\det(A) = 0$. Therefore, by applying Observation 1.4.3(b), we conclude that the given system has nonzero solutions. For any λ different from $-2, 1$, or 2, $\det(A) \neq 0$. Hence, again by applying Observation 1.4.3(b), we conclude that for any λ different from $-2, 1$, or 2, the given system has a unique solution.

Observation 1.4.4. Let $A = (a_{ij})_{n\times n}$ be an $n \times n$ invertible matrix. Then,

$$A^{-1} = \frac{\operatorname{adj}(A)}{\det(A)}.$$

Hence, Theorem 1.4.4 provides an another method of finding an inverse of a matrix (if its inverse exists). This is illustrated in the following example.

Example 1.4.8. Let us consider

$$A = \begin{bmatrix} 4 & 3 & 2 \\ 3 & 5 & 2 \\ 2 & 2 & 1 \end{bmatrix}.$$

Using the adjoint matrix, find the inverse of A (if it exists).

Solution procedure.

Step 1. To fulfill the goal of the problem, we imitate the procedure for finding an adjoint of the matrix in Example 1.4.5, and we obtain the following relationship (Example 1.4.6):

$$A\,\mathrm{adj}(A) = I\,\det(A).$$

Step 2. If the determinant of the matrix A is different from zero $[\det(A) \neq 0]$, then from Observation 1.4.4 we see that

$$A^{-1} = \frac{\mathrm{adj}(A)}{\det(A)}.$$

Step 3. In this case, from Example 1.4.3, $\det(A) = -1 \neq 0$. Now, from Examples 1.4.5 and 1.4.6, we substitute the values of $\det(A)$ and $\mathrm{adj}(A)$ into the above formula for the inverse of A, and we have

$$A^{-1} = \frac{\begin{bmatrix} 1 & 1 & -4 \\ 1 & 0 & -2 \\ -4 & -2 & 11 \end{bmatrix}}{(-1)} = \begin{bmatrix} -1 & -1 & 4 \\ -1 & 0 & 2 \\ 4 & 2 & -11 \end{bmatrix}.$$

Thus, we are able to find the inverse of the matrix by utilizing the concept of the adjoint of the given matrix.

Definition 1.4.4. Let **n**-tuple $(x_1, x_2, \ldots, x_j, \ldots, x_n)$ be in (1.12), and let it be denoted by $x = (x_j)_{n\times 1}$, where x_j's are numbers (real or complex). For real number x_j, the collection of all $x = (x_j)_{n\times 1}$ is denoted by R^n (complex x_j, C^n). The collection R^n or C^n satisfies the conclusions (A1)–(A4) and (S1)–(S4) of Theorem 1.3.1. Any nonempty collection $\mathcal{V}$ with two binary operations defined on $\mathcal{V}$ into itself satisfying (A1)–(A4) and (S1)–(S4) is called a vector space. Therefore, R^n or C^n are examples of vector spaces.

Illustration 1.4.4. Let $\mathcal{V} = C[[a,b], R]$ be a collection of all continuous functions defined on $[a,b]$ into R. From calculus, we know that (i) if f and g are continuous functions, then their sum (addition operation), defined by

$$(f+g)(x) = f(x) + g(x), \quad x \in [a,b],$$

is also continuous, and that (ii) for any real number α, αf (scalar multiplication operation), defined by

$$(\alpha f)(x) = \alpha f(x), \quad x \in [a,b],$$

is also a continuous function. Moreover, these two operations defined on $C[[a,b], R]$ into itself satisfy the properties (A1)–(A4) and (S1)–(S4) of Theorem 1.3.1. Therefore, by Definition 1.4.4, we conclude that $C\,[[a,b], R]$ is a vector space.

Observation 1.4.5. Let us consider

$$x_1 = (x_{i1})_{n\times 1}, \quad x_2 = (x_{i2})_{n\times 1}, \ldots, x_j = (x_{ij})_{n\times 1}, \ldots, x_m = (x_{in})_{n\times 1}.$$

By interchanging the role of A^j's with x_j's in (1.11) with scalar a_j's, (1.11) can be written as

$$x = a_1x_1 + a_2x_2 + \cdots + a_jx_j + \cdots + a_mx_m = \sum_{j=1}^{m} a_jx_j. \tag{1.21}$$

We note that x belongs to R^n or C^n, and x in (1.21) is called a linear combination of **n**-dimensional column vectors $x_1, x_2, \ldots, x_j, \ldots, x_m$ in R^n or C^n. The collection of all possible linear combinations of n-dimensional vectors $x_1, x_2, \ldots, x_j, \ldots, x_m$ in R^n or C^n is called the span of n-dimensional m vectors $x_1, x_2, \ldots, x_j, \ldots, x_m$ in R^n or C^n; it is denoted by

$$\text{Span}(\{x_1, x_2, \ldots, x_j, \ldots, x_m\}).$$

Theorem 1.4.5. *Let $x_1, x_2, \ldots, x_j, \ldots, x_m$ be n-dimensional vectors in R^n or C^n. Then,*

$$\text{Span}\left(\{x_1, x_2, \ldots, x_j, \ldots, x_m\}\right)$$

is a vector space (subspace of R^n or C^n).

Illustration 1.4.5. Let us consider a set of continuous functions,

$$f_0, f_1, f_2, \ldots, f_i, \ldots, f_{n-1},$$

defined by

$$f_0(t) = 1, \quad f_1(t) = t, \quad f_2(t) = t^2, \ldots, f_i(t) = t^i, \ldots, f_{n-1}(t) = t^{n-1},$$

for $t \in [a, b]$. Let us denote $\mathcal{S}$ by $\mathcal{S} = \{f_0, f_1, f_2, \ldots, f_i, \ldots, f_{n-1}\}$. It is clear that $\mathcal{S}$ is a nonempty subset of $C\,[[a, b], R]$. The span of $\mathcal{S}$ is

$$\text{Span}(\{f_0, f_1, f_2, \ldots, f_i, \ldots, f_{n-1}\}) = P_{n-1},$$

which is a collection of all polynomials of degree at most $n-1$. P_{n-1} is a vector subspace of $C\,[[a, b], R]$.

Observation 1.4.6. Let $\mathcal{S}$ be a set of all solutions to the homogeneous system (1.15), i.e. $Ax = 0$. Then, $\mathcal{S}$ is a vector space (subspace of R^n or C^n). It is called a solution space of the linear homogeneous system of algebraic Equations (1.15).

Example 1.4.9. For given $Ax = 0$, where A is given by

$$A = \begin{bmatrix} 6 & 4 & -1 \\ -3 & -2 & \frac{1}{2} \\ 2 & \frac{4}{3} & -\frac{1}{3} \end{bmatrix}, \quad x = \begin{bmatrix} x_1 \\ x_2 \\ x_3 \end{bmatrix}.$$

Find a solution space of the given homogeneous system of algebraic equations.

Solution procedure. We utilize the elementary reduction technique — the Gauss–Jordan method — to solve this algebraic system. The goal is to reduce the coefficient matrix A,

$$A = \begin{bmatrix} 6 & 4 & -1 \\ -3 & -2 & \frac{1}{2} \\ 2 & \frac{4}{3} & -\frac{1}{3} \end{bmatrix},$$

into an upper-triangular form. In the following, we imitate the procedure outlined in Example 1.3.15.

Step 1. For this purpose, we write the augmented coefficient matrix:

$$[A \mid b] = \left[\begin{array}{ccc|c} 6 & 4 & -1 & 0 \\ -3 & -2 & \frac{1}{2} & 0 \\ 2 & \frac{4}{3} & -\frac{1}{3} & 0 \end{array}\right], \quad \text{where } b = \begin{bmatrix} 0 \\ 0 \\ 0 \end{bmatrix}.$$

Step 2. We apply the following admissible elementary row operations, stated in step 2 of Example 1.3.15, to the augmented/partitioned matrix $[A \mid b]$:

$$[A \mid b] = \left[\begin{array}{ccc|c} 6 & 4 & -1 & 0 \\ -3 & -2 & \frac{1}{2} & 0 \\ 2 & \frac{4}{3} & -\frac{1}{3} & 0 \end{array}\right] \to E_3 \to \left[\begin{array}{ccc|c} 6 & 4 & -1 & 0 \\ 0 & 0 & 0 & 0 \\ 0 & 0 & 0 & 0 \end{array}\right] \to E_1 \to \left[\begin{array}{ccc|c} 1 & \frac{2}{3} & -\frac{1}{6} & 0 \\ 0 & 0 & 0 & 0 \\ 0 & 0 & 0 & 0 \end{array}\right].$$

Step 3. The algebraic system corresponding to the last augmented/partitioned matrix,

$$\left[\begin{array}{ccc|c} 1 & \frac{2}{3} & -\frac{1}{6} & 0 \\ 0 & 0 & 0 & 0 \\ 0 & 0 & 0 & 0 \end{array}\right],$$

is

$$x_1 + \frac{2}{3}x_2 - \frac{1}{6}x_3 = 0.$$

By solving for x_1, we have

$$x_1 = -\frac{2}{3}x_2 + \frac{1}{6}x_3.$$

The solution set for the given system of linear homogeneous system of algebraic equations is described by

$$\begin{aligned} S &= \left\{(x_1, x_2, x_3) \in R^3 : (x_1, x_2, x_3) = \left(-\frac{2}{3}x_2 + \frac{1}{6}x_3, x_2, x_3\right), x_2, x_3 \in R\right\} \\ &= \left\{(x_1, x_2, x_3) \in R^3 : (x_1, x_2, x_3) = \left(-\frac{2}{3}, 1, 0\right)x_2 + \left(\frac{1}{6}, 0, 1\right)x_3, x_2, x_3 \in R\right\} \\ &= \text{Span}\left(\left\{\left(-\frac{2}{3}, 1, 0\right), \left(\frac{1}{6}, 0, 1\right)\right\}\right). \end{aligned}$$

By the application of Theorem 1.4.5,

$$\text{Span}\left(\left\{\left(-\frac{2}{3},1,0\right),\left(\frac{1}{6},0,1\right)\right\}\right)$$

is the solution space of the given system.

Observation 1.4.7. Let

$$e_1 = (\delta_{i1})_{n\times 1},\quad e_2 = (\delta_{i2})_{n\times 1},\ldots,\quad e_j = (\delta_{ij})_{n\times 1},\ldots,\quad e_n = (\delta_{in})_{n\times 1}$$

be n-dimensional n vectors in R^n, where $\delta_{ij} = 0$ for $i \neq j$, and $\delta_{ij} = 1$ for $i = j$. We observe that $\text{Span}(\{e_1, e_2, \ldots, e_j, \ldots, e_n\}) = R^n$.

Definition 1.4.5. Let $x_1, x_2, \ldots, x_j, \ldots, x_m$ be in R^n. The m, n-dimensional column vectors $x_1, x_2, \ldots, x_j, \ldots, x_m$ are said to be linearly dependent if a system of linear homogeneous algebraic equations,

$$\begin{aligned} a_1x_1 + a_2x_2 + \cdots + a_jx_j + \cdots + a_mx_m &= 0,\\ \Phi\, a &= 0, \end{aligned} \tag{1.22}$$

has a nonzero (nontrivial) solution (at least one scalar $a_k \neq 0$), where $a = (a_j)_{n\times 1}$,

$$\Phi = (x_{ij})_{m\times n} = [x_1, x_2, \ldots, x_j, \ldots, x_m]. \tag{1.23}$$

On the other hand, n-dimensional column vectors $x_1, x_2, \ldots, x_j, \ldots, x_m$ are said to be linearly independent if these vectors are not linearly dependent, i.e. (1.22) has only a trivial solution ($a_1 = a_2 = \cdots = a_j = \cdots = a_m = 0$).

Example 1.4.10. We are given

$$x_1 = [1\ 1\ -1]^T,\quad x_2 = [1\ -1\ 1]^T,\quad x_3 = [-1\ 1\ 1]^T.$$

Determine whether these vectors in R^3 are linearly dependent or independent.

Solution procedure.

Step 1. Consider the linear combination of x_1, x_2, and x_3, and set it equal to zero as follows:

$$a_1x_1 + a_2x_2 + a_3x_3 = 0,$$

where a_1, a_2, and a_3 are unknown real numbers. By substituting the given vectors x_1, x_2, and x_3 and using a scalar multiplication to a vector and vector addition, the above equation can be rewritten as

$$\begin{aligned} a_1[1\ 1\ -1]^T + a_2[1\ -1\ 1]^T + a_3[-1\ 1\ 1]^T &= 0,\\ [a_1\ a_1\ -a_1]^T + [a_2\ -a_2\ a_2]^T + [-a_3\ a_3\ a_3]^T &= 0,\\ [a_1 + a_2 - a_3\ \ a_1 - a_2 + a_3\ \ -a_1 + a_2 + a_3]^T &= 0. \end{aligned}$$

Step 2. By using the concept of the equality of two matrices/vectors as in Definition 1.3.6, the last equation in step 1 can be rewritten as a system of algebraic equations,

$$a_1 + a_2 - a_3 = 0,$$
$$a_1 - a_2 + a_3 = 0,$$
$$-a_1 + a_2 + a_3 = 0,$$

and its matrix representation is $\Phi a = 0$, where Φ and a are

$$\Phi = \begin{bmatrix} 1 & 1 & -1 \\ 1 & -1 & 1 \\ -1 & 1 & 1 \end{bmatrix}, \quad a = \begin{bmatrix} a_1 \\ a_2 \\ a_3 \end{bmatrix}.$$

Step 3. Now, we imitate the procedure outlined in Example 1.4.7 to determine whether the system of linear algebraic equations in step 2 has a unique trivial or nonzero solution. For this purpose, we need to find the determinant of the coefficient matrix Φ. To find the determinant of the matrix Φ, we follow the procedure described in Example 1.4.3, and it is given by

$$\det(\Phi) = 1(-2) - 1(2) + (-1)(0) = -4.$$

From this and the application of Observation 1.4.3(b) $[\det(\Phi) = -4 \neq 0]$, we conclude that the linear homogeneous system of algebraic equations in step 2 has a unique trivial solution, i.e. $a_1 = a_2 = a_3 = 0$.

Step 4. From the conclusion in step 3 and Definition 1.4.5, we conclude that the given vectors in R^3 are linearly independent.

Theorem 1.4.6. *Let $x_1, x_2 \ldots, x_j, \ldots, x_n$ be n-dimensional vectors in R^n or C^n. The n vectors*

$$x_1, x_2, \ldots, x_j, \ldots, x_n$$

are linearly independent iff the determinant of the matrix

$$\Phi = (x_{ij})_{n \times n} = [x_1, x_2, \ldots, x_j, \ldots, x_n] \tag{1.24}$$

formed by these column vectors is different from zero, i.e. $\det(\Phi) \neq 0$.

Example 1.4.11. Let $x_1 = f_0$, $x_2 = f_1$ and $x_3 = f_2$ be vectors in $C[[a, b], R]$ in Illustration 1.4.5, where $f_0(t) = 1$, $f_1(t) = t$, $f_2(t) = t^2$. Determine whether these vectors in $C[[a, b], R]$ are linearly dependent or independent.

Solution procedure.

Step 1. Again, by following the procedure described in Example 1.4.10, we have

$$a_1 x_1 + a_2 x_2 + a_3 x_3 = 0,$$
$$a_1 f_0 + a_2 f_1 + a_3 f_2 = 0 \text{ (by substitution).}$$

The zero in the above equation means the "zero" function defined on $[a, b]$. a_1, a_2 and a_3 are unknown real numbers. From this we cannot go directly to step 2 in Example 1.4.10. The reason is that we do not have three algebraic equations with three unknowns. In other words, the number of equations must be equal to the number of unknown quantities. The following question arises: Under what conditions on functions f_0, f_1, and f_2 can one generate three equations? Of course, we have one equation. We need two more equations.

Step 2. To generate two more equations, we assume that the functions f_0, f_1, and f_2 are twice continuously differentiable on $[a, b]$. Under this assumption, we can differentiate the last equation with respect to the independent variable, and obtain two additional (in this case) algebraic equations:

$$\begin{aligned} a_1 f_0 + a_2 f_1 + a_3 f_2 &= 0, \\ a_1 f_0' + a_2 f_1' + a_3 f_2' &= 0' = 0, \\ a_1 f_0'' + a_2 f_1'' + a_3 f_2'' &= 0'' = 0, \end{aligned}$$

where f_0', f_1', and f_2' stand for the first derivatives of f_0, f_1, and f_2, respectively, and f_0'', f_1'', and f_2'' stand for the respective second derivatives. Of course, we have utilized the fact that the first and second derivatives of a "zero function" are zero functions on $[a, b]$.

Step 3. Now, we have three algebraic equations with three unknown numbers in step 1. We can either repeat steps 2 and 3 in the procedure outlined in Example 1.4.10 or apply Theorem 1.4.6. In this case,

$$\Phi = \begin{bmatrix} f_0 & f_1 & f_2 \\ f_0' & f_1' & f_2' \\ f_0'' & f_1'' & f_2'' \end{bmatrix}, \quad a = \begin{bmatrix} a_1 \\ a_2 \\ a_3 \end{bmatrix},$$

where the entries of the coefficient matrix Φ are functions, and a is an unknown constant vector,

$$\det(\Phi) = f_0(f_1' f_2'' - f_1'' f_2') - f_1(f_0' f_2'' - f_0'' f_2') + f_2(f_0' f_1'' - f_0'' f_1').$$

This is a determinant function. If $\det(\Phi)$ is different from zero on $[a, b]$, then f_0, f_1, and f_2 are linearly independent. On the other hand, if $\det(\Phi)$ is equal to zero on $[a, b]$, then f_0, f_1, and f_2 are linearly dependent on $[a, b]$. With regard to this example, we have

$$\Phi(t) = \begin{bmatrix} 1 & t & t^2 \\ 0 & 1 & 2t \\ 0 & 0 & 2 \end{bmatrix}, \quad \det(\Phi)(t) = (2) - t(0) + t^2(0) = 2, \quad \text{for } t \in [a, b].$$

Thus, $\det(\Phi)(t) = 2 \neq 0$. Therefore, f_0, f_1, and f_2 are linearly independent.

Note. The determinant of Φ is called the Wronskian of three functions. We further note that, in general, $\det(\Phi)$ may be zero even though the functions are linearly independent. In short, the nonvanishing property of the Wronskian is a sufficient condition that the given set of a finite number of functions is linearly independent.

Definition 1.4.6. Let $\{x_1, x_2, \ldots, x_j, \ldots, x_n\}$ be a subset of either R^n or C^n. $\{x_1, x_2, \ldots, x_j, \ldots, x_n\}$ is called a basis of R^n or C^n, if (1) the column vectors $x_1, x_2, \ldots, x_j, \ldots, x_n$ are linearly independent in R^n or C^n, and (2)

$$\text{Span}(\{x_1, x_2, \ldots, x_j, \ldots, x_n\}) = R^n$$

or C^n. Moreover, n is called the dimension of the span.

Theorem 1.4.7. *Let x_p be a solution to the system of linear nonhomogeneous algebraic equations* (1.12), *and let $\{x_1, x_2, \ldots, x_j, \ldots, x_m\}$ be a basis for the solution space $\mathcal{S}$ of the corresponding linear homogeneous algebraic equations* (1.15). *Then,*

$$x = x_c + x_p \tag{1.25}$$

is a solution to (1.12), *where x_c is a general solution to* (1.15) *defined by*

$$x_c = a_1 x_1 + a_2 x_2 + \cdots + a_j x_j + \cdots + a_m x_m. \tag{1.26}$$

1.4 Exercises

(1) Find $\det(A)$:

(a) $A = \begin{bmatrix} 1-\lambda & 2 & 0 \\ 1 & 2-\lambda & 1 \\ 1 & 1 & -\lambda \end{bmatrix}$ (b) $A = \begin{bmatrix} 1 & x_1 & x_1^2 \\ 1 & x_2 & x_2^2 \\ 1 & x_3 & x_3^2 \end{bmatrix}$

(2) Use the determinant to show that:

(a) AB is singular if either A or B is singular.
(b) $\det\left(AA^{-1}\right) = 1$.
(c) $\det\left(A^n\right) = (\det(A))^n$.

(3) Find those numbers λ for which $[1-\lambda\ 2\ 0]$, $[1\ 2-\lambda\ 1]$, and $[1\ 1\ -\lambda]$ are linearly dependent vectors in R^3.

(4) For given A, find a basis and the dimension of the solution space of $Ax = 0$. Describe the solution space geometrically:

(a) $A = \begin{bmatrix} 1 & -2 & -1 \\ 2 & -4 & -2 \\ 0 & 1 & 1 \end{bmatrix}$ (b) $A = \begin{bmatrix} 1 & 2 & -1 \\ 2 & 1 & 1 \\ 3 & 3 & 0 \end{bmatrix}$ (c) $A = \begin{bmatrix} 0 & 1 & 0 \\ 0 & 0 & 1 \\ 0 & 0 & 0 \end{bmatrix}$

(d) $A = \begin{bmatrix} 0 & 1 & 0 \\ 0 & 0 & 0 \\ 0 & 0 & 0 \end{bmatrix}$ (e) $A = \begin{bmatrix} 1 & 1 \\ 2 & 2 \end{bmatrix}$

(5) Find the adjoint matrix of A, and show that $A\,\text{adj}(A) = I\det(A)$:

$$\text{(a) } A = \begin{bmatrix} 2 & 1 & 0 \\ -1 & 1 & 1 \\ 0 & 2 & 3 \end{bmatrix} \quad \text{(b) } A = \begin{bmatrix} -5 & -10 & -20 \\ 5 & 5 & 10 \\ 2 & 4 & 9 \end{bmatrix} \quad \text{(c) } A = \begin{bmatrix} 1 & 1 \\ 2 & 2 \end{bmatrix}$$

(6) Are the matrices in Exercise 5 invertible? If the answer to the question is "yes," then determine its inverse.

1.5 Matrix Calculus

In this section, we briefly discuss the essentials of differential and integral calculus of matrix-valued functions and a function of matrices. Moreover, a few non-traditional results concerning matrix functions, in particular the determinant, are presented. These results provide a very useful tool for the study of differential equations.

Definition 1.5.1. Let $A: J \to R^{m\times n}$ be a relation defined by $A(t) = (a_{ij}(t))_{m\times n}$, for each t in J; J is a subset of R. A is called a matrix-valued function defined on J into $R^{m\times n}$, if entries a_{ij} are functions defined on J into R. The domain of A is J, and its range is in $R^{m\times n}$.

Example 1.5.1. Let $A: J \to R^{2\times 3}$ be a relation defined by

$$A(t) = \begin{bmatrix} t & 1 & \cos t \\ e^t & \ln t & \sinh t \end{bmatrix},$$

for each t in $J = (0, \infty)$. According to Definition 1.5.1, it is 2×3 matrix function. Its domain is $(0, \infty)$ and its range is $\mathcal{R}(A) = \{A(t) : t \in (0, \infty)\}$.

Definition 1.5.2. Let $A(t) = (a_{ij}(t))_{m\times n}$ be an $m \times n$ matrix function defined on an interval J. Let a be in R, and let $L = (l_{ij})_{m\times n}$ be an $m \times n$ matrix. The matrix L is said to be the limit of the matrix-valued function A if

$$\lim_{t\to a} A(t) = \left(\lim_{t\to a} a_{ij}(t)\right)_{m\times n} = (l_{ij})_{m\times n} = L,$$

i.e. the limit of the matrix function at $t = a$ is equal to the matrix corresponding to limits of all functions a_{ij} at $t = a$.

Example 1.5.2. Let $A: J \to R^{2\times 3}$ be a relation defined by

$$A(t) = \begin{bmatrix} t & 1 & \cos t \\ e^t & \ln t & \sinh t \end{bmatrix},$$

for each t in $J = (0, \infty)$. Let a be in $(0, \infty)$. From elementary calculus, $\lim_{t\to a} t = a$, $\lim_{t\to a} 1 = 1$, $\lim_{t\to a} \cos t = \cos a$, $\lim_{t\to a} e^t = e^a$, $\lim_{t\to a} \ln t = \ln a$, and $\lim_{t\to a} \sinh t = \sinh a$. From this and Definition 1.5.2, we conclude the 2×3 matrix

L, defined by

$$L = \begin{bmatrix} a & 1 & \cos a \\ e^a & \ln a & \sinh a \end{bmatrix},$$

is the limit of the 2×3 matrix function $A(t)$ at $t = a$, for a in $(0, \infty)$.

Observation 1.5.1. Let $A(t) = (a_{ij}(t))_{m\times n}$ be an $m \times n$ matrix-valued function defined on J, and let $\triangle t$ be an increment in t. The increment in the $m \times n$ matrix function A is the $m \times n$ matrix defined by

$$\begin{aligned} \triangle A(t) &= A(t + \triangle t) - A(t) = (a_{ij}(t + \triangle t))_{m\times n} - (a_{ij}(t))_{m\times n} \\ &= (a_{ij}(t + \triangle t) - a_{ij}(t))_{m\times n}. \end{aligned}$$

In the following, the definitions of continuity, differentiability, and integrability of matrix-valued functions are defined analogously to the elementary calculus of functions of real variables. This is summarized as follows:

Definition 1.5.3. Let $A(t) = (a_{ij}(t))_{m\times n}$ be an $m \times n$ matrix-valued function defined on J, and let a be in J.

(C) $\lim_{\triangle t\to 0}[A(a + \triangle t) - A(a)] = (\lim_{\triangle t\to 0} [a_{ij}(a + \triangle t) - a_{ij}(a)])_{m\times n} = (0)_{m\times n}$.

(D) $\lim_{\triangle t\to 0}\left[\frac{A(a+\triangle t)-A(a)}{\triangle t}\right] = \left(\lim_{\triangle t\to 0}\left[\frac{a_{ij}(a+\triangle t)-a_{ij}(a)}{\triangle t}\right]\right)_{m\times n}$. If this limit exists, then it is denoted by

$$\frac{d}{dt}A(t) = A'(t) = (a'_{ij}(t))_{m\times n} = \left(\frac{d}{dt}a_{ij}(t)\right)_{m\times n}.$$

(I) $\lim_{\mu(P)\to 0}\left[\sum_{l=1}^{k} A(t_l)\triangle t_l\right] = \left(\lim_{\mu(P)\to 0}\left[\sum_{l=1}^{k} a_{ij}(t_l)\triangle t_l\right]\right)_{m\times n}$, where

$$P : a = t_0 < t_1 < \cdots < t_{l-1} < t_l \cdots < t_k = b$$

is a partition of an interval $[a, b]$; $\triangle t_l = t_l - t_{l-1}$; and

$$\mu(P) = \max\{\triangle t_1, \triangle t_2, \ldots, \triangle t_l, \ldots, \triangle t_k\}$$

is called a norm or mesh of the partition P. If this limit exists, then it is denoted by

$$\int_a^t A(s)\,ds = \left(\int_a^t a_{ij}(s)\,ds\right)_{m\times n}.$$

Example 1.5.3. Let $A : J \to R^{2\times 3}$ be as in Example 1.5.1:

$$A(t) = \begin{bmatrix} t & 1 & \cos t \\ e^t & \ln t & \sinh t \end{bmatrix},$$

for each t in $J = (0, \infty)$. Let a be in $(0, \infty)$.

(C) From elementary calculus,

$$\lim_{\Delta t \to 0} (a + \Delta t) = a, \quad \lim_{\Delta t \to 0} 1 = 1,$$
$$\lim_{\Delta t \to 0} \cos(a + \Delta t) = \cos a, \quad \lim_{\Delta t \to 0} e^{(a+\Delta t)} = e^a,$$
$$\lim_{\Delta t \to 0} \ln(a + \Delta t) = \ln a, \quad \text{and} \quad \lim_{\Delta t \to 0} \sinh(a + \Delta t) = \sinh a.$$

From this and Definition 1.5.3(C), we conclude that the 2×3 matrix function $A(t)$ is continuous at $t = a$, for a in $(0, \infty)$. Moreover,

$$\lim_{\Delta t \to 0} A(a + \Delta t) = A(a) = \begin{bmatrix} a & 1 & \cos a \\ e^a & \ln a & \sinh a \end{bmatrix}.$$

(D) From elementary calculus, we have

$$\lim_{\Delta t \to 0} \left[\frac{(a + \Delta t) - a}{\Delta t} \right] = 1, \quad \lim_{\Delta t \to 0} \left[\frac{1 - 1}{\Delta t} \right] = 0,$$
$$\lim_{\Delta t \to 0} \left[\frac{\cos(a + \Delta t) - \cos a}{\Delta t} \right] = -\sin a, \quad \lim_{\Delta t \to 0} \left[\frac{e^{(a+\Delta t)} - e^a}{\Delta t} \right] = e^a,$$
$$\lim_{\Delta t \to 0} \left[\frac{\ln(a + \Delta t) - \ln a}{\Delta t} \right] = \frac{1}{a}, \quad \text{and} \quad \lim_{\Delta t \to 0} \left[\frac{\sinh(a + \Delta t) - \sinh a}{\Delta t} \right] = \cosh a.$$

From this and Definition 1.5.3 (D), we again conclude that the 2×3 matrix function $A(t)$ is differentiable at $t = a$, for a in $(0, \infty)$. Moreover,

$$\lim_{\Delta t \to 0} \left[\frac{A(a + \Delta t) - A(a)}{\Delta t} \right] = \left(\lim_{\Delta t \to 0} \left[\frac{a_{ij}(a + \Delta t) - a_{ij}(a)}{\Delta t} \right] \right)_{2\times 3}$$
$$= \begin{bmatrix} 1 & 0 & -\sin a \\ e^a & \frac{1}{a} & \cosh a \end{bmatrix}.$$

(I) From elementary calculus, we have

$$\int_a^t s\,ds = \left(\frac{s^2}{2} + C \right) \Big|_a^t = \frac{(t^2 - a^2)}{2}, \quad \int_a^t 1\,ds = (s + C)\Big|_a^t = (t - a),$$
$$\int_a^t \cos s\,ds = (\sin s + C)\Big|_a^t = (\sin t - \sin a), \quad \int_a^t e^s\,ds = (e^s + C)\Big|_a^t = e^t - e^a,$$
$$\int_a^t \ln s\,ds = (s \ln s - s + C)\Big|_a^t = (t \ln t - t - a \ln a + a),$$
$$\int_a^t \sinh s\,ds = (\cosh s + C)\Big|_a^t = (\cosh t - \cosh a).$$

From this and Definition 1.5.3 (I), we conclude that the indefinite integral of the 2×3 matrix function $A(t)$ over an interval $[a, t]$ is given by

$$\int_a^t A(s)ds = \left(\int_a^t a_{ij}(s)ds\right)_{2\times 3}$$
$$= \begin{bmatrix} \frac{(t^2-a^2)}{2} & (t-a) & (\sin t - \sin a) \\ e^t - e^a & (t\ln t - t - a\ln a + a) & (\cosh t - \cosh a) \end{bmatrix}.$$

In the following, we present the basic properties of the derivative and integral of matrix-valued functions.

Theorem 1.5.1. *Let A, B, x, and y be matrix-valued functions defined on J, and c be a scalar function. It is assumed that the matrices are appropriately defined to fulfill the following specified operations:*

(A1) $\frac{d}{dt}(A(t) + B(t)) = \frac{d}{dt}A(t) + \frac{d}{dt}B(t)$ *(addition rule)*;
(A2) $\frac{d}{dt}((cA)(t)) = c(t)\frac{d}{dt}A(t) + A(t)\frac{d}{dt}c(t)$, *where c is scalar function (product rule for scalar multiplication)*;
(P1) $\frac{d}{dt}(A(t)B(t)) = \frac{d}{dt}A(t)B(t) + A(t)\frac{d}{dt}B(t)$ *(product rule)*;
(P2) $\frac{d}{dt}(A(t)x(t)) = \frac{d}{dt}A(t)x(t) + A(t)\frac{d}{dt}x(t)$, *where x is an $n \times 1$ matrix function*;
(P3) $\frac{d}{dt}\left(y^T(t)x(t)\right) = \frac{d}{dt}y^T(t)x(t) + y^T(t)\frac{d}{dt}x(t)$;
(P4) $\frac{d}{dt}\left(x^T(t)x(t)\right) = \frac{d}{dt}x^T(t)x(t) + x^T(t)\frac{d}{dt}x(t) = 2x^T(t)\frac{d}{dt}x(t)$;
(I1) $\frac{d}{dt}\left(A^{-1}(t)\right) = -A^{-1}(t)\frac{d}{dt}A(t)A^{-1}(t)$, *if A is invertible*;
(T1) $\frac{d}{dt}\left(A^T(t)\right) = \left(\frac{d}{dt}A(t)\right)^T$ *(transpose rule)*.

Example 1.5.4. We are given

$$A(t) = \begin{bmatrix} t & \cos t \\ 1 & \sin t \end{bmatrix}.$$

Find

$$\frac{d}{dt}A^2(t).$$

Solution procedure. The goal of this problem is to find the first derivative of $A^2(t)$. We can either apply the formula (P1) in Theorem 1.5.1 or first find $A^2(t) = A(t)A(t)$ and then use the definition of the derivative of the matrix function. Here, we find

$$\frac{d}{dt}A^2(t)$$

by using the formula (P1), and we have

$$\begin{aligned}
\frac{d}{dt}A^2(t) &= \frac{d}{dt}A(t)A(t) + A(t)\frac{d}{dt}A(t) \\
&= \frac{d}{dt}\begin{bmatrix} t & \cos t \\ 1 & \sin t \end{bmatrix}\begin{bmatrix} t & \cos t \\ 1 & \sin t \end{bmatrix} + \begin{bmatrix} t & \cos t \\ 1 & \sin t \end{bmatrix}\frac{d}{dt}\begin{bmatrix} t & \cos t \\ 1 & \sin t \end{bmatrix} \\
&= \begin{bmatrix} 1 & -\sin t \\ 0 & \cos t \end{bmatrix}\begin{bmatrix} t & \cos t \\ 1 & \sin t \end{bmatrix} + \begin{bmatrix} t & \cos t \\ 1 & \sin t \end{bmatrix}\begin{bmatrix} 1 & -\sin t \\ 0 & \cos t \end{bmatrix} \\
&= \begin{bmatrix} t - \sin t & \cos t - \sin^2 t \\ \cos t & \cos t \sin t \end{bmatrix} + \begin{bmatrix} t & -t\sin t + \cos^2 t \\ 1 & -\sin t + \sin t \cos t \end{bmatrix} \\
&= \begin{bmatrix} 2t - \sin t & \cos t - \sin^2 t - t\sin t + \cos^2 t \\ 1 + \cos t & -\sin t + 2\sin t \cos t \end{bmatrix}.
\end{aligned}$$

Observation 1.5.2. Let $A(t) = (a_{ij}(t))_{m\times n}$ be an $m \times n$ differentiable matrix-valued function defined on J, and let Δt be an increment in t. The differential of the $m \times n$ matrix $A(t)$ is defined by

$$dA(t) = (da_{ij}(t)dt)_{m\times n} = A(t)dt = dt(a_{ij}(t))_{m\times n},$$

where $\Delta t = dt$. We note that the operation of the differential is linear. In fact,

(i) $d(A(t) + B(t)) = dA(t) + dB(t)$ (addition rule);
(ii) $d((cA)(t)) = c(t)dA(t) + A(t)dc(t)$, where c is the scalar function (product rule for scalar multiplication);
(iii) $d(A(t)B(t)) = dA(t)B(t) + A(t)dB(t)$ (product rule).

Observation 1.5.3. The higher derivatives of the matrix-valued function can be defined similarly. The calculus of several variables concepts can also be extended to matrix functions. For completeness and easy reference, we discuss partial derivatives, the gradient vector, and the Jacobean matrix.

(i) ***Gradient vector and chain rule.*** Let f be a real-valued composite function of nm real-valued functions

$$a_{11},\ldots,a_{1j},\ldots,a_{1n},\ldots,a_{i1},\ldots,a_{ij},\ldots,a_{in},\ldots,a_{m1},\ldots,a_{mj},\ldots,a_{mn}$$

of real variable t in J. By rearranging the a_{ij}'s, one can rewrite the real-valued function f as the composite function of either m, n-dimensional row vectors

$$A_1, A_2, \ldots, A_i, \ldots, A_m$$

or n, m-dimensional column vectors

$$A^1, A^2, \ldots, A^j, \ldots, A^n.$$

In short, f is a real-valued matrix function of $A(t)$ for t in J. Assume that the function f has partial derivatives, and functions

$$a_{11},\ldots,a_{1j},\ldots,a_{1n},\ldots,a_{i1},\ldots,a_{ij},\ldots,a_{in},\ldots,a_{m1},\ldots,a_{mj},\ldots,a_{mn}$$

are differentiable on J. From the calculus of several variables, we have

$$\begin{aligned}\frac{d}{dt}f(A(t)) &= \frac{d}{dt}f\left(a_{11},\ldots,a_{1j},\ldots,a_{1n},\ldots,a_{i1},\ldots,a_{ij},\ldots,\right.\\ &\quad \left.a_{in},\ldots,a_{m1},\ldots,a_{mj},\ldots,a_{mn}\right)\\ &= \frac{\partial}{\partial a_{11}}f\frac{d}{dt}a_{11}+\cdots+\frac{\partial}{\partial aij}f\frac{d}{dt}a_{ij}\\ &\quad +\cdots+\frac{\partial}{\partial a_{mn}}f\frac{d}{dt}a_{mn}\text{ (chain rule).}\end{aligned} \tag{1.27}$$

By rearranging the a_{ij}'s as the components of m, n-dimensional row vectors $A_1, A_2, \ldots, A_i, \ldots, A_m$, and applying Definition 1.3.10, the above chain rule formula can be written in various ways:

$$\begin{aligned}\frac{d}{dt}f(A) &= \frac{d}{dt}f\left(A_1,A_2,\ldots,A_i,\ldots,A_m\right)\\ &= \sum_{i=1}^{m}\frac{\partial}{\partial A_i}f\left(A_1,A_2,\ldots,A_i,\ldots,A_m\right)\frac{d}{dt}\left(A_i\right)^T\\ &= \sum_{i=1}^{m}\frac{d}{dt}A_i\left(\frac{\partial}{\partial A_i}f\left(A_1,A_2,\ldots,A_i,\ldots,A_m\right)\right)^T\\ &= \sum_{i=1}^{m}\frac{\partial}{\partial A_i}f(A)\left(\frac{d}{dt}A_i\right)^T = \sum_{i=1}^{m}\frac{d}{dt}A_i\left(\frac{\partial}{\partial A_i}f(A)\right)^T\\ &= \mathrm{tr}\left(\nabla f(A)\left[\frac{d}{dt}A\right]^T\right) = \mathrm{tr}\left(\frac{d}{dt}A[\nabla f(A)]^T\right),\end{aligned} \tag{1.28}$$

where T stands for "transpose of"; for any fixed integer i, $1 \le i \le m$, $A_i = (a_{ij})_{1\times n}$,

$$\frac{\partial}{\partial A}f(A) = \nabla f(A) = \left(\frac{\partial}{\partial a_{ij}}f(A)\right)_{m\times n}, \tag{1.29}$$

$$\begin{aligned}\frac{\partial}{\partial A_i}f(A) &= \nabla_i f(A) = \frac{\partial}{\partial A_i}f\left(A_1,A_2,\ldots,A_i,\ldots,A_m\right)\\ &= \left(\frac{\partial}{\partial a_{ij}}f(A)\right)_{1\times n}\end{aligned} \tag{1.30}$$

is the ith-row vector of $m \times n$ matrix $\nabla f(A)$; moreover,

$$\begin{aligned}(\nabla_i f(A))^T &= \left(\frac{\partial}{\partial a_{ij}}f(A)\right)_{n\times 1}\\ &= \left[\frac{\partial}{\partial a_{i1}}f(A),\ldots,\frac{\partial}{\partial a_{ij}}f(A),\ldots,\frac{\partial}{\partial a_{in}}f(A)\right]^T_{n\times 1},\end{aligned} \tag{1.31}$$

$$\nabla_i f(A) = \left[\frac{\partial}{\partial a_{i1}}f(A),\ldots,\frac{\partial}{\partial a_{ij}}f(A),\ldots,\frac{\partial}{\partial a_{in}}f(A)\right]_{1\times n}$$

is called the gradient vector of f with respect to the ith-row vector $A_i = (a_{ij})_{1\times n}$ or $(a_{i1}, a_{i2}, \ldots, a_{ij}, \ldots, a_{in})$.

Example 1.5.5. You are given: For $m = n$ and

$$f(A) = \sum_{j=1}^{n} a_{jj}.$$

In this case,

$$\frac{\partial}{\partial a_{ij}} f(A) = 0$$

for $i \neq j$, and

$$\frac{\partial}{\partial a_{jj}} f(A) = 1$$

for any fixed integer i, $1 \leq i \leq n$. Hence

$$\begin{aligned} \nabla_i f(A) &= [0, 0, \ldots, 1, \ldots, 0]_{1\times n} = (\delta_{ij})_{1\times n}, \\ \nabla f(A) &= \left(\frac{\partial}{\partial a_{ij}} f(A)\right)_{n\times n} = I. \end{aligned} \tag{1.32}$$

(ii) ***Jacobean matrix and chain rule.*** Now, we present a few formulas with regard to the $r \times 1$ matrix function $f = (f_k)_{r\times 1}$. We assume that for each k, $1 \leq k \leq r$, f_k is a real-valued composite function of nm real-valued functions

$$a_{11}, \ldots, a_{1j}, \ldots, a_{1n}, \ldots, a_{i1}, \ldots, a_{ij}, \ldots, a_{in}, \ldots, a_{m1}, \ldots, a_{mj}, \ldots, a_{mn}$$

of real variable t in J. Further, assume that the function f has partial derivatives, and

$$a_{11}, \ldots, a_{1j}, \ldots, a_{1n}, \ldots, a_{i1}, \ldots, a_{ij}, \ldots, a_{in}, \ldots, a_{m1}, \ldots, a_{mj}, \ldots, a_{mn}$$

are differentiable functions on J. From the calculus of several variables, we apply the definition of the derivative of the matrix function (Definition 1.5.3) to f, and obtain

$$\frac{d}{dt} f(A) = \left(\frac{d}{dt} f_k(A)\right)_{r\times 1}. \tag{1.33}$$

For each fixed k, $1 \leq k \leq r$, and for any fixed integer i, $1 \leq i \leq m$, $A_i = (a_{ij})_{1\times n}$, we apply the elementary formulas (1.27) and (1.28) to f_k, the kth component of f, and obtain

$$\frac{d}{dt} f_k(A) = \sum_{i=1}^{m} \frac{\partial}{\partial A_i} f_k(A) \left(\frac{d}{dt} A_i\right)^T = \sum_{i=1}^{m} \frac{d}{dt} A_i \left(\frac{\partial}{\partial A_i} f_k(A)\right)^T. \quad (1.34)$$

We substitute the expression in (1.34) for $\frac{d}{dt} f_k(A)$ into (1.33), and we get

$$\begin{aligned} \frac{d}{dt} f(A) &= \left(\sum_{i=1}^{m} \frac{\partial}{\partial A_i} f_k(A) \left(\frac{d}{dt} A_i\right)^T\right)_{r\times 1} \\ &= \left(\sum_{i=1}^{m} \frac{d}{dt} A_i \left(\frac{\partial}{\partial A_i} f_k(A)\right)^T\right)_{r\times 1} \\ &= \left(\operatorname{tr}\left(\frac{d}{dt} A \left[\nabla f_k(A)\right]^T\right)\right)_{r\times 1}. \end{aligned} \quad (1.35)$$

By using matrix multiplication and matrix addition, (1.35) can be written as

$$\begin{aligned} \frac{d}{dt} f(A) &= \sum_{i=1}^{m} \left(\frac{\partial}{\partial A_i} f_k(A) \left(\frac{d}{dt} A_i\right)^T\right)_{r\times 1} \\ &= \sum_{i=1}^{m} \left(\frac{d}{dt} A_i \left(\frac{\partial}{\partial A_i} f_k(A)\right)^T\right)_{r\times 1}. \end{aligned} \quad (1.36)$$

We note that for each i and k, $1 \le i \le m$ and $1 \le k \le r$,

$$\frac{\partial}{\partial A_i} f_k(A) = \nabla_i f_k(A) = \left[\frac{\partial}{\partial a_{i1}} f_k(A) \ldots, \frac{\partial}{\partial a_{i1}} f_k(A), \ldots, \frac{\partial}{\partial a_{in}} f_k(A)\right]_{1\times n}$$

and the ith term of $\frac{d}{dt} f(A)$ is

$$\begin{aligned} &\left(\frac{\partial}{\partial A_i} f_k(A(t)) \left(\frac{d}{dt} A_i\right)^T\right)_{r\times 1} \\ &= \left(\frac{d}{dt} A_i \left(\frac{\partial}{\partial A_i} f_k(A(t))\right)^T\right)_{r\times 1} \\ &= \begin{bmatrix} \frac{\partial}{\partial a_{i1}} f_1(A) & \frac{\partial}{\partial a_{i2}} f_1(A) & \cdots & \frac{\partial}{\partial a_{ij}} f_1(A) & \cdots & \frac{\partial}{\partial a_{in}} f_1(A) \\ \frac{\partial}{\partial a_{i1}} f_2(A) & \frac{\partial}{\partial a_{i2}} f_2(A) & \cdots & \frac{\partial}{\partial a_{ij}} f_2(A) & \cdots & \frac{\partial}{\partial a_{in}} f_2(A) \\ \cdots & \cdots & \cdots & \cdots & \cdots & \cdots \\ \frac{\partial}{\partial a_{i1}} f_k(A) & \frac{\partial}{\partial a_{i2}} f_k(A) & \cdots & \frac{\partial}{\partial a_{ij}} f_k(A) & \cdots & \frac{\partial}{\partial a_{in}} f_k(A) \\ \cdots & \cdots & \cdots & \cdots & \cdots & \cdots \\ \frac{\partial}{\partial a_{i1}} f_r(A) & \frac{\partial}{\partial a_{i2}} f_r(A) & \cdots & \frac{\partial}{\partial a_{ij}} f_r(A) & \cdots & \frac{\partial}{\partial a_{in}} f_r(A) \end{bmatrix} \begin{bmatrix} \frac{d}{dt} a_{i1} \\ \frac{d}{dt} a_{i2} \\ \cdots \\ \frac{d}{dt} a_{ij} \\ \cdots \\ \frac{d}{dt} a_{in} \end{bmatrix}. \end{aligned} \quad (1.37)$$

For each fixed integer i, $1 \leq i \leq m$, the $r \times n$ matrix

$$\left(\frac{\partial}{\partial a_{ij}} f_k(A)\right)_{r\times n}$$

is called the Jacobean matrix of f with respect to the ith-row vector $A_i = (a_{ij})_{1\times n}$ of the matrix function A and it is denoted by

$$\frac{\partial}{\partial A_i} f(A).$$

Example 1.5.6

(a) For $m, n = r$, and each fixed i, $1 \leq i \leq m$, $f_k(A) = \sum_{l=1}^{m} a_{lk}$. In this case, for each k, $1 \leq k \leq n$,

$$\frac{\partial}{\partial a_{ij}} f(A) = \begin{cases} 0, & \text{for } j \neq k, \\ 1, & \text{for } j = k, \end{cases}$$

for every $1 \leq i \leq m$. Hence, for each k and every i, $1 \leq i \leq m$ and $1 \leq k \leq n$,

$$\nabla_i f_k(A) = [0, 0, \ldots, 1, \ldots, 0]_{1\times n} = (\delta_{ik})_{1\times n}.$$

Moreover,

$$\frac{\partial}{\partial A_i} f(A) = I_{n\times n} \text{ identity matrix,} \tag{1.38}$$

where for any i, $1 \leq i \leq m$, $A_i = (a_{ij})_{1\times n}$.

(b) For $m = n = r = 2$,

$$\frac{\partial}{\partial A_1}\begin{bmatrix} a_{11} \\ a_{12}\end{bmatrix}^T = \begin{bmatrix} \nabla_1 a_{11} \\ \nabla_1 a_{12}\end{bmatrix}^T = \begin{bmatrix} 1 & 0 \\ 0 & 1\end{bmatrix}, \quad \frac{\partial}{\partial A_2}\begin{bmatrix} a_{11} \\ a_{12}\end{bmatrix}^T = \begin{bmatrix} \nabla_2 a_{11} \\ \nabla_2 a_{12}\end{bmatrix}^T = \begin{bmatrix} 0 & 0 \\ 0 & 0\end{bmatrix},$$
$$\frac{\partial}{\partial A_2}\begin{bmatrix} a_{21} \\ a_{22}\end{bmatrix}^T = \begin{bmatrix} \nabla_2 a_{21} \\ \nabla_2 a_{22}\end{bmatrix}^T = \begin{bmatrix} 1 & 0 \\ 0 & 1\end{bmatrix}, \quad \frac{\partial}{\partial A_1}\begin{bmatrix} a_{21} \\ a_{22}\end{bmatrix}^T = \begin{bmatrix} \nabla_1 a_{21} \\ \nabla_1 a_{22}\end{bmatrix}^T = \begin{bmatrix} 0 & 0 \\ 0 & 0\end{bmatrix}.$$

Theorem 1.5.2. *Let A, B, x, and y be matrix functions defined on J, and c be a scalar function. It is assumed that the matrices are appropriately defined to fulfill the following specified operations:*

(D1) $\int_a^t \frac{d}{ds} A(s)\, ds = \int_a^T A'(s)\, ds = \int_a^t dA(s)\, ds = A(t) - A(0)$;

(D2) $\frac{d}{dt} \int_a^t A(s)\, ds = A(t)$;

(A1) $\int_a^t (A(s) + B(s))\, ds = \int_a^t A(s)\, ds + \int_a^t B(s)\, ds$;

(A2) $\int_a^t (cA(s)) = c\int_a^t A(s)\, ds$, *where c is a constant scalar function;*

(P1) $\int_a^t AB(s)\, ds = A\int_a^t B(s)\, ds$, *where A is a constant matrix;*

(P2) $\int_a^t Ax(s)\,ds = A\int_a^t x(s)\,ds$, *where A is a constant matrix*;

(P3) $\int_a^t A(s)B\,ds = \int_a^t A(s)\,dsB$, *where B is a constant matrix*;

(T1) $\int_a^t A^T(s)\,ds = \left(\int_a^t A(s)ds\right)^T$.

By utilizing the discussions in Illustrations 1.3.1, 1.4.1, 1.4.2, and 1.4.3, we can introduce the first and second derivatives for the determinant of an $n \times n$ matrix function.

Illustration 1.5.1. In the derivation of the first and second derivatives of $\det(A(t))$, we use only row expansion of the determinant. Of course, everyone is also free to use the column expansion.

(i) For $n = 1$: Let $A(t) = (a_{ij}(t))_{1\times 1}$ be a differentiable matrix function. From Illustration 1.4.2(i), the determinant of $A(t)$ is $\det(A(t)) = a_{11}(t)$. This is exactly like the usual simple function. Hence, its first derivative, second derivative, and differential are

$$\frac{d}{dt}\det(A(t)) = \frac{d}{dt}W(A_1(t)) = a'_{11}(t),$$
$$\frac{d^2}{dt^2}\det(A(t)) = \frac{d^2}{dt^2}W(A_1(t)) = a''_{11}(t),$$

$\det(A) = dW(A_1) = a'_{11}dt$, respectively.

(ii) For $n = 2$: Let $A(t) = A = (a_{ij}(t))_{2\times 2}$ be a differentiable matrix function. From Illustration 1.4.2(ii), Definition 1.3.10, and notations in Observation 1.4.2, the various kinds of representations of the determinant of $A(t)$ are summarized as

$$\begin{aligned} W(A_1, A_2) = \det(A(t)) &= \sum_{j=1}^{2} a_{ij}(t)C_{ij}(t) \\ &= a_{11}(t)a_{22}(t) - a_{12}(t)a_{21}(t) = A_iC_i^T = C_iA_i^T \\ &= \sum_{i=1}^{2} a_{ij}(-1)^{i+j}A_k(ij)C_k^T(ij), \quad \text{for any } i = 1, 2. \end{aligned} \tag{1.39}$$

Elementary approach. By ordinary properties of the derivative (product, sum, and difference) and the definition of the determinant of a 2×2 matrix, the derivative of $\det(A(t))$ is given by

$$\begin{aligned} \frac{d}{dt}\det(A(t)) &= \frac{d}{dt}W(a_{11}, a_{12}, a_{21}, a_{22}) \\ &= \frac{d}{dt}[a_{11}(t)a_{22}(t) - a_{12}(t)a_{21}(t)] \end{aligned}$$

$$
\begin{aligned}
&= \frac{d}{dt}\left[a_{11}(t)a_{22}(t)\right] - \frac{d}{dt}\left[a_{12}(t)a_{21}(t)\right] \quad \text{(by addition rule)} \\
&= a_{11}(t)\frac{d}{dt}\left[a_{22}(t)\right] + \frac{d}{dt}\left[a_{11}(t)\right]a_{22}(t) \\
&\quad - a_{12}(t)\frac{d}{dt}\left[a_{21}(t)\right] - \frac{d}{dt}\left[a_{12}(t)\right]a_{21}(t) \quad \text{(by product rule)} \\
&= a_{11}(t)a'_{22}(t) + a'_{11}(t)a_{22}(t) - a_{12}(t)a'_{21}(t) - a'_{12}(t)a_{21}(t) \\
&= a'_{11}(t)a_{22}(t) - a'_{12}(t)a_{21}(t) + a_{11}(t)a'_{22}(t) - a_{12}(t)a'_{21}(t) \\
&= \sum_{i=1}^{2} a'_{1j}(t)C_{1j}(t) + \sum_{i=1}^{2} a'_{2j}(t)C_{2j}(t) \quad \text{(by notation)} \\
&= W\left(\frac{d}{dt}A_1, A_2\right) + W\left(A_1, \frac{d}{dt}A_2\right) \quad \text{(by Definition 1.4.2).}
\end{aligned} \tag{1.40}
$$

The above procedure uses the explicit knowledge of the determinant

$$
\det(A) = W\left(a_{11}, a_{12}, a_{21}, a_{22}\right) = a_{11}(t)a_{22}(t) - a_{12}(t)a_{21}(t)
$$

as a function of the entries of the matrix.

Calculus of multivariable approach: The first derivative formula (1.40) for $\det(A(t))$ can also be derived by using the determinant as the composite function of the entries of the matrix.

Using the representation (1.39) of the definition of the determinant of the 2×2 matrix as a composite function of entries of the matrix and its entries as functions of the independent variable t, we apply the formula (1.28) to the function defined in (1.39) to get

$$
\begin{aligned}
\frac{d}{dt}W\left(A_1, A_2\right) &= \frac{\partial}{\partial A_1}W\left(A_1, A_2\right)\frac{d}{dt}A_1^T + \frac{\partial}{\partial A_2}W\left(A_1, A_2\right)\frac{d}{dt}A_2^T \\
&= \sum_{i=1}^{2}\frac{\partial}{\partial A_i}W\left(A_1, A_2\right)\frac{d}{dt}A_i^T = \sum_{i=1}^{2}\frac{d}{dt}A_i\left(\frac{\partial}{\partial A_i}W(A_1, A_2)\right).
\end{aligned} \tag{1.41}
$$

From the expression of $W(A_1, A_2) = C_i A_i^T$ in (1.39) and using conclusions in Illustration 1.4.2(ii), we compute the gradient vectors of $W\left(A_1, A_2\right)$ with respect to vectors A_1 and A_2:

$$
\begin{aligned}
\frac{\partial}{\partial A_1}W\left(A_1, A_2\right) &= \frac{\partial}{\partial A_1}\left(C_i A_i^T\right), \quad \text{for any } i = 1, 2 \\
&= \frac{\partial}{\partial A_1}C_i A_i^T + C_i\frac{\partial}{\partial A_1}A_i^T \quad \text{[by Theorem 1.5.1(P3)]} \\
&= C_1\frac{\partial}{\partial A_1}A_1^T + \frac{\partial}{\partial A_1}C_1 A_1^T \quad \text{(by independent row expansion)}
\end{aligned}
$$

$$= C_1 \frac{\partial}{\partial A_1} A_1^T \quad \left(\text{by } C_1 \text{ independent of } A_1, \text{ i.e. } \frac{\partial}{\partial A_1} C_1 = 0\right)$$
$$= C_1 \quad [\text{by } (1.38)].$$

Similarly, the expression for

$$\frac{\partial}{\partial A_2} W(A_1, A_2)$$

is

$$\frac{\partial}{\partial A_2} W(A_1, A_2) = C_2.$$

By substituting these expressions into (1.41), using (1.39) and the definition of the determinant, we have

$$\begin{aligned} \frac{d}{dt} W(A_1, A_2) &= C_1 \frac{d}{dt} A_1^T + C_2 \frac{d}{dt} A_2^T \\ &= W\left(\frac{d}{dt} A_1, A_2\right) + W\left(A_1, \frac{d}{dt} A_2\right). \end{aligned} \tag{1.42}$$

We note that the calculus of the several variables approach provides a systematic procedure for computing the derivative of the determinant of the 2×2 matrix function A.

Let us discuss the computation of the second derivative of a 2×2 matrix function A. For this purpose, we apply the formula (1.28) to the functions defined in (1.42), and we have

$$\begin{aligned} \frac{d^2}{dt^2} W(A_1, A_2) &= \frac{d}{dt}\left[W\left(\frac{d}{dt} A_1, A_2\right) + W\left(A_1, \frac{d}{dt} A_2\right)\right] \\ &= \frac{d}{dt}\left(\sum_{i=1}^{2} C_i \frac{d}{dt} A_i^T\right) \\ &= \sum_{i=1}^{2} \frac{d}{dt}\left(C_i \frac{d}{dt} A^T\right) \quad [\text{by Theorem 1.5.1(A1)}] \\ &= \sum_{i=1}^{2} C_i \frac{d^2}{dt^2} A_i^T + \sum_{i=1}^{2} \frac{d}{dt} C_i \frac{d}{dt} A^T \quad [\text{by Theorem 1.5.1(P3)}]. \end{aligned} \tag{1.43}$$

We need to find a computational expression for $\frac{d}{dt} C_i$. For this purpose, we first examine the discussion in Observation 1.4.2, especially the conclusions. In this case,

$$\begin{aligned} C_{ij} &= (-1)^{i+j} \det(A_{ij}) = (-1)^{i+j} \sum_{l \neq j}^{2} a_{kl} C_{kl}(ij) \\ &= (-1)^{i+j} A_k(ij) C_k^T(ij) = (-1)^{i+j} C_k(ij) A_k^T(ij), \end{aligned} \tag{1.44}$$

where C_{ij}'s are functions of only A_k for $k \neq i$ and $i, k = 1, 2$. For $k \neq i$, $C_{kl}(ij)$'s are independent of $A_k(ij)$, and hence $C_{kl}(ij)$'s are independent of both row vectors A_i and A_k. With $n = 2$ and $C_{kl}(ij) = 1$, we have

$$\frac{\partial}{\partial A_k} C_{kl}(ij) = 0, \quad \frac{\partial}{\partial a_{kl}} C_{ij} = \begin{cases} (-1)^{i+j}, & \text{for } l \neq j, \\ 0, & \text{for } l = j, \end{cases}$$

and hence

$$\frac{\partial}{\partial A_k} C_i = \begin{bmatrix} (-1)^{i+1} & 0 \\ 0 & (-1)^{i+2} \end{bmatrix} \tag{1.45}$$

Now, for $k \neq i$, each $i, k = 1, 2$, from (1.36), we have

$$\begin{aligned} \frac{d}{dt} C_i &= \sum_{k \neq i}^{2} \left(\frac{\partial}{\partial A_k} C_i \left(\frac{d}{dt} A_k \right)^T \right) \\ &= \frac{\partial}{\partial A_k} C_i \left(\frac{d}{dt} A_k \right)^T . \end{aligned} \tag{1.46}$$

We substitute the expression (1.46) into (1.43), and we have

$$\begin{aligned} \frac{d^2}{dt^2} W(A_1, A_2) &= \sum_{i=1}^{2} C_i \frac{d^2}{dt^2} A_i^T + \sum_{i=1}^{2} \frac{d}{dt} C_i \frac{d}{dt} A_i^T \\ &= \sum_{i=1}^{2} C_i \frac{d^2}{dt^2} A_i^T + \sum_{i=1}^{2} \frac{\partial}{\partial A_k} C_i \left(\frac{d}{dt} A_k \right)^T \frac{d}{dt} A_i^T, \\ &\quad \text{for } k \neq 1, \text{ each } i, k = 1, 2 \\ &= \sum_{i=1}^{2} \frac{d^2}{dt^2} A_i C_i^T + 2 \frac{d}{dt} A_i \frac{\partial}{\partial A_k} C_i^T \left(\frac{d}{dt} A_k \right)^T \\ &= W\left(\frac{d^2}{dt^2} A_1, A_2 \right) + W\left(A_1, \frac{d^2}{dt^2} A_2 \right) + 2W\left(\frac{d}{dt} A_1, \frac{d}{dt} A_2 \right). \end{aligned} \tag{1.47}$$

(iii) For $n = 3$: Let $A(t) = A = (a_{ij}(t))_{3 \times 3}$ be the differentiable matrix function. From Illustration 1.4.3, Definition 1.3.10, and notations in Observation 1.4.2, the various kinds of representations of the determinant of $A(t)$ are summarized as

$$W(A_1, A_2, A_3) = \det(A(t)) = \sum_{j=1}^{3} a_{ij}(t) C_{ij}(t) = A_i C_i^T = C_i A_i^T, \tag{1.48}$$

for any $i = 1, 2, 3$.

Elementary approach. By ordinary properties of the derivative (product, sum, and difference) and the definition of the determinant of the 3×3 matrix, the derivative of $\det(A(t))$ is given by

$$\begin{aligned}
\frac{d}{dt}\det(A(t)) &= \frac{d}{dt} W\left(a_{11}, a_{12}, a_{13}, a_{21}, a_{22}, a_{23}, a_{31}, a_{32}, a_{33}\right) \\
&= \frac{d}{dt}\left[a_{11}(t)\det\left(A_{11}\right) - a_{12}(t)\det\left(A_{12}\right) + a_{13}(t)\det\left(A_{13}\right)\right] \\
&\quad \text{(by definition)} \\
&= \frac{d}{dt}\left[a_{11}(t)\det\left(A_{11}\right)\right] - \frac{d}{dt}\left[a_{12}(t)\det\left(A_{12}\right)\right] \\
&\quad + \frac{d}{dt}\left[a_{13}(t)\det\left(A_{13}\right)\right] \text{ (by sum rule)} \\
&= \frac{d}{dt}\left[a_{11}(t)\right]\det\left(A_{11}\right) + a_{11}(t)\frac{d}{dt}\left[\det\left(A_{11}\right)\right] - \frac{d}{dt}\det\left(A_{12}\right) \\
&\quad - a_{12}(t)\frac{d}{dt}\left[\det\left(A_{12}\right)\right] + \frac{d}{dt}\left[a_{13}(t)\right]\det\left(A_{13}\right) \\
&\quad + a_{13}(t)\frac{d}{dt}\left[\det\left(A_{13}\right)\right] \text{ (by product rule).} \qquad (1.49)
\end{aligned}$$

To compute the first derivative of the 3×3 matrix, we need to compute the first derivatives of C_{ij}, for $i, j = 1, 2, 3$. From the elementary approach to computation of the derivatives of 2×2 matrices, we have

$$\begin{aligned}
\frac{d}{dt}\det\left(A_{11}\right) &= \frac{d}{dt}\left[a_{22}a_{33} - a_{32}a_{23}\right] = W\left(\frac{d}{dt}A_2(11), A_3(11)\right) \\
&\quad + W\left(A_2(11), \frac{d}{dt}A_3(11)\right), \\
\frac{d}{dt}\det\left(A_{12}\right) &= \frac{d}{dt}\left[a_{21}a_{33} - a_{31}a_{23}\right] = W\left(\frac{d}{dt}A_2(12), A_3(12)\right) \\
&\quad + W\left(A_2(12), \frac{d}{dt}A_3(12)\right), \\
\frac{d}{dt}\det\left(A_{13}\right) &= \frac{d}{dt}\left[a_{21}a_{32} - a_{31}a_{22}\right] = W\left(\frac{d}{dt}A_2(13), A_3(13)\right) \\
&\quad + W\left(A_2(13), \frac{d}{dt}A_3(13)\right).
\end{aligned}$$

Substituting these derivatives in (1.49), we obtain the formula for the first derivative of the 3×3 matrix A:

$$\begin{aligned}
\frac{d}{dt}\det(A(t)) &= \frac{d}{dt} W\left(a_{11}, a_{12}, a_{13}, a_{21}, a_{22}, a_{23}, a_{31}, a_{32}, a_{33}\right) \\
&= \frac{d}{dt}a_{11}(t)\det\left(A_{11}\right) + a_{11}(t)
\end{aligned}$$

$$\times \left[W\left(\frac{d}{dt}A_2(11), A_3(11)\right) + W\left(A_2(11), \frac{d}{dt}A_3(11)\right)\right]$$

$$-\frac{d}{dt}\left[a_{12}(t)\right]\det\left(A_{12}\right) - a_{12}(t)$$

$$\times \left[W\left(\frac{d}{dt}A_2(12), A_3(12)\right) + W\left(A_2(12), \frac{d}{dt}A_3(12)\right)\right]$$

$$+\frac{d}{dt}\left[a_{13}(t)\right]\det\left(A_{13}\right) + a_{13}(t)$$

$$\times \left[W\left(\frac{d}{dt}A_2(13), A_3(13)\right) + W\left(A_2(13), \frac{d}{dt}A_3(13)\right)\right]$$

$$= \frac{d}{dt}a_{11}(t)\det\left(A_{11}\right) - \frac{d}{dt}\left[a_{12}(t)\right]\det\left(A_{12}\right) + \frac{d}{dt}\left[a_{13}(t)\right]\det\left(A_{13}\right)$$

$$+a_{11}(t)\left[W\left(\frac{d}{dt}A_2(11), A_3(11)\right)\right] - a_{12}(t)\left[W\left(\frac{d}{dt}A_2(12), A_3(12)\right)\right]$$

$$+a_{13}(t)\left[W\left(\frac{d}{dt}A_2(13), A_3(13)\right)\right] + a_{11}(t)\left[W\left(A_2(11), \frac{d}{dt}A_3(11)\right)\right]$$

$$-a_{12}(t)\left[W\left(A_2(12), \frac{d}{dt}A_3(12)\right)\right] + a_{13}(t)\left[W\left(A_2(13), \frac{d}{dt}A_3(13)\right)\right].$$

From the above expression, the definition, and the notation of the determinant of 3×3 matrices, we arrive at

$$\frac{d}{dt}\det(A(t)) = \frac{d}{dt}W\left(a_{11}, a_{12}, a_{13}, a_{21}, a_{22}, a_{23}, a_{31}, a_{32}, a_{33}\right)$$
$$= \frac{d}{dt}W\left(\frac{d}{dt}A_1, A_2, A_3\right) + W\left(A_1, \frac{d}{dt}A_2, A_3\right) + W\left(A_1, A_2, \frac{d}{dt}A_3\right). \tag{1.50}$$

The above procedure uses the explicit knowledge of the determinant as a function of entries of the matrix A,

$$\det(A) = W\left(a_{11}, a_{12}, a_{13}, a_{21}, a_{22}, a_{23}, a_{31}, a_{32}, a_{33}\right).$$

Calculus of multivariable approach. The calculus of the multivariable approach developed for a 2×2 matrix function is extended to the determinant of a 3×3 matrix function. Again, using the representation (1.48) of the definition of a determinant of a 3×3 matrix as a composite function of a matrix and its entries as functions of independent variable t, we apply the formula (1.28) to the function defined in (1.48) and obtain

$$\begin{aligned}\frac{d}{dt}W(A_1,A_2,A_3) &= \frac{\partial}{\partial A_1}W(A_1,A_2,A_3)\frac{d}{dt}A_1^T + \frac{\partial}{\partial A_2}W(A_1,A_2,A_3)\frac{d}{dt}A_2^T \\ &\quad + \frac{\partial}{\partial A_2}W(A_1,A_2,A_3)\frac{d}{dt}A_3^T \\ &= \sum_{i=1}^{3}\frac{\partial}{\partial A_i}W(A_1,A_2,A_3)\frac{d}{dt}A_i^T. \end{aligned} \tag{1.51}$$

From the expression of $W(A_1,A_2,A_3) = C_i A_i^T$ in (1.48), using conclusions from Illustration 1.4.3 and imitating the procedure outlined for the case of 2×2 matrix functions, the gradient vectors of $W(A_1,A_2,A_3)$ with respect to the vectors A_1, A_2, and A_3 are given by

$$\frac{\partial}{\partial A_i}W(A_1,A_2,A_3) = C_i\left(\text{by noting } \frac{\partial}{\partial A_i}C_{ij} = 0\right), \quad \text{for } i = 1,2,3. \tag{1.52}$$

This occurs because the ith-row expansion of $\det(A)$ depends on all rows, and C_{ij}'s depend on all rows except for the ith row, i.e. it is independent of the elements of the ith row of the matrix (see Observation 1.4.2 or Illustration 1.4.3), for any $i = 1,2,3$. From this and the definition of the determinant, the expression (1.51) reduces to

$$\begin{aligned}\frac{d}{dt}W(A_1,A_2,A_3) &= \sum_{i=1}^{3} C_i\frac{d}{dt}A_i^T \\ &= C_1\frac{d}{dt}A_1^T + C_2\frac{d}{dt}A_2^T + C_3\frac{d}{dt}A_3^T \\ &= W\left(\frac{d}{dt}A_1,A_2,A_3\right) + W\left(A_1,\frac{d}{dt}A_2,A_3\right) \\ &\quad + W\left(A_1,\frac{d}{dt}A_2,A_3\right). \end{aligned} \tag{1.53}$$

Once again, the above discussion shows that the calculus of several variables approach provides a systematic method for computing the derivative of the determinant of the 3×3 matrix function A.

Let us discuss the computation of the second derivative of the 3×3 matrix A. For this purpose, we apply the formula (1.28) to the function defined in (1.53) and obtain

$$\begin{aligned}\frac{d^2}{dt^2}W(A_1,A_2,A_3) &= \frac{d}{dt}\left[W\left(\frac{d}{dt}A_1,A_2,A_3\right)\right. \\ &\quad \left. + W\left(A_1,\frac{d}{dt}A_2,A_3\right) + W\left(A_1,\frac{d}{dt}A_2,A_3\right)\right]\end{aligned}$$

$$
\begin{aligned}
&= \frac{d}{dt}\left(\sum_{i=1}^{3} C_i \frac{d}{dt} A_i^T\right) \\
&= \sum_{i=1}^{3} \frac{d}{dt}\left(C_i \frac{d}{dt} A_i^T\right) \text{ [Theorem 1.5.1(A1)]} \\
&= \sum_{i=1}^{3} C_i \frac{d^2}{dt^2} A_i^T + \sum_{i=1}^{3} \frac{d}{dt} C_i \frac{d}{dt} A_i^T \text{ [Theorem 1.5.1(P3)].}
\end{aligned} \tag{1.54}
$$

To find $\frac{d}{dt} C_i$, we imitate the procedure for finding the first derivative of the determinant in the context of Observation 1.5.3, in particular (1.36):

$$
\frac{d}{dt} C_i = \sum_{k=1}^{3} \frac{\partial}{\partial A_k} C_i \left(\frac{d}{dt} A_k\right)^T, \quad \text{for any } k \neq i \quad \text{and} \quad i, k = 1, 2, 3. \tag{1.55}
$$

From the conclusions in Illustration 1.4.3 (or Observations 1.4.2 and 1.5.3), we recall the computation of C_{ij}'s and its dependence and/or independence on the rows of the matrix A in the context of the row expansion of the determinant. We then conclude that

$$
\frac{\partial}{\partial A_k} C_{kl}(ij) = 0, \quad \frac{\partial}{\partial a_{kl}} C_{ij} = \begin{cases} (-1)^{i+j} C_{kl}(ij), & \text{for } l \neq j, \\ 0, & \text{for } l = j, \end{cases}
$$

and hence

$$
\frac{\partial}{\partial A_k} C_i = \begin{cases} \left(\frac{\partial}{\partial a_{kl}} C_{ij}\right)_{3\times 3}, & \text{for any } k \neq i, \\ 0, & \text{for } k = i. \end{cases} \tag{1.56}
$$

From (1.55) and (1.56), (1.54) can be written as

$$
\begin{aligned}
\frac{d^2}{dt^2} W(A_1, A_2, A_3) &= \sum_{i=1}^{3} C_i \frac{d^2}{dt^2} A_i^T + \sum_{i=1}^{3} \sum_{k\neq i}^{3} \left(\frac{d}{dt} A_k\right) \frac{\partial}{\partial A_k} C_i \frac{d}{dt} A_i^T \\
&= \sum_{i=1}^{3} C_i \frac{d^2}{dt^2} A_i^T + \sum_{i=1}^{3} \sum_{k\neq i}^{3} \left(\frac{d}{dt} A_k\right) \frac{\partial}{\partial A_k} C_i \frac{d}{dt} A_i^T.
\end{aligned} \tag{1.57}
$$

For any $k \neq i$ and $i, k = 1, 2, 3$, we also note that the $\det(A(t))$ in (1.48) can also be written as

$$
W(A_1, A_2, A_3) = \det(A(t)) = A_k \frac{\partial}{\partial A_k} C_i A_i^T = A_k \frac{\partial^2}{\partial A_i \partial A_k} W A_i^T, \tag{1.58}
$$

where

$$
\frac{\partial}{\partial A_k} C_i = \frac{\partial}{\partial A_k}\left(\frac{\partial}{\partial A_i} W(A_1, A_2, A_3)\right)
$$

is independent of both row vectors A_i and A_k for $k \neq i$. The definition of the determinant, as in (1.52), Theorem 1.5.1 (T1), Illustration 1.4.3, Observation 1.4.2, and the definition of the product of matrices have all played a role in this deduction.

The presented argument in illustration 1.5.1 leads to the following result.

Theorem 1.5.3. *Let A be an $n \times n$ differentiable matrix function defined on J, and let*

$$\det(A) = W(A_1, \ldots, A_k, \ldots, A_n) = W\left(A^1, \ldots, A^k, \ldots, A^n\right)$$

be the determinant function defined in Definition 1.4.2. *Then*

$$\begin{aligned} \frac{d}{dt}\det(A) &= \frac{d}{dt}, W(A_1, \ldots, A_k, \ldots, A_n) \\ &= \sum_{i=1}^{n} C_i \frac{d}{dt} A_i^T \\ &= \sum_{k=1}^{n} W\left(A_1, \ldots, \frac{d}{dt}A_k, \ldots, A_n\right), \end{aligned} \tag{1.59}$$

$$\begin{aligned} d\det(A(t)) &= dW(A_1, \ldots, A_k, \ldots, A_n) \\ &= \sum_{k=1}^{n} dA_k \frac{\partial}{\partial A_k} W(A_1, \ldots, A_k, \ldots, A_n) \\ &= \sum_{k=1}^{n} W(A_1, \ldots, dA_k, \ldots, A_n). \end{aligned} \tag{1.60}$$

Proof. For the cases $n = 1, 2, 3$, Definitions 1.3.10 and 1.4.2, Observations 1.3.1, 1.4.1, 1.4.2, and 1.5.3, Illustration 1.5.1, and the principle of mathematical induction, we have

$$\frac{\partial}{\partial A_i} W = C_i, \quad \frac{\partial}{\partial A_i} C_i = 0, \quad \text{for any } i,\ 1 \leq i \leq n. \tag{1.61}$$

This occurs because the cofactor vector C_i is independent of the ith-row vector A_i of the matrix A, and $W(A_1, \ldots, A_i, \ldots, A_n)$ is independent of any row expansion of A. Hence,

$$\begin{aligned} \frac{d}{dt}\det(A) &= \sum_{i=1}^{n} \frac{\partial}{\partial A_i} W \frac{d}{dt} A_i^T = \sum_{k=1}^{n} C_i \frac{d}{dt} A_i^T \\ &= \sum_{i=1}^{n} W\left(A_1, \ldots, \frac{d}{dt}A_i, \ldots, A_n\right). \end{aligned} \tag{1.62}$$

This establishes the validity of (1.59). The validity of (1.60) comes from the above argument and the concept of the differential. □

Example 1.5.7. We are given

$$A(t) = \begin{bmatrix} \exp[2t] & \sin t & t \\ t^2 & \cos t & 1 \\ \exp[t] & t & \sin t \end{bmatrix}.$$

Find $\frac{d}{dt}\det(A)$.

Solution procedure. From the formula (1.59), for $n = 3$ we have

$$\begin{aligned} \frac{d}{dt}\det(A)(t) &= \frac{d}{dt}W(A_1, A_2, A_3) \\ &= W\left(\frac{d}{dt}A_1, A_2, A_3\right) + W\left(A_1, \frac{d}{dt}A_2, A\right) \\ &\quad + W\left(A_1, A_2, \frac{d}{dt}A_3\right). \end{aligned}$$

We need to compute the determinants of the following matrices:

$$\begin{bmatrix} \frac{d}{dt}A_1 \\ A_2 \\ A_3 \end{bmatrix} = \begin{bmatrix} 2\exp[2t] & \cos t & 1 \\ t^2 & \cos t & 1 \\ \exp[t] & t & \sin t \end{bmatrix},$$

$$\begin{bmatrix} A_1 \\ \frac{d}{dt}A_2 \\ A_3 \end{bmatrix} = \begin{bmatrix} \exp[2t] & \sin t & t \\ 2t & -\sin t & 0 \\ \exp[t] & t & \sin t \end{bmatrix},$$

$$\begin{bmatrix} A_1 \\ A_2 \\ \frac{d}{dt}A_3 \end{bmatrix} = \begin{bmatrix} \exp[2t] & \sin t & t \\ t^2 & \cos t & 1 \\ \exp[t] & 1 & \cos t \end{bmatrix}.$$

$$\begin{aligned} W\left(\frac{d}{dt}A_1, A_2, A_3\right) &= 2\exp[2t](\cos t \sin t - t) - \cos t\left(t^2 \sin t - \exp[t]\right) \\ &\quad + (t^3 - \cos t \exp[t]) \\ &= 2\exp[2t](\cos t \sin t - t) + t^2(t - \cos t \sin t), \\ W\left(A_1, \frac{d}{dt}A_2, A\right) &= -\exp[2t]\sin^2 t - 2t\sin^2 t + t\left(2t^2 + \exp[t]\sin t\right) \\ &= -\exp[2t]\sin^2 t + \exp[t]\, t\sin t + 2t\left(t^2 - \sin^2 t\right), \\ W\left(A_1, A_2, \frac{d}{dt}A_3\right) &= \exp[2t]\left(\cos^2 t - 1\right) - \sin t\left(t^2\cos t - \exp[t]\right) \\ &\quad + t\left(t^2 - \exp[t]\cos t\right) \end{aligned}$$

$$
\begin{aligned}
&= \exp[2t]\left(\cos^2 t - 1\right) + \exp[t](\sin t - t\cos t) \\
&\quad + t^2(t - \cos t \sin t), \\
\frac{d}{dt}\det(A)(t) &= 2\exp[2t](\cos t\sin t - t) + t^2(t - \cos t\sin t) \\
&\quad - \exp[2t]\sin^2 t + \exp[t]\, t\sin t + 2t\left(t^2 - \sin^2 t\right) \\
&\quad + \exp[2t]\left(\cos^2 t - 1\right) + \exp[t](\sin t - t\cos t) \\
&\quad + t^2(t - \cos t\sin t) \\
&= \exp[2t]\left(2\cos t\sin t - \sin^2 t + \cos^2 t - 1 - 2t\right) \\
&\quad + \exp[t](t\sin t + \sin t - t\cos t) \\
&\quad + (4t^3 - 2t^2\cos t\sin t) - 2t\sin^2 t.
\end{aligned}
$$

The following result is useful for the study of linear systems of differential equations.

Theorem 1.5.4. *Let A be an $n \times n$ twice differential matrix function defined on J, and let*

$$\det(A) = W(A_1, \ldots, A_k, \ldots, A_n) = W\left(A^1, \ldots, A^k, \ldots, A^n\right)$$

be the determinant function defined in Definition 1.4.2. *Then,*

$$
\begin{aligned}
\frac{d^2}{dt^2}\det(A) &= \frac{d^2}{dt^2} W(A_1, \ldots, A_k, \ldots, A_n) \\
&= \sum_{i=1}^{n} C_i \frac{d^2}{dt^2} A_i^T + \sum_{i=1}^{n}\sum_{k\neq i}^{n} \frac{d}{dt} A_k \frac{\partial}{\partial A_k} C_i \frac{d}{dt} A_i^T \\
&= \sum_{i=1}^{n} W\left(A_1, \ldots, \frac{d^2}{dt^2} A_i, \ldots, A_n\right) \\
&\quad + \sum_{i=1}^{n}\sum_{k\neq i}^{n} W\left(A_1, \ldots, \frac{d}{dt} A_i, \ldots, \frac{d}{dt} A_k, \ldots, A_n\right). \qquad (1.63)
\end{aligned}
$$

Proof. Based on the argument used in the proof of Theorem 1.5.3, Observations 1.4.1, 1.4.2, and 1.5.3, and Illustration 1.5.1(iii), from (1.59) and (1.28) we obtain expressions similar to (1.54), (1.56), and (1.57):

$$
\begin{aligned}
\frac{d^2}{dt^2}\det(A) &= \frac{d^2}{dt^2} W(A_1, \ldots, A_k, \ldots, A_n) \\
&= \sum_{i=1}^{n} C_i \frac{d^2}{dt^2} A^T + \sum_{i=1}^{n} \frac{d}{dt} C_i \frac{d}{dt} A_i^T, \qquad (1.64)
\end{aligned}
$$

$$\frac{d}{dt} C_i = \sum_{k=1}^{n} \frac{d}{dt} A_k \frac{\partial}{\partial A_k} C_i, \quad \text{for any } i, k = 1, 2, \ldots, n, \tag{1.65}$$

where

$$\frac{\partial}{\partial A_k} C_i = \begin{cases} \left(\frac{\partial}{\partial a_{kl}} C_i\right)_{n\times n}, & \text{for any } k \neq i, \\ 0, & \text{for } k = i, \end{cases} \tag{1.66}$$

$$\frac{\partial}{\partial A_k} C_{kl}(ij) = 0, \quad \frac{\partial}{\partial a_{kl}} C_{ij} = \begin{cases} (-1)^{i+j} C_{kl}(ij), & \text{for } l \neq j, \\ 0, & \text{for } l = j. \end{cases}$$

We substitute the expression (1.66) into (1.65) and then into (1.64). After that we derive a formula similar to (1.57), as follows:

$$\begin{aligned} \frac{d^2}{dt^2} \det(A) &= \frac{d^2}{dt^2} W(A_1, \ldots, A_k, \ldots, A_n) \\ &= \sum_{i=1}^{n} C_i \frac{d^2}{dt^2} A_i + \sum_{i=1}^{n} \sum_{k\neq i}^{n} \frac{d}{dt} A_k \frac{\partial}{\partial A_k} C_i \frac{d}{dt} A_i^T, \end{aligned} \tag{1.67}$$

for any $k \neq i$, $i, k = 1, 2, \ldots, n$. Recalling the representation of C_i as

$$\begin{gathered} C_i = \left((-1)^{i+j} A_k(ij) C_k^T(ij)\right)_{1\times n}, \\ \text{for any } k = 1, 2, \ldots, i-1, i+1, \ldots, n, \end{gathered} \tag{1.68}$$

and using (1.66), we have

$$\frac{\partial}{\partial A_k} C_i^T \left(\frac{d}{dt} A_k\right)^T = \left((-1)^{i+j} \sum_{k\neq i}^{n} \frac{d}{dt} A_k(ij) C_k^T(ij)\right)_{1\times n}. \tag{1.69}$$

From (1.67), (1.69), and the definition of the determinant and its variants, we obtain

$$\begin{aligned} \frac{d^2}{dt^2} \det(A) &= \frac{d}{dt} \left(\sum_{i=1}^{n} \frac{d}{dt} A_i \frac{\partial}{\partial A_i} W\right) \\ &= \sum_{k=1}^{n} C_i \frac{d^2}{dt^2} A_i^T + \sum_{i=1}^{n} \sum_{k\neq i}^{n} \frac{d}{dt} A_k \frac{\partial}{\partial A_k} C_i \frac{d}{dt} A_i^T \\ &= \sum_{i=1}^{n} C_i \frac{d^2}{dt^2} A^T + \sum_{i=1}^{n} \frac{d}{dt} a_{ij} \left[(-1)^{i+j} \sum_{k\neq i}^{n} C_k(ij) \frac{d}{dt} A_k^T(ij)\right] \end{aligned}$$

$$= \sum_{i=1}^{n} W\left(A_i, \ldots, \frac{d^2}{dt^2} A_i, \ldots, A_n\right)$$

$$+ \sum_{i=1}^{n} \sum_{k \neq i}^{n} W\left(A_1, \ldots, \frac{d}{dt} A_i, \ldots, \frac{d}{dt} A_k, \ldots, A_n\right). \tag{1.70}$$

This completes the proof of the theorem. □

In the following, we present a mean-value theorem for differential calculus and the Taylor polynomial theorem of degree 2 for a determinant function. These results will play a very important role in the study of differential equations.

Theorem 1.5.5 (generalized mean-value theorem). *Let A and X be $n \times n$ matrices. Let*

$$\det(A) = W(A) = W(A_1, \ldots, A_k, \ldots, A_n),$$
$$\det(X) = W(X) = W(X_1, \ldots, X_k, \ldots, X_n)$$

be the determinant of A and of X, respectively. Let L be an $n \times n$ matrix function defined on the interval $[0, 1]$ by $L(t) = A + t(X - A)$ for t in $[0, 1]$. Then

$$\det(X) - \det(A) = \sum_{i=1}^{n} (X_i - A_i) \int_0^1 C_i^T(L(t))\, dt. \tag{1.71}$$

Moreover,

$$\det(X) - \det(A) = \sum_{i=1}^{n} (X_i - A_i)\, C_i^T(A)$$

$$+ \sum_{i=1}^{n} (X_i - A_i) \int_0^1 C_i^T[L(t) - A]\, dt. \tag{1.72}$$

Proof. From Observation 1.4.2, we note that a determinant of an $n \times n$ matrix is the nth-degree homogeneous polynomial function of n^2 entries of the matrix as the independent variable. We know that every polynomial function in several independent variables is continuously differentiable. Hence, the determinant of any $n \times n$ matrix is continuously differentiable with respect to its independent variables. $L(t) = A + t(X - A)$ (a line segment joining A and X) for t in $[0, 1]$ is also continuously differentiable on $[0, 1]$. We define a function h defined on $[0, 1]$ into R as follows:

$$h(t) = \det(L(t)) = W(L(t)) = W(L(t)_1, \ldots, L_i(t), \ldots, L_n(t)).$$

h is a composite function of continuously differentiable functions, namely "det" and L. Now, Theorem 1.5.3 is applicable to $\det(L(t))$, and hence we have

$$h'(t) = \frac{d}{dt}\det(L(t)) = \sum_{i=1}^{n} \frac{d}{dt} L_i(t) C_i^T(L(t))$$
$$= \sum_{i=1}^{n} (X_i - A_i)\, C_i^T(L(t)) \quad \left[\text{by } \frac{d}{dt} L_i = (X_i - A_i)\right].$$

Now, we integrate both sides of the above expression from $t = 0$ to $t = 1$. This is possible because the expression is a continuous function on the interval $[0, 1]$. Hence,

$$\int_0^1 h'(t)\, dt = \int_0^1 \left[\sum_{i=1}^{n} (X_i - A_i)\, C_i^T(L(t))\right] dt,$$
$$h(1) - h(0) = \int_0^1 \left[\sum_{i=1}^{n} (X_i - A_i)\, C_i^T(L(t))\right] dt$$
$$= \sum_{i=1}^{n} (X_i - A_i) \int_0^1 C_i^T(L(t))\, dt \text{ [Theorem 1.5.2(A1), (P1)].}$$

This, along with the definition of the function h, yields

$$h(1) = \det(L(1)) = \det(X) \text{ (by } L(1) = A + 1(X - A) = X),$$
$$h(0) = \det(L(0)) = \det(A) \text{ (by } L(0) = A + 0(X - A) = A),$$

and hence

$$h(1) - h(0) = \det(X) - \det(A) = \sum_{i=1} (X_i - A_i) \int_0^1 C_i^T(L(t))\, dt.$$

This establishes the relation (1.71). The validity of the relation (1.72) results by adding and subtracting $\sum_{i=1}^{n} (X_i - A_i)\, C_i^T(A)$. This completes the proof of the theorem. □

Theorem 1.5.6 (Taylor's type formula). *Let us assume that all the hypotheses of Theorem* 1.5.5 *are satisfied. Let N be an $n \times n$ matrix function defined on* $[0,1] \times [0,1]$ *by $N(t,s) = A + st(X - A)$ for (t,s) in* $[0,1] \times [0,1]$. *Then,*

$$\det(X) = \det(A) + \sum_{i=1}^{n} (X_i - A_i)\, C_i^T(A)$$
$$+ \frac{1}{2!} \sum_{i=1}^{n} \sum_{k \neq i}^{n} (X_i - A_i) \frac{\partial}{\partial A_k} C_i^T(A)\, (X_k - A_k)^T$$
$$+ \sum_{i=1}^{n} \sum_{k \neq i}^{n} (X_i - A_i) \int_0^1 \left[\int_0^1 [0^{ik}(X - A)]\; dst\, (X_k - A_k)^T\right] dt, \tag{1.73}$$

where

$$\frac{\partial}{\partial A_k} C^{iT}(N(t,s)) - \frac{\partial}{\partial A_k} C^{iT}(A) = 0^{ik}(X-A);$$

$0^{ik}(X-A)$ *is independent of the ith and the kth row of the matrix* $X-A$ *and is bounded by the magnitude of* $X-A$.

Proof. We recall

$$\det(X) = \det(A) + \sum_{i=1}^{n} (X_i - A_i)\, C_i^T(A) + \sum_{i=1}^{n} (X_i - A_i) \int_0^1 \left[C_i^T(L(t)) - C_i^T(A)\right] dt. \tag{1.74}$$

Under the hypotheses of the theorem and by following the argument used in Theorem 1.5.5, $C_i^T(L(t)) - C_i^T(A)$ can be represented by

$$C_i^T(L(t)) - C_i^T(A) = \sum_{k\neq i} \int_0^1 \left[\frac{\partial}{\partial A_k} C_i^T(N(t,s))\, dst\, (X_k - A_k)\right], \tag{1.75}$$

where N is an $n\times n$ matrix function defined on $[0,1]\times[0,1]$ by $N(t,s) = A + st(X-A)$ for (t,s) $[0,1]\times[0,1]$. We substitute the expression (1.75) into (1.74), and we have

$$\begin{aligned}\det(X) &= \det(A) + \sum_{i=1}^{n} (X_i - A_i)\, C_i^T(A) + \sum_{i=1}^{n} (X_i - A_i) \\ &\quad \times \int_0^1 \left[\sum_{k\neq i}^{n} \int_0^1 \left[\frac{\partial}{\partial A_k} C_i^T(N(t,s))\, ds\right] t\, (X_k - A_k)^T\right] dt \\ &= \det(A) + \sum_{i=1}^{n} (X_i - A_i)\, C_i^T(A) + \sum_{i=1}^{n} \sum_{k\neq i}^{n} (X_i - A_i) \\ &\quad \times \int_0^1 \left[\int_0^1 \left[\frac{\partial}{\partial A_k} C_i^T(N(t,s))\, ds\right] t\, (X_k - A_k)^T\right] dt. \end{aligned} \tag{1.76}$$

Again, by adding and subtracting

$$\frac{1}{2!} \sum_{i=1}^{n} \sum_{k\neq i}^{n} (X_i - A_i) \frac{\partial}{\partial A_k} C_i^T(A)\, (X_k - A_k)^T$$

on the right-hand side of (1.76) and noting the fact that

$$\begin{aligned}&\frac{1}{2!} \sum_{i=1}^{n} \sum_{k\neq i}^{n} (X_i - A_i) \frac{\partial}{\partial A_k} C_i^T(A)\, (X_k - A_k)^T \\ &= \sum_{i=1}^{n} \sum_{k\neq i}^{n} (X_i - A_i) \int_0^1 \left[\int_0^1 \left[\frac{\partial}{\partial A_k} C_i^T(A)\right] t\, (X_k - A_k)^T\right] ds\, dt,\end{aligned}$$

(1.76) reduces to

$$\det(X) = \det(A) + \sum_{i=1}^{n} (X_i - A_i)\, C_i^T(A)$$
$$+ \frac{1}{2!} \sum_{i=1}^{n} \sum_{k \neq i}^{n} (X_i - A_i)\, \frac{\partial}{\partial A_k}\, C_i^T(A)\, (X_k - A_k)^T$$
$$+ \sum_{i=1}^{n} \sum_{k \neq i}^{n} (X_i - A_i) \int_0^1 \left[\int_0^1 [0^{ik}(X - A)]\, t\, (X_k - A_k)^T \right] ds\, dt,$$

where

$$\frac{\partial}{\partial A_k}\, C_i^T(N(t,s)) - \frac{\partial}{\partial A_k}\, C_i^T(A) = 0^{ik}(X - A).$$

This establishes the proof of the theorem. □

The simplest example of a scalar-valued matrix function is the determinant. In order to define more aspects of the matrix functions and its calculus concepts like the limit or continuity, we need a concept that describes "nearness" or "closeness" with regard to matrices. This concept centers around the "distance" between matrices. It is based on the definition of the distance between two points in a plane. We recall that the mathematical description of a plane is $R^2 = \{(a_{11}, a_{12}) : (a_{11}, a_{12}) = [a_{11}, a_{12}] = (a_{ij})_{1\times 2}$, all 1×2 matrices$\}$ and the distance between the two points $P\,(a_{11}, a_{12})$ and $Q\,(b_{11}, b_{12})$ in R^2 is defined by

$$|PQ| = \left((a_{11} - b_{11})^2 + (a_{12} - b_{12})^2 \right)^{\frac{1}{2}}$$
$$= \left(\sum_{i=1}^{1} \sum_{j=1}^{2} (a_{ij} - b_{ij})^2 \right)^{\frac{1}{2}} = \|A - B\|,$$

where $A = (a_{ij})_{1\times 2}$ and $B = (b_{ij})_{1\times 2}$. $\|A - B\|$ is the notation for the distance between two matrices A and B. By utilizing this definition and its interpretation, it is natural to define a distance between two matrices in $R^{m\times n}$.

Definition 1.5.4. Let $\mathcal{M}(m \times n) = R^{m\times n}$ be a collection of matrices of size $m \times n$. The distance between any two matrices $A = (a_{ij})_{m\times n}$ and $B = (b_{ij})_{m\times n}$ in $R^{m\times n}$ is defined by

$$\|A - B\| = \left(\sum_{i=1}^{m} \sum_{j=1}^{n} (a_{ij} - b_{ij})^2 \right)^{\frac{1}{2}}. \tag{1.77}$$

We note that it satisfies the following properties that are similar to the properties of the distance concept defined on R^2. They are as follows:

Theorem 1.5.7. *Let $\mathcal{M}(m \times n) = R^{m\times n}$ be a collection of matrices of size $m \times n$, and let A, B and C be arbitrary matrices in $\mathcal{M}(m \times n)$. The distance between any two matrices defined in* (1.77) *satisfies the following properties:*

(D1) $\|A - B\| \geq 0$ (*nonnegativity property*);
(D2) $\|A - B\| = 0$ *iff* $A = B$;
(D3) $\|A - B\| = \|B - A\|$ (*symmetric property*);
(D4) $\|A - B\| \leq \|A - C\| + \|C - A\|$ (*triangular inequality*).

Observation 1.5.4. We recall that the distance between the origin and any point in the plane is

$$|OP| = \left((a_{11} - 0)^2 + (a_{12} - 0)^2\right)^{\frac{1}{2}} = \left(\sum_{i=1}^{1}\sum_{j=1}^{2}(a_{ij} - 0)^2\right)^{\frac{1}{2}}$$

$$= \left(a_{11}^2 + a_{12}^2\right)^{\frac{1}{2}} = \left(\sum_{i=1}^{1}\sum_{j=1}^{2} a_{ij}^2\right)^{\frac{1}{2}} = \|A - 0\| = \|A\|.$$

From this and (1.77), the distance between $m \times n$ zero matrix $O = (0)_{m\times n}$ and any other $m \times n$ matrix $A = (a_{ij})_{m\times n}$ is given by

$$\|A - O\| = \left(\sum_{i=1}^{m}\sum_{j=1}^{n}(a_{ij} - 0)^2\right)^{\frac{1}{2}} = \left(\sum_{i=1}^{m}\sum_{j=1}^{n}(a_{ij}^2)\right)^{\frac{1}{2}} = \|A\|. \tag{1.78}$$

We further note that $\|A\|$ has the following properties that are similar to the properties of an absolute value of a real number:

(N1) $\|A\| \geq 0$ (nonnegative property);
(N2) $\|A\| = 0$ *iff* $A = O$;
(N3) $\|aA\| = |a|\|A\|$ (homogeneity property);
(N4) $\|A + B\| \leq \|A\| + \|B\|$ (triangular inequality).

Here A and B are arbitrary matrices in $R^{m\times n}$ and a is any scalar quantity. We note that $\|A\|$ is referred to as a norm of an $m \times n$ matrix A. Moreover, any distance and norm are functions defined on $R^{m\times n} \times R^{n\times m}$ and $R^{m\times n}$ into R_+, respectively, where $R_+ = [0, \infty)$, i.e. a set of nonnegative real numbers.

Definition 1.5.5. Let $f : R^{m\times n} \to R^{p\times q}$ be a relation. f is called a matrix function if, for each $m \times n$ matrix A in $R^{m\times n}$, the value of $f(A)$ is in $R^{p\times q}$ and it is uniquely determined by A.

Illustration 1.5.2. Let A be an $n \times n$ matrix, and let us define $\exp[A]$ as follows:

$$\exp[A] = I + \frac{A}{1!} + \frac{A^2}{2!} + \frac{A^3}{3!} + \cdots + \frac{A^l}{l!} + \cdots = \sum_{l=0}^{\infty} \frac{A^l}{l!}, \tag{1.79}$$

where A is an $n \times n$ square matrix, $0! = 1$, I is the $n \times n$ identity matrix, and $A^0 = I$. The $\exp[A]$ is well-defined because the right-hand series in (1.79) is based on the Maclaurin series for $\exp[t]$ in t on R. Moreover, the series in (1.79) converges because

$$\begin{aligned} \| \exp[A] \| &= \left\| I + \frac{A}{1!} + \frac{A^2}{2!} + \frac{A^3}{3!} + \cdots + \frac{A^l}{l!} + \cdots \right\| \\ &\leq \sum_{l=0}^{\infty} \left\| \frac{A^l}{l!} \right\| \quad \text{[Observation 1.5.4(N4)]} \\ &\leq \sum_{l=0}^{\infty} \frac{\|A^l\|}{l!} \quad \text{[Observation 1.5.4(N3)]} \\ &\leq \sum_{l=0}^{\infty} \frac{\|A\|^l}{j!}. \end{aligned} \tag{1.80}$$

By the comparison test [85], the series in (1.79) is convergent and its sum, $\exp[A]$, is a well-defined $n \times n$ matrix.

In order to appreciate and utilize the concept of a matrix function such as $\exp[A]$, we need to extend the concepts of limits and continuity of matrix functions.

Definition 1.5.6. Let $f : R^{m\times n} \to R^{p\times q}$ be a matrix function. In addition, let C be in $R^{m\times n}$ and let L be in $R^{p\times q}$. The matrix L in $R^{p\times q}$ is said to be the limit of the matrix function f at $X = C$, if $\lim_{X\to C} f(X) = L$. This statement means that for every $\epsilon > 0$, one must be able to determine a number $\delta(\epsilon, L, f) > 0$ such that $\|f(X) - L\| < \epsilon$, whenever $0 < \|X - C\| < \delta$. The matrix function f is said to be continuous at $X = C$ if: (a) $f(C)$ is defined, (b) $\lim_{X\to C} f(X) = L$ exists, and (c) $f(C) = L$.

Observation 1.5.5. From elementary calculus, we will use the sequence of a partial sum of a series. We define the sequence of a partial sum of a series in (1.79) as

$$S_k = \sum_{l=0}^{k} \frac{A^l}{l!}. \tag{1.81}$$

For each nonnegative integer $k \geq 0$, from Definitions 1.3.8, 1.3.9, and 1.3.11, and the repeated application of Theorem 1.3.1(A3), S_k is a well-defined $n \times n$ matrix. Moreover, by the principle of mathematical induction, the sequence in (1.81) is defined for every nonnegative integer $k \geq 0$ with $S_0 = I$. This is also an $n \times n$ matrix. Hence, we extend the convergence concept of a series of real numbers to the series of square matrices as follows: A series of matrices is said to be convergent if the sequence of partial sums of the given series converges, and its limit is the sum of the series. In the case of the series in (1.79), we have

$$\lim_{k\to\infty} S_k = \lim_{k\to\infty}\left[\sum_{l=0}^{k}\frac{A^l}{l!}\right] = \exp[A] = \sum_{l=0}^{\infty}\frac{A^l}{l!}. \tag{1.82}$$

In the following, we present a few algebraic properties of $\exp[A]$ which will be utilized to study systems of linear differential equations.

Theorem 1.5.8. *The matrix* exp *function possesses the following algebraic properties*:

(a) $\exp[O] = I$, *where* $O = (0)_{n\times n}$ *and* I *are* $n \times n$ *zero and identity matrices, respectively*;
(b) $\exp[-A]\exp[A] = \exp[A]\exp[-A] = I$, *i.e.* $\exp[-A]$ *is the inverse of* $\exp[A]$;
(c) $A\exp[A] = \exp[A]\,A$, *i.e. matrices* A *and* $\exp[A]$ *commute* (*Observation* 1.3.5);
(d) $\exp[A+B] = \exp[A]\exp[B]$ *if and only if* $AB = BA$ (*Observation* 1.3.5);
(e) $\frac{d}{dt}(\exp[At]) = A\exp[A] = \exp[At]\,A$.

Proof. The proof of (a) follows by substituting $A = O$ in (1.79). For the proof of (b), by substituting $-A$ for A and using (1.80)–(1.82), it follows that $\exp[-A]$ is well-defined. Moreover,

$$\exp[-A]\exp[A] = \left(I - \frac{A}{1!} + \frac{A^2}{2!} - \frac{A^3}{3!} + \cdots + (-1)^l\frac{A^l}{l!} + \cdots\right)$$
$$\times\left(I + \frac{A}{1!} + \frac{A^2}{2!} + \frac{A^3}{3!} + \cdots + \frac{A^l}{l!} + \cdots\right)$$

and the right-hand side expression can be multiplied term by term and then simplified. This simplification leads to the validity of (b). The proof of (c) follows by multiplication and regrouping through using Theorem 1.3.1 (A2). For the proof of (d), we consider

$$\begin{aligned}\exp[A+B] &= I + \frac{A+B}{1!} + \frac{(A+B)^2}{2!} + \frac{(A+B)^3}{3!} + \cdots + \frac{(A+B)^j}{l!} + \cdots\\ &= I + \frac{A+B}{1!} + \frac{A^2 + AB + BA + B^2}{2!}\\ &\quad + \frac{A^3 + A^2B + ABA + AB^2 + BA^2 + BAB + B^2A + B^3}{3!}\\ &\quad + \cdots + \frac{A^j + A^{j-1}B + \cdots + B^{j-1}A + B^j}{j!} + \cdots\end{aligned}$$

$$
\begin{aligned}
&= I + \frac{A+B}{1!} + \frac{A^2 + AB + AB + B^2}{2!} \\
&\quad + \frac{A^3 + A^2B + AAB + AB^2 + A^2B + ABB + AB^2 + B^3}{3!} \\
&\quad + \cdots + \frac{A^j + A^{j-1}B + \cdots + AB^{j-1} + B^j}{j!} \\
&\quad + \cdots \text{ (commutative property: } AB = BA) \\
&= I + \frac{A+B}{1!} + \frac{A^2 + 2AB + B^2}{2!} \\
&\quad + \frac{A^3 + 3A^2B + 3AB^2 + B^3}{3!} + \cdots \\
&\quad + \frac{A^j + jA^{j-1}B + \cdots + jAB^{j-1} + B^j}{j!} \\
&\quad + \cdots \text{ (by substitution)} \\
&= I + \frac{A}{1!} + \frac{A^2}{2!} + \frac{A^3}{3!} + \cdots + \frac{B}{1!} + \frac{2AB}{2!} + \frac{3A^2B}{3!} + \cdots \\
&\quad + \frac{B^2}{2!} + \frac{3AB^2}{3!} + \frac{B^3}{3!} \\
&\quad + \cdots \text{ (by rearranging the terms)} \\
&= \left(I + \frac{A}{1!} + \frac{A^2}{2!} + \frac{A^3}{3!} + \cdots\right) + \left(I + \frac{2A}{2!} + \frac{3A^2}{3!} + \cdots\right)\frac{B}{1!} \\
&\quad + \left(\frac{B^2}{2!} + \frac{3AB^2}{3!} + \cdots\right)\frac{B^2}{2!} + \frac{B^3}{3!} \\
&\quad + \cdots \text{ (by regrouping)} \\
&= \left(I + \frac{A}{1!} + \frac{A^2}{2!} + \frac{A^3}{3!} + \cdots\right) + \left(I + \frac{A}{1!} + \frac{A^2}{2!} + \cdots\right)\frac{B}{1!} \\
&\quad + \left(I + \frac{A}{1!} + \cdots\right)\frac{B^2}{2!} + \frac{B^3}{3!} \\
&\quad + \cdots \text{ (by simplifying)} \\
&= \left(I + \frac{A}{1!} + \frac{A^2}{2!} + \frac{A^3}{3!} + \ldots + \frac{A^l}{l!} + \cdots\right) \\
&\quad \times \left(I + \frac{B}{1!} + \frac{B^2}{2!} + \frac{B^3}{3!} + \cdots + \frac{B^l}{l!} + \cdots\right) \\
&= \exp[A]\exp[B].
\end{aligned}
$$

For the proof of (e), from (1.79) and the convergence of series we can differentiate each term with respect to t as follows:

$$\begin{aligned}
\frac{d}{dt}(\exp[At]) &= \frac{d}{dt}\left(I + \frac{At}{1!} + \frac{A^2t^2}{2!} + \frac{A^3t^3}{3!} + \cdots + \frac{A^lt^l}{l!} + \cdots\right) \\
&= \frac{d}{dt}(I) + \frac{d}{dt}\left(\frac{At}{1!}\right) + \frac{d}{dt}\left(\frac{A^2t^2}{2!}\right) + \frac{d}{dt}\left(\frac{A^3t^3}{3!}\right) \\
&\quad + \cdots + \frac{d}{dt}\left(\frac{A^lt^l}{l!}\right) + \cdots \\
&= \frac{A}{1!} + 2\frac{A^2t}{2!} + 3\frac{A^3t^2}{3!} + \cdots + l\frac{A^lt^{l-1}}{l!} \\
&\quad + \cdots \text{ [Theorem 1.5.1 (A2)]} \\
&= A + \frac{A^2t}{1!} + \frac{A^3t^2}{2!} + \cdots + \frac{A^lt^{l-1}}{(l-1)!} \\
&\quad + \cdots \text{ (by simplification)} \qquad (1.83) \\
&= A\left[I + \frac{At}{1!} + \frac{A^2t^2}{2!} + \cdots + \frac{A^{l-1}t^{l-1}}{(l-1)!}\right] \quad \text{[Theorem 1.3.1 (M2)]} \\
&= A\exp[At] \text{ [from (1.79)].} \qquad (1.84)
\end{aligned}$$

This, along with (1.83), yields

$$\begin{aligned}
\frac{d}{dt}(\exp[At]) &= \left[I + \frac{At}{1!} + \frac{A^2t^2}{2!} + \cdots + \frac{A^{l-1}t^{l-1}}{(l-1)!} + \cdots\right]A \quad \text{[Theorem 1.3.1 (M2)]} \\
&= \exp[At]\,A \text{ [from (1.79)].} \qquad (1.85)
\end{aligned}$$

From (1.84) and (1.85), the validity of the relation in (e) is proven and this completes the proof. □

Observation 1.5.6. Prior to the introduction of the differentiability of matrix functions, we recall and analyze the notion of the limit, continuity, and differentiability of a real-valued function f of real variables.

Limit. The statement $\lim_{x\to c} f(x) = L$ means that for every $\epsilon > 0$ one must be able to determine a number $\delta(\epsilon, L, f) > 0$ such that $|f(x) - L| < \epsilon$, whenever $0 < |x - c| < \delta$. In fact, this is really a test for the candidacy of the number L as the limit of f at $x = c$, and hence it requires *a priori* knowledge of the limit. We further observe that $\lim_{x\to c} f(x) = L$ means that $\lim_{x\to c}|f(x) - L| = 0$. Based on this observation and the above ϵ–δ test for the limit L of f at $x = c$, we can formulate another meaning of the limit concept. We define a function l as $l(x - c) = f(x) - L$. $\lim_{x\to c}|f(x) - L| = 0$ if and only if $\lim_{x\to c} l(x - c) = 0$. This suggests that $\lim_{x\to c} f(x) = L$ if and only if $f(x) = L + l(x-c)$ in the vicinity of c and $\lim_{x\to c} l(x-c) = 0$. A similar comment can be made with regard to continuity.

Differentiability. We recall the definition of the derivative of a real-valued function f of real variables at x:

$$f'(x) = \lim_{\triangle x\to 0}\left[\frac{f(x+\triangle x)-f(x)}{\triangle x}\right].$$

This limit can be rewritten as

$$\lim_{\triangle x\to 0}\left[G(\triangle x) - f'(c)\right] = 0,$$

where

$$G(\triangle x) = \frac{f(x+\triangle x)-f(x)}{\triangle x}.$$

We apply the above conclusion about the limit to $\lim_{\triangle x\to 0}\left[G(\triangle x) - f'(c)\right] = 0$. This leads to $\lim_{\triangle x\to 0}\left[G(\triangle x) - f'(c)\right] = 0$ if and only if $G(\triangle x) = f'(c) + l(\triangle x)$ in the vicinity of c and $\lim_{\triangle x\to 0} l(\triangle x) = 0$. This reduces to

$$G(\triangle x) = \frac{f(x+\triangle x)-f(x)}{\triangle x} = f'(c) + l(\triangle x)$$

in the vicinity of c and $\lim_{\triangle x\to 0} l(\triangle x) = 0$, which is equivalent to

$$f(x+\triangle x) - f(x) = f'(c)\triangle x + \triangle x l(\triangle x)$$

in the vicinity of c and $\lim_{\triangle x\to 0} l(\triangle x) = 0$. This statement provides an alternative definition to the derivative of f. In other words, a function is said to be differentiable at x if

$$f(x+\triangle x) - f(x) = f'(c)\triangle x + o(\triangle x),$$

where o is a function (a little o of) that satisfies the property $|\frac{o(z)}{z}| \to 0$ as $z \to 0$.

Observation 1.5.7. Let $f : R^{m\times n} \to R^{p\times q}$ be a matrix function. In addition, let X in $R^{m\times n}$ and let $\triangle X$ be an $m\times n$ increment matrix in X. The increment in the $p\times q$ matrix $Y = f(X)$ is the $p\times q$ matrix defined by

$$f(C+\triangle X) - f(C) = f'(C)\triangle X + o(\triangle X),$$

where $f'(C)$ is a linear map defined on $R^{m\times n} \to R^{p\times q}$; and o is a function defined on $R^{m\times n}$ into $R^{p\times q}$ that satisfies the property

$$\left\|\frac{o(z)}{z}\right\| \to 0 \text{ as } \|z\| \to 0.$$

We recall that the operation of a differential is linear. In the following, we present differential formulas for the product, quotient, and composition functions. Let us consider the following functions that are described by

$$dx = f(t,x)\,dt, \tag{1.86}$$

$$dy = h(t,y)\,dt. \tag{1.87}$$

Theorem 1.5.9 (differential formula — composition rule). *Let $V \in C[J \times R, R]$, and let $V(t,x)$ be a continuously differentiable function with respect to both of the variables t and x. Then*

$$dV(t,x(t)) = LV(t,x(t))\,dt, \tag{1.88}$$

where $x(t)$ is determined by (1.86)*, and L is a linear operator associated with* (1.86) *and defined by*

$$LV(t,x(t)) = \frac{\partial}{\partial t}V(t,x(t)) + f(t,x)\frac{\partial}{\partial x}V(t,x(t)). \tag{1.89}$$

Proof. We apply Theorem 1.5.6 to V to keep the terms up to the first-degree polynomial in Δt and Δx. For this purpose, we consider

$$\begin{aligned} V(t+\Delta t, x+\Delta x(t)) = V(t,x(t)) &+ \frac{\partial}{\partial t}V(t,x(t))\Delta t \\ &+ \Delta x(t)\frac{\partial}{\partial x}V(t,x(t)) + o(\Delta t, \Delta x(t)), \end{aligned}$$

which implies that

$$\begin{aligned} \Delta V &= V(t+\Delta t, x+\Delta x(t)) - V(t,x(t)) \\ &= \frac{\partial}{\partial t}V(t,x(t))\Delta t + \Delta x(t)\frac{\partial}{\partial x}V(t,x(t)) + o(\Delta t, \Delta x(t)). \end{aligned}$$

Using the concepts of the differential, i.e. $dt = \Delta t$,

$$dx(t) = f(t,x(t))\,dt + o(\Delta t),$$

and (1.89), the above expression reduces to

$$dV(t,x(t)) = \frac{\partial}{\partial t}V(t,x(t))\,dt + f(t,x(t))\frac{\partial}{\partial x}V(t,x(t))\,dt = LV(t,x(t))\,dt.$$

This validates (1.88). □

Definition 1.5.7. The formula

$$dV(t,x(t)) = \frac{\partial}{\partial t}V(t,x(t))\,dt + f(t,x(t))\frac{\partial}{\partial x}V(t,x(t))$$

is referred to as the differential of $V(t, x(t))$ along the vector field determined by (1.86).

Example 1.5.8. We are given $V(t, x) = x^2$. Find an expression for $dV(t, x(t))$ with respect to (1.86).

Solution procedure. We note that the given function $V(t, x) = x^2$ is a continuously differentiable function with respect to both t and x. In fact, it is constant with respect to the independent variable t. We apply Theorem 1.5.9 to the given function.

Step 1. We apply Theorem 1.5.9 (composition rule). First, we find $\frac{\partial}{\partial t} V(t, x)$ and $\frac{\partial}{\partial x} V(t, x)$. These partial derivatives are

$$\frac{\partial}{\partial t} V(t, x) = 0, \quad \frac{\partial}{\partial x} V(t, x) = 2x.$$

Step 2. We use the partial derivatives from step 1 to find $dV(t, x(t))$ in (1.88). In fact

$$dV(t, x(t)) = LV(t, x(t))\, dt = 2f(t, x(t))\, x(t)\, dt,$$

where $x(t)$ is determined by (1.86). This is the final expression for $dV(t, x(t))$ for the given $V(t, x)$.

Theorem 1.5.10 (differential formula — addition rule). *Let*

$$V_1, V_2 \in C[J \times R, R].$$

Assume that $V_1(t, z)$ and $V_2(t, z)$ are continuously differentiable functions with respect to t, x, and y. Define $V = V_1 + V_2$. Then

$$dV(t, x, y) = dV_1(t, x) + dV_2(t, y) \tag{1.90}$$

$$= [L_1 V_1(t, x) dt + L_2 V_2(t, y)]\, dt, \tag{1.91}$$

where the linear operators L_1 and L_2 associated with the differential equations (1.86) *and* (1.87) *are*

$$L_1 V_1(t, x) = \frac{\partial}{\partial t} V_1(t, x) + f(t, x) \frac{\partial}{\partial x} V_1(t, x), \tag{1.92}$$

$$L_2 V_2(t, y) = \frac{\partial}{\partial t} V_2(t, y) + h(t, y) \frac{\partial}{\partial y} V_2(t, y), \tag{1.93}$$

respectively.

Proof. First, by considering $x = [x, y]^T$ as a row vector (1×2 matrix) and using Observation 1.5.3, we have

$$\frac{\partial}{\partial t} V(t,x,y) = \frac{\partial}{\partial t} V_1(t,x) + \frac{\partial}{\partial t} V_2(t,y),$$

$$\frac{\partial}{\partial x} V(t,x,y) = \nabla V(t,x,y) = \left[\frac{\partial}{\partial x} V_1(t,x), \frac{\partial}{\partial y} V_2(t,y)\right].$$

By using (1.28) and (1.86) and imitating the proof of Theorem 1.5.9, the differential of $V(t,x,y)$ is represented by

$$dV(t,x,y) = \frac{\partial}{\partial t} V(t,x,y)\, dt + \frac{\partial}{\partial x} V(t,x,y)(dx)^T.$$

The rest of the proof can be completed by following the argument used in Theorem 1.5.9. □

Example 1.5.9. Let us consider $V_1(t,x) = \sin^2 x$ and $V_2(t,y) = e^y$. Find an expression for $dV(t, x(t), y(t))$ with respect to the solution processes of (1.86) and (1.87), where

$$V(t,x,y) = V_1(t,x) + V_2(t,y) = \sin^2 x + e^y.$$

Solution procedure. We note that the given functions $V_1(t,x) = \sin^2 x$ and $V_2(t,y) = e^y$ are continuously differentiable functions with respect to t and x. In fact, these functions are constant with respect to the independent variable t. We apply Theorem 1.5.10.

Step 1. We apply the argument and notations used in the proof of Theorem 1.5.10 (differential formula — addition rule) and compute

$$\frac{\partial}{\partial t} V(t,x,y) \text{ and } \frac{\partial}{\partial x} V(t,x,y).$$

With regard to $V(t,x,y) = \sin^2 x + e^y$ in this example,

$$\frac{\partial}{\partial t} V(t,x,y) = 0, \quad \frac{\partial}{\partial x} V(t,x,y) = [2 \sin x \cos x, e^y].$$

Step 2. Using the expressions in step 1, the formula (1.91) reduces to

$$\begin{aligned} dV(t,x,y) &= \frac{\partial}{\partial t} V(t,x,y)\, dt + \frac{\partial}{\partial x} V(t,x,y)(dx)^T \\ &= [2 \sin x \cos x, e^y]\,(dx)^T \\ &= 2 \sin x \cos x\, dx + e^y dy \\ &= [2 \sin x \cos x f(t,x) + e^y h(t,y)]\, dt. \end{aligned}$$

This gives us the desired expression for $dV(t,x,y)$ with respect to the solution processes of (1.86) and (1.87).

Theorem 1.5.11 (differential formula — product rule). *Let*

$$V_1, V_2 \in C\,[J \times R, R]\,.$$

Assume that $V_1(t,x)$ *and* $V_2(t,y)$ *are continuously differentiable functions with respect to* $t, x,$ *and* y. *Define* $V = V_1 V_2$. *Then*

$$dV(t,x,y) = V_2(t,y)\,dV_1(t,x) + V_1(t,x)\,dV_2(t,y) \tag{1.94}$$

$$= [V_2(t,y)\,L_1V_1(t,x) + V_1(t,x)\,L_2V_2(t,y)]\,dt, \tag{1.95}$$

where $L_1V_1(t,x)$ *and* $L_1V_2(t,x)$ *are as defined in* (1.92) *and* (1.93), *respectively.*

Proof. We compute dV by finding

$$\frac{\partial}{\partial t}V, \quad \frac{\partial}{\partial x}V.$$

From elementary calculus, we have

$$\frac{\partial}{\partial t}V(t,x,y) = V_2(t,y)\,\frac{\partial}{\partial t}\,V_1(t,x) + V_1(t,x)\,\frac{\partial}{\partial t}\,V_2(t,t) \text{ (by the product rule)},$$

$$\frac{\partial}{\partial x}V(t,x,y) = V_2(t,y)\,\frac{\partial}{\partial x_1}\,V_1(t,x) + V_1(t,x)\,\frac{\partial}{\partial x_2}\,V_2(t,y) \text{ (by the product rule)}.$$

By using these derivatives and following the argument used in the proof of Theorem 1.5.9, the differential of $V(t,x(t),y(t))$ is represented by

$$dV(t,x,y) = \left[V_2(t,y)\frac{\partial}{\partial t}V_1(t,x) + V_1(t,x)\frac{\partial}{\partial t}V_2(t,y)\right]dt$$
$$+ \left[V_2(t,y)\frac{\partial}{\partial x}V_1(t,x)\,dx + V_1(t,x)\frac{\partial}{\partial y}V_2(t,y)\,dy\right].$$

By regrouping the terms as the coefficients of $V_1(t,x)$ and $V_2(t,y)$, we finally have

$$dV(t,x,y) = V_2(t,y)\left[\frac{\partial}{\partial t}V_1(t,x) + f(t,x)\frac{\partial}{\partial x}V_1(t,x)\right]dt$$
$$+ V_1(t,x)\left[\frac{\partial}{\partial t}V_2(t,y) + h(t,y)\frac{\partial}{\partial y}V_2(t,y)dt\right]$$
$$= V_2\,(t,x_2)\,dV_1\,(t,x_1) + V_1\,(t,x_1)\,dV_2\,(t,x_2)\,.$$
$$\text{[by (1.92) and (1.93)].}$$

□

Example 1.5.10. You are given $V_1(t,x) = x^2$ and $V_2(t,y) = y^2$. Find $dV(t,x,y)$ with respect to the respective solution processes of (1.86) and (1.87), where

$$x^2y^2 = V(t,x,y) = V_1(t,y)V_2(t,y).$$

Solution procedure. We repeat the argument used in Theorem 1.5.9 and conclude that V_1 and V_2 are continuously differentiable functions with respect to both t and x. We now apply Theorem 1.5.11 (differential formula — product rule).

Step 1. To apply Theorem 1.5.11, we imitate the argument used in Theorem 1.5.10 to compute

$$\frac{\partial}{\partial t}V(t,x,y), \quad \frac{\partial}{\partial x}V(t,x,y).$$

With regard to $V(t,x,y)=x^2y^2$, in this example,

$$\frac{\partial}{\partial t}V(t,x,y)=0, \quad \frac{\partial}{\partial x}V(t,x,y)=[2xy^2,2x^2].$$

Step 2. Using the expressions in step 1, the formulas (1.90) and (1.91) reduce to

$$\begin{aligned}
dV(t,x,y) &= \frac{\partial}{\partial t}V(t,x,y)\,dt+\frac{\partial}{\partial x}V(t,x,y)\,(dx)^T\\
&= \left[2xy^2,2x^2y\right](dx)^T\\
&= 2xy^2dx+2x^2y\,dy\\
&= \left[2xy^2f(t,x)+2x^2yh(t,y)\right]dt\\
&= \left[y^22xf(t,x)+x^22yh(t,y)\right]dt.
\end{aligned}$$

This gives us the desired expression for $dV(t,x,y)$ with respect to the solution processes of (1.92) and (1.93).

Theorem 1.5.12 (differential formula — quotient rule). *Let*

$$V_1,V_2\in C[J\times R,R].$$

Assume that $V_1(t,x)$ and $V_2(t,y)$ are continuously differentiable functions with respect to t,x, and y. Define $V=\frac{V_1}{V_2}$, where $V_2\neq 0$. Then

$$dV(t,x,y)=\frac{V_2(t,y)\,dV_1(t,x)-V_1(t,x)\,dV_2(t,y)}{V_2^2(t,y)} \tag{1.96}$$

$$=\left[\frac{V_2(t,x_2)\,L_1V_1(t,x_1)-V_1(t,x_1)\,L_2V_2(t,x_2)}{V_2^2(t,x_2)}\right]dt, \tag{1.97}$$

where $L_1V_1(t,x_1)$ and $L_2V_2(t,x_2)$ are defined in (1.92) and (1.93), respectively.

Proof. The proof of this theorem follows by applying Theorem 1.5.11 to $V=V_1V_2^{-1}$ as the product of two functions, V_1 and V_2^{-1}. □

Example 1.5.11. You are given $V_1(t,x)=x^2$ and $V_2(t,y)=y$. Find $dV(t,x_1,x_2)$ with respect to the solution processes of (1.92) and (1.93), where

$$\frac{V_1(t,x)}{V_2(t,y)}=V(t,x,y).$$

Solution procedure. We note that the given functions $V_1(t,x)=x^2$ and $V_2(t,y)=y$ are continuously differentiable functions with respect to t, x and y. In fact, they are independent of the independent variable t. Moreover, $\frac{x^2}{y}=V(t,x,y)$ is a continuously differentiable function with respect to t, x, and y for all $y\neq 0$. We now apply Theorem 1.5.12 (differential formula — quotient rule).

Step 1. To apply Theorem 1.5.12, we imitate the argument used in Theorem 1.5.10

$$\frac{\partial}{\partial t}V(t,x,y),\quad \frac{\partial}{\partial x}V(t,x,y).$$

With regard to $V(t,x,y)=\frac{x^2}{y}$, in this example,

$$\frac{\partial}{\partial t}V(t,x,y)=0,\quad \frac{\partial}{\partial x}V(t,x,y)=\begin{bmatrix}\frac{2x}{y} & -\frac{x^2}{y^2}\end{bmatrix}.$$

Step 2. Using the expressions in step 1, the formulas (1.96) and (1.97) reduce to

$$\begin{aligned}
dV(t,x,y)&=\frac{\partial}{\partial t}V(t,x,y)\,dt+\frac{\partial}{\partial x}V(t,x,y)(dx)^T\\
&=\begin{bmatrix}\frac{2x}{y} & -\frac{x^2}{y^2}\end{bmatrix}(dx)^T\\
&=\frac{2x}{y}\,dx-\frac{x^2}{y^2}\,dy\\
&=\left[\frac{2x}{y}f(t,x)-\frac{x^2}{y^2}f(t,y)\right]dt\\
&=\left[\frac{2xyf(t,x)-x^2f(t,y)}{y^2}\right]dt.
\end{aligned}$$

This gives us the desired expression for $dV(t,x,y)$ with respect to the solution processes of (1.92) and (1.93).

1.5 Exercises

(1) Find $A^{-1}(t)$ for $A(t)=\begin{bmatrix}\cos t & \sin t\\ -\sin t & \cos t\end{bmatrix}$.
(2) Find $\frac{d}{dt}A(t)$ and $A^T(t)$, where $A(t)$ is as given in Exercise 1.
(3) Using Theorem 1.5.3, show that $\frac{d}{dt}\det(A(t))=0$, as in Exercise 1.
(4) Using Theorem 1.5.1, compute $\frac{d}{dt}A^{-1}(t)$ for $A(t)$ in Exercise 1.
(5) Compute $\int_a^t B(s)\,ds$, where $B(t)=\begin{bmatrix}t & t^2\\ 1 & \cos t\end{bmatrix}$.
(6) Find $\frac{d}{dt}\left[x^T(t)y(t)\right]$, where $x^T(t)=[t,\ln t,\cos t]$ and $y^T(t)=[t,\ln t,\cos t]$.
(7) Find $\frac{d}{dt}\left[x(t)y^T(t)\right]$, where $x(t)$ and $y(t)$ are as given in Exercise 6.
(8) Show that $\frac{d}{dt}\left[x(t)x^T(t)\right]=2x(t)\frac{d}{dt}\left[x^T(t)\right]$.
(9) Show that $\frac{d}{dt}A^2(t)=A(t)\frac{d}{dt}A(t)+\frac{d}{dt}A(t)A(t)$.

(10) Give an example to show that $2A(t)\frac{d}{dt}A(t) \neq A(t)\frac{d}{dt}A(t) + \frac{d}{dt}A(t)A(t)$. Justify your answer.

(11) Find the addition, product and quotient formulas with regard to the given differential equations with the specified functions:

(a) $V_1(t,x) = \sin^2 x$ and $V_2(t,x) = x^2$: $dx = 3x\,dt$;
(b) $V_1(t,x) = \exp[x]$ and $V_2(t,x) = \cosh x$: $dx = ax\,dt$;
(c) $V_1(t,x) = \sin^2 x$ and $V_2(t,x) = x^2$: $dx = -x\,dt$.

(12) Find the addition, product, and quotient formulas with regard to the given differential equations:

$$dx_1 = [\cos x_1 + \sin x_1]\,dt,$$

$$dx_2 = [-\sin x_2 + \cos x_2]\,dt,$$

by utilizing the following functions:

(a) $V_1(t,x) = \sin^2 x$ and $V_2(t,x) = x^2$;
(b) $V_1(t,x) = \exp[x]$ and $V_2(t,x) = \cosh x$;
(c) $V_1(t,x) = \cos^2 x$ and $V_2(t,x) = \sin^2 x$;
(d) $V_1(t,x) = \ln(1 + x^2)$ and $V_2(t,x) = \exp[x]$.

(13) Using Observation 1.5.3 and Corollary 1.5.2 and Theorem 1.5.3, find dA and $d(\det(A(x(t))))$, where

(a) $A(x(t)) = \begin{bmatrix} \cos x(t) & \sin x(t) \\ -\sin x(t) & \cos x(t) \end{bmatrix}$;

(b) $A(t) = \begin{bmatrix} 1 & \exp[x(t)] & t \\ \cos x(t) & t & 1 \\ t & \sin x(t) & \exp[x(t)] \end{bmatrix}$.

1.6 Notes and Comments

The authors do not claim that they have thoroughly searched for the original contributors of well-established and well-known ideas and results. Section 1.2 presents introductory tips for the readers who are interested in understanding and practicing a problem-solving process. It has been adapted from Ref. 69 and it is in the spirit of "teaching of mathematics" [105]. For further details, see Refs. 41,43 and 93. The content of Sections 1.3 and 1.4 is based on class notes of the second author [55] as well as Refs. 4,19 and 35. The basic classical multivariate calculus material in Section 1.5 has been derived from Refs. 85 and 87, and the class notes [55]. The development of results about determinant functions (Theorems 1.5.3–1.5.6) is new [53]. The remaining results and concepts are based on advanced differential equations [15, 19] and analysis [120].

Chapter 2

First-Order Differential Equations

2.1 Introduction

In this chapter, mathematical modeling, procedures for solving first-order scalar linear differential equations, and their fundamental conceptual analysis are developed. Prior to the presentation of the technical procedures and the concepts, an attempt is made to dispel any doubts or to answer any questions that are frequently asked by students. These include: Why is this course required? Why should I learn this material? How will this help me?

The mathematical modeling in Section 2.2 takes a proactive approach to motivate the student. Section 2.3 deals with first-order differential equations whose solutions can be directly found by the methods of integration. This class of first-order differential equations is referred to as integrable differential equations. Moreover, the mathematical models of laminar blood flow in an artery and the motion of particles in the air are presented to illustrate the usage of this class of differential equations. Section 2.4 is devoted to first-order scalar homogeneous differential equations. The eigenvalue-type method is utilized to solve this class of differential equations with both constant and variable coefficients. This approach is motivated by observing the fact that the problems of solving linear scalar differential equations are analogous to the problems of solving linear scalar algebraic equations. By integrating the knowledge of the derivatives of exponential functions and the concept of solution of a scalar differential equation, the problem of finding a solution to a linear scalar differential equation is reduced to the problem of solving linear algebraic equations. The step-by-step procedures for finding the general solutions and the solutions of initial value problems are logically and clearly outlined. Various examples and illustrations are utilized to better describe the procedures and the usefulness of the differential equations. Section 2.5 deals with first-order nonhomogeneous scalar differential equations. The method of variation of constants parameters is used to solve both the general and the initial value problems. This method is very powerful, as it provides a conceptual procedure for solving very complex differential equation problems in a systematic way. The usefulness of this class of differential equations in the mathematical modeling of single-species dynamic processes, population dynamics, the well-known and well-recognized illustrations of single-species processes, Newton's law of cooling, diffusion processes under a controlled environment, and cell membrane and central nervous system dynamics is also presented. The optional Section 2.6 begins with ten natural questions. Only three questions can be answered at this time (a better understanding of the material is required to answer the remaining questions). An attempt is made to provide rigorous justifications for the validity of conceptual algorithms (theorems, lemma, or corollary). In addition, it provides a foundation for answering the remaining questions posed at the beginning of the section. Again this section is optional, but interested students can explore it with the help of their instructor.

2.2 Mathematical Modeling

This section is devoted to the development of deterministic mathematical models for dynamic processes in the biological, chemical, engineering, medical, physical, and social sciences. This development is based on a theoretical and experimental setup, the fundamental laws of science and engineering, and basic information about dynamic processes.

Deterministic Modeling Procedure 2.2.1. First, for the development of mathematical models of dynamic processes, we use a conceptually common description of processes in the sciences and engineering known as the "state" of a given system. Some examples are: "distance" traveled by an object in a physical process, "concentration" of a chemical substance in a chemical process, "number of species" in a biological process, and "price" of a commodity/service in a sociological process. It is measured by a scalar/vector quantity.

Let $x(t)$ be a state of a system at a time t. The state of the system is observed over an interval $[t, t+\Delta t]$, where Δt is a small increment in t. Without loss of generality, we assume that Δt is positive. The process is observed in a controlled environment, and the data are collected according to the laws of science and engineering. Let $x(t_0) = x(t)$, $x(t_1), x(t_2), \ldots, x(t_k), \ldots, x(t_n) = x(t+\Delta t)$ be experimentally observed state data set of a system at $t_0 = t$, $t_1 = t+\tau, t_2 = t+2\tau, \ldots, t_k = t+k\tau, \ldots, t_n = t+\Delta t = t+n\tau$ over the inverval $[t, t+\Delta t]$, where n belongs to $\{1, 2, 3, \ldots\}$ and $\tau = \frac{\Delta t}{n}$ and $\Delta x(t_k) = x(t_k) - x(t_{k-1})$. These observations are made under the following conditions:

DCM 1. The observations of the state of the system are made at

$$t_1, t_2, \ldots, t_k, \ldots, t_n$$

under the controlled environment.

DCM 2. The state dynamic of the system follows its own laws (if any). It is assumed that it is evolving under the closed environment without any uncertainties. The state of the system is observed over every time subinterval of length τ.

DCM 3. For each $k = 1, 2, \ldots, n$, it is assumed that the state is either increased by $\Delta x(t_k)$ (the positive change in the state) or decreased by $\Delta x(t_k)$ (the negative change in the state). We refer to $\Delta x(t_k)$ as a microscopic/local experimental or knowledge-based observed increment to the state of the system over the subinterval of length τ.

DCM 4. Depending on the state dynamic law(s) of the process, it is assumed that $\Delta x(t_k)$ may depend on the state of the system $x(t)$, the time t, and the length of the subinterval τ.

In short, the initial state and n experimental or knowledge-based changes $\Delta x(t_k)$ in the state,

$$\begin{aligned} x(t_0) &= x(t), \\ x(t_1) - x(t_0) &= Z_1, \\ x(t_2) - x(t_1) &= Z_2, \\ &\ldots\ldots\ldots\ldots, \\ x(t_k) - x(t_{k-1}) &= Z_k, \\ &\ldots\ldots\ldots\ldots, \\ x(t_n) - x(t_{n-1}) &= Z_n, \end{aligned} \tag{2.1}$$

are observed at t_0 and over the respective subintervals

$$[t_0, t_1], [t_1, t_2], \ldots, [t_{k-1}, t_k], \ldots, [t_{n-1}, t_n]$$

of the given interval $[t, t + \Delta t]$ of length Δt.

From (2.1), the final state, $x(t + \Delta t) = x(t_n)$, of the process is expressed by

$$x(t + \Delta t) = x(t) + \sum_{i=1}^{n} Z_i, \tag{2.2}$$

where $\sum_{i=1}^{n} Z_i$ is called an aggregate increment to the given state $x \equiv x(t)$ of the system at the given time t over the interval $[t, t + \Delta t]$ of length Δt.

In this case, the aggregate change of the state of the system $x(t + \Delta t) - x(t)$ under n observations of the system over the given interval $[t, t + \Delta t]$ of length Δt is described by

$$x(t + \Delta t) - x(t) = n \frac{\sum_{i=1}^{n} Z_i}{n} = \frac{\Delta t}{\tau} S_n, \tag{2.3}$$

where

$$S_n = \frac{1}{n} \sum_{i=1}^{n} Z_i.$$

S_n is the sample average of the state aggregate incremental data. Under the assumption of the smallness of τ, a large n with $\tau n = \Delta t$, and the nature of the system [finiteness of $x(t + \Delta t) - x(t)$], we have

$$\lim_{\tau \to 0+} \left[\frac{S_n}{\tau} \right] = C, \tag{2.4}$$

where C may depend on time and the initial state $x(t)$ of the system. From DCM 3 and (2.4), (2.3) reduces to

$$\begin{aligned} x(t + \Delta t) - x(t) &= \text{(total amount of positive increment)} \\ &\quad - \text{(total amount of negative increment)} \\ &\approx C\Delta t. \end{aligned} \tag{2.5}$$

Thus, the aggregate change of the state of the system $x(t + \Delta t) - x(t)$ in (2.5) under the controlled environment over the given interval $[t, t + \Delta t]$ of length Δt is interpreted as the average/expected change of the state of the system $C\Delta t$.

If Δt is very small, then its differential is $dt = \Delta t$, and from (2.5) the differential dx of the state is defined by

$$dx = C\, dt. \tag{2.6}$$

The expression (2.6) is called a differential equation. It is usually denoted by

$$\frac{dx}{dt} = C, \tag{2.7}$$

where C is called the instantaneous rate of change of the state of the system.

Illustration 2.2.1 (economic dynamics). This modeling procedure can be applied to formulate mathematical models in a variety of dynamic processes in the social sciences, namely economics, management, and information sciences. Let E be a collection of economic entities, such as an asset, information, product, or service. Here, a state of the system $x(t)$ stands for either a value or a specific rate of a value (per capita rate: the rate per unit time per unit size of the entity) at a time t in $J = [a, b] \subseteq R$. The specific rate of the entity is observed over an interval of $[t, t + \Delta t]$, where Δt is a small increment in t. Without loss of generality, we assume that Δt is positive. Its mathematical description can be made by using the above model,

$$x(t + \Delta t) - x(t) \approx C\Delta t,$$

which implies that

$$dx = C\, dt.$$

We note that if $x(t)$ is the specific rate of the economic entity at time t, then C is called a measure of the specific rate (per capita growth/decay rate) of the entity at a time t; if $x(t)$ is a value of an entity at a time t, then C is a rate of the growth/decay of the entity at the time t.

Illustration 2.2.2 (decay/growth dynamic processes). The above-discussed modeling procedure provides a basis for mathematical modeling of several decay/growth dynamic processes in biological, chemical, compartmental, pharmacological, physical, and social sciences. Here, a state of the system $x(t)$ stands for either a size or mass of an entity or a specific rate of a size or mass of an entity at time t in $J = [a, b]$, $a, b \in R$ and $J \subseteq R$, where the entity can be any biological, chemical, medical, or social science, such as substance, population, charge, or ionizing beam of particles. The specific rate of a **mass or size (or mass or size)** of an entity is observed over an interval of $[t, t + \Delta t]$, where Δt is a small increment in t. Moreover, increments $\Delta x(t_k)$ are either purely positive or negative. Here, the size may be a volume, weight, diameter, number, etc. By following the modeling procedure, we have

$$x(t + \Delta t) - x(t) \approx C\Delta t,$$

which implies that

$$dx = C\,dt.$$

As before (Illustration 2.2.1), we note that if $x(t)$ is the specific rate of an entity at a time t, then C is called a measure of the specific rate [per capita growth $(C > 0)$/decay $(C < 0)$ rate] of the mass or size of the entity at a time t; if $x(t)$ is the mass or size of an entity at a time t, C is called a measure of the growth/decay of the mass or size of the entity at a time t.

Illustration 2.2.3 (birth and death processes). The modeling process can also be utilized to formulate mathematical models for dynamic processes that are composed of both birth and death processes in the biological, chemical, compartmental, physical, and social sciences. As before, $x(t)$ is as described in Illustration 2.2.2. Due to the nature of the increment, the specific rate of mass or size (or a mass or size) of an entity needs to be decomposed into two parts: (i) birth–growth and (ii) death–decay. The increments corresponding to these birth and death processes are modeled separately. In fact, by following the argument used in Illustration 2.2.2, increments to the birth and death processes are described by

$$\begin{aligned} B(t + \Delta t) - B(t) &\approx C_b\Delta t, \\ D(t + \Delta t) - D(t) &\approx -C_d\Delta t \end{aligned}$$

over an interval, respectively. Hence, the net change in $x(t)$ over an interval of length Δt is:

$$\begin{aligned} x(t + \Delta t) - x(t) &= B(t + \Delta t) - B(t) + D(t + \Delta t) - D(t) \\ &\approx (C_b - C_d)\,\Delta t. \end{aligned}$$

Moreover,

$$dx = C\,dt.$$

Here, $C = C_b - C_d$ is called the intrinsic growth rate.

Illustration 2.2.4 (diffusion processes). The above modeling procedure in the context of Fick's diffusion law [31, 113] is frequently utilized to formulate a mathematical model for dynamic processes in the biological, biochemical, biophysical, chemical, compartmental, epidemiological, pharmacological, physical, physiological, social sciences and systems under a limited resources/capacity/environmental conditions/control mechanism. Let D be a collection of a diffusible type of entities in the biological, chemical, medical, physical, and social sciences, such as chemical substance, microscopic species, or charge. Again, here, a state of the system $x(t)$ stands for either a biomass or mass or a specific rate of biomass or mass (per capita growth/decay rate) of an entity at a time t in $J = [a, b]$, $a, b \in R$ and $J \subseteq R$. The specific rate of biomass or mass (or a biomass or mass) is observed over an interval of $[t, t + \triangle t]$, where $\triangle t$ is a small increment in t. Further details of the dynamic model is left as an exercise for the reader.

Observation 2.2.1. We further emphasize that based on experimental observations, information, and basic scientific laws/principles in the biological, chemical, engineering, medical, physical, and social sciences, we infer that in general the magnitude of the microscopic or local increment depends on both the initial time t and the initial state $x(t) \equiv x$ of a system. As a result of this and in general, the rate coefficient (C) defined in (2.4) need not be an absolute constant. It may depend on both the initial time t and the initial state $x(t) \equiv x$ of the system, as long as their dependence on t and x is very smooth. As a result of this, (2.5) reduces to

$$x(t + \triangle t) - x(t) \approx C(t, x)\triangle t, \tag{2.8}$$

where $C(t, x)$ is also referred to as the rate of the state of the system on the interval of length $\triangle t$. Moreover, the differential equation (2.6) reduces to

$$dx = C(t, x)\,dt. \tag{2.9}$$

The following examples justify the scope and the significance of Observation 2.2.1 in the modeling of dynamic processes in the biological, chemical, engineering, medical, physical, and social sciences.

Example 2.2.1 (economic dynamics). Let us consider Illustration 2.2.1. By utilizing the notations, definitions, and conditions outlined in Deterministic Modeling Procedure 2.2.1, we modify the conditions DCM 3 and DCM 4 to incorporate the dependence of the initial state x and time t. For example, $\triangle x(t_k) \equiv x(t)\triangle x(t_k)$.

By imitating the argument used in Deterministic Modeling Procedure 2.2.1, we arrive at

$$x(t + \triangle t) - x(t) \approx Cx(t)\triangle t,$$

which implies that

$$dx = Cx\,dt.$$

Here, we assume that C is constant. Under the above conditions, Cx is called the rate of change of the value, and we refer to C as the specific rate (per capita growth/decay rate) of change of the value of the asset (information, product, service) at the time t over an interval of length $\triangle t$.

Example 2.2.2 (decay dynamic processes). Let us consider Illustration 2.2.2. By using the notations, definitions, and conditions outlined in Illustration 2.2.2, we modify Illustration 2.2.2. From Example 2.2.1, we have $\triangle x(t_k) \equiv -x(t)\triangle x(t_k)$. With this and by following the argument used in Example 2.2.1, we obtain

$$x(t + \triangle t) - x(t) \approx -Cx(t)\triangle t,$$

which implies that

$$dx = -Cx\,dt.$$

Here we assume that C is constant. Under the above conditions, Cx is called the rate of decay of the amount (mass, size) of the substance (population, charge, ionizing beam of particles, etc.). We refer to C as the specific rate (per capita decay rate) of the amount or mass (size of the entity) at the time t over an interval of length $\triangle t$.

Example 2.2.3 (birth and death processes). Let us consider Illustration 2.2.3. We use all the notations, definitions, and conditions outlined in Illustration 2.2.3 to incorporate the dependence of the initial state x with regard to both decay and growth processes analogous to decay processes in Example 2.2.2,

$$x(t + \triangle t) - x(t) \approx (C_b - C_d)\,x(t)\triangle t,$$

and hence

$$dx = Cx\,dt,$$

where $C = C_b - C_d$ is the per capita intrinsic growth rate.

Example 2.2.4 (diffusion processes). Let us consider Illustration 2.2.4. We use all of the notations, definitions, and conditions outlined in Illustration 2.2.4 and incorporate the dependence of the initial state x with regard to diffusion processes in the various sciences and under limited resources (capacity, environmental conditions, control mechanism, etc.). By following the argument used in Example 2.2.2, we set

$$\triangle x(t_k) \equiv x(t)\,(1 - bx(t))\,\triangle x(t_k);$$

moreover,

$$x(t+\triangle t)-x(t)=\bar{r}x(t)(1-bx(t))\triangle t,$$

which implies that

$$dx=\bar{r}x(1-bx)\,dt,$$

where $C=\bar{r}x(1-bx)$ is called a Pearl–Verhulst logistic growth rate of the entity and $\bar{r}(1-bx)$ is the per capita growth rate. Further details of this example are given in Chapter 3.

Observation 2.2.2

(i) From (2.5) and (2.8), we further remark that

$$\frac{x(t+\triangle t)-x(t)}{\triangle t}\approx C, \tag{2.10}$$

$$\frac{x(t+\triangle t)-x(t)}{\triangle t}\approx C(t,x). \tag{2.11}$$

In addition to the interpretations about $C/C(t,x)$, we shed light on their existence in the context of controlled dynamic processes.

(ii) From (2.6) and (2.9), we have

$$\frac{dx}{dt}=C, \tag{2.12}$$

$$\frac{dx}{dt}=C(t,x), \tag{2.13}$$

respectively.

(iii) The left-hand side expression in (2.10)/(2.11) is the average (time average) rate of change of the state of the system under controlled (deterministic) environmental perturbations over a time interval of length $\triangle t$. The right-hand side expression is the rate of change of the state of the system at the given time t and the state x. In short, the evolution of a dynamic system under controlled environmental perturbations is determined by a predictable/deterministic rate. Moreover, it is characterized by $C/C(t,x)$.

The presented models provide motivation for studying the following type of first-order differential equation:

$$dx=f(t,x)\,dt, \tag{2.14}$$

where dx is differential, and f is defined and continuous on $[a,b]\times R$ into R.

This chapter deals with the methods of solving first-order differential equations.

2.2 Exercises

(1) ***Compound interest problem*** [6, 113]. We are given the following: (i) Let $r(t)$ be a constantly adjustable annual interest rate function per dollar per year at a time t (in years) with a specific rate $\bar{r}$. (ii) Let $x(t) \equiv x$ be the number of dollars in a savings account at a time t (in years). Show that

(a) $dr(t) = \bar{r}\,dt$ (b) $dx = r(t)x\,dt$

(2) ***Cell growth problem*** [6, 113]. You are given the following: (i) Let $k(t)$ be a constantly adjustable absolute growth rate function per unit mass per unit time at a time t. (ii) Let $x(t) \equiv x$ be the amount of mass at a time t (in an appropriate unit of time). Show that

(a) $dk(t) = \bar{k}\,dt$ (b) $dx = k(t)x\,dt$

(3) ***Population growth problem*** [6, 104, 113]. We assume the following conditions: (i) Let $\beta(t)$ be a constantly adjustable birth rate function per individual per unit time at a time t (in years), i.e., $\beta(t)$ be a constantly adjustable specific birth rate. (ii) Let $x(t)$ be the size of the species at a time t (in an appropriate unit of time). Show that

(a) $d\beta(t) = \bar{\beta}\,dt$ (b) $dx = \beta(t)x\,dt$

(4) ***Radiation problem*** [1, 6, 113]. An ionizing beam of particles consists of protons, neutrons, deuterons, electrons, γ-ray quanta, etc. It is known that high polymers, namely protein or nucleic acids hit by an ionizing beam, may be irreversibly altered. In short, the polymers are damaged. Let us assume that N_0 is the number of undamaged molecules of a specific chemical compound. These molecules are in the cell, and are assumed to be susceptible to radiation. Let n be the number of ionizing particles which cross the unit area of the target. This n is called a "dose of radiation." We are given the following: (i) Let $\lambda(n)$ be a constantly adjustable rate function of decay of undamaged molecules per undamaged molecule per unit dose at a dose level n (number of ionizing per particles unit area of the target), i.e., let $\lambda(n)$ be considered to be a constantly adjustable specific decay rate of an undamaged molecular population of a chemical compound. (ii) Let $N(n) \equiv N$ be the number of undamaged molecules after the exposure to radiation at the level of dose n. Show that

(a) $d\lambda(n) = \bar{\lambda}\,dn$ (b) $dN = \lambda(n)\,N\,dn$

(5) ***Compartmental problem*** [6, 46, 113]. Let S be a solute. It is distributed in a fluid compartment with a fixed volume V_s. The solute S is removed (cleared) from the compartment by removing a constant fraction of the fluid in the compartment per unit time (V_r volume per unit time). However, no additional amount of S is added to the fluid compartment. Under these stated conditions, the concentration of S in the fluid compartment will decrease. A particular case of this general situation is the "washout" of a solute from a well-mixed compartment by a steady inflow of the pure solvent through the compartment, and

the removal of the solute by a steady constant rate of turnover. We are given the following: (i) Let $k(t)$ be a constantly adjustable clearance/volume distribution (turnover) rate function per unit amount of S per unit time at a time t. (ii) Let $Q(t) \equiv Q$ be the amount of the mass at the time t (in an appropriate unit of time). Show that

(a) $dk(t) = \bar{k}\,dt$ (b) $dQ = -k(t)Q\,dt$

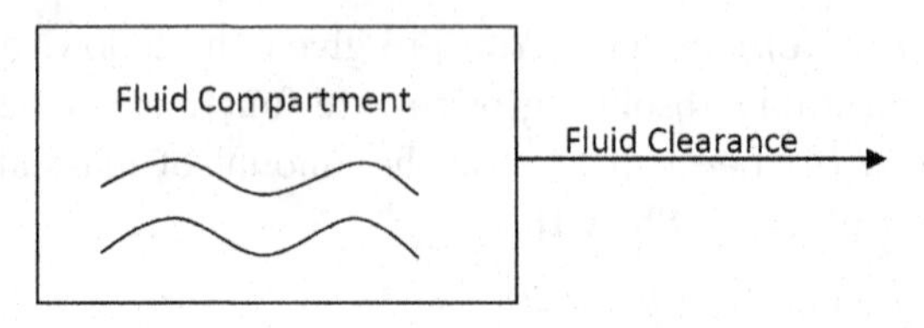

Fig. 2.1

(6) ***Electric circuits problem*** [6, 40, 113]. Let us consider a capacitor of fixed capacitance C (farads) carrying an initial charge Q_0 (coulombs). In an RC circuit the voltage difference $\triangle E$ [electromotive force/voltage E (volts)] across the capacitor causes the current I (amperes) to flow through a fixed resistance R (ohms). In this circuit, let $Q(t)$ and $\triangle E(t)$ be the charge on the capacitor and the voltage difference, respectively, at any time t (in seconds). It is known that $C\triangle E(t) = Q(t)$. By Ohm's law, we have $\triangle E = I(t)R$. From the definition of current, $I(t) = \frac{d}{dt}Q(t)$. In the RC circuit, we have $I(t) = -\frac{d}{dt}Q(t)$. You are given the following: Let $k(t) = \frac{1}{CR(t)}$ be a constantly adjustable rate of electric charge (turnover rate) function per unit coulomb per unit time at a time t. Show that

(a) $dk(t) = \bar{k}\,dt$ (b) $dQ = -k(t)Q\,dt$

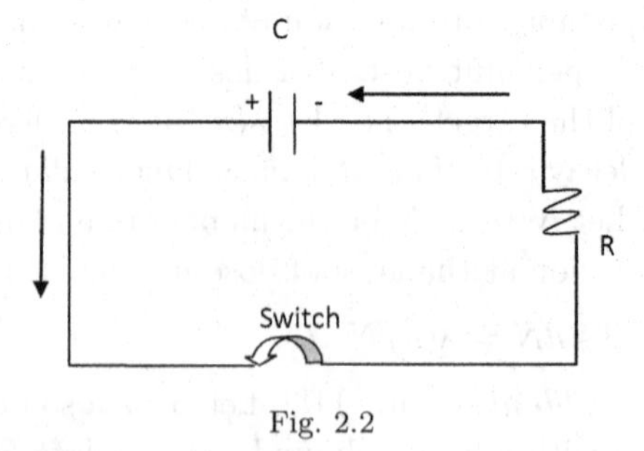

Fig. 2.2

2.3 Integrable Equations

In this section, we find a solution process for a very simple form of first-order deterministic differential equation. The direct methods of integration will be used to solve a very simple class of first-order deterministic differential equations. These types of differential equations are called integrable differential equations.

2.3.1 *General problem*

Let us consider the following first-order differential equation:

$$dx = f(t)\,dt, \tag{2.15}$$

where x is a state variable and f is a continuous function defined on R into R; the rate function f is independent of the state variable x (dependent variable) of a dynamic system.

In this subsection, our goal is to discuss a procedure for finding a general solution to (2.15). This provides a basis for solving real-world problems.

Definition 2.3.1. Let $J = [a, b]$, for $a, b \in R$, and hence $J \subseteq R$ be an interval. A solution to a first-order deterministic differential equation of the type (2.15) is a deterministic process or function x defined on J into R such that it satisfies (2.15) on J in the sense of deterministic calculus [85, 87]. In short, the usual (elementary calculus) differential $dx(t)$ of x is equal to the right-hand side expression in (2.15).

Example 2.3.1. Verify that $x(t) = t + 5$ is a solution to the following differential equation: $dx = dt$. Moreover, for any arbitrary constant c, show that $x(t) = t + c$ is also a solution to the given differential equation.

Solution procedure. We apply the deterministic differential calculus to $x(t) = t+5$, and we have $dx(t) = dt$. This shows that the given function $x(t) = t + 5$ satisfies the given differential equation. Furthermore, $x(t) = t + c$ also satisfies the same differential equation. This is due to the fact that the differential of a constant is zero. Therefore, by Definition 2.3.1, the given functions are solutions to the given differential equation.

Example 2.3.2. Given that $x(t) = \sin t + c$, where c is an arbitrary constant, prove that $x(t)$ is a solution process for the differential equation $dx = \cos t\, dt$.

Solution procedure. We apply the deterministic differential calculus to $x(t) = \sin t + c$, and obtain

$$\begin{aligned} dx(t) &= d(\sin t + c) \quad \text{(from the given expression)} \\ &= d(\sin t) + dc \quad \text{(by the sum rule for the differential)} \\ &= \cos t\, dt \quad \text{(by computation of the differential).} \end{aligned}$$

This proves that the given function satisfies the differential equation.

2.3.2 *Procedure for finding a general solution*

The procedure for finding a general solution to the differential equation (2.15) is described below. It basically depends on methods of integration in deterministic calculus [85].

Step 1. A solution x to (2.15) is simply determined by finding the antiderivative of a function f with respect to t. In fact, by integrating this function with respect to t, we have

$$x(t) = c + \int f(t)\,dt, \tag{2.16}$$

where c is an arbitrary constant of integration (deterministic real number). The integral in (2.16) is a usual Cauchy–Riemann integral [85]. It can be computed by employing the methods of deterministic integration, whenever it is possible.

Step 2. A general solution to (2.15) is defined by

$$x(t) = c + \int f(t)\,dt, \tag{2.17}$$

where c is an arbitrary constant as defined in (2.16). The solution in (2.17) is called a general solution to (2.15).

In the following, several examples are provided to illustrate the above-described procedure.

Example 2.3.3. Consider the following differential equation: $dx = t\,dt$. Find a general solution to the given differential equation.

Solution procedure. Here, $f(t) = t$ is the coefficient of dt in (2.15). In this case, a solution is described by the general antiderivative of $f(t) = t$:

$$x(t) = c + \int t\,dt = c + \frac{t^2}{2}.$$

This completes step 1. In view of step 2, we conclude that

$$x(t) = c + \frac{t^2}{2}$$

is the general solution of Example 2.3.3. This completes the above-described procedure.

In the following subsection, we discuss a procedure for finding a particular solution to a deterministic differential equation. This idea leads to the formulation of an initial value problem in the study of differential equations and its applications. Moreover, illustrations are provided to exhibit the usefulness of the concept for solving real-world problems.

2.3.3 *Initial value problem*

Let us formulate an initial value problem for the differential equation (2.15). We consider

$$dx = f(t)\,dt, \quad x(t_0) = x_0. \tag{2.18}$$

Here x and f are as defined in (2.15); t_0 is in J; x_0 is a real number. (t_0, x_0) is called an initial data or initial condition. In short, x_0 takes values in R, and it is the value of the solution function at $t = t_0$ (the initial/given time). Moreover, x_0 is an initial (given) state of a dynamic process. The problem of finding a solution of (2.18) is referred to as the initial value problem (IVP). Its solution is represented by $x(t) = x(t, t_0, x_0)$ for $t \geq t_0$ and $t, t_0 \in J$.

Definition 2.3.2. A solution to the IVP (2.18) is a real-valued deterministic process or function x defined on J into R such that $x(t)$ and its differential $dx(t)$ satisfy the scalar differential equation in (2.18), and $x(t)$ satisfies the given initial condition (t_0, x_0). In short: (i) $x(t)$ is a solution to (2.15) (Definition 2.3.1) and (ii) $x(t_0) = x_0$.

Example 2.3.4. Verify that $x(t) = t + c$ is a solution to the following differential equation: $dx = dt$. Then determine a value of the constant c so that $x(t)$ is the solution to the IVP: $dx = dt$, $x(0) = 5$.

Solution procedure. By imitating the procedure described in Example 2.3.1, we conclude that $x(t) = t + c$ is the general solution to the given differential equation. We need to find a constant c so that $x(t) = t + c$ is the solution to the given IVP: $dx = dt$, $x(0) = 5$. For this purpose, we substitute $t = 0$ in $x(t) = t + c$, and obtain $x(0) = 0 + c$. We know by Definition 2.3.2 that $x(0) = 5$. Hence, $5 = 0 + c$. Now, we solve for c, and obtain $c = 5$. Finally, we substitute for $c = 5$ into the expression $x(t) = t + c$, and we have $x(t) = t + 5$. This is the desired solution to the given IVP.

Example 2.3.5. Determine the value of a constant c so that $x(t) = \sin t + c$ is the solution to the IVP: $dx = \cos t\, dt$, $x(\pi) = 0$.

Solution procedure. By imitating the procedure described in Example 2.3.2, we conclude that $x(t) = \sin t + c$ is a solution to

$$dx = \cos t\, dt.$$

We follow the procedure outlined in Example 2.3.4 to determine the value of the constant c. Again, for this purpose, we substitute $t = \pi$ into $x(t) = \sin t + c$, and we have $x(\pi) = \sin \pi + c$. We are given $x(\pi) = 0$. Therefore, we have $0 = \sin \pi + c = 0 + c = c$. Hence, $c = 0$. This completes the goal of the problem.

2.3.4 *Procedure for solving the IVP*

In the following, we use the procedure for finding the solution function of the first-order scalar differential equation (2.15) in Subsection 2.3.2. Now, we briefly summarize the procedure for finding a solution to the IVP (2.18). The procedure is as follows:

Step 1. The computed nontrivial solution in (2.17) can also be represented as follows:

$$x(t) = c + \int_a^t f(s)\,ds, \tag{2.19}$$

where $a, t \in [a, b] = J$, for $a, b \in R$. Here, c is an arbitrary constant of integration. The general solution in (2.19) of (2.15) is defined (in view of the continuity of the rate function on its respective domain of definition).

Step 2. In this step, we need to find an arbitrary constant c in (2.19). For this, we utilize the given data (t_0, x_0), and solve the scalar algebraic equation

$$x(t_0) = c + \int_a^{t_0} f(s)\,ds = x_0, \tag{2.20}$$

for any given t_0 in J. This equation has a unique solution, and it is given by

$$\begin{aligned} c &= x(t_0) - \int_a^{t_0} f(s)\,ds \\ &= x_0 - \int_a^{t_0} f(s)\,ds. \end{aligned} \tag{2.21}$$

Of course, c depends on f, x_0, t_0, and a.

Step 3. The solution to the IVP (2.18) is determined by substituting the expression of c in (2.21) into (2.19). Hence, we have

$$\begin{aligned} x(t) &= c + \int_a^t f(s)\,ds \quad \text{[from (2.19)]} \\ &= x_0 - \int_a^{t_0} f(s)\,ds + \int_a^t f(s)\,ds \quad \text{[from (2.21)]} \\ &= x_0 + \int_a^t f(s)\,ds - \int_a^{t_0} f(s)\,ds \quad \text{(by rearrangement)} \\ &= x_0 + \int_{t_0}^t f(s)\,ds. \quad \text{(by the integral properties).} \end{aligned} \tag{2.22}$$

The solution to the IVP (2.18) is denoted by $x(t) = x(t, t_0, x_0)$, and is called a particular solution to (2.15).

Now, a couple of illustrations are presented to exhibit the usefulness of the concept of the IVP and its applications.

Illustration 2.3.1 (laminar blood flow in an artery [6, 95, 113]). By considering a piece of an artery or of a vein as a cylindrical tube with constant radius R and length l, the velocity of the blood flow in the artery or vein is described by

$$dv = -2kr\,dr, \quad v(R) = 0, \quad v(0) = kR^2,$$

where r is the distance from any point of the blood from the axis; $k = P/4\eta l$; l, R, and r are measured in centimeters; P stands for the pressure difference between the two ends of the tube (dyne/cm^2 = cm^{-1}g s^{-2}) in the CGS system; η is the inner friction of the blood at the artery wall, and is called the coefficient of viscosity of the blood measured in poise (cm^{-1}g s^{-2}). The speed is highest along the center axis of the tube. It is assumed that the velocity increases from zero at the wall toward the center, and it is maximum at the center.

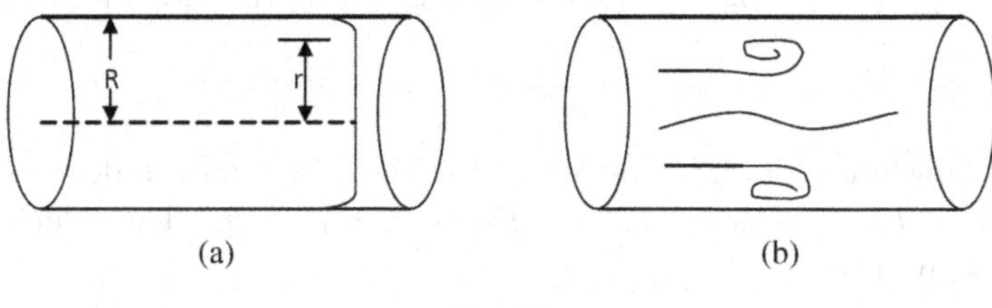

Fig. 2.3

By using the deterministic integral and the given initial and boundary conditions, we have

$$\begin{aligned} v(r) &= v(0) - 2k\int_0^r s\,ds \\ &= v(0) - kr^2 \\ &= kR^2 - kr^2, \end{aligned}$$

and hence

$$v(r) = k(R^2 - r^2), \text{ for } 0 \le r \le R.$$

This describes the velocity profile of the blood flow through the circular-cylinder-like artery or vein with constant radius R and length l. This function $v(r)$ the law of laminar flow was experimentally discovered by a French physiologist and physician, Jean Louis Poiseuille (1799–1869).

Example 2.3.6 (vertical motion of a rock [40]). A person standing on a cliff throws a rock upward. It is observed that after 2 s, the rock is at the maximum height in feet, and then after 5 s, it hits the ground. The motion of the rock is based on the gravitational force. Find the velocity and the position of the rock at time t, for the given initial velocity V_0 and position s_0 at the initial time t_0.

Solution procedure. Let $s(t)$ be a position of the rock at a time t. The motion of the rock is based on the gravitational force. Therefore, it is described by the

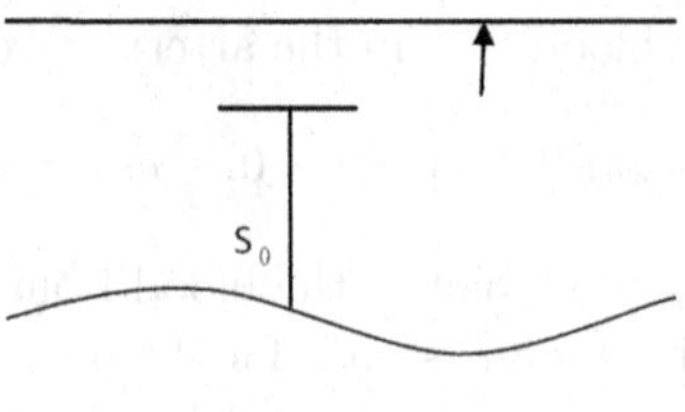

Fig. 2.4

following differential equation:

$$dV(t) = -g\,dt,$$

where $V(t) = \frac{d}{dt}\,s(t)$, $g \equiv 32\,\text{ft/s}^2$ ($9.81\,\text{m/s}^2$). By using the integral, we have

$$V(t) = \frac{d}{dt}\,s(t) = -gt + c,$$

where c is a constant of integration. We utilize the initial data to determine c, and it is given by $V(t)_{t=t_0} = V_0 = -gt_0 + c$. Hence, $c = V_0 + gt_0$. The velocity $V(t)$ at any time t is given by

$$\begin{aligned} V(t) = \frac{d}{dt}\,s(t) &= -gt + V_0 + gt_0 \\ &= -g\,(t - t_0) + V_0. \end{aligned}$$

The position of the rock at any time t is given by integrating the above expression over an interval $[t_0, t]$, and we obtain

$$\begin{aligned} \int_{t_0}^{t} V(u)\,du = \int_{t_0}^{t} \frac{d}{dt}\,s(u)\,du &= \int_{t_0}^{t} [-g\,(u - t_0) + V_0]\,du \\ = s(u)\,|_{t_0}^{t} &= \left[-\frac{1}{2}g\,(u - t_0)^2 + V_0 u\right]\,|_{t_0}^{t}, \end{aligned}$$

and hence

$$s(t) = s(t_0) - \frac{1}{2}\,g\,(t - t_0)^2 + V_0\,(t - t_0)\,.$$

Thus, the velocity and position of the rock are given by

$$\begin{aligned} V(t) &= -g\,(t - t_0) + V_0, \\ s(t) = s(t_0) &- \frac{1}{2}\,g(t - t_0)^2 + V_0(t - t_0), \end{aligned}$$

respectively.

Illustration 2.3.2 (Weber–Fechner law [6, 18]). The Weber–Fechner law describes that the increment to the reaction for equal increments in a stimulus decreases as the magnitude of the stimulus increases. Let R be the response number of nerve impulses emitted per second by a sense organ (eyes, nose, etc.), and S be the magnitude of the stimulus (intensity of the light, smell, etc.). In addition, let $\triangle R$ and $\triangle S$ be increments to the response and the stimulus functions, respectively. From Fechner's law we have the mathematical model

$$\Delta R = k\,\frac{\Delta S}{S},$$

which implies that

$$dR = k\,\frac{dS}{S},$$

where k is some positive constant. This is an integrable type of differential equation. Its solution is given by using one of the methods of integration. In this case, we have

$$R(S) = k \ln S + c,$$

where c is the constant of integration. This expression is termed the Weber–Fechner law. This is the general solution to the derived differential equation. For the initial value problem, if $R(S_0) = 0$. This means that S_0 is the response detection threshold. Thus, the response is described by

$$\begin{aligned} R(S) &= k \ln S - k \ln S_0 \\ &= k \ln\left(\left|\frac{S}{S_0}\right|\right). \end{aligned}$$

We note that the parameters k and S_0 depend on the type of stimulus and the individual. The curve of $R(S)$ describes the law proposed by Fechner.

2.3 Exercises

(1) Verify by substitution that each given function (process) is a solution to the given differential equation:

(a) $x(t) = \cos 2t$, $dx = -2\sin 2t\,dt$ (b) $x(t) = t^4$, $dx = 4t^3 dt$

(2) Find the general solution to the following differential equations:

(a) $dx = \sec^2 t\,dt$ (b) $dx = \frac{2t}{1+t^2}\,dt$ (c) $dx = \ln(t+1)\,dt$

(d) $dx = t^3 dt$ (e) $dx = \frac{1}{1+t^2}\,dt$ (f) $dx = \sin t\,dt$

(g) $dx = \ln(t+1)\,dt$ (h) $dx = \sec^2 t\,dt$ (i) $dx = \tan t\,dt$

(3) Show that the vertical distance traveled by a falling body is given by

$$s(t) = \frac{1}{2}\,gt^2,$$

where t is time (measured in seconds), s is the vertical distance traveled by the body, and $g \approx 32\,\mathrm{ft/s^2}$ ($9.81\,\mathrm{m/s^2}$) at the surface of the Earth.

(4) In reference to Poiseuille's law, Illustration 2.3.1, let us consider arterial blood with its concentration of O_2-bound hemoglobin. For humans, its viscosity is less than that of venous blood, on the average $\eta = 0.027$ poise (Diem, 1962, p. 548). The blood is flowing through a wide arterial capillary of length $l = 2\,\text{cm}$ and $R = 8 \times 10^{-3}\,\text{cm}$. The difference in pressure between the two ends of the vessel is $P = 4 \times 10^3\,\text{dyne/cm}^{-2}$ (3 mm mercury). Find the velocity $v(r)$.

(5) In a metabolic experiment, it is observed that the rate of change of the speed of glucose metabolism is -0.03 per hour. Show that the dynamic of glucose metabolism is described by:
   (a) $dr(t) = -0.06\,dt$, $r(0) = 0$, where $r(t) = \frac{d}{dt}M(t)$, with $M(t)$ being the mass of glucose at time t hour.
   (b) Find the following: (i) $r(t)$, (ii) $r(t)$ at $t = 1$ and 3.
   (c) Moreover, given the hemoglobin A1C test, the lower end of the reference interval is given as $M(0) = 4.5$. Show that

$$M(t) = 4.5 - 0.03t^2.$$

   (d) Find the following: (i) $M(t)$, (ii) $M(t)$ at $t = 1$ and 3.

(6) In reference to Exercises 2.2, find continuously adjustable specific rates:
   (i) (a) with $r(0) = 0$ (compound interest problem)
   (ii) (a) with $k(0) = 0$ (cell growth problem)
   (iii) (a) with $\beta(0) = 0$ (population growth problem)
   (iv) (a) with $\lambda(0) = 0$ (radiation problem)
   (v) (a) with $k(0) = 0$ (compartmental problem)
   (vi) (a) with $k(0) = 0$ (electric circuit problem)

(7) The Weber–Fechner law often ceases to describe the experimental data. However, the Loewenstein equation is better in some cases. The Loewenstein equation [18, 89] is given by

$$dR = k\frac{dS}{(1+kS)^2}, \quad R(0) = 0,$$

where k is a constant. Determine the Loewenstein equation.

(8) Let F be a cumulative distribution function (CDF) of a random variable X. For $x \in R$, let e be a function defined by the expected value of the random variable of X for $X \leq x$, and it is denoted by $e(x)$. In addition, let Δx be an increment to x, and let $e(x + \Delta x) - e(x)$ be a change of expected value of X over an interval $[x, x + \Delta x]$ for $\Delta x > 0$ (or $[x + \Delta x, x]$ for $\Delta x < 0$). For Δx sufficiently small, show that

$$de(x) = xdF(x) \text{ if and only if } e(x + \Delta x) - e(x) = \int_x^{x+\Delta x} sdF(s).$$

(9) Prove or disprove the following statement: If $E[X]$ is finite, then $E[X] = \lim_{x\to\infty} e(x)$.

(10) We are given a straight thin metallic rod of length $x \in [0, +\infty)$. It is perpendicular to an axis and its foot is on the axis of rotation. For $s \in [0, x]$, let m be the mass of an object over an interval $[0, s]$. In addition, let $I(x)$ be the moment of inertia of the rod about the axis of rotation. Show that

$$dI(x) = x^2 dm(x) \text{ if and only if } I(x + \triangle x) - I(x) = \int_x^{x+\triangle x} s^2 dm(s).$$

2.4 Linear Homogeneous Equations

In this section, we utilize the eigenvalue type of method to compute a solution to a first-order linear scalar homogeneous differential equation. This idea is efficient and useful for solving linear first-order scalar homogeneous differential equations. This method can be considered to be an alternative to the integrating factor approach to solving deterministic differential equations.

2.4.1 *General problem*

Let us consider the following first-order linear scalar homogeneous differential equation:

$$dx = f(t)\, x\, dt, \tag{2.23}$$

where x is a state of the process and f is a continuous function defined on an interval $J \subseteq R$, where J is as defined before.

In this subsection, our goal is to discuss a procedure for finding a general solution to (2.23). This provides a basis for analyzing, understanding, and solving problems in the biological, chemical, physical, and social sciences.

Definition 2.4.1. Let $J = [a, b]$ for $a, b \in R$ and hence $J \subseteq R$ be an interval. A solution process for the first-order linear scalar deterministic differential equation (2.23) is a function x defined on J into R such that it satisfies (2.23) on J in the sense of elementary calculus. In short, if we substitute a function $x(t)$ and its differential $dx(t)$ into (2.23), then the equation remains valid for all t in J.

Example 2.4.1. Verify that $x(t) = 5e^{at}$ is a solution to the differential equation $dx = ax\, dt$, where a is any given real number. Moreover, for any arbitrary constant c, show that $x(t) = ce^{at}$ is also a solution to the given differential equation.

Solution procedure. We apply deterministic differential calculus to $x(t) = 5e^{at}$, and we compute the following differential:

$$\begin{aligned} dx(t) &= d\left(5e^{at}\right) \quad \text{(from the given expression)} \\ &= 5ae^{at} dt \quad \text{(by properties of the differential and the chain rule)} \\ &= ax\, dt \quad \text{(by substitution for } 5e^{at}\text{)}. \end{aligned}$$

This shows that the given function $x(t) = 5e^{at}$ satisfies the given differential equation. Furthermore, by repeating the above argument, one can show that $x(t) = ce^{at}$, and it also satisfies the same differential equation. We note that the "5" is replaced by an arbitrary constant c. From Definition 2.4.1, we conclude that the given functions are solutions to the given differential equation.

2.4.2 *Procedure for finding a general solution*

We introduce an eigenvalue–eigenvector-like approach to solving linear homogeneous scalar differential equations. Of course, this may sound complicated. However, the underlying ideas are very simple, and it is motivated by the knowledge of solving linear algebraic equations. First, we observe that the study of linear scalar differential equations is analogous to the study of linear scalar algebraic equations. Therefore, the problem of finding a solution to a linear homogeneous scalar differential equation reduces to the problem of reducing to and then solving a corresponding linear scalar algebraic equation. The details are as follows:

Step 1. Let us seek the following form of solution to (2.23):

$$x(t) = \exp\left[\int_a^t \lambda(s)ds\right] c, \tag{2.24}$$

where c is an unknown constant (real number), λ is an unknown continuous function defined on J into R, and $a, t \in J$.

Case 1. We note that if $c = 0$ (for any λ), then $x(t)$ in (2.24) is a trivial solution [zero solution, i.e. zero process (function) on J] to (2.23). In this case, the candidate for a solution defined in (2.24) is always the trivial solution to (2.23).

Case 2. Therefore, our major goal is to seek a nontrivial solution (nonzero solution) to (2.23). For this purpose, we need to find the unknown function λ and the unknown nonzero constant c in (2.24). This is achieved in the following:

Step 2. By applying the deterministic differential (scalar version of Theorems 1.5.9 and 1.5.11 and the second fundamental theorem of calculus [85]) to $x(t)$ in (2.24), we have

$$\begin{aligned} dx(t) &= d\exp\left[\int_a^t \lambda(s)ds\right] c \quad \text{[from (2.24)]} \\ &= \exp\left[\int_a^t \lambda(s)ds\right] c\, d\left[\int_a^t \lambda(s)ds\right] \quad \text{(by the chain rule)} \\ &= \lambda(t)\exp\left[\int_a^t \lambda(s)ds\right] c\, dt \quad \text{(by the second fundamental theorm of calculus).} \end{aligned} \tag{2.25}$$

Therefore, under the assumption that x is the solution to (2.23), and from Definition 2.4.1, we conclude that $x(t)$ in (2.24) and $dx(t)$ in (2.25) satisfy the differential equation (2.23). From this discussion, we have

$$dx = \exp\left[\int_a^t \lambda(s)ds\right] \lambda(t)c\,dt = f(t)\exp\left[\int_a^t \lambda(s)ds\right] c\,dt. \tag{2.26}$$

From (2.26), we infer that for any nonzero constant c and unknown function λ,

$$\exp\left[\int_a^t \lambda(s)ds\right] \lambda(t)\,dt = f(t)\exp\left[\int_a^t \lambda(s)ds\right] dt. \tag{2.27}$$

For any integrable function λ, we note that

$$\exp\left[\int_a^t \lambda(s)ds\right] \neq 0 \text{ on } R.$$

Therefore, from (2.27) and the method of undetermined coefficients, we have

$$\lambda(t) = f(t). \tag{2.28}$$

Thus, the unknown function is determined.

Step 3. From step 2, and recalling the fact that the arbitrary constant c is nonzero, a solution x to (2.23) is determined. In fact, from (2.28), (2.24) reduces to

$$x(t) = \exp\left[\int_a^t f(s)ds\right] c, \tag{2.29}$$

for any given arbitrary constant c. This completes the procedure for finding a solution to (2.23). For easy reference, this procedure is summarized in Fig. 2.5.

Observation 2.4.1

(i) The solution expression of (2.23) in (2.29) can also be written as

$$\begin{aligned} x(t) &= \exp\left[\int_a^t f(s)ds\right] c \\ &= \Phi(t)c, \end{aligned} \tag{2.30}$$

where

$$\Phi(t) = \exp\left[\int_a^t f(s)ds\right], \tag{2.31}$$

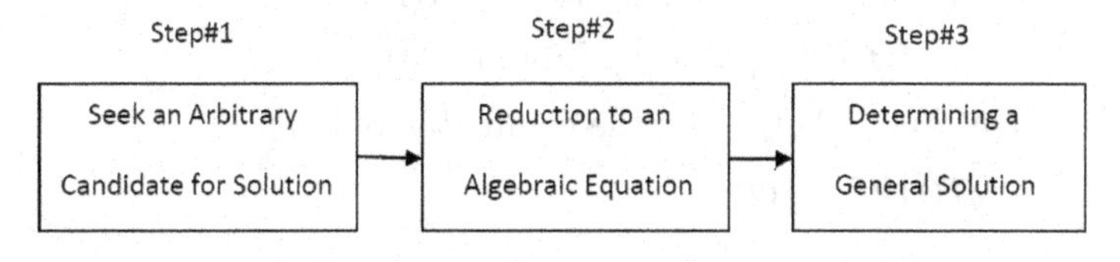

Fig. 2.5 Flowchart for finding general solution.

with the nonzero arbitrary constant c as defined in step 2. Based on this, the previous considerations, and the discussion, we are ready to draw the following conclusions:

(ii) We observe that $\Phi(t)$ is a nonzero function for any $a, t \in J$ and $t \geq a$. $\Phi(t)$ has an algebraic inverse. In fact,

$$(\Phi(t))^{-1} = \exp\left[-\int_a^t f(s)ds\right]. \tag{2.32}$$

(iii) The solution $x(t)$ in (2.30) is called the general solution to the deterministic differential equation (2.23).

(iv) Moreover, $\Phi(t)$ defined in (2.31) depends on the rate function f in the differential equation (2.23) and $a \in J$. In short, $\Phi(t)$ is uniquely determined by the rate function and $a \in J$.

(v) In particular, for $c = 1$ and from (2.30), we conclude that $\Phi(t)$ is also a nontrivial solution to the differential equation (2.23). From (2.30), we further conclude that any other solution to (2.23) is determined by a scalar multiple of $\Phi(t)$. Therefore, $\Phi(t)$ is called the general fundamental solution to deterministic differential equation (2.23). In fact,

$$d\Phi(t) = f(t)\Phi(t)\, dt. \tag{2.33}$$

Theorem 2.4.1. *Let Φ be the fundamental solution to* (2.23). *Then, Φ is invertible in the algebraic sense, and it satisfies the following first-order linear differential equation:*

$$d\Phi^{-1}(t) = -f(t)\Phi^{-1}(t)\, dt. \tag{2.34}$$

Proof. From Observation 2.4.1(ii), the fundamental solution Φ to (2.23) is algebraically invertible. The algebraic inverse is denoted by Φ^{-1}. By using the product rule (Theorem 1.5.11), we compute the following differential:

$$d\left(\Phi\Phi^{-1}\right) = d\Phi\Phi^{-1} + \Phi d\Phi^{-1} = d1 = 0.$$

From this and (2.33), we have

$$\begin{aligned}
d\Phi^{-1} &= -\Phi^{-1}[d\Phi\Phi^{-1}] \quad (\text{by solving the above equation for } d\Phi^{-1}) \\
&= -\Phi^{-1}d\Phi\Phi^{-1} \quad [\text{by Theorem 1.5.1(I1)}] \\
&= -\Phi^{-1}[f(t)\Phi(t)dt]\Phi^{-1} \quad [\text{from (2.33)}] \\
&= \Phi^{-1}\left[-f(t)\, dt\right] \quad (\text{by the commutative law}) \\
&= -f(t)\Phi^{-1}(t)\, dt.
\end{aligned}$$

This completes the proof of the theorem. □

Observation 2.4.2. From Theorem 2.4.1, we observe that $\Phi^{-1}(t)$ is a fundamental solution to the following differential equation:

$$dy = -yf(t)\,dt. \tag{2.35}$$

This differential equation is called the adjoint to (2.23). This idea is utilized to study systems of differential equations as well as the basic properties of scalar differential equations. Moreover, $\Phi^{-1}(t)$ is denoted by $\Psi(t)$.

Example 2.4.2. We are given $dx = fx\,dt$, where f is a nonzero constant. Find the fundamental and general solutions to the given differential equation.

Solution procedure. The goal is to find the general solution to the given differential equation. Let us compare the coefficient of dt of a given differential equation with the corresponding coefficient of the differential equation (2.23). Here $f(t) \equiv f$ is a constant function.

Now, by imitating steps 2 and 3, we have $\lambda(t) = f$. From this information, the representation of the solution in step 3 is

$$x(t) = \exp\left[f(t-a)\right]c,$$

where c is an arbitrary constant. From Observation 2.4.1(iii), we conclude that this expression is the general solution. Moreover, from (2.31), its fundamental solution is

$$\Phi(t) = \exp[f(t-a)],$$

which is our desired result.

Example 2.4.3. Find a general solution to $dx = -fx\,dt$, where f is a nonzero constant.

Solution procedure. We can follow the mechanical argument used in the solution process of Example 2.4.2, and obtain the fundamental and general solutions to the given differential equation:

$$\begin{aligned}\Phi(t) &= \exp[-f(t-a)],\\ x(t) &= \exp\left[-f(t-a)\right]c,\end{aligned}$$

respectively.

Example 2.4.4. Find the general solutions to $\sqrt{1-t^2}\,dx = \sin^{-1}t\,x\,dt$.

Solution procedure. The goal is to find the general solution to the given differential equation. Let us compare the coefficients of dt of the given differential equation with the corresponding coefficients of the differential equation (2.23). For this purpose we need to rewrite the given differential equation in a "standard form of (2.23)":

$$dx = \frac{\sin^{-1} t}{\sqrt{1-t^2}} x\,dt.$$

Here, $f(t) = \frac{\sin^{-1} t}{\sqrt{1-t^2}}$. Now, by imitating steps 2 and 3, we have

$$\lambda(t) = \frac{\sin^{-1} t}{\sqrt{1-t^2}}.$$

From this information, the solution of step 2, and Observation 2.4.1(iii), the general solution is

$$x(t) = \exp\left[\int_a^t \frac{\sin^{-1} s}{\sqrt{1-s^2}}\,ds\right] c.$$

This is our desired solution to the given differential equation.

In the following subsection, we discuss a procedure for finding a particular solution to a linear homogeneous deterministic differential equation. This idea leads to the formulation of an initial value problem in the study of differential equations and their applications. Moreover, illustrations are provided to exhibit the usefulness of the concept.

2.4.3 *Initial value problem*

Let us formulate an initial value problem for a first-order linear homogeneous differential equation of the type (2.23). We consider

$$dx = f(t)\,x\,dt, \quad x(t_0) = x_0. \tag{2.36}$$

Here, x and f are as defined in (2.23); t_0 is in J; x_0 is a real number. (t_0, x_0) is called an initial data or initial condition. In short, x_0 belongs to R, and is the value of the solution function at $t = t_0$ (the initial/given time). Moreover, x_0 is an initial (given) state of a dynamic process. The problem of finding a solution to (2.36) is referred to as the initial value problem (IVP). Its solution is represented by $x(t) = x(t, t_0, x_0)$ for $t \geq t_0$ and $t, t_0 \in J$.

Definition 2.4.2. A solution to the IVP (2.36) is a real-valued process (function) x defined on J into R such that $x(t)$ and its deterministic differential $dx(t)$ satisfy the scalar differential equation in (2.36) and any given initial condition (t_0, x_0). In short, (i) $x(t)$ is a solution to (2.23) (Definition 2.4.1) and (ii) $x(t_0) = x_0$.

Example 2.4.5. Verify that $x(t) = 3\exp[-(t - t_0)]$ is a solution to the following IVP: $dx = -xdt$, $x(t_0) = 3$.

Solution procedure. To show the given function to be a solution of the given differential, we find its differential:

$$
\begin{aligned}
dx(t) &= d\left(3\exp\left[-\left(t - t_0\right)\right]\right)\\
&= 3\exp\left[-\left(t - t_0\right) d\left(-\left(t - t_0\right)\right)\right] \quad \text{(by the chain rule)}\\
&= 3\exp\left[-\left(t - t_0\right)\right]\left(-dt\right)\\
&= -3\exp\left[-\left(t - t_0\right)\right] dt \quad \text{(by simplification)}\\
&= -x(t)\, dt \quad \text{(by substitution)}.
\end{aligned}
$$

This shows that $x(t)$ and $dx(t)$ satisfy the given differential equation. To complete the solution process for the given problem, we verify that the given solution satisfies the given initial condition. For this purpose, we substitute $t = t_0$ into $x(t)$ and simplify

$$
\begin{aligned}
x(t_0) &= 3\exp\left[-\left(t_0 - t_0\right)\right] \quad \text{(by evaluation)}\\
&= 3\exp[0] \quad \text{(by simplification)}\\
&= 3(1) = 3 \quad \text{(by definition of an exponential function)}.
\end{aligned}
$$

This shows that the given solution satisfies the given initial data. Therefore, by Definition 2.4.2, we conclude that the given process $x(t)$ is the solution to the given IVP. This is our desired goal.

2.4.4 *Procedure for solving the IVP*

In the following, we utilize the procedure for finding a solution to the first-order linear scalar homogeneous differential equation (2.23) in Subsection 2.4.2. The problem of finding a solution to the IVP (2.36) is summarized below:

Step 1. From Observation 2.4.1 and (2.29) or (2.30), the general solution to (2.23) is

$$x(t) = \exp\left[\int_a^t f(s)ds\right] c = \Phi(t)c,$$

where $a, t \in J \subseteq R$, and $\Phi(t)$ is as defined in (2.31). Here, c is an arbitrary constant.

Step 2. In this step, we need to find an arbitrary constant c in (2.30). For this purpose, we utilize the given initial data (t_0, x_0), and solve the scalar algebraic equation

$$x(t_0) = x_0 = \Phi(t_0)c, \tag{2.37}$$

for any given t_0 in J. Since

$$\exp\left[\int_a^{t_0} f(s)ds\right] \neq 0$$

for any continuous function f, the algebraic equation (2.37) has a unique solution. This solution is given by

$$c = \Phi^{-1}(t_0)x_0 = \exp\left[-\int_a^{t_0} f(s)ds\right] x_0, \tag{2.38}$$

which depends on f, x_0, t_0, and a.

Step 3. The solution to the IVP (2.36) is determined by substituting the expression for c in (2.38) into (2.30). Hence, we have

$$\begin{aligned} x(t) &= \Phi(t)c \quad [\text{from } (2.30)] \\ &= \Phi(t)\Phi^{-1}(t_0)x_0 \quad [\text{from } (2.38)] \\ &= \exp\left[\int_{t_0}^{t} f(s)ds\right] x_0 \quad (\text{from properties of the integral}) \\ &= \Phi(t, t_0)x_0 \quad (\text{by notation}), \end{aligned} \tag{2.39}$$

where

$$\begin{aligned} \Phi(t, t_0) &= \Phi(t)\Phi^{-1}(t_0) \equiv \Phi(t)\Phi^{-1}(t_0) \equiv \Phi(t, t_0) \\ &= \exp\left[\int_{t_0}^{t} f(s)ds\right]. \end{aligned} \tag{2.40}$$

Thus, the solution $x(t) = x(t, t_0, x_0)$ to the IVP (2.36) is represented by

$$x(t) = x(t, t_0, x_0) = \exp\left[\int_{t_0}^{t} f(s)ds\right] x_0 = \Phi(t, t_0)x_0.$$

This solution is called a particular solution to (2.23). $\Phi(t, t_0)$, defined in (2.40), is called a normalized fundamental solution to (2.23), because $\Phi(t, t_0)$ has an algebraic inverse [$\Phi(t, t_0) \neq 0$ on J], and $\Phi(t_0, t_0) = 1$. For easy reference, this procedure is summarized in Fig. 2.6.

Now, a few illustrations and numerical examples are presented to exhibit the usefulness of the concept of the IVP and its solution.

Illustration 2.4.1 (photochemical reaction [31, 113]). Under the assumptions of the following controlled ideal experimental conditions: (a) let A be a surface area cm^2 of a uniform absorbing material; (b) let N_0 be a number of uniformly distributed monochromatic photons (a photon is unit of radiant energy moving with the velocity

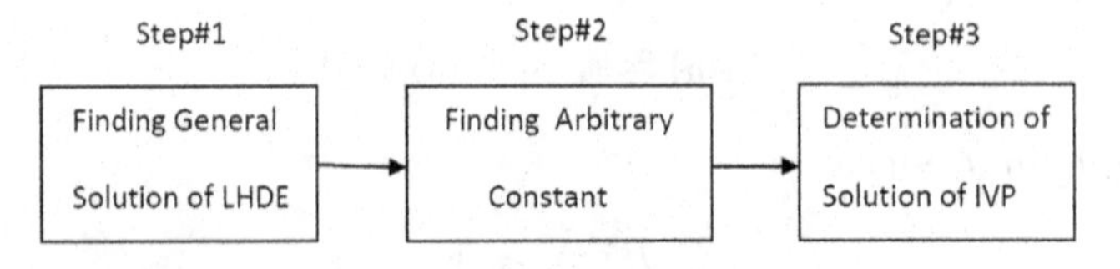

Fig. 2.6 Flowchart for finding particular solution.

of light) perpendicular to the surface area A in (a); and (c) it is assumed that there is no reflection or scattering of the photons.

Let us denote by $N(x)$ the number of uniformly distributed monochromatic photons entering in the direction perpendicular to the surface in the layer lying x cm below the surface, and by $\triangle N(x) = N(x+\triangle x) - N(x)$ the number of uniformly distributed monochromatic photons absorbed while passing through an infinitesimal layer $\triangle x$ lying x cm below the surface. Under the above-stated conditions, it is known that the number of photons,

$$\triangle n(x) = \frac{\triangle N(x)}{A},$$

absorbed per second per square centimeter below the surface is directly proportional to the product of the thickness $\triangle x$ and the number of photons,

$$n(x) = \frac{N(x)}{A},$$

per second per square centimeter incident upon the layer lying x cm below the surface. That is to say,

$$\triangle n(x) = -k\, n(x) \triangle x,$$

where k is an appropriate constant of proportionality, and k is called the coefficient of absorption or the specific rate of absorption. The absorption of the photons in the layer of thickness $\triangle x$ is equivalent to the decrease in the number of photons that hit the layer lying x cm below the surface. Thus, for very small $\triangle x$, we have

$$dn(x) = -kn(x)dx, \quad n_0 = n(0) = \frac{N_0}{A}.$$

This is an IVP. By following the procedure for solving the IVP (2.36), we obtain

$$n(x) = n_0 \exp[-kx].$$

This is equivalent to Bouguer's law, [31, 113], and it is commonly known as Lambert's law [31, 113].

$$I(x) = I_0 \exp[-kx],$$

where $I(x)$ denotes the intensity of a light beam at a depth of x units below the absorbing surface. It is known that the intensity of the light beam is directly proportional to the photon flux:

$$n(x) = \frac{N(x)}{A}.$$

Remark 2.4.1 [6,37]. We note that the Bouguer–Lambert law is applicable to any homogeneous transparent substance, such as glass, Plexiglas, liquids, and thin layers viewed under the microscope. Furthermore, light waves and other electromagnetic waves, namely x-rays and gamma rays, behave similarly.

Example 2.4.6. Using Illustration 2.4.1, find the depth of the layer to absorb: (i) 50% and (ii) 90% of the photons.

Solution procedure. From Illustration 2.4.1, the number of photons $n(x)$ per second per square centimeter incident upon the layer lying x cm below the surface is given by

$$n(x) = \exp[-kx]\, n_0,$$

and hence

$$\frac{n(x)}{n_0} = \exp[-kx].$$

By using the fact that the function "ln" is an inverse function of "exp," the above presented mathematical expression can be written as

$$\ln\left[\frac{n(x)}{n_0}\right] = -kx.$$

Therefore,

$$x = -\frac{1}{k}\ln\left[\frac{n(x)}{n_0}\right].$$

By using the above expression for x, we have

$$x_1 = -\frac{1}{k}\ln\left[\frac{\frac{1}{2}n_0}{n_0}\right] = -\frac{1}{k}\ln\left[\frac{1}{2}\right] = \frac{1}{k}\ln[2].$$

Similarly,

$$x_2 = \frac{1}{k}\ln\left[\frac{10}{9}\right].$$

If we either know the value of the constant k or can determine the constant k, then the depths x_1 and x_2 are completely determined.

Illustration 2.4.2 (chemical kinetics [6, 132]). Under the well-mixed and constant temperature conditions, the decomposition of a gaseous nitrogen pentoxide (N_2O_5) into nitrogen oxide (NO_2) and oxygen (O_2) is as follows:

$$2N_2O_5 \to 4NO_2 + O_2.$$

Let $c(t)$ be the concentration of gaseous nitrogen pentoxide measured in moles per liter at a time t. It is known that under constant temperature the reaction rate per mole per liter per unit of time (the specific reaction rate) is constant. It is also known that in the deoxidation process of gaseous nitrogen pentoxide, the number of molecules of gaseous nitrogen pentoxide decreases. By the law of first-order chemical reaction, we have

$$\triangle c(t) = -k\, c(t)\triangle t,$$

where k is a constant of proportionality, and k is called the specific reaction rate of deoxidation of gaseous nitrogen pentoxide. Thus for very small Δt, we have

$$dc(t) = -kc(t)\,dt, \quad c_0 = c(0).$$

This is an IVP. By following the procedure for solving the IVP (2.36), we obtain

$$c(t) = c_0 \exp\left[-kt\right].$$

Under the controlled experimental conditions, this concentration solution is in good agreement with the experimental results.

Illustration 2.4.3 (Otto Frank model — arterial pulse–diastolic phase [21, 25]). This illustration is based on a model by Otto Frank. He made an attempt to establish the pressure–time relationship in the aorta. By following Frank, the aorta is represented by volume, and its capacity depends on the pressure. The term "Windkessel" (elastic container connected to a tube with a finite resistance to flow) represents the peripheral resistance. There are two phases:

APM 1 (diastolic phase). During this phase, the rate of inflow from the heart to the aorta is zero.

APM 2 (systolic phase). In this phase, blood enters the aorta from the heart. The model of this phase will be discussed in Section 2.5.

APM 3. In the diastolic phase, it is assumed that: (a) there is a linear relationship between the volume and the pressure; (b) Poiseuille's law is obeyed in this case.

Let P be the pressure in the Windkessel, and let V be the volume of blood (fluid). From APM 3(a), we have

$$dP = \alpha dV,$$

where α is constant.

If we assume that the pressure at the distal end (venous side) is zero, then Poiseuille's law is applicable, and we have

$$dV = -\frac{P}{\beta}\,dt, \text{ where } \beta = \frac{8l\eta}{\pi r^4}$$

and β is constant. It represents the resistance to blood flow, and the minus sign signifies that the greater the pressure, the greater the decrease in the volume per unit time in the Windkessel.

By using the above two expressions, we eliminate the V, and we have

$$dP = -\frac{\alpha}{\beta} P dt, \quad P(0) = P_0.$$

Now, we utilize the procedure in Subsection 2.4.4 and obtain

$$P(t) = P_0 \exp\left[-\frac{\alpha}{\beta} t\right].$$

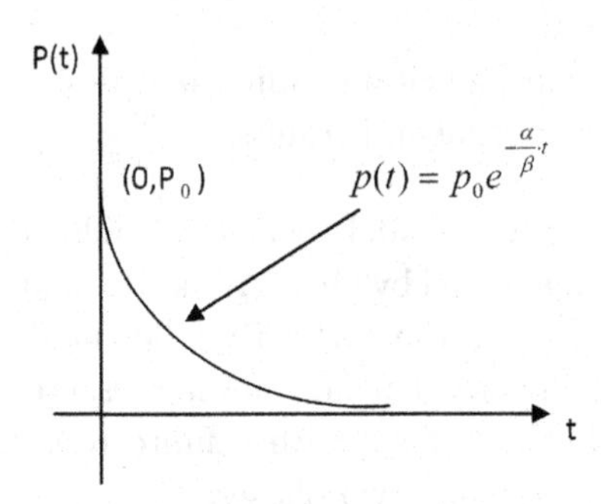

Fig. 2.7

This describes the pressure–time curve during the diastolic phase. Moreover, it signifies the pressure drop in the aorta during the diastolic period.

$$\frac{\beta}{\alpha} = \tau$$

is called the time constant.

2.4 Exercises

(1) Verify by substitution that each given process/function is a solution to the given differential equation:

(a) $x(t) = \exp[-3t]\, c$, $dx = -3x\, dt$
(b) $x(t) = \exp[3t]\, c$, $dx = 3x\, dt$
(c) $x(t) = \exp\left[\frac{1}{2} t\right] c$, $dx = \frac{1}{2} x\, dt$
(d) $x(t) = \left(1 + t^2\right) c$, $dx = \frac{2t}{1+t^2}\, x\, dt$

(2) Find the general and fundamental solutions to the following differential equations:

(a) $dx = -2x\, dt$
(b) $dx = -x\, dt$
(c) $dx = 2x\, dt$
(d) $dx = bx\, dt$

(e) $dx = -bx\,dt$
(f) $dx = \cos tx\,dt$

(3) Find the general solution to the following differential equations:

(a) $(1+t^2)\,dx = t^2 x\,dt$
(b) $(90+t)\,dx = 3x\,dt$
(c) $\sqrt{1+t^2}\,dx = (1+t^2)\,x\,dt$
(d) $dx = \tan t2x\,dt$
(e) $(1+\sin^4 t)\,dx = \sin^2 t \sin 2tx\,dt$
(f) $dx = \ln(1+t)\,x\,dt$

(4) Solve the following initial value problems:

(a) $dx = 4x\,dt$, $x(0) = 2$
(b) $dx = 2x\,dt$, $x(0) = 3$
(c) $(90+t)\,dx = 3x\,dt$, $x(0) = 90$
(d) $dx = \cos tx\,dt$, $x(0) = 1$

(5) Due to the technological changes in the measurement of sensory perception, the Weber–Fechner law (Illustration 2.3.2) was extended by S. Steven. Steven's law states that the rate of change of response per unit stimulus is directly proportional to the response per unit stimulus. (a) Derive the differential equation and (b) obtain Steven's power law [18].

(6) Let F be a cumulative distribution function (cdf) of a random variable X. For $x \in R$, let e be the function defined by the expected value of a random variable over an interval $(-\infty, x]$, and it is denoted by $e(x)$. Solve the following initial value problems:

(a) $de = -ke\,dF(x)$, $e_0 = e(0)$
(b) $de = x\,e\,dF(x)$
(c) $de = ke\,dF(x)$, $e_0 = e(0)$

(7) Let F be a male cumulative age distribution function (cdf). For $x = 25$ and the initial value problems in Exercise 6, find an expression for the expected age $e(25)$ of a male with an initial expected age of $e_0 = 5$.

(8) We are given a straight thin metallic rod of length $x \in [0, +\infty)$. It is perpendicular to an axis and its foot is on the axis of rotation. For $s \in [0, x]$, let m be the mass of an object over an interval $[0, s]$. In addition, let $I(x)$ be the moment of inertia of the rod about the axis of rotation. Solve the following initial value problems:

(a) $dI = -KI\,dm(x)$, $I_0(1) = 1$
(b) $dI = xI\,dm(x)$, $I_0(1) = 1$

2.5 Linear Nonhomogeneous Equations

In this section, we utilize the method of variation of constant parameters to compute a particular solution to first-order linear nonhomogeneous scalar differential equations. This method is efficient and useful for solving first-order linear nonhomogeneous scalar differential equations. Moreover, it can be an alternative to the integrating factor approach as well as the method of undetermined coefficients for solving deterministic differential equations.

2.5.1 *General problem*

Let us consider the following first-order linear nonhomogeneous scalar differential equation:

$$dx = [f(t)x + p(t)]\,dt, \tag{2.41}$$

where x is a state variable, and f and p are continuous functions defined on an interval $J \subseteq R$ into R.

In this subsection, our goal is to discuss a procedure for finding a general solution to (2.41). [The concept of the solution process defined in Definition 2.4.1 is directly applicable to (2.41).] The details are left to the reader. The problem of finding a solution provides a basis for analyzing, understanding, and solving problems in the biological, chemical, physical, and social sciences.

2.5.2 *Procedure for finding a general solution*

To find a general solution to (2.41), we imitate the conceptual ideas of finding a general solution to the linear nonhomogeneous algebraic equations described in Theorem 1.4.7. The basic ideas are: (i) to find a general solution x_c to the first-order linear homogeneous scalar differential equation corresponding to (2.41), (ii) to find a particular solution x_p to (2.41), and (iii) to set up a candidate $x = x_c + x_p$ for testing the general solution to (2.41). Here, we introduce the method of variation of constant parameters to find a particular solution x_p to the differential equation (2.41). This is a very powerful and elegant approach to finding a particular solution to (2.41).

Step 1. First, we find a fundamental solution to the first-order linear homogeneous scalar differential equation corresponding to (2.41), i.e

$$dx = f(t)x\,dt. \tag{2.42}$$

We note that this scalar differential equation is exactly the same as (2.23). Therefore, we use steps 1–3 described in the procedure in Subsection 2.4.2 for finding a general solution to (2.23) in Section 2.4, and determine a fundamental solution to (2.42) (Observation 2.4.1):

$$\Phi(t) = \Phi(t) = \exp\left[\int_a^t f(s)ds\right]. \tag{2.43}$$

Step 2. (method of variation of constant parameters). A particular solution x_p to (2.4.1) is a function that satisfies the differential equation (2.41). To find a particular solution to (2.41), we define a function as

$$x_p(t) = \Phi(t)c(t), \quad c(a) = c_a, \tag{2.44}$$

where $\Phi(t)$ is determined in step 1, and $c(t)$ is an unknown function with $c(a) = c_a$ for $a \in J$. Our goal of finding a particular solution $x_p(t)$ to (2.41) is equivalent to finding an unknown function $c(t)$ in (2.44) passing through c_a at $t = a$. For this purpose, we assume that $x_p(t)$ is a solution to (2.41). From this assumption, using the product formula (Theorem 1.5.11), and the fact that $\Phi(t)$ is the fundamental solution to (2.42), we have

$$dx_p(t) = d\Phi(t)c(t) + \Phi(t)\,dc(t). \tag{2.45}$$

By using the algebraic inverse $\Phi^{-1}(t)$ of $\Phi(t)$, we solve for $dc(t)$, and obtain

$$\begin{aligned} dc(t) &= \Phi^{-1}(t)[dx_p(t) - d\Phi(t)c(t)] \quad \text{[from (2.45)]} \\ &= \Phi^{-1}(t)\,[[f(t)x_p(t) + p(t)]\,dt - d\Phi(t)c(t)] \quad (x_p \text{ is the solution}) \\ &= \Phi^{-1}(t)\,[[f(t)x_p(t) + p(t)]\,dt - f(t)\Phi(t)c(t)dt] \quad \text{[by (2.33) and (2.44)]} \\ &= \Phi^{-1}(t)p(t)\,dt \quad \text{(by simplification).} \end{aligned} \tag{2.46}$$

We recall that the problem of finding a particular solution $x_p(t)$ defined in (2.44) reduces to the problem of finding the unknown function $c(t)$ in (2.44) passing through c_a at $t = a$. This problem further reduces to the problem of solving the following IVP:

$$dc(t) = \Phi^{-1}(t)p(t)\,dt, \quad c(a) = c_a. \tag{2.47}$$

By following the procedure in Subsection 2.3.4 (the IVP described in Section 2.3), the solution to the IVP (2.47) is given by

$$c(t) = c_a + \int_a^t \Phi^{-1}(s)p(s)\,ds. \tag{2.48}$$

We substitute the expression for $c(t)$ in (2.48) into (2.44), and we have

$$\begin{aligned} x_p(t) &= \Phi(t)\left[c_a + \int_a^t \Phi^{-1}(s)p(s)ds\right] \\ &= \Phi(t)c_a + \int_a^t \Phi(t)\Phi^{-1}(s)p(s)\,ds. \end{aligned} \tag{2.49}$$

Step 3. From step 1 and Observation 2.4.1, for any arbitrary constant c the function defined by $x_c(t) = \Phi(t)c$ is the general solution to (2.42). From the conceptual ideas described in Theorem 1.4.7, we define a function $x(t) = x_c(t) + x_p(t)$, where $x_p(t)$

is determined in step 2. Now, we claim that the function $x(t) = x_c(t) + x_p(t)$ is a general solution to (2.41). For this purpose, we consider

$$\begin{aligned} x(t) &= x_c(t) + x_p(t) \\ &= \Phi(t)c + \Phi(t)c_a + \int_a^t \Phi(t)\Phi^{-1}(s)p(s)\,ds \quad \text{[from (2.43) and (2.49)]} \\ &= \Phi(t)(c + c_a) + \int_a^t \Phi(t)\Phi^{-1}(s)p(s)\,ds \quad \text{(by regrouping)} \\ &= \Phi(t)c + \int_a^t \Phi(t)\Phi^{-1}(s)p(s)\,ds \quad \text{(by notation)} \\ &= \Phi(t)\left[c + \int_a^t \Phi^{-1}(s)p(s)\,ds\right] \quad \text{[by Theorem 1.3.1(M2)]}, \end{aligned} \tag{2.50}$$

where $c = c + c_a$ is an arbitrary constant due to the fact that c and c_a are arbitrary constants. Thus, $\Phi(t)c$ is also a general solution to (2.42). We show that $x(t)$ is a general solution to (2.41):

$$\begin{aligned} dx(t) &= d\left(\Phi(t)\left[c + \int_a^t \Phi^{-1}(s)p(s)\,ds\right]\right) \quad \text{[from (2.50)]} \\ &= d\Phi(t)\left[c + \int_a^t \Phi^{-1}(s)p(s)\,ds\right] \\ &\quad + \Phi(t)d\left[c + \int_a^t \Phi^{-1}(s)p(s)\,ds\right] \quad \text{(by Theorem 1.5.11)} \\ &= d\Phi(t)\left[c + \int_a^t \Phi^{-1}(s)p(s)\,ds\right] \\ &\quad + \Phi(t)d\left[\int_a^t \Phi^{-1}(s)p(s)\,ds\right] \quad \text{(by Theorem 1.5.10).} \end{aligned} \tag{2.51}$$

We recall that

$$d\Phi(t) = f(t)\Phi(t)dt \quad \text{[from (2.33)]},$$

$$d\left[\int_a^t \Phi^{-1}(s)p(s)ds\right] = \Phi^{-1}(t)p(t)\,dt \quad \text{(by the fundamental theorem [85]).}$$

We substitute these expressions in (2.51), and simplify by using calculus as

$$\begin{aligned} dx(t) &= [f(t)\Phi(t)\,dt]\left[c + \int_a^t \Phi^{-1}(s)p(s)\,ds\right] + \Phi(t)\Phi^{-1}(t)p(t)\,dt \\ &= f(t)\Phi(t)\left[c + \int_a^t \Phi^{-1}(s)p(s)\,ds\right]dt \\ &\quad + \Phi(t)\Phi^{-1}(t)p(t)\,dt \quad \text{(by rearranging)} \end{aligned}$$

$$= \left[f(t)\Phi(t) \left[c + \int_a^t \Phi^{-1}(s)p(s)\, ds \right] + p(t) \right] dt \quad \text{(by simplifications)}$$

$$= f(t)x(t)dt + p(t)\, dt \quad [\text{from } x(t) \text{ in } (2.50)]. \tag{2.52}$$

This shows that

$$x(t) = x_c(t) + x_p(t)$$

is indeed a solution to (2.41). Moreover, from (2.30) the general solution to (2.41) is expressed by

$$\begin{aligned}
x(t) &= \Phi(t) \left[c + \int_a^t \Phi^{-1}(s)p(s)\, ds \right] \quad [\text{from } (2.50)] \\
&= \Phi(t)c + \int_a^t \Phi(t)\Phi^{-1}(s)p(s)\, ds \quad [\text{by Theorem 1.3.1(M2)}] \\
&= \Phi(t)c + \int_a^t \Phi(t,s)p(s)\, ds \quad [\text{from } (2.40)] \qquad (2.53) \\
&= \exp\left[\int_a^t f(s)\, ds \right] c \\
&\quad + \int_a^t \exp\left[\int_s^t f(u) du \right] p(s)\, ds \quad [\text{from } (2.30) \text{ and } (2.40)].
\end{aligned}$$

For easy reference, this procedure is summarized in Fig. 2.8.

Observation 2.5.1

(i) From (2.31) (Observation 2.4.1), we note that $x_c(t) = \Phi(t)c$ is the general solution to (2.42), where $\Phi(t)$ is the fundamental solution to (2.42), and c is an arbitrary constant. $x_c(t) = \Phi(t)c$ is also referred to as the complementary solution to (2.41). However, to find a particular solution to (2.41) in step 2, we assume that $x_p(t) = \Phi(t)c(t)$ is a particular solution to (2.41), where $c(t)$ is an unknown function. The basic structure of the complementary and the particular solution is the same. In the case of the particular solution, we treated the constant c in the complementary solution as an unknown function of independent variable t as a parameter. Because of this fact, the method of finding a particular solution to (2.41) is called the method of variation of constant parameters.

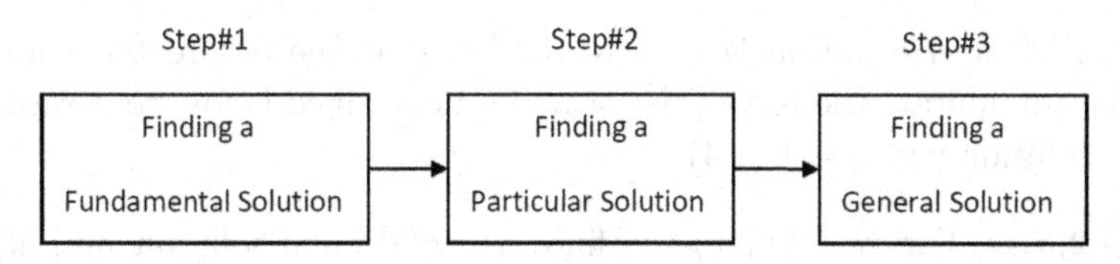

Fig. 2.8 Flowchart for finding general solution.

(ii) Instead of using a general fundamental solution in (2.44), one can use the normalized fundamental solution $\Phi(t, t_0)$ in (2.40) which corresponds to (2.42), and compute a particular solution to (2.41).

(iii) We further observe that the method of variation of constant parameters determines the general solution to (2.41). For instance, if $c_a = 0$ in (2.44), then the particular solution (2.49) reduces to

$$\begin{aligned} x_p(t) &= \Phi(t) \int_a^t \Phi^{-1}(s)p(s)\, ds \\ &= \int_a^t \Phi(t)\Phi^{-1}(s)p(s) \quad \text{[by Theorem 1.3.1(M2)]} \\ &= \int_a^t \Phi(t,s)p(s) \quad \text{[from (2.40)]}. \end{aligned} \tag{2.54}$$

(iv) If c_a in (2.47) is any arbitrary constant c, then the particular solution in (2.49) is truly a general solution to (2.41). In other words,

$$\begin{aligned} x_p(t) &= \Phi(t) \left[c + \int_a^t \Phi^{-1}(s)p(s)ds \right] \\ &= \Phi(t)c + \int_a^t \Phi(t,s)p(s)\, ds. \end{aligned} \tag{2.55}$$

This solution is equivalent to the general solution described in (2.53).

Observation 2.5.2. In the following, we present an alternative procedure for finding a particular solution to (2.41). From the definition of a particular solution in (2.44) and Observation 2.4.1(ii), we have

$$c(t) = \Phi^{-1}(t)x_p(t). \tag{2.56}$$

Now, by using Theorems 1.5.1 and 2.4.1 and by assuming that $x_p(t)$ is a particular solution to (2.41), we have

$$\begin{aligned} dc(t) &= d\Phi^{-1}(t)x_p(t) + \Phi^{-1}(t)\, dx_p(t) \quad \text{(by Theorem 1.5.11)} \\ &= -f(t)\Phi^{-1}(t)x_p(t)\, dt \\ &\quad + \Phi^{-1}(t)\left[f(t)x_p(t) + p(t)\right] dt \quad \text{(by substitution)} \\ &= \Phi^{-1}(t)p(t)\, dt \quad \text{(by simplification)}. \end{aligned} \tag{2.57}$$

This differential equation is exactly the same as the differential equation in (2.47). One can imitate the rest of the procedure described before to determine the particular solution process of (2.41).

Example 2.5.1. For $dx = (fx+p)\, dt$, find: (i) a particular solution, and (ii) a general solution to the given differential equation.

Solution procedure. Let us compare the coefficient of "dt" of the given differential equation with the corresponding coefficient of a differential equation in (2.41). Here $f(t) \equiv f$ and $p(t) \equiv p$ are constant functions.

From step 1 of finding a particular solution to the given differential equation and Example 2.4.2, a fundamental solution to the corresponding first-order linear homogeneous scalar differential equation,

$$dx = fx\,dt,$$

is given by

$$\Phi(t) = \Phi(t) = \exp[f(t-a)].$$

By applying the method of variation of constant parameters and Observation 2.5.1(iii), the particular solution to the given differential equation is

$$\begin{aligned} x_p(t) &= \Phi(t)\int_a^t \Phi^{-1}(s)p\,ds \\ &= p\int_a^t \exp\left[f(t-s)\right]ds \quad \text{(by substitution)} \\ &= -\frac{p}{f}\left[1-\exp[f(t-a)]\right] \quad \text{(by integration)}. \end{aligned}$$

Moreover, from (2.53) the general solution to the given differential equation is

$$\begin{aligned} x(t) &= x_c(t) + x_p(t) \\ &= \exp\left[f(t-a)\right]c + \frac{p}{f}\left(\exp\left[f(t-a)\right]-1\right). \end{aligned}$$

This is our desired result.

Example 2.5.2. Find a general solution to $dx = (-fx + p)\,dt$.

Solution procedure. To avoid repetitiveness, we just follow the mechanical argument used in the solution of Example 2.5.1. In this example, $f(t) \equiv -f$ and $p(t) \equiv p$ are constant functions. Moreover, from Example 2.4.3 the fundamental solution corresponding to the homogeneous differential equation

$$dx = -fxdt$$

is

$$\Phi(t) = \Phi(t) = \exp\left[-f(t-a)\right].$$

By using the same steps outlined in the solution procedure of Example 2.5.1, the particular and the general solution to the given differential are represented by

$$\begin{aligned} x_p(t) &= \Phi(t)\int_a^t \Phi^{-1}(s)p\,ds = p\int_a^t \exp\left[-f(t-s)\right]ds \quad \text{(by substitution)}, \\ x(t) &= x_c(t) + x_p(t) \end{aligned}$$

$$= \exp[-f(t-a)]\,c + p\int_a^t \exp[-f(t-s)]\,ds$$
$$= \exp[-f(t-a)]\,c + \frac{p}{f}\,(1-\exp[-f(t-a)]) \quad \text{[by (2.53)]}.$$

In the following subsection, we discuss a procedure for finding a particular solution to a linear nonhomogeneous deterministic differential equation. This idea leads to the formulation of an IVP in the study of differential equations and its applications. Moreover, illustrations are provided to exhibit the usefulness of the concept of solving real-world problems.

2.5.3 *Initial value problem*

Let us formulate an IVP for a first-order linear nonhomogeneous differential equation of the type (2.41). We consider

$$dx = [f(t)x + p(t)]\,dt, \quad x(t_0) = x_0. \tag{2.58}$$

2.5.4 *Procedure for solving the IVP*

In the following, we utilize the procedure for finding a solution to the first-order linear nonhomogeneous scalar differential equation (2.41) described in Subsection 2.5.2 for the problem of finding the solution to the IVP (2.58). Now, we briefly summarize the procedure for finding a solution to the IVP (2.58).

Step 1. By following the procedure for finding a general solution to (2.41), we have

$$\begin{aligned} x(t) &= \Phi(t)\left[c + \int_a^t \Phi^{-1}(s)p(s)ds\right] \\ &= \Phi(t)c + \left[\int_a^t \Phi(t,s)p(s)ds\right], \end{aligned} \tag{2.59}$$

where $a, t \in J \subseteq R$, $\Phi(t)$ is as defined in (2.43), and $\Phi(t,s) = \Phi(t)\Phi^{-1}(s)$. Here, c is an arbitrary constant.

Step 2. In this step, we need to find an arbitrary constant c in (2.59). For this purpose, we utilize the given initial data (t_0, x_0), and solve the linear scalar algebraic equation

$$x(t_0) = \Phi(t_0)\left[c + \int_a^{t_0} \Phi^{-1}(s)p(s)ds\right] = x_0, \tag{2.60}$$

for any given t_0 in J. Since

$$\exp\left[\int_a^{t_0} f(s)ds\right] \neq 0$$

for any continuous function f, the algebraic equation (2.60) has a unique solution, given by

$$c = \Phi^{-1}(t_0)x_0 - \int_a^{t_0} \Phi^{-1}(s)p(s)\,ds. \tag{2.61}$$

Step 3. The solution to the IVP (2.58) is determined by substituting the expression of (2.61) into (2.59). Hence, we have

$$\begin{aligned}
x(t) &= \Phi(t)c + \int_a^t \Phi(t,s)p(s)\,ds \quad \text{[from (2.59)]} \\
&= \Phi(t)\left[\Phi^{-1}(t_0)x_0 - \int_a^{t_0} \Phi^{-1}(s)p(s)ds\right] \\
&\quad + \int_a^t \Phi(t,s)p(s)\,ds \quad \text{[from (2.61)]} \\
&= \Phi(t,t_0)x_0 - \int_a^{t_0} \Phi(t,s)p(s)\,ds \\
&\quad + \int_a^t \Phi(t,s)p(s)\,ds \quad \text{[by Theorem 1.3.1 (M2) and (2.40)]} \\
&= \Phi(t,t_0)x_0 + \int_{t_0}^t \Phi(t,s)p(s)ds \quad \text{[by integral properties and grouping)},
\end{aligned} \tag{2.62}$$

where

$$\Phi(t,t_0) = \Phi(t)\Phi^{-1}(t) \equiv \Phi(t)\Phi^{-1}(t_0) = \exp\left[\int_{t_0}^t f(s)ds\right]. \tag{2.63}$$

Thus, the solution $x(t) = x(t,t_0,x_0)$ to the IVP (2.58) in (2.62) is also represented by

$$\begin{aligned}
x(t) &= x(t,t_0,x_0) \\
&= \exp\left[\int_{t_0}^t f(s)dsx_0\right] + \int_{t_0}^t \exp\left[\int_s^t f(u)du\right] p(s)\,ds.
\end{aligned} \tag{2.64}$$

The solution (2.64) is referred to as a particular solution to (2.41). For easy reference, this procedure is summarized in Fig. 2.9.

Now, some numerical examples are given to demonstrate the procedure. In addition, a few illustrations are presented to exhibit the usefulness of the concept of the IVP and its applications.

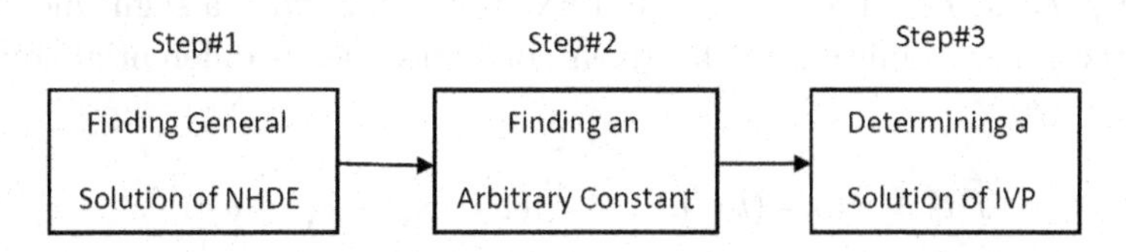

Fig. 2.9 Flowchart for finding particular solution.

Example 2.5.3. We are given $dx = (fx + p)\, dt$, $x(t_0) = x_0$. Find the solution to the given IVP for any constants f and p.

Solution procedure. From step 1 and Example 2.5.1, the general solution to the given nonhomogeneous differential equation is described by

$$x(t) = \exp\left[f(t-a)\right] c + \frac{p}{f}\left(\exp\left[f(t-a)\right] - 1\right).$$

By applying step 2 of the procedure in Subsection 2.5.4, the unknown constant c is determined by

$$x_0 = \exp\left[f\left(t_0 - a\right)\right] c + \frac{p}{f}\left(\exp\left[f\left(t_0 - a\right)\right] - 1\right),$$

which can be solved for c, and we get

$$c = \exp\left[f\left(a - t_0\right)\right] x_0 + \frac{p}{f}\left(\exp\left[f\left(a - t_0\right)\right] - 1\right).$$

Now, we substitute the expression of c into the above general solution. After algebraic simplification, we obtain

$$\begin{aligned}
x(t) &= \exp\left[f(t-a)\right]\left(\exp\left[f\left(a-t_0\right)\right] x_0 + \frac{p}{f}\left(\exp\left[f\left(a-t_0\right)\right] - 1\right)\right) \\
&\quad + \frac{p}{f}\left(\exp[f(t-a) - 1]\right) \quad \text{(by substitution of } c\text{)} \\
&= \exp\left[f(t-a)\right] \exp\left[f\left(a-t_0\right)\right] x_0 + \frac{p}{f}\left(\exp\left[f(t-a)\right] - 1]\right) \\
&\quad + \frac{p}{f}\left(\exp\left[f(t-a)\right] \exp\left[f\left(a-t_0\right)\right] - \exp[f(t-a)]\right) \quad \text{(by multiplication)} \\
&= \exp\left[f\left(t-t_0\right)\right] x_0 + \frac{p}{f}\left(\exp\left[f(t-a)\right] - 1\right) \\
&\quad + \frac{p}{f}\left(\exp\left[f\left(t-t_0\right)\right] - \exp\left[f(t-a)\right]\right) \quad \text{(by simplification)} \\
&= \exp\left[f\left(t-t_0\right)\right] x_0 + \frac{p}{f}\left(\exp\left[f\left(t-t_0\right)\right] - 1\right) \quad \text{(by simplification).}
\end{aligned}$$

This is the desired solution to the given IVP.

Example 2.5.4. Solve the IVP $dx = (-fx + p)\, dt$, $x(t_0) = x_0$, for any given constants f and p.

Solution procedure. From step 1 and Example 2.5.2 (with a slight modification in sign), the general solution to the given nonhomogeneous differential equation is represented by

$$x(t) = \exp\left[-(t-a)\right] c + \frac{p}{f}\left(1 - \exp\left[-f(t-a)\right]\right).$$

By using the same argument from Example 2.5.3, we obtain

$$x(t) = \exp\left[-f\left(t - t_0\right)\right] x_0 + \frac{p}{f}\left(1 - \exp\left[-f\left(t - t_0\right)\right]\right).$$

This is the desired solution to the given IVP.

Illustration 2.5.1 (birth/death and immigration process [6, 55]). From Example 2.2.3, we have

$$dx = (fx + p)\, dt, \quad x(t_0) = x_0,$$

where x stands for the number of species in a given population. Here $C(t, x) = fx+p$, f is the intrinsic growth rate, p is the expected (deterministic) legal immigration rate of the population, and x_0 is the initial size of the population at the initial time t_0.

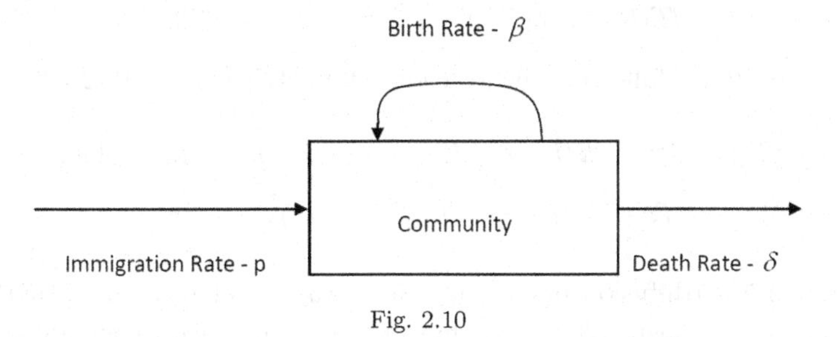

Fig. 2.10

By following the discussion of solving the IVP in Example 2.5.3, we have

$$x(t) = \exp\left[f\left(t - t_0\right)\right] x_0 + \frac{p}{f}\left(\exp\left[f\left(t - t_0\right)\right] - 1\right).$$

This expression describes the size of the population under a controlled environment. From this, we can analyze the effects of legal immigration on the size of the population.

Illustration 2.5.2 (Newton's law of cooling [6, 27, 55]). Let T_s be the constant temperature of the surrounding medium. Let T_0 be the initial temperature of a body at an initial time t_0 without an internal heating system. Let $T(t)$ be the temperature of the body at a time t for $t \geq t_0$. For $\Delta t \neq 0$, under the property of the transfer of

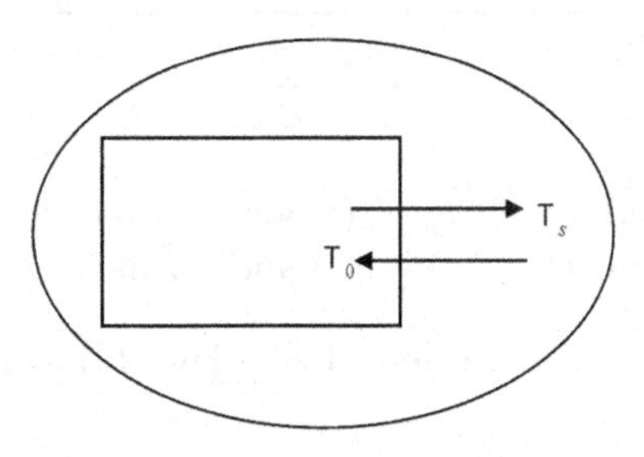

Fig. 2.11

heat from a body of higher temperature to a body of lower temperature, the sign change in the temperature $T(t+\Delta t) - T(t)$ of the body is described by

(a) If $T_0 - T_s < 0$, then $T(t+\Delta t) - T(t)$ and Δt have the same sign;
(b) If $T_0 - T_s > 0$, then $T(t+\Delta t) - T(t)$ and Δt have the opposite sign.

Under ideal experimental conditions, Newton's law of cooling is applicable. Therefore, we have

$$T(t+\Delta t) - T(t) = k\,(T_s - T(t))\,\Delta t,$$

where k is a constant of proportionality and is called the coefficient of cooling (or the specific rate of cooling). Depending on the situation, $T_0 - T_s < 0$ or $T_0 - T_s > 0$, the surrounding medium or the body is losing the heat. Therefore, for small Δt,

$$dT(t) = -k\,(T(t) - T_s)\,dt, \quad T(t_0) = T_0.$$

This is an IVP. By following the procedure for solving the IVP (2.58), we obtain

$$\begin{aligned} T(t) &= \exp\left[-k\,(t-t_0)\right] T_0 - \frac{kT_s}{k}\left(\exp\left[f\,(t-t_0)\right] - 1\right) \\ &= T_s + \exp\left[-k\,(t-t_0)\right](T_0 - T_s). \end{aligned}$$

Illustration 2.5.3 (diffusion process [31, 55, 113]). Let us assume that a cell of constant volume V is suspended in a homogeneous liquid with a constant concentration C of a solute S. Let $c = c(t)$ be the concentration of the solute S inside the cell and uniformly distributed over the cell. Due to the diffusion process, the molecules from the solute will enter the cell from the surrounding liquid, but there will also be molecules from the solute that will leave the cell. Thus, there is flow of molecules through the cell membrane in both directions. The net flow is determined by the sign of the change in concentration $c(t+\Delta t) - c(t)$ of the solute in the cell, namely

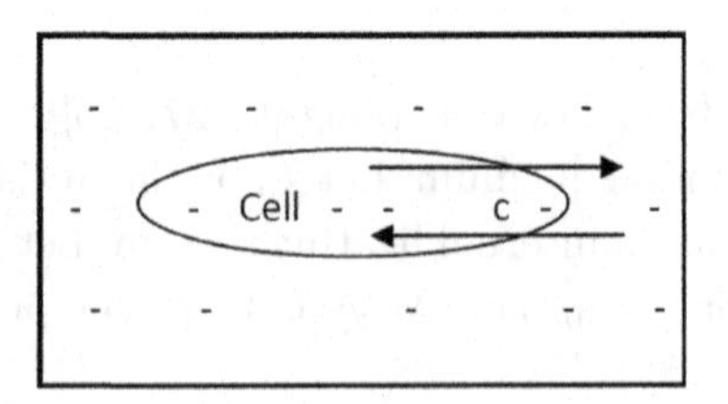

Fig. 2.12

(a) If $(C - c(t)) > 0$, then $c(t+\Delta t) - c(t)$ and Δt have the same sign;
(b) If $(C - c(t)) < 0$, then $c(t+\Delta t) - c(t)$ and Δt have the opposite sign.

Under ideal experimental conditions, Fick's law of the diffusion process is applicable. Then, we have

$$c(t+\Delta t) - c(t) = \frac{kA}{V}\,(C - c(t))\,\Delta t,$$

where A is the surface area of the cell membrane, V is the volume of the cell, and k is a constant of proportionality and is called the diffusion coefficient. This constant is determined by the condition, structure, and thickness of the membrane. It is called the permeability of the membrane for a given solute. Depending on the initial condition, $c_0 - C < 0$ or $c_0 - C > 0$, the cell or surrounding plasma is receiving the solute. Thus, for small $\triangle t$,

$$dc(t) = -\frac{kA}{V}(c(t) - C)\,dt, \quad c_0(t_0) = c_0.$$

This is an IVP. By following the procedure for solving the IVP (2.58), we obtain

$$\begin{aligned} c(t) &= \exp\left[-\frac{kA}{V}(t - t_0)\right] c_0 - C\left(\exp\left[-\frac{kA}{V}(t-t_0)\right] - 1\right) \\ &= C + \exp\left[-\frac{kA}{V}(t - t_0)\right](c_0 - C). \end{aligned}$$

Illustration 2.5.4 (cell membrane [31, 39, 110]). The body fluid is full of positive- and negative charged ions. Due to the ionic permeability of the membrane as well as the difference in ionic composition between the intra- and extracellular fluids, the potential difference across the resting membrane is negative. These types of differences are controlled by a sodium–potassium exchange system. The ionic permeability of the membrane depends on: (i) the opening and closing of the ionic channels across the membrane (ionic mobility across the membrane), (ii) the membrane thickness, and (iii) the partition coefficient between the aqueous extra- and intracellular fluids and the membrane. It is easier to measure the current carried by the ion than the ionic charge. As a result of this, it is natural to consider membrane conductance rather than membrane permeability.

We consider the RC circuit model (Exercise 6, Section 2.2) as a prototype model for the cell membrane. Here, the membrane acts like a resistance (the reciprocal of membrane conductance) and a capacitance in parallel. From Exercise 6, Section 2.2, we have

$$dI_a = -\frac{1}{C_m} I_a\,dt, \quad I_a(t_0) = I_{a0},$$

where a is the ionic species which can pass through the channel;

$$I(t) = -\frac{d}{dt}Q(t);$$

$C_m \triangle E(t) = Q(t)$; $\triangle E_m$ is the voltage difference [electromotive force/voltage E (volts)] across the membrane; C_m is the membrane capacity per unit area (C_m is assumed to be constant); $\triangle E(t) = I(t)R_m$ (Ohm's law) with

$$R_m = \frac{1}{P_a}, \quad P_a = \frac{\mu\beta RT}{DF},$$

where R is the universal gas constant, T the absolute temperature, F Faraday's constant, D the thickness of the homogeneous membrane, μ the ionic mobility, β the partition coefficient, and

$$\frac{\triangle E}{D}$$

the potential constant;

$$I_a = P_a \frac{F^2 \triangle E(t)}{RT} \left[\frac{[a]_0 - [a]_i \exp\left[-\frac{F\triangle E}{RT}\right]}{1 - \exp\left[-\frac{F\triangle E}{RT}\right]} \right].$$

Here, $[a]_0$ and $[a]_i$ are ionic concentrations outside and inside the cell, respectively.

Illustration 2.5.5 (central nervous system [39, 110]). The central nervous system (CNS) depends on (i) the transmission of excitation from one neuron to another at the synapse, and (ii) the structural arrangement of the numerous neurons. The intensity of an individual impulse in an axon is constant and independent of how the axon has been stimulated. This constant impulse arriving at a synapse either generates another constant impulse in the higher order or remains ineffective. The axons or nerve fibers are pathways for transmitting the stimuli from one part of the CNS to another. The pathways that transmit the stimuli from the different sensory organs to the brain are called afferent pathways. Those that transmit the stimuli from the brain to the different end organs are called efferent pathways. There are also certain pathways from one part of the brain to another. These pathways are referred to as first-order and second-order pathways. The intensity of the excitation (or inhibition) in a pathway depends on both the intensity of the stimulus and the number of excited (or inhibited) nerve fibers. In the region of the brain where one pathway terminates and another originates (sensory and motor area), there are a large number of neurons. These neurons form closed electric circuits (cycles). The protoplasm contains a large number of different ions, both cations (positive electric charge) and anions (negative electric charge). The stimulus to the nerve cell generates an electric current that causes the cations to move to the cathode and the anions to move to the anode. This leads to the excitation (or inhibition) of the nerve cell at the cathode.

It is assumed that the constant stimuli intensity rate I of excitation (or inhibition) in a pathway per unit time is measured in the electric current. Let $\epsilon = \epsilon(t)$ be the concentration of the "exciting" cations and $\iota = \iota(t)$ be the concentration of the "inhibiting" cations near the cathode at a time t. Based on this understanding, there is a threshold for the excitation/inhibition process. Let us denote this threshold value by c. Using the threshold value c, the excitation and the inhibition state of cations are described by

$$\frac{\epsilon}{\iota} \geq c \text{ excitation}, \quad \frac{\epsilon}{\iota} < c \text{ inhibition}.$$

Let ϵ and ι, respectively, be the concentrations at rest for the exciting cations and for the inhibiting cations. The dynamic motion of cations and anions, can be described by the decay and birth processes (Example 2.2.3). It is represented by

$$\epsilon(t + \triangle t) - \epsilon(t) = AI\triangle t - a\epsilon(t)\triangle t,$$

where A, a, and I are positive constants. From this, we have

$$d\epsilon(t) = (AI - a\epsilon(t))\,dt, \quad \epsilon(0) = \epsilon_0.$$

By using the same argument, the mathematical model for the inhibition process is given by

$$d\iota(t) = (BI - b\iota(t))\,dt, \quad \iota(0) = \iota_0,$$

where B, b, and I are positive constants. By following the procedure of Example 2.5.4, we have

$$\epsilon(t) = \exp[-at]\,\epsilon_0 + \frac{AI}{a}(1 - \exp[-at]),$$
$$\iota(t) = \exp[-bt]\,\iota_0 + \frac{BI}{b}(1 - \exp[-bt]).$$

Illustration 2.5.6 (Otto Frank model — arterial pulse–systolic phase [21, 25]). This is a continuation of Illustration 2.4.3. In this phase, the blood is pumped into the aorta (Windkessel). Due to the contraction of the heart, dV (differential of volume of the blood) is the algebraic sum of two factors, namely the total inflow (from the heart) and the outflow (through the peripheral resistance) over the interval of length dt. Therefore, by Poiseuille's law, the outflow is

$$-\frac{P}{\beta}\,dt.$$

Let us assume that the volume of blood entering the Windkessel during the interval of length dt is $e(t)\,dt$:

Net change in volume = Volume change due to inflow
− volume change due to outflow,

$$dV = e(t)\,dt - \frac{P}{\beta}\,dt,$$
$$dP = \alpha dV.$$

By using the above two expressions, we eliminate V, and we have

$$dP = \left[-\frac{\alpha}{\beta}P + \alpha e(t)\right]dt, \quad P(0) = P_0.$$

Now, we utilize the procedure in Subsection 2.5.4, and we obtain

$$P(t) = P_0 \exp\left[-\frac{\alpha}{\beta}t\right] + \alpha\int_0^t \exp\left[-\frac{\alpha}{\beta}(t-s)\right]e(s)\,ds.$$

This describes the pressure–time curve during the systolic phase. Moreover, it signifies the pressure drop in the aorta during the diastolic period. The function $e(t)$ is called the forcing quantity. We note that this function is unknown.

If we assume that the outflow rate is approximately $e(t) = A\sin Bt$, then $P(t)$ is expressed as

$$
\begin{aligned}
P(t) &= P_0 \exp\left[-\frac{\alpha}{\beta}t\right] + \alpha\int_0^t \exp\left[-\frac{\alpha}{\beta}(t-s)\right] A\sin Bs\,ds \\
&= P_0 \exp\left[-\frac{\alpha}{\beta}t\right] + \alpha A\exp\left[-\frac{\alpha}{\beta}t\right]\int_0^t \exp\left[\frac{\alpha}{\beta}s\right]\sin Bs\,ds \\
&= P_0 \exp\left[-\frac{\alpha}{\beta}t\right] + \frac{\alpha A}{\left(\frac{\alpha}{\beta}\right)^2 + B^2}\exp\left[-\frac{\alpha}{\beta}t\right] \\
&\quad \times \left[\exp\left[\frac{\alpha}{\beta}s\right]\left(\frac{\alpha}{\beta}\sin Bs - B\cos Bs\right)\Big|_0^t\right] \\
&= P_0 \exp\left[-\frac{\alpha}{\beta}t\right] + \frac{\alpha A}{\left(\frac{\alpha}{\beta}\right)^2 + B^2}\left[\frac{\alpha}{\beta}\sin Bt - B\cos Bt\right] \\
&\quad + \frac{\alpha AB}{\left(\frac{\alpha}{\beta}\right)^2 + B^2}\exp\left[-\frac{\alpha}{\beta}t\right] \\
&= \left[P_0 + \frac{\alpha AB}{\left(\frac{\alpha}{\beta}\right)^2 + B^2}\right]\exp\left[-\frac{\alpha}{\beta}t\right] \\
&\quad + \frac{\alpha A}{\left(\frac{\alpha}{\beta}\right)^2 + B^2}\left[\frac{\alpha}{\beta}\sin Bt - B\cos Bt\right].
\end{aligned}
$$

This completes the description of the pressure–time curve in the systolic phase.

Example 2.5.5. We consider the RC electric circuit of Exercise 6, Section 2.2. To this circuit, an external voltage $E_e(t)$ is applied. (a) Find a differential equation, (b) define a physical standard form of the differential equation, (c) find the solution to this example IVP, and find (d) the solution to the particular forms [40] (i) $E_e(t) = E_0$, a constant, (ii) $E_e(t) = E_0 \sin\omega t$, (iii) $E_0\cos\omega t$.

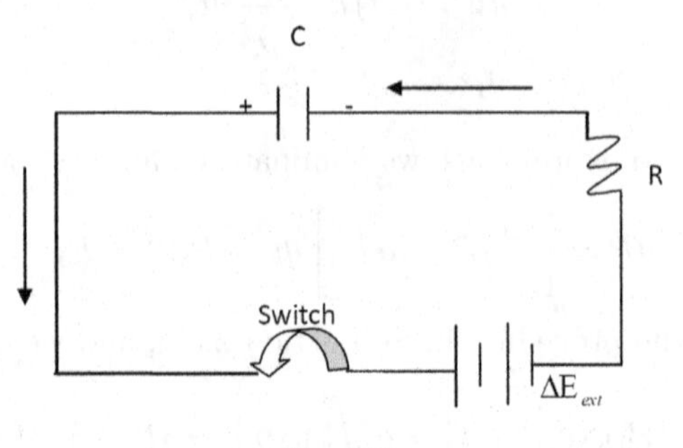

Fig. 2.13

Solution procedure. In this case, the dynamic of the electric charge is described (by the following first-order nonhomogeneous equation with constant coefficient) as

$$dQ(t) = \left[-kQ(t) + \frac{E_e(t)}{R}\right] dt, \quad Q(0) = Q_0,$$

where

$$Q, R, k = \frac{1}{RC}, \quad Q_0,$$

as defined in Exercise 6, Section 2.2. $E_e(t)$ is an external voltage applied to the circuit. We write this equation in the form

$$\tau dQ(t) = [-Q(t) + CE_e(t)]\, dt, \quad Q(0) = Q_0,$$

where

$$\frac{1}{k} = \tau.$$

The above differential equation is called the physical standard form. This completes parts (a) and (b). The term $E_e(t)$ is called the forcing term. By applying the procedure in Subsection 2.5.4, the solution to the IVP given by

$$Q(t) = Q_0 \exp\left[-\frac{t}{\tau}\right] + \int_0^t \exp\left[-\frac{1}{\tau}(t-s)\right] \frac{E_e(s)}{R}\, ds.$$

This completes part (c). To solve part (d), we consider (i). In this case, the above solution reduces to

$$\begin{aligned}
Q(t) &= Q_0 \exp\left[-\frac{t}{\tau}\right] + \int_0^t \exp\left[-\frac{1}{\tau}(t-s)\right] \frac{E_0}{R}\, ds \\
&= Q_0 \exp\left[-\frac{t}{\tau}\right] + \frac{E_0}{R} \exp\left[-\frac{t}{\tau}\right] \int_0^t \exp\left[\frac{1}{\tau}s\right] ds \\
&= Q_0 \exp\left[-\frac{t}{\tau}\right] + \frac{E_0}{R} \exp\left[-\frac{t}{\tau}\right] \left[\exp\left[\frac{1}{\tau}s\right] |_0^t\right] \\
&= Q_0 \exp\left[-\frac{t}{\tau}\right] + \frac{E_0}{R}\left(1 - \exp\left[-\frac{t}{\tau}\right]\right).
\end{aligned}$$

In the case of (ii), we have

$$\begin{aligned}
Q(t) &= Q_0 \exp\left[-\frac{t}{\tau}\right] + \frac{E_0}{R} \int_0^t \exp\left[-\frac{1}{\tau}(t-s)\right] \sin \omega s\, ds \\
&= Q_0 \exp\left[-\frac{t}{\tau}\right] + \frac{CE_0}{1+\omega^2\tau^2} \exp\left[-\frac{t}{\tau}\right] \left[\exp\left[\frac{s}{\tau}\right] (\sin \omega s - \tau\omega \cos \omega s) |_0^t\right] \\
&= Q_0 \exp\left[-\frac{t}{\tau}\right] + \frac{CE_0}{1+\omega^2\tau^2} \exp\left[-\frac{t}{\tau}\right] \left[\exp\left[\frac{t}{\tau}\right] (\sin \omega t - \tau\omega \cos \omega t) + \tau\omega\right]
\end{aligned}$$

$$= Q_0 \exp\left[-\frac{t}{\tau}\right] + \frac{CE_0}{1+\omega^2\tau^2}\left[\sin \omega t - \tau\omega \cos \omega t + \tau\omega \exp\left[-\frac{t}{\tau}\right]\right]$$

$$= Q_0 \exp\left[-\frac{t}{\tau}\right] + \frac{CE_0}{1+\omega^2\tau^2}\left[\sin \omega t + \tau\omega\left(\exp\left[-\frac{t}{\tau}\right] - \cos \omega t\right)\right].$$

Similarly, by using the above process, the solution to (d)(iii) is

$$Q(t) = Q_0 \exp\left[-\frac{t}{\tau}\right] + \frac{E_0}{R}\int_0^t \exp\left[-\frac{1}{\tau}(t-s)\right] \cos \omega s \, ds$$

$$= Q_0 \exp\left[-\frac{t}{\tau}\right] + \frac{CE_0}{1+\omega^2\tau^2} \exp\left[-\frac{t}{\tau}\right]\left[\exp\left[\frac{s}{\tau}\right](\cos \omega s + \tau\omega \sin s\omega s)\,|_0^t\right]$$

$$= Q_0 \exp\left[-\frac{t}{\tau}\right] + \frac{CE_0}{1+\omega^2\tau^2} \exp\left[-\frac{t}{\tau}\right]\left[\exp\left[\frac{t}{\tau}\right](\cos \omega t + \tau\omega \sin \omega t) - \tau\omega\right]$$

$$= Q_0 \exp\left[-\frac{t}{\tau}\right] + \frac{CE_0}{1+\omega^2\tau^2}\left[\cos \omega t + \tau\omega \sin \omega t - \tau\omega \exp\left[-\frac{t}{\tau}\right]\right].$$

We draw a few conclusions that shed light on the nature of the solution. For a large value of t,

$$\exp\left[-\frac{t}{\tau}\right]$$

is almost zero. The terms

$$Q_0 \exp\left[-\frac{t}{\tau}\right], \quad \tau\omega \exp\left[-\frac{t}{\tau}\right]$$

are called transient terms, and the term

$$\frac{CE_0}{1+\omega^2\tau^2}\left[\cos \omega t + \tau\omega \sin \omega t - \tau\omega \exp\left[-\frac{t}{\tau}\right]\right]$$

is called a forced response. This gives rise to

$$Q(t) = \frac{CE_0}{1+\omega^2\tau^2}[\cos \omega t + \tau\omega \sin \omega t]$$

$$= \frac{CE_0\sqrt{1+\omega^2\tau^2}}{1+\omega^2\tau^2}\left[\frac{1}{\sqrt{1+\omega^2\tau^2}} \cos \omega t + \frac{\tau\omega}{\sqrt{1+\omega^2\tau^2}} \sin \omega t\right]$$

$$= \frac{CE_0}{\sqrt{1+\omega^2\tau^2}}\left[\frac{1}{\sqrt{1+\omega^2\tau^2}} \cos \omega t + \frac{\tau\omega}{\sqrt{1+\omega^2\tau^2}} \sin \omega t\right]$$

$$= \frac{CE_0}{\sqrt{1+\omega^2\tau^2}}[\cos\theta \cos \omega t + \sin\theta \sin \omega t]$$

$$= \frac{CE_0}{\sqrt{1+\omega^2\tau^2}} \cos(\omega t - \theta),$$

where

$$\cos\theta = \frac{1}{\sqrt{1+\omega^2\tau^2}}, \quad \sin\theta = \frac{\tau\omega}{\sqrt{1+\omega^2\tau^2}},$$

and $\tan^{-1}(\tau\omega) = \theta$.

At this point, we introduce a few more definitions in the theory of vibration. The above expression represents a simple harmonic oscillation [40].

If $\theta = 0$, then we get

$$Q(t) = \frac{CE_0}{\sqrt{1+\omega^2\tau^2}} \cos\omega t,$$

and if $\theta = -\frac{\pi}{2}$, then we have

$$Q(t) = \frac{CE_0}{\sqrt{1+\omega^2\tau^2}} \sin\omega t.$$

The expression

$$\frac{CE_0}{\sqrt{1+\omega^2\tau^2}}$$

is called the amplitude of the vibration. We note that

$$\cos(\omega t - \theta) = \cos(\omega t + 2\pi - \theta) = \cos\left(\omega\left(t + \frac{2\pi}{\omega}\right) - \theta\right).$$

This shows that the value of Q at t and $t + \frac{2\pi}{\omega}$ are the same for any t. Hence

$$T = \frac{2\pi}{\omega}$$

is called the period of the vibration. The constant ω is called the circular frequency. The frequency in physics [40] is defined by

$$T = \frac{1}{\nu}.$$

The relationship between the frequency in physics and circular frequency is

$$\omega = 2\pi\nu.$$

$\omega t - \theta$ is called the phase angle. $-\theta$ is the phase constant, but if θ is positive, then it is called the phase lag.

2.5 Exercises

(1) Find the fundamental and general solutions to the following differential equations:

(a) $dx = (-2x+3)\,dt$ (b) $dx = (2x-3)\,dt$
(c) $dx = (2x+3)\,dt$ (d) $dx = (bx-q)\,dt$
(e) $dx = (-bx+p)\,dt$ (f) $dx = \left(\sqrt{b}\,x - q\right)dt$

(2) Find the general solution to the following differential equations:

(a) $(90+t)\,dx = (3x+p)\,dt$ (b) $\sin t\,dx = (\cos tx + \sin 2t)\,dt$
(c) $dx = \tan t\,(2x - b\cos t)\,dt$ (d) $\frac{1+t}{1+t^2}\,dx = (x+\sin 2t)\,dt$

(3) Solve the following initial value problems:
(a) $dx = (5-4x)\,dt$, $x(0)=2$
(b) $dx = \sqrt{6}\,(x-1)\,dt$, $x(0)=2$
(c) $dx = \left(2x+\sqrt{6}\,p\right)dt$, $x(0)=3$
(d) $(9+t)\,dx = (3x+9+t)\,dt$, $x(0)=1$

(4) At 10 A.M., a thermometer reading 75°F is taken outdoors, where the temperature is 30°F. At 10:05, the thermometer reads 50°F. Under ideal conditions, find the thermometer reading at any time: (a) t min, (b) $t = 10$ min, and (c) $t = 50$ min.

(5) Under the ideal experimental conditions, the amount of wheat production reaches a maximum level of C units. It has been observed that the rate of change of yield of wheat per unit amount of fertilizer is directly proportional to $C - y$, where y stands for the amount of yield of wheat at the level of the dose x of the application of the fertilizer. Find an expression for the amount of wheat y at a level x of the dose of the application of the fertilizer knowing that $y(x_0) = y_0$.

(6) Write the following differential equations in the physical standard form, and find: (i) a general solution, (ii) a solution to the initial value problem, and (iii) the amplitude, period, frequency, phase angle, and physical frequency (if possible).

(a) $dQ = [-4Q + 3\exp[at]]\,dt$, $Q(0) = Q_0$
(b) $dI = [-2I + 4\cos 3t]dt$, $I(0) = I_0$

(7) Given the RL electric circuit with an externally applied voltage $E_e(t)$: (a) Derive the differential equations with regard to the current; (b) write the derived differential equations in the physical standard form; (c) solve the IVP, and (d) find the solution to the particular forms (i) $E_e(t) = E_0$, a constant, (ii) $E_e(t) = E_0 \sin\omega t$, (iii) $E_0\cos\omega t$, and (iv) determine the amplitude, period,

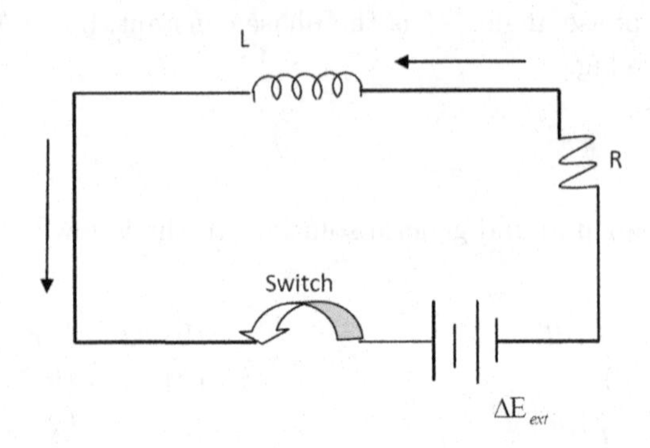

Fig. 2.14

frequency, phase angle, and the physical frequency with regard to the solution to the differential equation with the forcing term: $E_e(t) = E_0 \sin \omega t$.

(8) ***Continuous intravenous injection of glucose*** [21]. Let e be the rate of infusion (mg/min) of glucose in the blood. Let Q be the quantity (mg) in the blood, and let $c = \frac{Q}{V}$ be the concentration of the glucose in the blood, where V stands for the volume of distribution. The injected glucose is eliminated at a rate proportional to the concentration in the blood. Find: (a) the glucose concentration in the blood knowing that $Q(0) = Q_0$; (b) the glucose concentration in the absence of the infusion; (c) k and V knowing that $e = 297$ mg/min experimentally observed:

$$c(t) = 53.8\,(1 - \exp[-0.0519t])\,.$$

(9) ***H. C. Burger model — heart function test*** [21]. The goal of this problem is to find oxygen debt (x). This is a clinically important problem. During exercise the body is able to incur an oxygen debt. This means that the muscles are able to work anaerobically. But in the recovery state, oxygen is needed to replenish the energy stores (glucose). With greater muscular effort, more oxygen will be needed to replenish the energy stores that were used during effort. It is assumed that the oxygen debt is proportional to the work done (w). The oxygen debt also depends upon the oxygen supply [amount of oxygen (O) taken through the lungs]. It is assumed that the oxygen consumption in the muscles is ignored. Show: (a) $dx = \alpha dw - dO$, (b) if the extra oxygen uptake per second (in the lung) is proportional to the oxygen debt at that instant, then

$$dx = [\alpha P(t) - kx]\,dt,$$

where

$$P(t) = \frac{dw}{dt}.$$

(c) Write the differential equation in (b) in the physical standard form, (d) formulate the initial value problem for (b) and solve it, (e) find the solutions corresponding to the problem (d) for the cases (i) $P(t) = 0$, (ii) $P(t) = \beta$ constant, and (f) from the solution in (e) and using the measurable quantity

$$\frac{dO}{dt}\left[\frac{dO}{dt} = kx, \text{ in particular } \ln\left(\frac{dO}{dt}\right) = \ln\left(\frac{dO}{dt}\right)|_{t=0} - kt\right],$$

one can find k.

(10) Let F be a cumulative distribution function (cdf) of a random variable X. For $x \in R$, let e be the function defined by the expected value of the random variable of X over an interval $(-\infty, x]$, and it is denoted by $e(x)$. Solve the following initial value problems:

(a) $de = [xe + p(x)]\,dF(x)$, $e_0 = e(0)$

(b) $de = [-ke + p(x)]\,dF(x)$, $e_0 = e(0)$

(11) Let F be a male cumulative age distribution function (cdf). For $x = 25$ and the initial value problems in Exercise 10, find an expression for the expected age $e(25)$ of a male with an initial expected age of $e_0 = 5$.
(12) We are given a straight thin metallic rod of length $x \in [0, +\infty)$. It is perpendicular to an axis and its foot is on the axis of rotation. For $s \in [0, x]$, let m be the mass of an object over an interval $[0, s]$. In addition, let $I(x)$ be the moment of inertia of the rod about the axis of rotation. Solve the following initial value problems:

(a) $dI = (-KI + p)\, dm(x)$, $I(1) = 1$
(b) $dI = (KI + p(x))\, dm(x)$, $I(1) = 1$

2.6 Fundamental Conceptual Algorithms and Analysis

In Sections 2.3–2.5, we presented the computational procedures (methods) for finding solutions to linear scalar differential equations. The presented techniques provide the basis for addressing and raising several important questions in the modeling of dynamic processes in the biological, chemical, engineering, medical, physical, and social sciences. Some examples are:

(1) Are linear differential equations adequate mathematical models of dynamic processes?
(2) Prior to finding a solution to a differential equation, how do we know that it has a solution?
(3) Are there more than one solutions to a given differential equation?
(4) Are the methods of integration sufficient for finding a closed-form solution to any differential equation?
(5) Are the procedures for solving differential equations in Sections 2.4 and 2.5 applicable to any other types of differential equations?
(6) Do we really need a closed-form solution to a differential equation?
(7) In the absence of a closed-form solution to a differential equation, is the modeling of the dynamic process wasteful?
(8) Without a closed-form solution, is there any type of qualitative knowledge about a solution process shedding light on the dynamic processes?
(9) Is it feasible to study the qualitative properties of solutions to differential equations without the closed-form knowledge about them?
(10) Are there any basic universal qualitative properties of solutions to differential equations?

The answers to a few of these questions are outlined in this section. In particular, the three most basic questions are 2, 3, and 10. These questions generate the three most important problems in the theory of differential equations: (i) the existence problem, (ii) the uniqueness problem, and (iii) the fundamental properties of solution processes.

The presented procedures for finding solutions to IVPs in Sections 2.4 and 2.5 justify the existence of solutions to IVPs for first-order linear differential equations. This can be summarized via the following result. The existence result and the results presented thereafter play very important roles for both the theoretical and the practical points of view, particularly in the study of (a) time-varying (nonstationary) and time-invariant (stationary) nonlinear scalar, (b) linear time-varying systems, (c) nonhomogeneous systems, and (d) nonlinear systems of differential equations.

Theorem 2.6.1 (existence theorem). *Let us consider the IVP*

$$dy = [f(t)y + p(t)]\,dt, \quad y(t_0) = y_0, \tag{2.65}$$

where dx stands for the deterministic differential of a state variable; f and p are any given continuous functions defined on an interval $J = [a, b]$ $(J \subseteq R)$ into R; initial data/conditions $(t_0, y_0) \in J \times R$; and y_0 is a real number. Then the IVP (2.65) *has a solution $y(t, t_0, y_0) = y(t)$ through the given initial data/conditions (t_0, y_0) for $t \geq t_0$, t_0, and t in J. Moreover,*

$$y(t) = \Phi(t, t_0)y_0 + \int_{t_0}^{t} \Phi(t, s)p(s)\,ds, \tag{2.66}$$

where

$$\Phi(t, t_0) \equiv \Phi(t, t_0) = \exp\left[\int_{t_0}^{t} f(s)\,ds\right] \tag{2.67}$$

is the fundamental solution process of (2.23), *i.e. the first-order linear homogeneous scalar differential equation corresponding to* (2.65):

$$dx = f(t)x\,dt, \quad x(t_0) = y_0. \tag{2.68}$$

Proof. The justifications for the conclusions of the existence theorem follow from the computational procedures described in Sections 2.4 and 2.5, Observations 2.4.1, 2.5.1, and 2.5.2, and Theorem 2.4.1. Moreover, the solution expression (2.66) is the byproduct of these procedures. □

Observation 2.6.1

(i) We assume that the IVP (2.65) has a solution. Let us denote it by

$$y(t) = y(t, t_0, y_0).$$

From the definition analogous to Definition 2.4.2 of the IVP (2.68)/(2.36), the function $y(t)$ and its differential must satisfy both the differential equation (2.65) and the given initial conditions/data (t_0, y_0) for $t \geq t_0$, t_0 and t in J, i.e.

$$dy(t) = [f(t)y(t) + p(t)]\,dt, \quad y(t_0) = y_0. \tag{2.69}$$

From the elementary integral [85], the solution $y(t)$ to the initial problem (2.69) is equivalent to the following integral equation:

$$y(t) = y_0 + \int_{t_0}^{t} [f(s)y(s) + p(s)]\, ds. \tag{2.70}$$

We recall that the integral term on the right-hand side of (2.70) is in the sense of Riemann–Cauchy calculus [85].

(ii) We remark that the solution to the IVP (2.65) has two representations/expressions: (a) a computationally explicit (2.66) and (b) conceptually implicit (2.70).

(iii) Under the continuity assumption on the rate coefficients f and p, other standard assumptions, and without the knowledge about the exact expression of the solution, we can, alternatively, establish the existence of the solution process of the IVP (2.65). In fact, by using the approximation procedure

$$y_n(t) = \begin{cases} y_0 \text{ for } t_0 \le t \le t_0 + \dfrac{\gamma}{n}, \\ y_0 + \displaystyle\int_{t_0}^{t-\frac{1}{n}} [f(s)y_n(s) + p(s)]\, ds, \end{cases}$$

for

$$t_0 + \frac{k\gamma}{n} \le t \le t_0 + \frac{(k+1)\gamma}{n},$$
$$k = 1, 2, \ldots, n-1, \quad \gamma = b - a,$$

one can establish the existence of the solution to the IVP (2.65). This procedure utilizes an additional conceptual aspect of the rate function

$$C(t, y) = f(t)y + p(t),$$

namely the condition

$$|C(t, y)| \le M|y| + N \quad \text{(linear growth condition [15, 42, 83, 84])}, \tag{2.71}$$

where M and N are some nonnegative real numbers. Of course, it is obvious that

$$C(t, y) = f(t)y + p(t)$$

in (2.65) satisfies the growth condition (2.71).

(iv) Finally, we observe that (2.66) can be rewritten as

$$y(t) = x(t) + \int_{t_0}^{t} \Phi(t, s)p(s)\, ds, \tag{2.72}$$

where $x(t)$ is the solution to the IVP (2.68) through (t_0, y_0), i.e.

$$x(t) = x(t, t_0, y_0).$$

Our experience of finding a solution to the IVP (2.65) is based on the construction of a general fundamental solution (Observation 2.4.1). The construction of a general fundamental solution depends on finding an unknown function in (2.24) with an arbitrary constant and a choice of "a" in J. As we know, the choice of an unknown constant as well as "a" in J depends on the person solving the problem. As a result of this, the process of finding a general fundamental solution process of (2.23) or (2.68) is not unique. Therefore, the presented solution procedure does not address the question of the uniqueness of the solution to the IVP. In short, a general fundamental solution to (2.23) or (2.68) is not uniquely determined. Theorem 2.6.1 does not guarantee the uniqueness of the solution process of (2.65). The following result further justifies these statements.

Lemma 2.6.1. *Let* Φ_1 *be fundamental solution to* (2.23). Φ_2 *is any other general fundamental solution process of* (2.23) *if and only if* $\Phi_2(t) = \Phi_1(t)C$ *on* J *for some nonzero constant* C.

Proof. First, we assume that $\Phi_1(t)$ is a given general fundamental solution to (2.23), and $\Phi_2(t)$ is any general fundamental solution to (2.23). From the definition of the general fundamental solution process of (2.23) [Observation 2.4.1(v)], we recall that Φ_1 and Φ_2 will differ only by the choice of "a" in J. First, we prove that $\Phi_2 = \Phi_1 C$ on J for some nonzero constant number C. For this, we define $\Psi = \Phi_1^{-1}\Phi_2$ on J. This is defined in view of Observation 2.4.1(ii) and the definition of the general fundamental solution [Observation 2.4.1(v)]. Our goal reduces to showing that Ψ is a nonzero constant number. For this, we compute the deterministic differential of Ψ:

$$\begin{aligned} d\Psi &= d\left(\Phi_1^{-1}\Phi_2\right) \quad \text{(by definition of } \Psi\text{)} \\ &= d\left(\Phi_1^{-1}\right)\Phi_2 + \Phi_1^{-1}d\Phi_2 \quad \text{(by Theorem 1.5.11)} \\ &= -\Phi_1^{-1}f(t)\Phi_2 dt + \Phi_1^{-1}f(t)\Phi_2\, dt \quad \text{[by Theorem 2.4.1 and (2.33)]} \\ &= 0 \quad \text{(final simplification).} \end{aligned} \tag{2.73}$$

From (2.73), we conclude that $\Psi(t) = C$ is constant on J (because the differential of a function is zero if and only if the function is constant). By substitution and simplification, we conclude that $\Phi_2 = \Phi_1 C$ on J. Now, it remains for us to show that Ψ is nonzero on J. Given that $\Phi_2 \neq 0 \neq \Phi_1$, $\Phi_2 = \Phi_1 C$, and from elementary algebra, we conclude that $C \neq 0$. Therefore, C is a nonzero constant number. This completes the proof of the "if" part.

To prove the "only if" part, we start with $\Phi_2 = \Phi_1 C$, where C is a nonzero constant number and show that Φ_2 is a general fundamental solution process of (2.23). We find the differential of both sides of the expressions, and obtain

$$\begin{aligned} d\Phi_2 &= d\Phi_1 C \quad \text{[by Theorem 1.5.1(P1)]} \\ &= [f(t)\Phi_1 dt]\, C \quad \text{[by substitution from (2.33)]} \\ &= f(t)\Phi_1 C dt \quad \text{(by regrouping)} \\ &= f(t)\Phi_2 dt \quad \text{(by substitution for } \Phi_1 C\text{),} \end{aligned}$$

and hence $\Phi_1 C$ is a solution to (2.23). Since $\Phi_2 = \Phi_1 C \neq 0$ for $C \neq 0$, we conclude that Φ_2 is the general fundamental solution to (2.23). This completes the proof of the theorem. □

Observation 2.6.2

(i) Obviously, two different linear homogeneous scalar differential equations cannot have the same general fundamental solution. In fact, from (2.33), we note that $f(t)dt = d\Phi(t)\Phi^{-1}(t)$. This is true due to Observation 2.4.1.

(ii) Let us assume that Φ_1 and Φ_2 are two given general fundamental solutions to (2.23). Let $t_0 \in J$ be a given. From Lemma 2.6.1, we have $\Phi_2(t) = \Phi_1(t)C$ on J. For $t_0 \in J$, we have $\Phi_2(t_0) = \Phi_1(t_0)C$. This implies that $C = \Phi_1^{-1}(t_0)\Phi_2(t_0)$. From this, we obtain $\Phi_2(t) = \Phi_1(t)\Phi_1^{-1}(t_0)\Phi_2(t_0)$, which implies that

$$\Phi_2(t)\Phi_2^{-1}(t_0) = \Phi_1(t)\Phi_1^{-1}(t_0)$$

in view of Observation 2.4.1. From this and using the notation of a normalized fundamental solution to (2.68) or (2.23), we have

$$\Phi(t, t_0) \equiv \Phi_2(t, t_0) = \Phi_1(t, t_0). \tag{2.74}$$

This shows that the normalized fundamental solution to (2.68) or (2.23) is uniquely determined by f and t_0. The normalized fundamental solution $\Phi(t, t_0)$ to (2.68) or (2.23) defined in (2.74) is referred to as the normalized fundamental solution to (2.68) or (2.23) at $t = t_0$.

Below, we present a few algebraic properties of the normalized fundamental solution to (2.68) or (2.23). The proof of the result is based on the observed properties of a general fundamental solution, Observation 2.4.1, Theorem 2.4.1, and differential/integral calculi [85].

Lemma 2.6.2. *Let $\Phi(t, t_0) \equiv \Phi(t, t_0)$ and $\Psi(t, t_0) \equiv \Psi(t, t_0)$ be normalized fundamental solutions to* (2.68) *and* (2.35), *respectively, at $t = t_0$. Let $\Phi_1(t, t_1)$ be a normalized fundamental solution to* (2.23) *at $t = t_1$. Then, for all t_0, t_1, s, and t in J,*

$$\text{(a)} \quad \Psi(t, t_0)\Phi(t, t_0) = 1 = \Phi(t_0, t)\Phi(t, t_0), \textit{for } t_0 \leq t, \tag{2.75}$$

where

$$\Psi(t, t_0) = \Phi^{-1}(t, t_0) = \Phi(t_0, t), \textit{for } t_0 \leq t; \tag{2.76}$$

$$\text{(b)} \quad \Phi_1(t, t_1) = \Phi(t, t_0)\Phi_1(t_0, t_1), \textit{for } t_0 \leq t; \tag{2.77}$$

$$\text{(c)} \quad \Phi(t, t_0) = \Phi(t, s)\Phi(s, t_0), \textit{for } t_0 \leq t; \tag{2.78}$$

$$\text{(d)} \quad \Phi(t, s) = \Phi(t, t_0)\Psi(s, t_0) = \Phi(t, t_0)\Phi(t_0, s), \textit{for } t_0 \leq t; \tag{2.79}$$

$$\text{(e)} \quad d_s\Phi(t, s) = -f(s)\Phi(t, s)\, ds, \tag{2.80}$$

where $d_s\Phi(t, s)$ is the deterministic differential with respect to s for fixed t.

Proof. To prove (a), by using the differential we compute

$$\begin{aligned} d\left(\Psi(t,t_0)\Phi(t,t_0)\right) &= d\Psi(t,t_0)\Phi(t,t_0) \\ &\quad + \Psi(t,t_0)d\Phi(t,t_0) \quad \text{(by Theorem 1.5.11)} \\ &= -f(t)\Psi(t,t_0)dt\Phi(t,t_0) \\ &\quad + \Psi(t,t_0)f(t)\Phi(t,t_0)dt \quad \text{[from (2.33) and (2.34)]} \\ &= 0 \quad \text{(final simplification).} \end{aligned}$$

This establishes $\Psi(t,t_0)\Phi(t,t_0) = c$, where c is an arbitrary constant. This is due to the fact that if the differential of a function is zero on an interval J for $t \geq t_0$, then the function is constant on J. Moreover, $\Psi(t,t_0)$ and $\Phi(t,t_0)$ are the normalized fundamental solutions (2.68) and (2.35), respectively, at $t = t_0$. Therefore,

$$\Psi(t,t_0)\Phi(t,t_0) = c = \Psi\left(t_0,t_0\right)\Phi\left(t_0,t_0\right) = 1. \tag{2.81}$$

This shows that $\Psi(t,t_0)$ is the algebraic inverse of $\Phi(t,t_0)$, and it is denoted by $\Phi\left(t_0,t\right)$. This statement is equivalent to other notations in (2.76). In view of these notations and (2.81), we have

$$\Psi(t,t_0)\Phi(t,t_0) = 1 = \Phi\left(t_0,t\right)\Phi(t,t_0), \text{ for } t_0 \leq t.$$

This completes the proof (2.75).

To prove (b), from Lemma 2.6.1 and Observation 2.6.2(ii) we have $\Phi_1\left(t,t_1\right) = \Phi(t,t_0)C$. By using the assumption of this lemma and $t = t_1$, we get $C = \Phi_1\left(t_0,t_1\right)$. After substitution, we obtain

$$\Phi_1\left(t,t_1\right) = \Phi(t,t_0)\Phi_1\left(t_0,t_1\right), \text{ for } t_0 \leq t.$$

This establishes the relation (2.77).

To prove (c), for $t_0 \leq t$ and any s in J, we utilize the elementary algebraic properties and the nature of $\Phi(t,t_0)$, and obtain

$$\begin{aligned} \Phi(t,t_0) &= \Phi(t)\Phi^{-1}(t_0) \quad \text{[by the definition in (2.40)]} \\ &= \Phi(t)\Phi^{-1}(s)\Phi(s)\Phi^{-1}(t_0) \quad \text{[by the existence of } \Phi^{-1}(s)\text{]} \\ &= \Phi(t,s)\Phi\left(s,t_0\right) \quad \text{[by the definitions in (2.40) and (2.75)].} \end{aligned}$$

This completes the proof of (2.78).

For the proof of (d), from (2.78) and solving for $\Phi(t,s)$ we have

$$\begin{aligned} \Phi(t,s) &= \Phi(t,t_0)\Phi^{-1}(s,t_0) \quad \text{[by (2.78)]} \\ &= \Phi(t,t_0)\Psi\left(s,t_0\right) \quad \text{[by the notation (2.76)]} \\ &= \Phi(t,t_0)\Phi\left(t_0,s\right), \text{ for } t_0 \leq t \quad \text{[by the notation (2.76)].} \end{aligned}$$

This proves (d).

For the proof of (e), we apply the differential (Theorem 1.5.11) to both sides with respect to s to the expression in (2.79), and obtain

$$\begin{aligned} d_s\Phi(t,s) &= \Phi(t,t_0)d_s\Psi(s,t_0) \\ &= \Phi(t,t_0)\left[-f(s)\right]\Psi(s,t_0)\,ds \quad \text{[from (2.34)]} \\ &= -\Phi(t,t_0)f(s)\Psi(s,t_0)\,ds \\ &\quad \text{[by Theorem 1.3.1(M2) and commutativity]} \\ &= -\Phi(t,s)f(s)\,ds \quad \text{[by (2.79)]} \\ &= -f(s)\Phi(t,s)\,ds \quad \text{[by Theorem 1.5.8(c)].} \end{aligned}$$

This proves and completes the proof of the lemma. □

Observation 2.6.3

(i) We must emphasize that the proof of Lemma 2.6.2 depends heavily on algebraic reasoning and the closed-form representation of $\Phi(t,t_0)$ in (2.67).
(ii) An alternative proof of these expressions in Lemma 2.6.2 can be given by utilizing a conceptual approach. In fact, the proofs of parts (a), (c), and (e) will be illustrated later in the section.
(iii) We further remark that (2.34) provides another simple calculus-based approach to finding the inverse of the general fundamental solution to (2.23).

Prior to the presentation of the uniqueness result and for the sake of completeness, let us define the concept of the uniqueness of the solution to (2.65).

Definition 2.6.1. Let $y_1(t) = y_1(t,t_0,y_0)$ be a solution to the IVP (2.65) for $t \geq t_0$ and t, t_0 in J, and let $y_2(t) = y_2(t,t_0,y_0)$ be any other solution to (2.65). $y_1(t,t_0,y_0)$ is said to be the unique solution to the IVP (2.65) if and only if

$$y_1(t,t_0,x_0) = y_2(t,t_0,y_0)$$

for all $t \geq t_0$ and t, t_0 in J.

We are now ready to present the uniqueness result for the solution to (2.65).

Theorem 2.6.2 (uniqueness theorem). *Assume that the hypotheses of Theorem* 2.6.1 *have been satisfied. Let* $y(t) = y(t,t_0,y_0)$ *be a solution to the IVP* (2.65). *Then* $y(t,t_0,x_0)$ *is a unique solution to* (2.65) *through* $(t_0,y_0) \in J \times R$ *for* $t \geq t_0$ *and* t, t_0 *in* J.

Proof. We assume that the IVP (2.65) has two solutions. Let $y_1(t) = y_1(t,t_0,y_0)$ and $y_2(t) = y_2(t,t_0,y_0)$ be the two solutions through the same initial data (t_0,y_0) corresponding to the two normalized fundamental solutions $\Phi_1(t,t_0)$ and $\Phi_2(t,t_0)$ to (2.68) (at $t = t_0$), respectively. Our goal is to show that $y_1(t) = y_2(t)$, i.e. $y_1(t,t_0,y_0) = y_2(t,t_0,y_0)$ for $t \geq t_0$ and t, t_0 in J. From Observation 2.6.2, we note that $\Phi_2(t, t_0) = \Phi_1(t, t_0)$ for $t \geq t_0$, and hence $\Phi_2(t, t_0)y_0 = \Phi_1(t, t_0)y_0$ is valid for $t \geq t_0$, and t, t_0 in J. From Theorem 2.6.1, we have

$$y_1(t) = \Phi_1(t, t_0)y_0 + \int_{t_0}^{t} \Phi_1(t, s)p(s)\, ds,$$

$$y_2(t) = \Phi_2(t, t_0)y_0 + \int_{t_0}^{t} \Phi_2(t, s)p(s)ds,$$

where

$$\Phi_2(t, t_0) = \Phi_1(t, t_0) = \exp\left[\int_{t_0}^{t} [f(s)]\right].$$

The expressions on the right-hand of $y_1(t)$ and $y_2(t)$ are exactly the same. Therefore, $y_1(t, t_0, y_0) = y_2(t, t_0, y_0)$ for $t \geq t_0$ and t, t_0 in J. This establishes the uniqueness of the solution process of the IVP (2.65), which is our desired result. □

Observation 2.6.4

(i) Under the continuity assumption on the rate coefficients f and p, and other standard assumptions, without the knowledge about the exact expression of the solution, we can establish the existence and uniqueness of the solution process of the IVP (2.65). In fact, by using the approximation procedure

$$y_n(t) = \begin{cases} y_0 \text{ for } n = 0, \text{ on } t_0 \leq t \leq \gamma, \\ y_0 + \displaystyle\int_{a}^{t} [f(s)y_{n-1}(s) + p(s)]\, ds, \end{cases} \tag{2.82}$$

for $n \geq 1$, we can establish the existence and uniqueness of the solution to the IVP (2.65). In addition to the linear growth condition (2.71), this procedure utilizes an additional conceptual aspect of the rate function $C(t, y) = f(t)y + p(t)$, namely the condition

$$|C(t, x) - C(t, y)| \leq L|x - y| \text{ (Lipschitz condition [15, 42, 83, 84])}, \tag{2.83}$$

where L is a positive real number. Again, we can easily verify that the function $C(t, y) = f(t)y + p(t)$ in (2.65) satisfies the Lipschitz condition (2.83).

(ii) The uniqueness theorem (Theorem 2.6.2) conceptualizes the solution $y(t, t_0, y_0) = y(t)$ to the IVP (2.65) as a function of three variables (t, t_0, y_0). From the definition of a function, for given (t, t_0, y_0), the output $y(t, t_0, y_0)$ [the value of the solution at (t, t_0, y_0)] is uniquely determined under the conditions of the uniqueness theorem. This conceptualization plays a very important role in the area of mathematical modeling and the advanced study of differential equations. Moreover, the notation $y(t, t_0, y_0)$ is just a natural symbol for the solution to (2.65).

(iii) We also observe that the proofs of the relations (2.77) and (2.78) can be reproduced by applying Theorem 2.6.2. The details are left to the reader.

In the following, we present a few results that shed light on the algebraic properties of the solution to the IVPs corresponding to the linear homogeneous scalar differential equations (2.34) and (2.68) or (2.35). These properties are used in studying the fundamental properties of the solution to the IVP (2.65). Moreover, these results provide an auxiliary conceptual tool and the insight for undertaking the study of both linear and nonlinear systems of differential equations.

Lemma 2.6.3 (principle of superposition). *Let* $x_1(t, t_0, x_0) = x_1(t)$ *and* $x_2(t, t_0, x_0) = x_2(t)$ *be the solutions to the IVP* (2.68) *[or* (2.35)*] through* (t_0, x_0) *for* $t \geq t_0$ *and* t, t_0 *in* J. *Let* c_1 *and* c_2 *be any two scalar real/complex numbers. Then*

$$c_1 x_1(t, t_0, x_0) + c_2 x_2(t, t_0, x_0) \tag{2.84}$$

is a solution to (2.68) *[or* (2.35)*] through the initial data* $(t_0, c_1 x_0 + c_2 x_0)$ *for* $t \geq t_0$ *and* t, t_0 *in* J.

Proof. Let $x_1(t) = x_1(t, t_0, x_0)$ and $x_2(t) = x_2(t, t_0, x_0)$ be solutions to the IVP (2.68). From Definition 2.4.2 of the solution to the IVP (2.68) [or (2.35)], we note that for $t = t_0$ the expression $c_1 x_1 (t_0, t_0, x_0) + c_2 (t_0, t_0, x_0)$ reduces to $c_1 x_0 + c_2 x_0$. Furthermore, we have the existence of differentials $dx_1(t) = dx_1(t, t_0, x_0)$ and $dx_2(t) = dx_2(t, t_0, x_0)$. Thus, from Theorems 1.5.10 and 1.5.11 and noting the fact that a differential of the constant is zero, we have

$$\begin{aligned}
&d\left(c_1 x_1(t, t_0, x_0)\right) + d\left(c_2 x_2(t, t_0, x_0)\right) \quad \text{(differential of the expression)}\\
&\quad = c_1 dx_1(t, t_0, x_0) + c_2 x_2(t, t_0, x_0) \quad \text{(by Theorem 1.5.11)}\\
&\quad = c_1 \left[f(t) x_1(t) dt\right] + c_2 \left[f(t) x_2(t) dt\right] \quad \text{(by substitution)}\\
&\quad = \left[f(t) c_1 x_1(t) dt\right] + \left[f(t) c_2 x_2(t) dt\right] \quad \text{(by laws of real numbers)}\\
&\quad = f(t) \left[c_1 x_1(t) + c_2 x_2(t)\right] dt \quad \text{(by the linearity of rate functions)}\\
&\quad = f(t) \left[c_1 x_1(t, t_0, x_0) + c_2 x_2(t, t_0, x_0)\right] dt \quad \text{(by solution notation).}
\end{aligned}$$

This shows that $c_1 x_1(t, t_0, x_0) + c_2 x_2(t, t_0, x_0)$ and $d\left(c_1 x_1(t, t_0, x_0)\right) + d(c_2 x_2 (t, t_0, x_0))$ satisfy the IVP (2.68). This establishes the conclusion of the theorem. □

Observation 2.6.5

(i) From Illustration 1.4.4, Observation 1.4.5, and the conclusion of Lemma 2.6.3, we infer that an arbitrary linear combination of solutions to (2.68) is also a solution to the IVP (2.68).

(ii) In fact, from Observation 1.4.5, Theorem 1.4.5, and Illustration 1.4.5, we conclude that the span $S\left(\{x_1, x_2\}\right)$ of solutions x_1 and x_2 to (2.68) is a vector subspace of $C\left[[t_0, b], R\right]$.

(iii) From Definition 1.4.5, the dimension of the subspace $S(\{x_1, x_2\})$ is only one arbitrary constant. In fact, it is a linear combination of a single function.

The following result shows that the normalized fundamental solution $\Phi(t, t_0)$ to (2.68) or (2.35) possesses certain basic algebraic properties as a function for each fixed (t, t_0) for $t \geq t_0$ and t, t_0 in J.

Lemma 2.6.4. *Let $x(t) = x(t, t_0, x_0)$ be a solution to the IVP (2.68) through (t_0, x_0) for $t \geq t_0$ and t, t_0 in J. For fixed (t, t_0) the transformation/mapping of $\Phi(t, t_0)$ is defined on R into R by*

$$\Phi(t, t_0)x_0 = x(t, t_0, x_0),$$

for x_0 in R. Then, for every x_0, x_1, x_2 and α in R,

$$\Phi(t, t_0)(x_1 + x_2) = \Phi(t, t_0)x_1 + \Phi(t, t_0)x_2, \tag{2.85}$$

$$\Phi(t, t_0)(\alpha x_0) = \alpha\Phi(t, t_0)x_0. \tag{2.86}$$

Proof. From the notation, we note that

$$\Phi(t, t_0)(x_1 + x_2) = x(t, t_0, x_1 + x_2),$$

$$\Phi(t, t_0)x_1 + \Phi(t, t_0)x_2 = x(t, t_0, x_1) + x(t, t_0, x_2).$$

Moreover, $x(t, t_0, x_1 + x_2)$, $x(t, t_0, x_1)$, and $x(t, t_0, x_2)$ are solutions to the IVP (2.68) through $(t_0, x_1 + x_2)$, (t_0, x_1) and (t_0, x_2), respectively. We apply Lemma 2.6.3 with $c_1 = c_2 = 1$ to $x(t, t_0, x_1)$ and $x(t, t_0, x_2)$, and conclude that

$$x(t, t_0, x_1) + x(t, t_0, x_2)$$

is also a solution to (2.68) with the initial data $(t_0, x_1 + x_2)$. Moreover, by the definition of the solution to the IVP (2.68) through $(t_0, x_1 + x_2)$, it is denoted by $x(t, t_0, x_1 + x_2)$. In summary, $x(t, t_0, x_1 + x_2)$ and

$$x(t, t_0, x_1) + x(t, t_0, x_2)$$

are solutions to the IVP (2.68) through $(t_0, x_1 + x_2)$. Therefore, by the application of Theorem 2.6.2, these two solutions must be equal on J, i.e.

$$x(t, t_0, x_1 + x_2) = x(t, t_0, x_1) + x(t, t_0, x_2).$$

This completes the proof of (2.85). Based on the above argument, the proof of (2.86) can be reformulated analogously. The details are left to the reader. □

Observation 2.6.6

(i) We observe that the mapping/transformation that satisfies the properties described in (2.85) and (2.86) is called a linear transformation/mapping. For fixed (t, t_0), the normalized fundamental solution $\Phi(t, t_0)$ to (2.68)/(2.35) at $t = t_0$ is a linear transformation defined on R into itself. In fact, it is a polynomial function of degree 1.

(ii) We further note that the proof of Lemma 2.6.4 is trivial in the light of the algebra of real numbers (Theorem 1.3.1). However, we choose to use the uniqueness theorem. This line of argument has greater universal appeal for incorporating the study beyond the study of linear scalar differential equations.

In the following, we present a relationship between the solution to the linear homogeneous scalar differential equation (2.68) and its adjoint differential equation (2.35).

Lemma 2.6.5. *Let* $x(t) = \Phi(t, t_0)x_0$ *and* $z(t) = x_0\Phi(t, t_0)$ *be a solution to the IVP (2.68) and the corresponding IVP associated with its adjoint differential equation (2.35) through* (t_0, x_0), *i.e.*

$$dz = -zf(t)\,dt, \quad z(t_0) = x_0, \tag{2.87}$$

respectively, where $\Phi(t, t_0)$ *and* $\Psi(t, t_0)$ *are the normalized fundamental solutions to (2.68) and (2.87) for* $t \geq t_0$ *and* t, t_0 *in* J. *Then*

$$z(t)x(t) = c, \tag{2.88}$$

where c *is a constant number. Moreover,* $c = x_0^2$ *and*

$$\Psi(t, t_0)\Phi(t, t_0) = I, \tag{2.89}$$

for all $t \geq t_0$ *and* t, t_0 *in* J.

Proof. By imitating the proof of Lemma 2.6.2(a), we compute

$$\begin{aligned} d(z(t)x(t)) &= dz(t)x(t) + z(t)\,dx(t) \quad \text{(by Theorem 1.5.11)} \\ &= -f(t)z(t)x(t)\,dt \\ &\quad + z(t)f(t)x(t)\,dt \quad \text{(by substitution)} \\ &= 0 \quad \text{(by final simplification).} \end{aligned}$$

This establishes $z(t)x(t) = c$, where c is a constant real number. This is due to the fact that if the differential of a function is zero on an interval J, then the process is constant on J. This completes the proof of (2.88). The proof of $c = x_0^2$ follows from the fact that

$$\begin{gathered} z(t_0)x(t_0) = c = x_0^2, \\ \Phi(t_0, t_0) = 1 = \Psi(t_0, t_0), \end{gathered}$$

and

$$z(t)x(t) = x_0\Psi(t, t_0)\Phi(t, t_0)x_0 = x_0x_0,$$

for any x_0 in R. Finally, from

$$x_0\Psi(t, t_0)\Phi(t, t_0)x_0 = x_0x_0$$

we have

$$\Psi(t, t_0)\Phi(t, t_0) = 1.$$

This completes the proof of (2.89) as well as the proof of the lemma. □

Observation 2.6.7

(i) The conclusions of Lemma 2.6.5 provide a relationship between the normalized fundamental solutions to (2.68) and (2.87). Moreover, Lemma 2.6.5 provides an alternative conceptual proof of Lemma 2.6.2(a).
(ii) In addition, Lemma 2.6.5, conceptually, confirms that the solution provided by Theorem 2.4.1 is indeed the algebraic inverse of a fundamental solution to (2.68). Of course, Theorem 2.4.1 provides an analytical method for finding an inverse of any fundamental solution to (2.68).

In the following, we provide an analytical (conceptual) and alternative proof of Lemma 2.6.2(c) that was based on an elementary algebraic approach. This result provides a very general argument for establishing the validity of the relation (2.78). Moreover, the presented argument provides some conceptual reasoning for undertaking the study of further complex properties of solutions to more general differential equations.

Lemma 2.6.6. *Let $\Phi(t, t_0)$ be a normalized fundamental matrix solution to* (2.68) *at $t = t_0$, and $x(t, t_0, x_0)$ be the solution to the IVP* (2.68) (t_0, x_0). *Then, for all $t \geq t_0$ and t, t_0 in J,*

$$\text{(a) } x(t, s, x(s, t_0, x_0)) = x(t, t_0, x_0), \tag{2.90}$$

$$\text{(b) } \Phi(t, s)\Phi(s, t_0) = \Phi(t, t_0). \tag{2.91}$$

Proof. To prove (a) and (b), from the definitions of the solution to the IVP (2.68) and the equality of two functions (or two transformations/mappings), it is enough to show that $\Phi(t, s)\Phi(s, t_0)x_0 = \Phi(t, t_0)x_0$ for every x_0 in R. We recall that $x(t) = x(t, t_0, x_0) = \Phi(t, t_0)x_0$ and $x(s) = x(s, t_0, x_0) = \Phi(s, t_0)x_0$. Then we consider

$$\begin{aligned}
\Phi(t, s)\Phi(s, t_0)x_0 &= \Phi(t, s)(\Phi(s, t_0)x_0) \quad \text{(by grouping)} \\
&= \Phi(t, s)x(s, t_0, x_0) \quad \text{(by substitution)} \\
&= x(t, s, x(s, t_0, x_0)) \quad \text{(by definition and notation)} \\
&= x(t, s, x(s)) \quad \text{(by notation of solution)}.
\end{aligned}$$

We note that $x(t, t_0, x_0)$ and $x(t, s, x(s, t_0, x_0))$ are solutions to (2.68) through (t_0, x_0) and $(s, x(s)) = (s, x(s, t_0, x_0))$, respectively. Furthermore, for $s = t$, $x(s) = x(s, t_0, x_0)$ and $x(t, s, x(s)) = x(t, s, x(s, t_0, x_0))$ reduce to $x(t) = x(t, t_0, x_0)$ and $x(t, t, x(t)) = x(t, t, x(t, t_0, x_0))$, respectively, and hence by Definition 2.4.2 we have $x(t, t_0, x_0) = x(t, t, x(t, t_0, x_0))$. Therefore, both of the solutions $x(s, t_0, x_0)$ and

$x\left(t, s, x\left(s, t_0, x_0\right)\right)$ pass through the point $(t, x(t)) = \left(t, x\left(t, x_0, x_0\right)\right)$. Thus, by the uniqueness of the solution to (2.68), we have $x\left(t, s, x\left(s, t_0, x_0\right)\right) = x(t, t_0, x_0)$. This establishes the validity of (2.90). Moreover, $x\left(t, s, x\left(s, t_0, x_0\right)\right) = x(t, t_0, x_0)$ is true for every x_0 in R. Hence, $\Phi(t, s)\Phi\left(s, t_0\right) x_0 = \Phi(t, t_0)x_0$ for every x_0 in R. From this together with the definition of the equality of two functions, we conclude that $\Phi(t, t_0) = \Phi(t, s)\Phi\left(s, t_0\right)$. This completes the proof of (2.91) and the proof of the theorem. □

A very important byproduct of the uniqueness theorem is that the knowledge of the solution at a future time depends only on the knowledge of the initial data, and it is independent of its knowledge between the initial time and the future time. This is demonstrated by Lemma 2.6.6 in the context of the solution to the IVP (2.68). The following result extends this idea to the IVP (2.65).

Lemma 2.6.7. *Let the hypotheses of Theorem* 2.6.2 *be satisfied. For* $t_0 \le s \le t$, *let* $y(t) = y(t, t_0, y_0)$ *and let* $y(t, s, y(s)) = y\left(t, s, y(s, t_0, y_0)\right)$ *be the solutions to* (2.65) *through* (t_0, y_0) *and* $(s, y(s))$, *respectively. Then*

$$y(t) = y(t, t_0, y_0) = y(t, s, y(s, t_0, y_0)), \textit{ for } t_0 \le s \le t. \tag{2.92}$$

Proof. For $t_0 \le s \le t$, let $y(t, t_0, y_0)$ and $y(t, s, y(s))$ be solutions to (2.65) through (t_0, y_0) and $(s, y(s)) = (s, y(s, t_0, y_0))$, respectively. From the representation of $y(t) = y(t, t_0, y_0)$ in (2.66) and Lemma 2.6.2(c) or Lemma 2.6.6, we have

$$\begin{aligned}
y(s) &= \Phi\left(s, t_0\right) y_0 + \int_{t_0}^{s} \Phi(s, u)p(u)\, du, \\
y(t) &= \Phi(t, t_0)y_0 + \int_{t_0}^{t} \Phi(t, u)p(u)\, du \\
&= \Phi(t, s)\Phi\left(s, t_0\right) x_0 + \int_{t_0}^{s} \Phi(t, u)u(u)\, du \\
&\quad + \int_{s}^{t} \Phi(t, u)p(u)\, du \quad \text{[by Lemma 2.6.2(c) and properties of integrals]} \\
&= \Phi(t, s)\left[\Phi\left(s, t_0\right) y_0 + \int_{t_0}^{s} \Phi(s, u)p(u) du\right] \\
&\quad + \int_{s}^{t} \Phi(t, u)p(u)\, du \quad \text{[by Lemma 2.6.2(c) and regrouping]} \\
&= \Phi(t, s)y(s) + \int_{s}^{t} \Phi(t, u)p(u)\, du \quad [\text{by } y(s) = y(s, t_0, y_0)] \\
&= y(t, s, y(s)) \quad \text{(by notation)} \\
&= y\left(t, s, y(s, t_0, y_0)\right) \quad [\text{by notation } (s, y(s, t_0, y_0))].
\end{aligned}$$

This establishes the relation (2.92) and completes the proof. □

Observation 2.6.8

(i) We emphasize that the proof of Lemma 2.6.7 is based on the explicit solution to (2.65), the algebraic features of the normalized fundamental solution to (2.68), the computational representation of the solution to (2.65) in (2.66), and the properties of elementary integrals.

(ii) One can also use the integral representation of the solution in (2.70) to (2.65) and the properties of integrals to establish the validity of Lemma 2.6.7. In fact,

$$
\begin{aligned}
y(t) &= y_0 + \int_{t_0}^{t} [f(u)y(u) + p(u)]\, du \\
&= y_0 + \int_{t_0}^{s} [f(u)y(u) + p(u)]\, du \\
&\quad + \int_{s}^{t} [f(u)y(u) + p(u)]\, du \quad \text{(by integral properties)} \\
&= y(s) + \int_{s}^{t} [f(u)y(u) + p(u)]\, du \quad \text{(by the solution representation)} \\
&= y(t, s, y(s)) \quad \text{(by notation)} \\
&= y(t, s, y(s, t_0, y_0)) \quad [\text{by notation of } (s, y(s, t_0, y_0))].
\end{aligned}
$$

We note that we do not require the explicit knowledge of the solution to (2.65).

(iii) The validity of Lemma 2.6.7 also follows from reasoning similar to what was presented in the proof of Lemma 2.6.6 and the application of Theorem 2.6.2. In fact, we observe that $y(t, s, y(s)) = y(t, s, y(s, t_0, y_0))$ (by notation and definition of the solution). For $s = t$, we have

$$
\begin{aligned}
y(t, s, y(s, t_0, y_0)) &= y(t, t, y(t, t_0, y_0)) \quad \text{(by Definition 2.4.2)} \\
&= y(t, t_0, y_0).
\end{aligned} \tag{2.93}
$$

This shows us that the solutions $y(t, t_0, y_0)$ and $y(t, s, y(s, t_0, y_0))$ are equal at $s = t$. Thus, by the uniqueness of Theorem 2.6.2,

$$y(t, t_0, y_0) = y(t, s, y(s, t_0, y_0))$$

is valid for $t_0 \le s \le t$.

In the following, we show that the solution to (2.65) has additional one of the fundamental properties. This statement is illustrated by the following results.

Theorem 2.6.3 (continuous dependence of the solution on initial conditions). *Let the hypotheses of Theorem 2.6.2 be satisfied. Then the solution process $y(t) = y(t, t_0, y_0)$ is continuous with respect to (t, t_0, y_0) for $t \ge t_0$ and t, t_0 in J. In particular, it is continuous with respect to the initial conditions (data).*

Proof. First, we note that by virtue of the solution, $y(t, t_0, y_0)$ is continuous in t for each (t_0, y_0). We need to examine the continuity with respect to (t_0, y_0) for each t. For this purpose, we examine the solution in (2.66). From the expression of $y(t) = y(t, t_0, y_0)$ in (2.66), it is clear that $y(t)$ is a linear function of y_0. That is to say, $y(t)$ is a first-degree polynomial function in y_0 for each fixed (t, t_0) for $t \geq t_0$ and t, t_0 in J. Elementary calculus notes that every polynomial function is continuous. Therefore, we conclude that $y(t)$ is continuous with respect to y_0 for each fixed (t, t_0) for $t \geq t_0$ and t, t_0 in J. To prove the continuity of the solution $y(t, t_0, y_0)$ to (2.65) with respect to t_0 for fixed (t, y_0), we need to examine the continuity of functions in the two terms on the right-hand side of (2.66). We note that the continuity of the function $\Phi(t, t_0)y_0$ in the first term with respect to t_0 follows from Lemma 2.6.2(e). This occurs since $d_{t_0}\Phi(t, t_0)$ satisfies the adjoint differential equation (2.87). This implies that $\Phi(t, t_0)$ is continuous in t_0, and hence the function $\Phi(t, t_0)y_0$ in the first term is continuous with respect to t_0. The continuity of the function on the second term on the right-hand side with respect to t_0 follows from the continuity of the rate coefficient functions of differential equations and their properties of the indefinite integral as a function of t_0. Finally, from these arguments, we conclude that the solution $y(t, t_0, y_0)$ is continuous with respect to the initial data (t_0, y_0). □

We present below a result that exhibits the differentiability of the solution to (2.65) with respect to all three variables (t, t_0, y_0). Prior to this, we will prove a result concerning the differentiability of the solution to (2.68) with respect to all three variables (t, t_0, x_0).

Lemma 2.6.8. *Let $x(t, t_0, x_0)$ be the solution process of the IVP (2.68) through (t_0, x_0). Then, for all $t \geq t_0$ and t, t_0 in J,*

(a) $\frac{\partial}{\partial x_0} x(t) = \frac{\partial}{\partial x_0} x(t, t_0, x_0)$ *exists and it satisfies the differential equation*

$$dx = f(t)x\,dt, \quad \frac{\partial}{\partial x_0} x(t_0) = 1, \tag{2.94}$$

where

$$\frac{\partial}{\partial x_0} x(t, t_0, x_0)$$

stands for a deterministic partial derivative of the solution $x(t, t_0, x_0)$ to (2.68) with respect to x_0 at (t, t_0, x_0) for fixed (t, t_0);

(b) $\frac{\partial}{\partial t_0} x(t) = \frac{\partial}{\partial t_0} x(t, t_0, x_0)$ *exists and it satisfies the differential equation*

$$dx = f(t)x\,dt, \quad \frac{d}{dt_0} x(t_0) = z_0, \tag{2.95}$$

where

$$\partial_{t_0} x(t, t_0, x_0) = \frac{\partial}{\partial t_0} x(t, t_0, x_0)dt_0$$

stands for the partial differential of the solution $x(t, t_0, x_0)$ *to* (2.68) *with respect to* t_0 *at* (t, t_0, x_0) *for fixed* (t, x_0); *its initial differential condition* $\partial x(t_0)$

$$\left(\frac{dx_0(t_0)}{dt_0} = \frac{\partial}{\partial t_0} x(t_0) = z_0 \right)$$

satisfies the following differential equation:

$$\frac{\partial}{\partial t_0} x\left(t_0, t_0, x_0\right) dt_0 \equiv \partial x(t_0) = -f(t_0) x_0 \, dt_0 \tag{2.96}$$

or

$$\frac{d}{dt_0} x(t_0) \equiv \frac{d}{dt_0} x\left(t_0, t_0, x_0\right) = z_0 = -f(t_0) x_0.$$

Proof. From Lemma 2.6.4, the solution $x(t, t_0, x_0) = \Phi(t, t_0) x_0$ to the IVP (2.68) through (t_0, x_0) is a linear function of x_0. That is to say, $x(t, t_0, x_0)$ is a first-degree polynomial in x_0 for each fixed (t, t_0) for $t \geq t_0$ and t, t_0 in J. The elementary calculus tells us that every polynomial function is continuously differentiable. Therefore, the solution $x(t) = x(t, t_0, x_0)$ is differentiable with respect to x_0 for each fixed (t, t_0). In fact, we have

$$\frac{\partial}{\partial x_0} x(t) = \frac{\partial}{\partial x_0} x(t, t_0, x_0) = \Phi(t, t_0). \tag{2.97}$$

Moreover,

$$\frac{\partial}{\partial x_0} x(t, t_0, x_0)$$

is a continuous function in x_0, because it is independent of x_0 (i.e. constant) for each fixed (t, t_0) for $t \geq t_0$ and t, t_0 in J. Moreover, from (2.97), it is the normalized fundamental solution to (2.68). This validates statement (a).

Now, we will prove statement (b). Again, from the definition of the solution $x(t, t_0, x_0) = \Phi(t, t_0) x_0 = \Phi(t) \Phi^{-1}(t_0) x_0$ to the IVP (2.68), Theorem 1.5.11, and applying the formula (Theorem 1.5.11), we have

$$\begin{aligned} \partial_{t_0} x(t, t_0, x_0) &= \Phi(t) d_{t_0} \Phi^{-1}(t_0) x_0 \quad \text{[from (2.40)]} \\ &= \Phi(t)[\Phi^{-1}(t_0)[-f(t_0)]] dt_0 \quad \text{[from (2.35)]} \\ &= \Phi(t, t_0)[-f(t_0)] x_0 \, dt_0 \quad \text{(by simplification)} \\ &= \Phi(t, t_0) \partial x(t_0) \quad \text{[from (2.96)]}, \end{aligned} \tag{2.98}$$

where $\partial_{t_0} x(t, t_0, x_0)$ denotes the partial differential of the solution $x(t, t_0, x_0)$ to (2.68) with respect to t_0 at (t, t_0, x_0); $\partial x(t_0)$ is as defined in the lemma and satisfies the differential equation (2.96).

From (2.96), it is clear that $\partial_{t_0} x(t, t_0, x_0)$ satisfies the IVP (2.95) with $\partial x(t_0) = \partial_{t_0} x\,(t_0, t_0, x_0)$, which satisfies the differential equation (2.96). □

Observation 2.6.9

(i) From the proof of Lemma 2.6.8, we observe that

$$\frac{\partial}{\partial x_0}\,x(t, t_0, x_0)$$

is independent of x_0, i.e.

$$\frac{\partial}{\partial x_0}\,x(t, t_0, x_0) = \Phi(t, t_0).$$

It is obvious that

$$\frac{\partial}{\partial x_0}\,x(t, t_0, x_0)$$

is continuous in (t, t_0, x_0). Moreover, its second partial derivative of $x(t, t_0, x_0)$ with respect to x_0 for fixed (t, t_0)

$$\left(\frac{\partial^2}{\partial x_0^2}\,x(t, t_0, x_0)\right)$$

exists, and

$$\frac{\partial^2}{\partial x_0^2}\,x(t, t_0, x_0) = 0.$$

(ii) By using (i) and Lemma 2.6.8, we introduce the partial differentials of the solution $x(t, t_0, x_0)$ to (2.68) with respect to x_0 and t_0 at (t, t_0, x_0):

$$\partial x_0 x(t, t_0, x_0) = \frac{\partial}{\partial x_0}\,x(t, t_0, x_0)dx_0 = \Phi(t, t_0)dx_0, \tag{2.99}$$

$$\partial t_0 x(t, t_0, x_0) = \frac{\partial}{\partial t_0}\,x(t, t_0, x_0)\partial x_0(t_0) = \Phi(t, t_0)\partial x_0(t_0), \tag{2.100}$$

where dx_0 is a differential with respect to x_0 of $x(t, t_0, x_0)$ at t_0, i.e.

$$dx(t_0) = dx_0, \tag{2.101}$$

and

$$\frac{\partial}{\partial t_0}\,x\,(t_0, t_0, x_0)\,dt_0 = \partial x(t_0)$$

is the partial differential of $x(t, t_0, x_0)$ with respect to t_0 for fixed (t_0, x_0) at t_0. Note the difference between dx_0 and $\partial x(t_0)$.

(iii) Using the notations (2.99) and (2.100), we compute the mixed second partial derivatives of $x(t, t_0, x_0)$. Therefore, the mixed partial differentials of the solution $x(t, t_0, x_0)$ to (2.68) are given by

$$\begin{aligned}
\partial^2_{t_0 x_0} x(t, t_0, x_0) &= \partial_{t_0} (\partial_{x_0} x(t, t_0, x_0)) \quad \text{(by the partial differential)} \\
&= \partial_{t_0} \left(\frac{\partial}{\partial x_0} x(t, t_0, x_0) dx_0 \right) \quad \text{[from (2.99)]} \\
&= \Phi(t, t_0)[-f(t_0)] dt_0\, dx_0 \quad \text{[from (2.80)]} \\
&= -\Phi(t, t_0) f(t_0) dt_0\, dx_0 \quad \text{[from (2.101)]}, \qquad (2.102) \\
\partial^2_{x_0 t_0} x(t, t_0, x_0) &= \partial_{x_0} (\partial_{t_0} x(t, t_0, x_0)) \quad \text{(by the partial differential)} \\
&= \partial_{x_0} (\Phi(t, t_0) \partial x(t_0)) \quad \text{[from (2.100)]} \\
&= \Phi(t, t_0) dx_0 \, (\partial x(t_0)) \quad \text{[from Theorem 1.5.1(P2)]} \\
&= \Phi(t, t_0) \left[-f(t_0)\right] dt_0\, dx_0 \quad \text{[from (2.96)]} \\
&= \Phi(t, t_0)[-f(t_0)] dt_0\, dx_0 \quad \text{(by notation)} \\
&= -\Phi(t, t_0) f(t_0) dt_0\, dx_0 \quad \text{[from (2.101)]}. \qquad (2.103)
\end{aligned}$$

Hence, we conclude that its mixed second derivatives exist

$$\left(\frac{\partial^2}{\partial_{x_0} \partial_{t_0}} x(t, t_0, x_0) \right)$$

$$\left(\frac{\partial^2}{\partial_{x_0} \partial_{t_0}} x(t, t_0, x_0) = -\Phi(t, t_0) f(t_0) = \frac{\partial^2}{\partial_{x_0} \partial_{t_0}} x(t, t_0, x_0) \right)$$

and are equal. Moreover, it is the solution to the adjoint differential equation (2.87) with the initial data $-f(t_0)$.

(iv) We further observe that Lemma 2.6.8(b) provides a conceptual proof of Lemma 2.6.2(e). In fact, from (2.96) and (2.98), we have

$$\begin{aligned}
&\frac{\partial}{\partial x_0} (\partial_{t_0} x(t, t_0, x_0)) \\
&= \frac{\partial}{\partial x_0} (\Phi(t, t_0) \partial x(t_0)) \quad \text{[from (2.98)]} \\
&= \frac{\partial}{\partial x_0} (\Phi(t, t_0)[-f(t_0)] dt_0 x_0) \quad \text{[from (2.96)]} \\
&= \Phi(t, t_0) \left([-f(t_0)] dt_0 \frac{\partial}{\partial x_0} x_0 \right) \quad \text{[from Theorem 1.5.1(P2)]} \\
&= \Phi(t, t_0) ([-f(t_0)] dt_0) \quad \left(\text{from } \frac{\partial}{\partial x_0} x_0 = I_{n \times n} \right).
\end{aligned}$$

On the other hand,

$$\partial_{t_0}\left(\frac{\partial}{\partial x_0}x(t,t_0,x_0)\right) = \partial_{t_0}\left(\Phi(t,t_0)\right) \quad \text{[from (2.97)]}$$

$$= \partial_{t_0}\Phi(t,t_0) \quad \text{(by notation).}$$

From the above discussion and (iii), we validate (2.80).

Theorem 2.6.4 (differentiability of solutions with respect to the initial conditions). *Assume that the hypotheses of Theorem 2.6.2 are satisfied. Let $y(t,t_0,y_0)$ be the solution to the IVP (2.65). Then, for all $t \geq t_0$ and t, t_0 in J,*

(a) $\frac{\partial}{\partial y_0}y(t) = \frac{\partial}{\partial y_0}y(t,t_0,y_0)$ *exists and it satisfies the IVP (2.94);*
(b) $\frac{\partial}{\partial t_0}y(t) = \frac{\partial}{\partial t_0}y(t,t_0,y_0)$ *exists and it satisfies the differential equation*

$$dx = f(t)x\,dt, \quad \partial_{t_0}y\left(t_0,t_0,y_0\right) = dy(t_0), \tag{2.104}$$

where $\partial y(t_0)$ satisfies the following differential equation:

$$dy(t_0) = \left[\left(-f(t_0)\right)y_0 - p(t_0)\right]dt_0 \tag{2.105}$$

or

$$\frac{d}{dt_0}y(t_0) = z_0 = -f(t_0)y_0 - p(t_0).$$

Proof. From the direct observation of the solution to the IVP (2.65) in (2.66), we conclude that $y(t,t_0,y_0)$ is a first-degree polynomial in y_0. Because of Observation 2.6.1(iv), this first term is indeed $x(t,t_0,y_0) = \Phi(t,t_0)y_0$. By following the argument used in the proof of Lemma 2.6.8, we conclude that

$$\frac{\partial}{\partial y_0}y(t,t_0,y_0)$$

exists, and it is equal to $\Phi(t,t_0)$, where $\Phi(t,t_0)$ is the normalized fundamental solution to (2.68). As a result of this,

$$\frac{\partial}{\partial y_0}y(t,t_0,y_0) = \Phi(t,t_0) = \frac{\partial}{\partial y_0}x(t,t_0,y_0),$$

and it satisfies the corresponding (variational differential equation) first-order linear homogeneous scalar differential equation (2.68) with the initial data $(t_0,1)$. Hence,

$$\frac{\partial}{\partial y_0}y(t,t_0,y_0)$$

is the solution to the IVP (2.94).

To prove (b), again from the solution (2.72) to the IVP (2.65),

$$y(t,t_0,y_0) = x(t,t_0,y_0) + \int_{t_0}^{t}\Phi(t,s)p(s)\,ds. \tag{2.106}$$

The first term $x(t,t_0,y_0) = \Phi(t,t_0)y_0$ on the right-hand side has a differential with respect to t_0. Moreover, from the application of Lemma 2.6.8, in particular (2.97), we can conclude that

$$\partial_{t_0} x(t,t_0,y_0) = \Phi(t,t_0)\partial x(t_0),$$

where

$$\partial x(t_0) = -f(t_0)y_0\, dt_0.$$

By using Lemma 2.6.2(d) and applying the fundamental theorem of integral calculus [85] to the second term on the right-hand side of (2.106), we get

$$\begin{aligned}
&\partial_{t_0}\left(\int_{t_0}^{t} \Phi(t,s)p(s)ds\right) \\
&\quad = \Phi(t,a)\partial_{t_0}\left(\int_{t_0}^{t} \Psi(s,a)p(s)\, ds\right) \\
&\quad = \Phi(t,a)\Psi(t_0,a)\left[-p(t_0)\right] dt_0 \quad \text{(by partial differential)} \\
&\quad = -\Phi(t,t_0)p(t_0)dt_0 \quad \text{(by simplification)}.
\end{aligned} \tag{2.107}$$

From the above discussion, with (2.106) and (2.107) we have

$$\begin{aligned}
\partial_{t_0} y(t,t_0,y_0) &= \partial_{t_0} x(t,t_0,y_0) \\
&\quad + \partial_{t_0}\left(\int_{t_0}^{t} \Phi(t,s)p(s)\, ds\right) \quad \text{(by Theorem 1.5.10)} \\
&= \Phi(t,t_0)\left[-f(t_0)y_0 dt_0\right] \\
&\quad + \Phi(t,t_0)\left[-p(t_0)\right] dt_0 \quad \text{[from (2.107)]} \\
&= \Phi(t,t_0)\left[-f(t_0)y_0 - p(t_0)\right] dt_0 \quad \text{(by regrouping)} \\
&= \Phi(t,t_0)\partial y(t_0),
\end{aligned} \tag{2.108}$$

where $dy(t_0)$ is as defined in (2.105), i.e.

$$\partial y(t_0) = \left[-f(t_0)y_0 - p(t_0)\right] dt_0, \tag{2.109}$$

and it satisfies (2.104).

From the representation of the solution to the IVP (2.65) in (2.66), it is clear that $\partial_{t_0} y(t,t_0,y_0)$ is the solution to the IVP (2.104). This completes the proof of the theorem. □

Observation 2.6.10. We note that the proof of Theorem 2.6.4 can be reconstructed by utilizing the conceptual aspect of Lemma 2.6.8. For the sake of simplicity, we chose to prove Theorem 2.6.4 by employing various algebraic properties coupled with the solution representation in (2.66).

Now we present a very important result that connects the solution to (2.65) with the corresponding homogeneous differential equation (2.68). This result is the Alekseev-type variation of the constant parameter formula [2]. It provides a conceptual tool and an insight for developing the results concerning linear/nonlinear and time-varying (time-invariant and scalar) systems of differential equations. In addition, it provides an alternative method for investigating the qualitative properties of more complex differential equations.

Theorem 2.6.5 (variation of constant formula). *Let the hypotheses of Theorem* 2.6.2 *be satisfied. Let* $y(t) = y(t, t_0, y_0)$ *and* $x(t) = x(t, t_0, y_0)$ *be the solutions to the IVPs* (2.65) *and* (2.68) *through the initial data* (t_0, y_0), *respectively. Then*

$$y(t, t_0, y_0) = x(t, t_0, y_0) + \int_{t_0}^{t} \Phi(t, s)p(s)\, ds. \tag{2.110}$$

Proof. From the assumptions of the theorem, for $t_0 \leq s \leq t$, we define a solution to the IVP (2.68) through $(s, y(s))$, as follows:

$$x(t, s, y(s, t_0, y_0)) = x(t, s, y(s)), \tag{2.111}$$

where $y(s) = y(s, t_0, y_0)$, a solution to the IVP (2.65) through (t_0, y_0). From (2.111) and the definition of the IVP, we observe that, for $s = t_0$,

$$x(t, t_0, y(t_0, t_0, y_0)) = x(t, t_0, y_0). \tag{2.112}$$

From the application of Lemma 2.6.8, the differential formula (Theorem 1.5.11), the chain rule, and Observation 2.6.9 with respect to s for $t_0 \leq s \leq t$, we have

$$\begin{aligned} d_s x(t, s, y(s)) &= \partial_{t_0} x(t, s, y(s)) + \partial_{x_0}(t, s, y(s)) \quad \text{(by Theorem 1.5.10)} \\ &= \Phi(t, s)\partial y(s) + \Phi(t, s)\, dy(s) \quad \text{[from (2.99) and (2.100)]} \\ &= \Phi(t, s)[-f(s)y(s)\, ds] \\ &\quad + \Phi(t, s)[f(s)y(s) + p(s)]\, ds \quad \text{[by (2.101) and (2.105)]} \\ &= \Phi(t, s)p(s)\, ds \quad \text{(by simplifying).} \end{aligned} \tag{2.113}$$

By integrating (2.113) on both sides with respect to s from t_0 to t, using Lemma 2.6.6, and the notations and the uniqueness (Theorem 2.6.2) of the solution to (2.68), we get

$$x(t, t, y(t, t_0, y_0)) - x(t, t_0, y(t_0)) = y(t, t_0, y_0) - x(t, t_0, y_0) = \int_{t_0}^{t} \Phi(t, s)p(s)\, ds,$$

which can be written as

$$y(t, t_0, y_0) = x(t, t_0, y_0) + \int_{t_0}^{t} \Phi(t, s)p(s)\, ds.$$

This completes the proof of the theorem. Moreover, from (2.66) and (2.72), we have

$$y(t, t_0, y_0) = \Phi(t, t_0)y_0 + \int_{t_0}^{t} \Phi(t, s)p(s)\, ds. \tag{2.114}$$

□

Observation 2.6.11

(i) From the conclusion of Theorem 2.6.5, it is obvious that Theorem 2.6.5 provides an alternative conceptual approach to the closed-form representation of the solution to (2.65). In fact, one can compare the solution expressions on the right-hand sides of (2.72) or (2.66) and (2.110) or (2.114). These expressions are exactly the same. This representation is based on the conceptual knowledge of the corresponding homogeneous differential equation.
(ii) We further observe that the solution representation of (2.65) in either (2.66) or (2.72) was derived based on the computational knowledge of the corresponding homogeneous scalar differential equation (2.23) or (2.68). In particular, the smoothness (conceptual) properties of the solution with respect to its initial data (conditions) outlined in Observation 2.6.9 were utilized.
(iii) For the validity of Theorem 2.6.5, we merely needed the existence of the solution to (2.65).

2.7 Notes and Comments

The focus of this chapter and of future chapters relate to the biological, chemical, engineering, medical, physical, and social sciences published in journals or textbooks and from related conference agendas. The authors' effort to expand on this stems from the second author's class and interdisciplinary research experience, multidisciplinary research collaborations with his mathematical, engineering, and biological science colleagues, and his classroom experiences from the past 45 years. One of the main objectives is to prepare the undergraduate students and interdisciplinary readers to meet the demands and challenges that exist in today's world.

The development of the deterministic modeling procedure in Section 2.2 is new. It is based on the elementary descriptive statistical approach and is a modified form of the classical modeling process in engineering and the sciences.

The examples, exercises, and illustrations are based on the second author's classroom experiences. The method of solving integrable differential equations is the topic of Section 2.3. The development of Illustration 2.3.1 (laminar blood flow in an artery [6, 39, 95, 113]), Illustration 2.3.2 (Weber–Fechner law [6, 18]), and Example 2.3.6 (vertical motion of rock [40]) is outlined. To further the reader's interest, Exercise 4 (Poiseuille's law [6, 22]), Exercise 7 (Loewenstein equation [18, 89]), Exercise 5 (glucose metabolism [6, 113]), and several original exercises (8–10) are provided.

An eigenvalue and eigenvector–type approach and its systematic use to solve the first-order linear scalar homogeneous differential equations are developed in Section 2.4. The concepts of fundamental, general, and particular solutions are introduced and systematically determined. Furthermore, several numerical and applied examples are presented to illustrate these concepts. In addition, several illustrations and examples, namely Illustrations 2.4.1 (photochemistry [30, 31, 113]), 2.4.2 (chemical kinetics [6, 30, 31, 86, 99, 132]), and 2.4.3 (Otto Frank model: arterial pulse–diastolic phase [21, 39, 25]), and Example 2.4.6 [6, 30, 31, 113] are constructed. Many numerical exercises, as well as the applied Exercise 5 (Stevens' power law [18, 17]) and other new exercises (6–8), are given at the end of Section 2.4.

Section 2.5 deals with the method of variation of parameters for solving first-order linear nonhomogeneous differential equations. Various numerical and applied examples are presented to illustrate the method and concepts. In addition to the classical applied Illustrations 2.5.1 (birth–death and immigration process [6, 55, 94, 104, 129]), 2.5.2 (Newton's law of cooling [6, 27, 40, 55, 107]), 2.5.3 (diffusion process [30, 31, 39, 113]), 2.5.4 (cell membrane [31, 39, 55, 113, 109, 110]), 2.5.5 (central nervous system [39, 55, 109, 111]), and 2.5.6 (Otto Frank model: arterial pulse–systolic phase [21, 25, 39]), and Example 2.5.5 (RC electric circuit [40, 113]) are outlined to meet the interests of diverse readers/students. Several numerical and applied exercises, in particular Exercise 7 (RL electric circuit [40, 113]), Exercise 8 (continuous intravenous injection glucose [21]), Exercise 9 (H. C. Burger model — the heart function test [21]), and new exercises (10–12), are given at the end of the section.

Section 2.6 outlines the basis and scope of conceptual algorithms. This material is based on Refs. 7,15,19,23,42,55,83,84 and the second author's classroom experience.

Chapter 3

First-Order Nonlinear Differential Equations

3.1 Introduction

The development of the mathematical modeling and procedures for solving first-order nonlinear scalar differential equations is the objective of this chapter. The problem of finding a solution to nonlinear scalar differential equations depends on the level of the student's knowledge about solving linear scalar differential equations and the recent ideas about advanced nonlinear differential equations.

Section 3.2 deals with the mathematical modeling of dynamic processes. Several dynamic processes in the biological, chemical, engineering, medical, physical, and social sciences are outlined to illustrate the scope of modeling. We introduce an innovative method of solving nonlinear scalar differential equations in Section 3.3. It is a five-step general algorithm for finding either an implicit or an explicit form of a solution process for a class of first-order nonlinear scalar differential equations. This problem-solving algorithm is called the energy or Lyapunov function method. By employing a nonlinear function or transformation, a nonlinear scalar differential equation is reduced to a scalar solvable differential equation. Finally, either an implicit or an explicit form of the solution to the given scalar differential equation is obtained. Section 3.4 focuses on a subclass of first-order scalar nonlinear differential equations that are directly reduced to integrable differential equations. The subclass of differential equations outlined in Section 3.4 includes the subclass of exact differential equations and the differential equations reducible to exact differential equations, as special cases. Section 3.5 deals with a subclass of first-order scalar nonlinear differential equations that are reducible to scalar linear nonhomogeneous differential equations. The scope of this method is exhibited by showing that the variable separable differential equations, the homogeneous differential equations, and the essentially time-invariant differential equations are reducible to integrable differential equations in Sections 3.6, 3.7, and 3.9, respectively. Also, the Bernoulli-type differential equations are reduced to linear nonhomogeneous scalar differential equations in Section 3.8. Moreover, the mathematical models for enzymatic reactions, acid-catalyzed hydrolysis, concurrent chemical reactions, single-species ecosystems, liquid leakage problems, etc. are detailed in Section 3.6. To note the importance of homogeneous differential equations, mathematical models of the relative growth allometry and the dynamics of swarms are also presented in Section 3.7. The Solow's economic growth model and Richard Levins' model of population extinction are exhibited in Section 3.8.

3.2 Mathematical Modeling

This section is formulated in the spirit of Section 2.2. We briefly outline the development of the deterministic mathematical modeling for dynamic processes in the biological, chemical, engineering, medical, physical, and social sciences. This development is based on a theoretical experimental setup, the fundamental laws of science and engineering, and dynamic processes (including the existing deterministic mathematical models). An attempt is made to relate the diverse and apparently different phenomena with some conceptual and/or computational frameworks and ideas which will show another model-building approach. Several illustrations, examples, and observations are presented to enhance the learning experience.

***Deterministic Modeling Procedure* 3.2.1.** First, we will present an elementary modeling procedure for developing a mathematical model. The basic ideas are given below.

Let $x(t)$ be a state of a dynamic system at time t. The state of the system is observed over an interval $[t, t+\Delta t]$ of length of magnitude Δt, with Δt being a small increment in t. Without loss of generality, we assume that Δt is positive. The dynamic process is assumed to be evolving under the controlled environment. Let $x(t_0) = x(t)$, $x(t_1), x(t_2), \ldots, x(t_k), \ldots, x(t_n) = x(t+\Delta t)$ be experimentally observed state data of a system at $t_0 = t$, $t_1 = t+\tau$, $t_2 = t+2\tau, \ldots, t_k = t + k\tau, \ldots, t_n = t+\Delta t = t+n\tau$ over the interval $[t, t+\Delta t]$, where n belongs to $\{1, 2, 3, \ldots\}$ and $\tau = \frac{\Delta t}{n}$ and $\Delta x(t_k) = x(t_k) - x(t_{k-1})$. Here, $x(t_0) = x(t) \equiv x$.

From the above discussion, the microscopically observed changes of the state of the system,

$$
\begin{aligned}
x(t_0) &= x(t),\\
x(t_1) - x(t_0) &= Z_1(t_0, x, \tau),\\
x(t_2) - x(t_1) &= Z_2(t_1, x_1, \tau),\\
&\ldots\ldots\ldots\ldots\ldots\ldots,\\
x(t_k) - x(t_{k-1}) &= Z_k(t_{k-1}, x_{k-1}, \tau),\\
&\ldots\ldots\ldots\ldots\ldots\ldots,\\
x(t_n) - x(t_{n-1}) &= Z_n(t_{n-1}, x_{n-1}, \tau),
\end{aligned}
\tag{3.1}
$$

are observed at t_0 and over n subintervals

$$[t_0, t_1], [t_2, t_3], \ldots, [t_{k-1}, t_k], \ldots, [t_{n-1}, t_n]$$

of the given interval $[t, t + \triangle t]$ of length $\triangle t$.

By following the given argument used in Section 2.2, the final state

$$x(t + \triangle t) = x(t_n)$$

of the system is given by

$$\begin{aligned} x(t + \triangle t) - x(t) &= \sum_{k=1}^{n} Z_k(t_{k-1}, x_{k-1}, \tau) \\ &= \frac{\triangle t}{\tau}\left[\frac{\sum_{k=1}^{n} Z_k(t_{k-1}, x_{k-1}, \tau)}{n}\right] \quad \text{(from } \tau \text{ and } n\text{)} \\ &= \frac{S_n}{\tau}\triangle t \quad \text{(by notation)}, \end{aligned} \tag{3.2}$$

where

$$S_n = \frac{1}{n}\left[\sum_{k=1}^{n} Z_k(t_{k-1}, x_{k-1}, \tau)\right].$$

S_n is the sample average of the state aggregate increment data. Under the assumption of smallness of τ, a large n with $\tau n = \triangle t$, and the nature of the system [finiteness of $x(t + \triangle t) - x(t) = x(t_n) - x(t_0)$], we have

$$\lim_{\tau \to 0^+}\left[\frac{S_n}{\tau}\right] = C(t, x).$$

From this discussion, (3.1)–(3.2), and imitating the argument used in Section 2.2, we have

$$x(t + \triangle t) - x(t) \approx C(t, x)\triangle t.$$

This yields the differential equation

$$dx = C(t, x)dt \quad \text{(from the concept of the differential)}, \tag{3.3}$$

where $C(t, x)$ is the rate coefficient. The differential equation (3.3) is a mathematical model of the dynamic process under a controlled environment.

Next, we present a few illustrations and examples of mathematical modeling of dynamic processes in the chemical, biological, engineering, medical, physical, and social sciences.

Illustration 3.2.1 (chain reaction: Rice–Herzfeld mechanism [30, 31, 86, 99]). A large number of organic chemical reactions (particularly the decomposition types) have simple orders. They are like elementary processes. During the process of decomposition, (a) the presence of organic free radicals are observed, and (b) the dissociation of a carbon–carbon (C–C) bond requires more than 80 kcal mole^{-1}, but the activation energy for most of the reactions requires much less than this. These contradictory observations have been explained by the Rice–Herzfeld mechanism:

Initiation process. In this process, free radicals are formed from normal reactant molecules. This is caused by either chemical reaction or thermal decomposition/ absorption of radiation. We now exhibit the decomposition of the molecule M into two radicals by the dissociation of a C–C bond:

$$M \xrightarrow{k_1} {}^{\cdot}R_1 + {}^{\cdot}R_1'.$$

Propagation process. In this process, the reactant molecules are converted into product molecules without any change in the total number of free radicals. However, there could be a change in the type of free radical molecule. In the current illustration, the radical ${}^{\cdot}R_1$ initiates the chain, but the ${}^{\cdot}R_1'$ is not involved in the propagation:

$$ {}^{\cdot}R_1 + M \xrightarrow{k_2} {}^{\cdot}R_2 + R_1H,$$

$$ {}^{\cdot}R_2 \xrightarrow{k_3} {}^{\cdot}R_1 + M'.$$

In the reaction k_2, ${}^{\cdot}R_1$ reacts with M and extracts a hydrogen atom and generates the radical ${}^{\cdot}R_2$. In the reaction k_3, ${}^{\cdot}R_2$ breaks at a C–C bond and regenerates the radical ${}^{\cdot}R_1$.

Termination process. In this final process, the generated free radicals recombine to yield normal molecules. Either the regenerated ${}^{\cdot}R_1$ in the propagation process initiates a new sequence or the chain termination takes place:

$$ {}^{\cdot}R_1 + {}^{\cdot}R_1 \xrightarrow{k_{4a}} P_1,$$

$$ {}^{\cdot}R_2 + {}^{\cdot}R_2 \xrightarrow{k_{4b}} P_2,$$

$$ {}^{\cdot}R_1 + {}^{\cdot}R_2 \xrightarrow{k_{4c}} P_3.$$

In the present discussion, we consider the molecular decomposition processes that lead to the first termination process. In this case, we note that the radical ${}^{\cdot}R_1$ propagates the chain by a bimolecular process, and ${}^{\cdot}R_2$ is assumed to be of the first-order.

By utilizing the law of mass action, we find the mathematical description of the above single-species chemical reaction system. For this purpose, let us denote the concentration of a substance S by $[S]$ (the number of moles of S per unit volume $= [S]$) and the instantaneous rate of change of concentration (velocity) by $\frac{d}{dt}[S]$:

Step 1. Find $\frac{d}{dt}[M]$. The rate of change of $\frac{d}{dt}[M]$ is determined by

$$M \xrightarrow{k_1} {}^{\cdot}R_1 + {}^{\cdot}R_1', \quad {}^{\cdot}R_1 + M \xrightarrow{k_2} {}^{\cdot}R_2 + R_1H.$$

M is decomposed by these two processes, and its net rate of decomposition is given by

$$-\frac{d}{dt}[M] = k_1[M] + k_2[M]\,[{}^{\cdot}R_1] = [M]\,(k_1 + k_2\,[{}^{\cdot}R_1]).$$

Step 2. Find $\frac{d}{dt}[{}^{\cdot}R_1]$. The rate of change of $\frac{d}{dt}[{}^{\cdot}R_1]$ is determined by

$$M \xrightarrow{k_1} {}^{\cdot}R_1 + {}^{\cdot}R_1',$$

$${}^{\cdot}R_1 + M \xrightarrow{k_2} {}^{\cdot}R_2 + R_1H,$$

$${}^{\cdot}R_2 \xrightarrow{k_3} {}^{\cdot}R_1 + M',$$

$${}^{\cdot}R_1 + {}^{\cdot}R_1 \xrightarrow{k_{4a}} P_1.$$

${}^{\cdot}R_1$ is actually formed by two processes: $M \xrightarrow{k_1} {}^{\cdot}R_1 + {}^{\cdot}R_1'$ and ${}^{\cdot}R_2 \xrightarrow{k_3} {}^{\cdot}R_1 + M'$. Therefore, the rate of formation of ${}^{\cdot}R_1$ is

$$\frac{d}{dt}[{}^{\cdot}R_1] = k_1[M] + k_3\,[{}^{\cdot}R_2],$$

and ${}^{\cdot}R_1$ is decomposed by two processes: ${}^{\cdot}R_1 + M \xrightarrow{k_2} {}^{\cdot}R_2 + R_1H$ and ${}^{\cdot}R_1 + {}^{\cdot}R_1 \xrightarrow{k_{4a}} P_1$. Therefore, the rate of decomposition of ${}^{\cdot}R_1$ is

$$-\frac{d}{dt}[{}^{\cdot}R_1] = k_2\,[{}^{\cdot}R_1]\,[M] + k_{4a}\,[{}^{\cdot}R_1]^2.$$

Thus, the net rate of change of ${}^{\cdot}R_1$ is

$$\frac{d}{dt}[{}^{\cdot}R_1] = k_1[M] + k_3\,[{}^{\cdot}R_2] - k_2\,[{}^{\cdot}R_1]\,[M] - k_{4a}\,[{}^{\cdot}R_1]^2.$$

Step 3. Find $\frac{d}{dt}[{}^{\cdot}R_2]$. The rate of change of $\frac{d}{dt}[{}^{\cdot}R_2]$ is determined by

$${}^{\cdot}R_1 + M \xrightarrow{k_2} {}^{\cdot}R_2 + R_1H \text{ and } {}^{\cdot}R_2 \xrightarrow{k_3} {}^{\cdot}R_1 + M'.$$

${}^{\cdot}R_2$ is formed by one process: ${}^{\cdot}R_1 + M \xrightarrow{k_2} {}^{\cdot}R_2 + R_1H$. Therefore, the rate of formation of ${}^{\cdot}R_2$ is

$$\frac{d}{dt}[{}^{\cdot}R_2] = k_2\,[{}^{\cdot}R_1]\,[M],$$

and ${}^{\cdot}R_2$ is decomposed by one process: ${}^{\cdot}R_2 \xrightarrow{k_3} {}^{\cdot}R_1 + M'$. Therefore, the rate of decomposition of ${}^{\cdot}R_2$ is

$$-\frac{d}{dt}[{}^{\cdot}R_2] = k_3\,[{}^{\cdot}R_2].$$

Thus, the net rate of change of ${}^{\cdot}R_2$ is

$$\frac{d}{dt}[{}^{\cdot}R_2] = k_2\,[{}^{\cdot}R_1]\,[M] - k_3\,[{}^{\cdot}R_2].$$

The mathematical description of the Rice–Herzfeld decomposition [99] of the single chemical species is summarized by

$$\begin{aligned}
\frac{d}{dt}[M] &= -[M]\,(k_1 + k_2\,[{}^{\cdot}R_1]),\\
\frac{d}{dt}\,[{}^{\cdot}R_1] &= k_1[M] + k_3\,[{}^{\cdot}R_2] - k_2\,[{}^{\cdot}R_1]\,[M] - k_{4a}\,[{}^{\cdot}R_1]^2,\\
\frac{d}{dt}\,[{}^{\cdot}R_2] &= k_2\,[{}^{\cdot}R_1]\,[M] - k_3\,[{}^{\cdot}R_2].
\end{aligned} \tag{3.4}$$

This is a three-dimensional system of nonlinear differential equations. In order to compare the mathematical model with the existing one, we need to use the standard stationary state (a concept introduced by Bodenstein in 1913) approximation scheme (referred to as the stationary state hypothesis [86, 99, 125]) to determine the concentrations of $[{}^{\cdot}R_1]$ and $[{}^{\cdot}R_2]$. For this purpose, we set

$$\frac{d}{dt}\,[{}^{\cdot}R_1] = 0 = \frac{d}{dt}\,[{}^{\cdot}R_2],$$

and solve for $[{}^{\cdot}R_1]$ and $[{}^{\cdot}R_2]$. Hence,

$$\begin{aligned}
k_1[M] + k_3\,[{}^{\cdot}R_2] - k_2\,[{}^{\cdot}R_1]\,[M] - k_{4a}\,[{}^{\cdot}R_1]^2 &= 0,\\
k_2\,[{}^{\cdot}R_1]\,[M] - k_3\,[{}^{\cdot}R_2] &= 0.
\end{aligned}$$

By adding these algebraic equations, we have

$$k_1[M] - k_{4a}\,[{}^{\cdot}R_1]^2 = 0,$$

which implies that $k_{4a}\,[{}^{\cdot}R_1]^2 = k_1[M]$, and hence

$$[{}^{\cdot}R_1] = \left(\frac{k_1[M]}{k_{4a}}\right)^{\frac{1}{2}}.$$

We substitute this in the first equation in (3.4), and obtain

$$\begin{aligned}
d[M] &= -[M]\left(k_1 + k_2\left(\frac{k_1[M]}{k_{4a}}\right)^{\frac{1}{2}}\right)dt\\
&= -\left[k_1[M] + \frac{k_2 k_1}{\sqrt{k_1 k_{4a}}}[M]^{\frac{3}{2}}\right]dt \quad \text{(by algebraic simplification)}.
\end{aligned} \tag{3.5}$$

Example 3.2.1 ([86, 99, 132]). The Rice–Herzfeld mechanism of a chain reaction is exhibited by the thermal decomposition of acetaldehyde (CH_3CHO). The thermal decomposition of acetaldehyde products are two free radicals, namely $CH_3^{\cdot}$ and

$CHO^{\cdot}$ (chain carriers). In this case the three terminal processes are

$$\begin{aligned}
&\text{Initiation:} && CH_3CHO \xrightarrow{k_1} CH_3^{\cdot} + CHO^{\cdot};\\
&\text{Propagation:} && CH_3^{\cdot} + CH_3CHO \xrightarrow{k_2} CH_3^{\cdot} + CH_4 + CO;\\
&\text{Termination:} && CH_3^{\cdot} + CH_3^{\cdot} \xrightarrow{k_{4a}} C_2H_6,\\
& && CHO^{\cdot} + CHO^{\cdot} \xrightarrow{k_{4b}} CH_4 + CO,
\end{aligned}$$

and the overall reaction is

$$CH_3CHO \rightarrow CH_4 + CO,$$

with traces of C_2H_6 and H_2. The rate of reaction is determined at the propagation stage of the reaction. Our main goal is to find a mathematical model for the dynamic of the overall reaction rate. For this purpose, we introduce the comparable notations and follow the same reasoning outlined in Illustration 3.2.1.

We set $[M] = [CH_3CHO]$, $[{}^{\cdot}R_1] = [CH_3^{\cdot}]$, $[{}^{\cdot}R_1'] = [CHO^{\cdot}]$, $[{}^{\cdot}R_2] = [CH_4] + [CO]$, $[P_1] = [C_2H_6]$, $[P_2] = [CH_4] + [CO]$, and employing the argument used in Illustration 3.2.1 (under first terminal process), the system of differential equations (3.4) reduces to

$$\begin{aligned}
&\frac{d}{dt}[M] = -[M]\,(k_1 + k_2\,[{}^{\cdot}R_1]),\\
&\frac{d}{dt}[{}^{\cdot}R_1] = k_1[M] - k_2\,[{}^{\cdot}R_1]\,[M] + k_2\,[{}^{\cdot}R_1]\,[M] - k_{4a}\,[{}^{\cdot}R_1]^2,\\
&\frac{d}{dt}[{}^{\cdot}R_1'] = k_1[M] - k_{4b}\,[{}^{\cdot}R_1']^2,\\
&\frac{d}{dt}[{}^{\cdot}R_2] = k_2[M]\,[{}^{\cdot}R_1].
\end{aligned}$$

Our main goal is to derive a system of differential equations that describes the above reaction. Therefore, we only need to focus our attention on the first, second, and fourth differential equations in the above-presented list. Hence, we have

$$\begin{aligned}
&\frac{d}{dt}[M] = -[M]\,(k_1 + k_2\,[{}^{\cdot}R_1]),\\
&\frac{d}{dt}[{}^{\cdot}R_1] = k_1[M] - k_{4a}\,[{}^{\cdot}R_1]^2,\\
&\frac{d}{dt}[{}^{\cdot}R_2] = k_2[M]\,[{}^{\cdot}R_1].
\end{aligned}$$

By applying the steady state hypothesis to the second differential equations, we have

$$k_1[M] - k_{4a}\,[{}^{\cdot}R_1]^2 = 0,$$

and we solve for $[\dot{}R_1]$. The algebraic solution is

$$[\dot{}R_1] = \left(\frac{k_1}{k_{4a}}[M]\right)^{\frac{1}{2}} = \left(\frac{k_1}{k_{4a}}\right)^{\frac{1}{2}} [M]^{\frac{1}{2}}.$$

The rates of formation of the product and decomposition of the reactant are described by

$$d[M] = -[M]\left(k_1 + k_2\sqrt{\frac{k_1}{k_{4a}}}\,[M]^{\frac{1}{2}}\right) dt,$$

$$d[\dot{}R_2] = k_2\sqrt{\frac{k_1}{k_{4a}}}\,[M]^{\frac{3}{2}}\, dt.$$

The remaining comments can be made directly by using Illustration 3.2.1.

Illustration 3.2.2 (enzymatic reaction process [38, 86, 99, 125, 132]). The basic pieces for building inanimate or animate objects are atoms and molecules. Each elementary chemical reaction in living cells generates the change in one, two, or at most a few molecules at a time. In particular, exergonic (energy-yielding) and endergonic (energy-requiring) chemical reaction processes play significant roles in the cell growth and reproduction processes. Of course, it takes energy to build a large molecule from small (elementary) molecules. For example, the work has to be done to build complex structures of proteins, nucleic acids, cell membranes, etc., from basic chemical elements. The living cells have a unique ability to utilize these types of chemical reactions in their environment. The activities of living organisms are governed by molecules that are formed by enzymes, genes, chromosomes, etc. In particular, the growth processes in biological sciences are highly complex chemical reaction processes.

Based on the Briggs–Haldane steady state approach (1925 [125]), we briefly outline a very simple enzymatic reaction process. First, we recall some basic chemical terms. A phenomenon is called catalysis occurs when one of the participating substances in the chemical reaction: (c_1) either increases or decreases the chemical reaction, (c_2) keeps its concentration constant in the chemical reaction, (c_3) cannot alter the enthalpy change, the free energy change, or the equilibrium constant of reaction, and (c_4) is called either an accelerator or an inhibitor. The participating substance that obeys the above-cited four conditions (c_1, c_2, c_3, c_4) is called a catalyst. Moreover, when one of the products or participating substances plays the role of a catalyst, the product or the participating substance (reactant) is called an autocatalyst.

Now we introduce a few notations and a brief description about an enzymatic reaction process. Let us denote a reactant (also called a substrate) and an enzyme by S and E, respectively. The role of the enzyme as a catalyst is played by the existence of an intermediate substance known as a enzyme–substrate complex, ES. The substrate ES is the binding of S with E at the substrate binding site (the active site or catalytic site). This anchored substrate ES undergoes its dissociation and formation process of E and P. More precisely, the enzymatic reaction process (ERP) is based on the following well-known assumptions:

ERP 1. The enzyme is a catalyst (Berzelius, 1835–1838).

ERP 2. The formation of the enzyme–substrate complex is very rapid (A. J. Brown, 1902).

ERP 3. Only a single substrate and a single enzyme–substrate complex are enveloped. The enzyme–substrate complex is decomposed into a free enzyme and product.

ERP 4. Shortly after the start of the reaction, the enzyme–substrate complex will build up to a near-constant or "steady state" (a concept introduced by Bodenstein, 1913). This means that after the initial presteady period, the formation and the decomposition rates of the enzyme–substrate complex are the same (B. E. Briggs and J. B. S. Haldane, 1925).

ERP 5. The formation of the enzyme–substrate complex does not change the concentration of the substrate (the number of moles of S per unit volume $= [S]$). This means that the substrate concentration is much larger than the concentration of the enzyme.

ERP 6. The overall rate of the reaction is limited by the dissociation of the enzyme–substrate complex to form a free enzyme and product.

ERP 7. The reaction rate is measured during the early stage of the reaction, so that the reverse reaction is insignificant.

The chemical mechanism of the single enzyme and the single substrate reaction is described by

$$E + S \rightleftarrows ES \rightarrow E + P,$$
$$E + P \rightarrow ES \quad \text{(neglected due to assumption ERP 7)}.$$

For the derivation of the Michaelis–Menten rate equation, we need to introduce further notations:

Let $[S]$, $[E]$, $[ES]$, and $[P]$ be concentrations of the substrate (S), enzyme (E), enzyme–substrate complex (ES), and product (P), respectively. Let v be an initial rate of change of concentration, the instantaneous rate of change of concentration (velocity), $\frac{d}{dt}[P]$ (or $-\frac{d}{dt}[S]$) of the given substrate. k_p stands for the rate constant for the breakdown of ES to $E + P$. k_1 and k_{-1} stand for the rate constants for the formation of ES from $E + S$ and the dissociation of ES to $E + S$, respectively.

To find the steady state and the reaction rate of the above enzymatic reaction, as in Illustration 3.2.1, we need to derive a system of differential equations with regard to $[E]$, $[ES]$, $[S]$, and $[P]$. For this purpose, we consider the following:

Step 1. Find $\frac{d}{dt}[E]$. The rate of change of $\frac{d}{dt}[E]$ is determined by

$$ES \to E+S,\ ES \to E+P,\ E+S \to ES.$$

E is formed by two processes: $ES \to E+S$ and $ES \to E+P$. Therefore, the rate of formation of E is

$$\frac{d}{dt}[E] = k_{-1}[ES] + k_p[ES] = (k_{-1}+k_p)[ES],$$

and E is anchored by one process: $E+S \to ES$. Therefore, the binding rate of E is

$$-\frac{d}{dt}[E] = k_1[S][E].$$

Thus, the net rate of change of E is

$$\frac{d}{dt}[E] = -k_1[S][E] + (k_{-1}+k_p)[ES]. \tag{3.6}$$

Step 2. Find $\frac{d}{dt}[ES]$. The rate of change of $\frac{d}{dt}[ES]$ is determined by

$$E+S \to ES,\ E+P \to ES,\ ES \to E+S \text{ and } ES \to E+P.$$

ES is really formed by one process, $E+S \to ES$, and $E+P \to ES$ is negligible due to assumption ERP 7. Therefore, the rate of formation of ES is

$$\frac{d}{dt}[ES] = k_1[S][E],$$

and ES is decomposed by two processes: $ES \to E+S$ and $ES \to E+P$. Therefore, the rate of decomposition of ES is

$$-\frac{d}{dt}[ES] = k_{-1}[ES] + k_p[ES].$$

Thus, the net rate of change of ES is

$$\frac{d}{dt}[ES] = k_1[S][E] - (k_{-1}+k_p)[ES]. \tag{3.7}$$

Step 3. Find $\frac{d}{dt}[S]$. The rate of change of $\frac{d}{dt}[S]$ is determined by

$$ES \to E+S \quad \text{and} \quad E+S \to ES.$$

S is formed by one process: $ES \to E+S$. Therefore, the rate of formation of S is

$$\frac{d}{dt}[S] = k_{-1}[ES],$$

and S is decomposed by one process: $E + S \to ES$. Therefore, the rate of decomposition of S is

$$-\frac{d}{dt}[S] = k_1[S][E].$$

Thus, the net rate of change of S is

$$\frac{d}{dt}[S] = -k_1[S][E] + k_{-1}[ES]. \tag{3.8}$$

Step 4. Find $\frac{d}{dt}[P]$. The rate of change of $\frac{d}{dt}[P]$ rate determined by

$$ES \to E + P.$$

P is formed by one process: $ES \to E+P$. Therefore, the rate of formation of P is

$$\frac{d}{dt}[P] = k_p[ES]. \tag{3.9}$$

This is the net rate of change of P.

The mathematical model of a single-species and single enzymatic reaction process is summarized by

$$\begin{aligned}
\frac{d}{dt}[E] &= -k_1[S][E] + (k_{-1} + k_p)[ES], \\
\frac{d}{dt}[ES] &= k_1[S][E] - (k_{-1} + k_p)[ES], \\
\frac{d}{dt}[S] &= -k_1[S][E] + k_{-1}[ES], \\
\frac{d}{dt}[P] &= k_p[ES].
\end{aligned} \tag{3.10}$$

The steady state of the above enzymatic reaction is determined by the concentrations of the intermediate substrate $[ES]$ and $[E]$. This means that

$$\frac{d}{dt}[ES] = 0 = \frac{d}{dt}[E].$$

From this consideration, (3.6), and (3.7), we have

$$\begin{aligned}
-k_1[S][E] + (k_{-1} + k_p)[ES] &= 0, \\
k_1[S][E] - (k_{-1} + k_p)[ES] &= 0.
\end{aligned}$$

By solving these equations, we obtain

$$k_1[S][E] = (k_{-1} + k_p)[ES]$$

and hence

$$[ES] = \frac{k_1[S]}{k_{-1} + k_p}[E]. \tag{3.11}$$

From assumption ERP 7 and the total enzyme concentration, $[E^*]$ at the steady state of the chemical reaction is a sum:

$$[E^*] = [E] + [ES] \quad \text{(by the conservation of sites).} \tag{3.12}$$

From (3.11) and (3.12), we have

$$[E^*] - [ES] = [E] = \frac{[ES]\,(k_{-1} + k_p)}{k_1[S]},$$

which implies that

$$\frac{([E^*] - [ES])[S]}{[ES]} = \frac{k_{-1} + k_p}{k_1} = K_m \quad \text{(by algebraic simplification).} \tag{3.13}$$

Then K_m can be called the steady state constant instead of the equilibrium constant. The difference between the steady state (K_m) and the equilibrium state constant ($\frac{k_{-1}}{k_1}$) is that $\frac{k_p}{k_1}$ vanishes when $k_p \ll k_{-1}$.

By dividing both sides of the initial rate of change of concentration in (3.9) by $[E^*]$, we have

$$\begin{aligned}
\frac{v}{[E^*]} &= \frac{k_p[ES]}{[E^*]} \\
&= \frac{k_p[ES]}{[E] + [ES]} \quad \text{[from (3.12)]} \\
&= \frac{k_p \frac{k_1[S]}{k_{-1}+k_p}[E]}{[E] + \frac{k_1[S]}{k_{-1}+k_p}[E]} \quad \text{[from (3.11)]} \\
&= \frac{k_p \frac{k_1[S]}{k_{-1}+k_p}}{1 + \frac{k_1[S]}{k_{-1}+k_p}} \quad \text{(by the cancelation law)} \\
&= \frac{\frac{k_p k_1[S]}{k_{-1}+k_p}}{\frac{(k_{-1}+k_p)+k_1[S]}{k_{-1}+k_p}} \quad \text{(by simplifying)} \\
&= \frac{k_p k_1[S]}{(k_{-1} + k_p) + k_1[S]} \quad \text{(by simplifying)} \\
&= \frac{k_p[S]}{\frac{k_{-1}+k_p}{k_1} + [S]} \quad \text{(by simplifying)} \\
&= \frac{k_p[S]}{K_m + [S]} \quad \text{[from (3.13)].}
\end{aligned}$$

Hence,

$$v = \frac{k_p\,[E^*]\,[S]}{K_m + [S]}. \tag{3.14}$$

This is the Michaelis–Menten rate equation under the Briggs–Haldane approach [125]. For $[E^*] = [ES]$, from (3.9), $v_{\max} = k_p[E^*]$. Under this change, (3.14) reduces to

$$v = \frac{v_{\max}[S]}{K_m + [S]} = \frac{v_{\max}[S]}{K_m + [S]}. \tag{3.15}$$

This is the Henry–Michaelis–Menten rate equation under the Briggs–Haldane approach.

From (3.9), (3.14), and (3.15), we conclude that

$$\frac{d}{dt}[P] = \frac{k_p[E^*][S]}{K_m + [S]}, \tag{3.16}$$

$$\frac{d}{dt}[P] = \frac{v_{\max}[S]}{K_m + [S]}. \tag{3.17}$$

On the other hand,

$$\frac{d}{dt}[P] = -\frac{d}{dt}[S], \tag{3.18}$$

where $[P] = [S_0] - [S]$, and $[S_0]$ stands for the initial concentration of the substrate. This suggests that the computation of the product $[P]$ is done by calculating the substrate $[S]$ at a time t. Therefore, from (3.16), (3.17), and (3.18), we have

$$d[S] = -\frac{k_p[E^*][S]}{K_m + [S]}\,dt, \quad [S(t_0)] = [S_0], \tag{3.19}$$

$$d[S] = -\frac{v_{\max}[S]}{K_m + [S]}\,dt, \quad [S(t_0)] = [S_0]. \tag{3.20}$$

Illustration 3.2.3 (ecological process [55, 104]). The biological processes have a natural tendency to maintain their steady (equilibrium) states, or the gradual adjustment to their states under environmental changes. The birth and death processes in the biological systems play a significant role. For the development of a mathematical model of single-species ecological processes (SEPs) in biological sciences, we make the following assumptions:

SEP 1. All the members of the given single-organism/species community are capable of reproducing their offspring. However, in the case of bisexual species, only females are able to reproduce their offspring with the assumption that there is no shortage of males. Let $N(t)$ be the size of the species/organisms in the community at a time t.

SEP 2. The birth and death processes are independent of age. The intrinsic rate of natural increase is: [birth rate − death rate]. It is denoted by $\alpha = \beta - \delta$, where β and δ are the birth and death rates per member of the community.

SEP 3. From SEP 1 and SEP 2, one can formulate a deterministic mathematical model of a single-organism/species community (Thomas Malthus, 1798):

$$dN = \alpha N dt, \quad N(t_0) = N_0. \tag{3.21}$$

The growth of the species can be treated as unrestricted exponential growth for $\alpha > 0$, or purely decaying exponential decay for $\alpha < 0$. By applying the method of finding the solution procedure in Subsection 2.4.4, we have, $N(t) = \exp\left[\alpha(t-t_0)\right] N_0$.

SEP 4. The model in (3.21) is the Malthusian growth model. This suggests that the intraspecies effects are ignored. This is feasible provided that the population size of the community is small and the resources are unlimited. However, the growth of population is restricted due to finite resources. Therefore, it is reasonable to think that the crowding affects the intrinsic rate. Unrestricted growth causes increase in demand for food/space and other basic resources. Hence, it leads to decreased fecundity or even starvation/death. To overcome this situation, one can formulate the following mathematical model:

$$dN = \alpha(\kappa - N)dt, \quad N(t_0) = N_0, \tag{3.22}$$

where $\kappa > 0$. This represents the saturation level of the population and is called the carrying capacity of the environment (α is defined in SEP 3). At the beginning, we noted that the growth rate of N is almost linear. By applying the method of finding the solution procedure in Subsection 2.5.4, we have $N(t) = \kappa + \exp\left[-\alpha(t-t_0)\right](\kappa - N_0)$.

SEP 5. In order to obtain a biologically more suitable deterministic mathematical model, we combine the two modeling approaches described in SEP 3 and SEP 4. This leads to the well-known deterministic mathematical model

$$dN = \alpha N(\kappa - N)dt, \quad N(t_0) = N_0, \tag{3.23}$$

where α and κ are as defined in SEP 4. The basic idea about the formulation of this model is based on the law of mass action. Here, N is the size of the population at time t, and $\kappa - N$ characterizes the amount of resources available for generating the growth of the population at time t. The rate of change in the population is determined by the product of these quantities. This differential equation is termed the Pearl–Verhulst logistic equation. The solution process of (3.23) is $\kappa N_0(N_0 + (\kappa - N_0)\exp[-\alpha\kappa(t-t_0)])^{-1}$. For details see Example 3.6.2.

Example 3.2.2 (population genetics [34, 49, 118, 119, 121]). The genetic population mathematical models provide knowledge about the microscopic scale of population dynamics. In a single-species process, the number of different possible genotypes ("genotype" refers to the genetic makeup of an individual determined by the totality of genetic factors) is far greater than the number of individuals in the single-species community. Each member's genotype is determined by a large number of genes. Many of the genes have more than one possible allele ("allele" refers to a particular form of gene). In genetic population dynamics (GPD), the genes in the same allele (form) are capable of producing their own type of genes except when mutation occurs. It is assumed [34, 49, 119] that:

GPD 1. At one particular locus (the location of a gene on a chromosome) for which there are two possible alleles in the population.

GPD 2. The frequency (relative proportion) of an allele in the gene population changes due to Wright's systematic evolutionary pressures.

GPD 3. By Wright (1949), the first source consists of three "systematic evolutionary pressures": (a) mutation — influence of the presence of two alleles; (b) migration — exchange of individuals with different genetic structures; (c) selection — density-dependent effects of the progeny due to differences in survival, mating, and fertility.

Now, we introduce a few notations about the above-described gene population system. Let A_1 and A_2 be alleles at one locus in the population. Let the generations be overlapping in continuous time t. Let x be the frequency of the A_1 allele. Let s be the selective advantage of A_1 over A_2. Overall, the genes are not neutral. The frequency x satisfies the following logistic differential equation:

$$dx = sx(1 - x)dt, \quad x(t_0) = x_0. \tag{3.24}$$

Illustration 3.2.4 (epidemiological process [5, 55, 136]). Most epidemiological phenomena are very complex. Mathematical models provide information about the communicable diseases to the various health departments for planning and decision-making processes.

For simplicity, we assume that the population is divided into two disjoint groups of individuals, and the spread of the infection is presumed to satisfy the following conditions; [55, 136]

EDP 1 (infective and susceptible groups). The infective group, i.e. those individuals who are capable of transmitting the disease to others. The infective population size is denoted by I. The susceptible group, i.e. those individuals who are not infective but who are capable of contracting the disease and becoming infective. The susceptible population size is denoted by S.

EDP 2. A single infected individual is introduced into a population of equally susceptible individuals. Through contact, the disease will spread.

EDP 3. An individual that is once infected will remain this way during the process. There is no removal.

EDP 4. The total population size is constant. Hence, from EDP 1, $N = S + I$, where N is the total size of the population.

EDP 5. From EDP 2, we can say that the larger the population (infective/susceptible), the greater the number of interactions in the population. Therefore, the growth rate of the number of infectives I is proportional to I as well as $S = N - I$. This leads to the following differential equation:

$$dI = \alpha I(N - I), \quad I(0) = 1, \tag{3.25}$$

where α is a positive constant of proportionality.

EDP 6. We note that the mathematical model in EDP 5 is purely a deterministic one. This model makes sense provided that assumptions EDP 2 and EDP 4 are strictly feasible. If the total population is uniformly distributed in the community/region (well-mixed/homogeneous), then assumption EDP 2 is valid. If the infected community is completely closed to other communities, and if the measurement system of counting $I(t)$ is perfect, then EDP 4 is feasible.

Illustration 3.2.5 (economic growth model: Solow [130]). An age-old economic problem is the relationship between capital accumulation and employment. In particular, under what conditions is the growth in the income with continuous full employment possible, and are these conditions sufficient to maintain the stability of the economy? The answers to these questions are provided by the neoclassical aggregate growth model of Robert M. Solow [130, 131]. This model economy considers a single composite commodity. This commodity can either be consumed currently or accumulate as capital. The accumulated capital and the supply of homogeneous labor are the inputs to the production output. Here, the output is considered to be the net output after any capital depreciation. The Harrod–Domar consistency conditions [28, 130] are as follows:

EGM 1. The labor force is a constant fraction of the population. The per capita growth rate of the population is constant. It is independent of any economic forces.

EGM 2. Part of the instant output is consumed, and the remainder is saved and invested. It is assumed that the rate of saving per unit of the net output is constant. This constant reflects a behavior parameter in any capitalist economy.

EGM 3. The technology of the model economy is described by two constant coefficients. These constants are the number of workers per unit output and the amount of capital per unit output. These are fixed constant coefficients.

Let Y, K, and L denote the net production output, the accumulated capital, and the supply of homogeneous labor to the single composite commodity macroeconomics system, respectively. From assumption EGM 1, we have

$$\frac{1}{L}\frac{d}{dt}L = n = \frac{1}{\alpha L}\frac{d}{dt}(\alpha L),$$

where α is any nonzero constant $(0 < \alpha \leq 1)$, αL stands for the available supply of labor. n is constant, and it stands for the per capita growth rate of the population or labor force. The presented expression exhibits the labor force as a constant fraction of the population that grows at the same growth rate. Furthermore, from the above relation, we have

$$\frac{d}{dt}L = nL. \tag{3.26}$$

From EGM 2, we obtain

$$\frac{\text{rate of saving}}{\text{number of units of net output}} = \frac{1}{Y}\frac{d}{dt}K = s,$$

which implies that

$$\frac{d}{dt}K = sY, \tag{3.27}$$

where s is constant and stands for the rate of saving per unit of the net output. From EGM 3, we have

$$\begin{aligned} \frac{\text{number of workers}}{\text{number of units of net output}} &= \frac{L}{Y} = a, \\ \frac{K}{Y} = k &= \frac{\text{amount of capital}}{\text{number of units of net output}}, \end{aligned} \tag{3.28}$$

where a and k are fixed constants that are independent of time. Furthermore, we consider

$$\frac{\text{amount of capital}}{\text{number of workers}} = \frac{K}{L} = \frac{\frac{K}{Y}}{\frac{L}{Y}} = \frac{k}{a} \quad \text{[from (3.28)]}.$$

This shows that

$$K = \frac{k}{a}L, \quad \frac{d}{dt}K = \frac{k}{a}\frac{d}{dt}L. \tag{3.29}$$

Now, we consider

$$\begin{aligned} \frac{s}{k} &= \frac{\frac{\frac{d}{dt}K}{Y}}{\frac{K}{Y}} \quad \text{[from (3.27) and (3.28)]} \\ &= \frac{\frac{d}{dt}K}{K} \quad \text{(by simplification)} \\ &= \frac{\frac{k}{a}\frac{d}{dt}L}{\frac{k}{a}L} \quad \text{[from (3.29)]} \\ &= \frac{\frac{d}{dt}L}{L} \quad \text{(by simplification)} \\ &= \frac{nL}{L} \quad \text{[from (3.26)]} \\ &= n \quad \text{(by simplification)}. \end{aligned} \tag{3.30}$$

The answers to the proposed questions are given by Harrod and Domar. In fact, (3.30) is a necessary and sufficient condition for steady state economic growth. Moreover, $s = nk$. This means that the saving rate (s) is the product of the capital/output ratio and the rate of growth of the labor force.

In general, if the constants s, n, and k are given, and are independent, then the validity of their relationship in (3.30) is just an accident. Furthermore, if k is a variable, it is possible that k can take the value $\frac{s}{n}$. This idea has led to the modification of the above Harrod–Domar model (3.30). This modification was introduced by Solow. Solow's model assumes that all the assumptions of Harrod and Domar (EGM 1 and EGM 2) are satisfied except for the crucial assumption EGM 3 [the constancy of the capital/output ratio k]. This assumption is replaced by the following:

EGM 4. It is assumed that the economic production is subject to the constant returns to scale in the two homogeneous factors of production, namely capital and labor. This means that production possibilities are described by a production relation that shows that the net output per worker is uniquely determined for each capital per worker. Moreover, it means that if both capital and labor are doubled or reduced by one-half, then the net production will be doubled or reduced by one-half. In short, the neoclassical aggregate model in its simplest version assumes that the production function is homogeneous of degree 1, and it exhibits an unlimited substitutability property between capital and labor (the constant-returns-to-scale production function).

Let f be a production function of a single composite commodity. We assume that it satisfies assumption EGM 4. We recall the unlimited substitutability. This means that in order to produce any given output from zero (excluded/included) to infinity, the capital per worker must be defined. Mathematically speaking, the interval $(0, \infty)/[0, \infty)$ is a subset of the range of the production function f. Moreover, from EGM 4, we have

$$Y = f(K, L) = f\left(L\frac{K}{L}, L\right) = Lf\left(\frac{K}{L}, 1\right). \tag{3.31}$$

From (3.31), we obtain

$$\frac{Y}{L} = f\left(\frac{K}{L}, L\right) = f(r, 1), \tag{3.32}$$

where $r = \frac{K}{L}$ is the capital per worker, and $\frac{Y}{L}$ is the net output per labor. Moreover, from (3.27) and (3.31), we have

$$\frac{d}{dt}K = sY = sf(K, L). \tag{3.33}$$

We note that $k = \frac{K}{Y}$ is the capital/output ratio. The output per unit capital $\frac{Y}{K}$ is the reciprocal of the capital/output ratio and is described by

$$\frac{1}{k} = \frac{Y}{K} = \frac{Lf(\frac{K}{L}, 1)}{K} = \frac{f(\frac{K}{L}, 1)}{\frac{K}{L}} = \frac{f(r, 1)}{r}, \quad r \neq 0. \tag{3.34}$$

Our main goal is to find k in (3.34) for the given constants n, s, and the function f in (3.31). For this purpose, we differentiate $r = \frac{K}{L}$ on both sides, and we obtain

$$\begin{aligned}
dr = d\left[\frac{K}{L}\right] &= \frac{LdK - KdL}{L^2} \quad \text{(from the quotient rule of the differential)} \\
&= \frac{Lsf(K, L) - nKL}{L^2} dt \quad \text{[from (3.26) and (3.33)]} \\
&= \frac{L^2 sf(\frac{K}{L}, 1) - nKL}{L^2} dt \quad \text{[from (3.31)]} \\
&= \left[\frac{L^2 sf(\frac{K}{L}, 1)}{L^2} - \frac{nKL}{L^2}\right] dt \quad \text{(by simplification)} \\
&= \left[sf(r, 1) - n\frac{K}{L}\right] dt \quad \text{(by simplification)} \\
&= sf(r, 1)dt - nr\,dt \quad \text{(by substitution).}
\end{aligned} \tag{3.35}$$

The relation $f(r, 1)$ describes the total product curve as a function of capital per worker. Moreover, it gives the output per worker as the function of capital per worker. The differential equation (3.35) describes the change of the capital/labor ratio over the time interval of length $\Delta t \approx dt$. It is equal to the difference between the increment of capital and the increment of labor. If the change of the capital/labor ratio over a time interval of length $\Delta t \approx dt$ is zero, i.e.

$$sf(r, 1) - nr = 0, \tag{3.36}$$

then the capital/labor ratio is constant ($r = \frac{K}{L}$ is constant). This implies that the growth rate of capital is equal to the labor force rate, i.e.

$$\frac{LdK - KdL}{L^2} = 0 = \frac{\frac{d}{dt}K}{K} - \frac{\frac{d}{dt}L}{L}.$$

Moreover, under the unlimited substitutability property between capital and labor of the production function, one can solve the algebraic equation (3.36), so that

$$\frac{s}{n} = \frac{r^*}{f(r^*, 1)}. \tag{3.37}$$

Under suitable conditions on f one can establish the stability of r^*. Further discussion requires more advanced knowledge of differential equations.

Example 3.2.3 (Cobb–Douglas function [130]). The production function is defined by

$$Y = K^a L^{1-a} = f(K, L),$$

where a is a parameter, and $a < 1$, $a \neq 0$. This is a first-degree homogeneous function ($K^a L^{1-a} = L[\frac{K}{L}]^a$) and admits the unlimited substitutability property between capital and labor [$[0, \infty) \subseteq R(f)$ = the range of f, where $f(r) = Lr^a$]. The Cobb–Douglas function satisfies this property. Moreover, it is a concave function for $0 < a < 1$. Hence, the marginal productivity of capital rises indefinitely as the capital per worker ratio decreases, i.e.

$$\frac{d}{dr} f(r) = aLr^{a-1} \to \infty,$$

as $r \to 0^+$. By following the discussion of Illustration 3.2.5, we arrive at

$$dr = (sr^a - nr)\, dt,$$

and it is clear that r^* in (3.37) is uniquely determined by $sr^a - nr = 0$ and, hence,

$$r^* = \left(\frac{n}{s}\right)^{\frac{1}{a-1}} = \left(\frac{s}{n}\right)^{\frac{1}{b}},$$

where $b = 1 - a$.

This concludes the discussion with regard to the Cobb–Douglas production function as an example of Illustration 3.2.5.

3.2 Exercises

(1) Use Illustration 3.2.1. Find: (i) the mathematical description of the Rice–Herzfeld decomposition of the single chemical species, (ii) the steady states of decomposition of the single chemical species, (iii) the steady states of intermediate substrates, and (iv) the instantaneous rate of change of concentration (velocity), $\frac{d}{dt}[M]$, by using the following:

(a) the terminal reaction (k_{4b}) (b) the terminal reaction (k_{4c})

(2) Let $[A_0] = a$ and $[B_0] = b$ be the given concentration of n-amyl fluoride ($n - C_5H_{11}F$) and of sodium ethoxide ($NaOC_2H_5$), and $a \neq b$. It is known that one mole of n-amyl fluoride reacts with exactly one mole of sodium ethoxide to form one mole of sodium fluoride and one mole of n-amyl ethoxide. The reaction mechanism is described by

$$n - C_5H_{11}F + NaOC_2H_5 \to NaF + n - C_5H_{11}OC_2H_5.$$

Let $x \equiv x(t)$ be the number of moles of n-amyl fluoride that have reacted with the moles of sodium ethoxide. We note that a and b decrease by the same number of moles. This reaction requires the collision of molecules of n-amyl fluoride with molecules of sodium ethoxide. Find an expression for $x(t)$ at any time t.

(3) ***Unimolecular reaction*** [132]. Let M be a normal molecule, and let M^* be its activated molecule. The problem of the origin of the activation energy in a unimolecular gas reaction is well known. A reaction in the gas phase is assumed to be unimolecular, provided that it: (i) follows the first-order rate law, (ii) is homogeneous, (iii) is not a chain reaction, and (iv) changes order from one to two at a pressure of a few millimeters. In 1923, F. A. Lindemann recognized that it is possible for normal molecules to receive their energy of activation by collision. The possible time delay between the activation and reaction processes is due to the deactivation of activated molecules by the collision with the normal molecules. The Lindemann mechanism may be described as follows:

$$\begin{aligned}
&M + M \to M^* + M \quad &&\text{(activation: reaction rate } k_1\text{)},\\
&M^* + M \to M + M \quad &&\text{(deactivation: reaction rate } k_2\text{)},\\
&M^* \to P &&\text{(product formation: reaction rate } k_3\text{)},
\end{aligned}$$

where M and M^* stand for a unimolecular reactant and a unimolecular activated molecule, respectively.

(a) Derive a system of differential equations that represents the given Lindemann mechanism.
(b) Determine the reaction rate of M at any pressure.
(c) Using the reaction rate in (b), justify the fact that the order of reaction changes from one to two.

4. ***Fixed proportion — Harrod–Domar model*** [130]. Let a stand for the number of units of capital for producing a unit of output, and let b signify the number of workers needed to produce a unit of output. Of course, a unit of output can be produced with more capital and/or labor than this. Let us define a production function f as

$$Y = f(K, L) = \min\left(\frac{K}{a}, \frac{L}{b}\right).$$

(a) Show that the production function f satisfies the neoclassical conditions, i.e. it is homogeneous of degree 1 and it has an unlimited substitutability property between capital and labor.
(b) (i) Find the domain and range of f and (ii) make a sketch of f.
(c) Show that $dr = \left[s \min\left(\frac{r}{a}, \frac{1}{b}\right) - nr\right] dt$, where $r = \frac{K}{L}$, and s and n are as defined in Illustration 3.2.5.

5. A family of functions is defined by $Y = f(K, L) = (aK^p + L^p)^{1/p}$ for $p > 0$.

(a) Show that f is a homogeneous function of degree 1.
(b) For what values of p does f satisfy the unlimited substitutability property?

(c) For $p = \frac{1}{2}$, show that $Y = f(K, L) = a^2 K + 2a\sqrt{KL} + L$.

(d) Show that $dr = s(A\sqrt{r} + 1)(B\sqrt{r} + 1)dt$, where $A = a - \sqrt{n/s}$, $B = a + \sqrt{n/s}$, and s and n are as defined in Illustration 3.2.5.

3.3 Energy Function Method

In this section, we present a very general conceptual algorithm for finding the solution process of a first-order nonlinear differential equation. The method seeks an energy function associated with a given dynamic process. By knowing the existence of a solution process, we assume that there is an energy function associated with a given dynamic system. The basic ideas are: (1) to seek an unknown energy function, (2) to associate a simpler differential equation with an unknown energy function and the original nonlinear differential equation, (3) to determine an energy function and the rate functions of a simpler differential equation in the context of a conceptually simpler differential equation and the original nonlinear differential equation, and (4) to find a representation of a solution to the original differential equation in the context of the energy function and the solution process of a simpler differential equation. We note that during the reduction process (to a simpler differential equation), the energy and rate functions of a simpler differential equation are determined. A solution to an original nonlinear differential equation is recasted in the context of the energy function and the solution of easily solvable differential equations like: (a) a directly integrable differential equation, (b) a first-order linear differential equation and (c) nonlinear differential equations.

3.3.1 *General problem*

Let us consider the following first-order nonlinear differential equation:

$$dx = f(t, x)dt, \tag{3.38}$$

where f is a continuous function defined on $J \times R$ into R, $J = [a, b] \subseteq R$.

Let us present a definition of the solution process of (3.38):

Definition 3.3.1. Let $J = [a, b]$ for $a, b \in R$, and hence $J \subseteq R$ be an interval. A function x defined on J into R is said to be a solution process of the first-order differential equation (3.38) if it satisfies the following conditions: (a) $x(t)$ is a real number and (b) the function x and its differential dx satisfy (3.38) on J. In short, the differential dx of x is equal to the right-hand side expression in (3.38).

Now, without a proof, we present a result that provides sufficient conditions for the existence and the uniqueness of the solution process corresponding to the differential equation (3.38).

Theorem 3.3.1. *Assume that f is a continuous function defined on $J \times R$ into $R, J = [a, b]$. Further assume that*

($H_{3.3}$): *the rate function f in (3.38) satisfies the conditions*

$$|f(t, x)| \leq K(1 + |x|) \quad (\textit{growth condition}),$$

$$|f(t, x) - f(t, y)| \leq L|x - y| \quad (\textit{Lipschitz condition}),$$

for all (t, x), $(t, y) \in J \times R$, where K and L are some positive numbers. Then, the IVP

$$dx = f(t, x)dt, \quad x(t_0) = x_0 \tag{3.39}$$

has a unique solution, $x(t) = x(t, t_0, x_0)$, through (t_0, x_0) for $t \geq t_0$, $t, t_0 \in J$ for a given real number x_0.

In this subsection, our goal is to discuss a general procedure for finding a representation (explicit or implicit) of a general solution to (3.38). This provides a basis for solving various real-world dynamic problems.

3.3.2 *Procedure for finding a general solution representation*

The procedure for finding a representation of a general solution to the differential equation (3.38) is described below. A method of the energy/Lyapunov function is used to determine (a closed-form) representation (explicit or implicit) of a solution to (3.38).

Step 1 (seeking an energy function). Let us assume that $x(t)$ is a solution process of (3.38). To find an explicit/implicit representation of $x(t)$ as a function of t, we seek an unknown function $V(t, x)$ (energy/Lyapunov function) defined on $J \times R$ into R possessing the following properties:

(a) $V(t, x)$ is continuous on $J \times R$ into $R (V \in C[J \times R, R])$;
(b) For $(t, x) \in J \times R$, $V(t, x)$ is monotonic in x for each t;
(c) V is continuously differentiable with respect to t and x;
(d) For each $t \in J$, $V(t, x)$ has an inverse function $E(t, x)$ defined on $J \times R$ into R, i.e. $V(t, E(t, x)) = x = E(t, V(t, x))$.

Step 2 [differential of the energy function along the differential equation (3.38)]. We find a differential of $V(t, x)$ along a vector field determined by the differential equation (3.38). We simply assume that (3.38) has a solution process $x(t)$ on J with an unknown closed-form representation. We recall that our main goal is to find an explicit/implicit representation (if possible) of $x(t)$ on J. For this purpose, we apply the differential formula [Theorem 1.5.9 to $V(t, x(t))$] with respect to the solution process of (3.38), and we have

$$dV(t, x(t)) = LV(t, x(t))dt, \tag{3.40}$$

where $x(t)$ is the solution process of (3.38), and L is a linear differential operator associated with (3.38) that is defined by

$$LV(t,x(t)) = \frac{\partial}{\partial t}V(t,x(t)) + f(t,x(t))\frac{\partial}{\partial x}V(t,x(t)). \tag{3.41}$$

Step 3 (idea of a simpler differential equation). We denote $m(t) = V(t,x(t))$ as a composite function of $V(t,x)$ and $x(t)$. With this notation, from (3.40) and (3.41), we make a conceptual choice of a convenient function F so that the process $m(t)$ satisfies the following differential equation:

$$dm = F(t,m)dt, \tag{3.42}$$

where

$$F(t,V(t,x)) = \frac{\partial}{\partial t}V(t,x) + f(t,x)\frac{\partial}{\partial x}V(t,x). \tag{3.43}$$

The condition (3.43) allows us to choose an unknown energy function in step 1(a) depending on the conceptual choice of the convenient function F in (3.42). "Convenient" means convenient enough to find an explicit/implicit (if possible) form of the solution process $x(t)$ of (3.38) for the class of rate functions f as large as possible. The choice of the convenient function F in (3.42) satisfying the condition (3.43) depends on the class of rate functions f and the ability of the problem-solver to find a closed-form solution to (3.42). Moreover, the condition (3.43) leads to the determination of a class of differential equations (3.38). The differential equation (3.42) is referred to as the reduced differential equation.

Step 4 (determination of energy and convenient choice of functions). Depending on the class of functions determined by the condition (3.43), step 3 allows us to determine: (i) an unknown energy function described in step 1 and (ii) the conceptual convenient choice function F in (3.42). In short, an energy function and a reduced differential equation (3.42) are determined with respect the class of nonlinear differential equations determined in step 3.

Step 5 [determination of a representation of the solution to (3.38)]. The energy function and reduced differential equation, determined in step 4 with respect to the class of nonlinear differential equations in step 3, are used to find a representation of the solution to (3.38). This is achieved by using the closed-form solution $m(t)$ to (3.42) and the property (d) of the energy function $V(t,x)$ in step 1 (if it exists). The recovery of the original solution process is given by

$$x(t) = E(t,m(t)) = E(t,V(t,x(t))). \tag{3.44}$$

This is the desired closed-form solution representation of a given solution $x(t)$ to (3.38).

The differential equation (3.42) is supposed to be solvable in a closed form. After finding its solution $m(t)$, we follow step 5 described in the procedure to find a solution representation of the solution $x(t)$ to (3.38). The following flowchart summarizes the steps described in the energy function method:

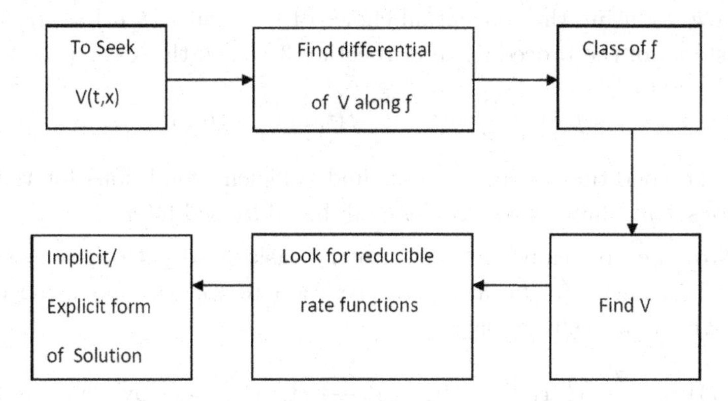

Fig. 3.1 Flowchart: Energy function method.

3.4 Integrable Reduced Equations

In this section, we make a particular choice of a "convenient function" F in (3.42) as follows: $F(t, m) = p(t)$, where p is an unknown continuous function. We assume that p satisfies (3.43). Using this "conceptual and convenient choice of function," we determine the class of differential equations (3.38) as per step 3 of the procedure in Subsection 3.3.2. From the knowledge of this class, we then determine an energy/Lyapunov function and the expressions for the conceptual choice of the function p. In short, the energy function and the rate function F depend on (3.38). The determined class of differential equations is reducible to an integrable differential equation. The method of solving the differential equations of Section 2.3 is needed to find a solution representation of (3.38).

3.4.1 *General problem*

Let us consider a reduction problem for a broad class of nonlinear differential equations (3.38). This class of differential equations is reducible to the following type of first-order integrable reduced differential equation:

$$dm = p(t)dt, \tag{3.45}$$

where p is a continuous function defined on R into R; the rate function p is independent of state variable m (dependent variable) of a dynamic system.

3.4.2 *Procedure for finding a general solution representation*

To fulfill the goal of finding an expression for the solution process of (3.38), we describe a brief procedure: (IDE_1) to determine a class of rate function f in (3.38), (IDE_2) to find the energy/Lyapunov function, and (IDE_3) to look for the convenient choice of the function p in the context of the class and the given rate function f in (3.38).

Step 1. We mirror steps 1 and 2 outlined in the procedure in Subsection 3.3.2, and arrive at (3.40) and (3.41).

Step 2. Now, by using the conceptual choice of function $F(t,m) = p(t)$ in (3.45), we repeat step 3 of the procedure in Subsection 3.3.2. In this case, (3.43) becomes

$$p(t) = \frac{\partial}{\partial t} V(t,x) + f(t,x)\frac{\partial}{\partial x} V(t,x). \tag{3.46}$$

Under the condition (3.46), we first find sufficient conditions for the class of rate functions f in (3.38). This problem will be addressed later.

Step 3. Now, we are ready to enter into the task of step 4 of the procedure in Subsection 3.3.2. Let $x(t)$ be any given solution to (3.38). Under the condition (3.46) and Theorem 1.5.9, we have

$$\begin{aligned} dV(t,x(t)) &= \left[\frac{\partial}{\partial t}V(t,x(t)) + f(t,x(t))\frac{\partial}{\partial x}V(t,x(t))\right]dt \quad \text{(by Theorem 1.5.9)} \\ &= p(t)dt. \quad \text{[from (3.46)]}. \end{aligned} \tag{3.47}$$

Therefore, we obtain

$$V(t,x(t)) = \int dV(t,x(t)) = \int p(t)dt + c, \tag{3.48}$$

where c is a constant of integration.

From (3.48), we note that the energy function $V(t,x)$ is determined by the choice function $p(t)$. Moreover, the general solution process of (3.45) is given by

$$m(t) = V(t,x(t)) = \int_a^t p(s)\,ds + c. \tag{3.49}$$

Step 4. Here, the class of differential equations (3.38) is determined by the conceptual condition (3.46). The energy function in (3.49) is utilized to find a representation of the solution process of (3.38). In fact, one can solve (3.49) for $x(t)$ by knowing the general solution $m(t)$ to (3.45). This completes the procedure for finding the solution representation of (3.38). The following flowchart describes the process of integrable differential equation reduction:

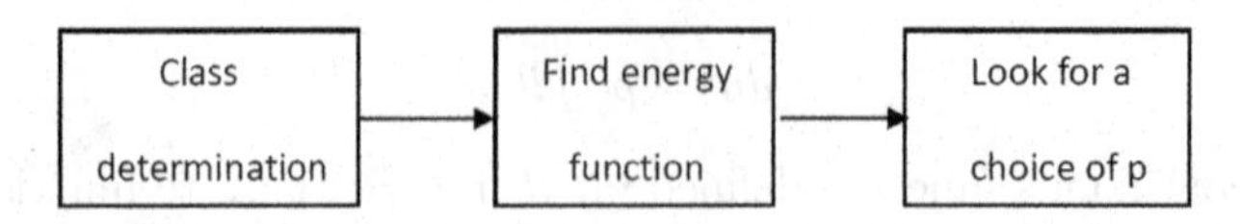

Fig. 3.2 Flowchart: Energy function and integrable equation.

Observation 3.4.1

(i) In the solution procedure presented in Subsection 3.4.2, step 2 was not able to explicitly determine the class of differential equations in (3.38). This was due to the fact that (3.38) is the most general nonlinear differential equation. We will limit our discussion to the following type of rate representation function:

$$f(t,x) = -\frac{M(t,x)}{N(t,x)}, \tag{3.50}$$

where $M(t,x)$ and $N(t,x)$ are continuous functions. From (3.46) and (3.50), we have

$$p(t) = \frac{\partial}{\partial t}V(t,x) - \frac{u(t,x)M(t,x)}{u(t,x)N(t,x)}\frac{\partial}{\partial x}V(t,x), \tag{3.51}$$

where $p(t)$ and $V(t,x)$ are unknown functions, $u(t,x)$ is an arbitrary continuous function, and $u(t,x) \neq 0$. Our goal is to find $p(t)$ and $V(t,x)$. For this purpose, we define

$$V(t,x) = \int u(t,x)N(t,x)dx, \tag{3.52}$$

and assume that

$$\frac{\partial}{\partial t}V(t,x) = \frac{\partial}{\partial t}\left(\int u(t,x)N(t,x)dx\right)$$

exists. We choose (if possible) the function $u(t,x)$, so that the expression

$$\frac{\partial}{\partial t}\left(\int u(t,x)N(t,x)dx\right) - u(t,x)M(t,x) \tag{3.53}$$

is only a function of t. This implies that

$$\frac{\partial}{\partial x}\left(\frac{\partial}{\partial t}\left(\int u(t,x)N(t,x)dx\right) - u(t,x)M(t,x)\right) = 0. \tag{3.54}$$

This condition determines the class of functions $f(t,x)$ defined in (3.50). Of course, we note that this class of functions depends on the existence of $u(t,x)$. From (3.52) and (3.53), (3.51) reduces to

$$\begin{aligned}
p(t) &= \frac{\partial}{\partial t}V(t,x) - \frac{u(t,x)M(t,x)}{u(t,x)N(t,x)}\frac{\partial}{\partial x}V(t,x)\\
&= \frac{\partial}{\partial t}\left(\int u(t,x)N(t,x)dx\right)\\
&\quad - \frac{u(t,x)M(t,x)}{u(t,x)N(t,x)}u(t,x)N(t,x) \quad \text{(by substitution)}\\
&= \frac{\partial}{\partial t}\left(\int u(t,x)N(t,x)dx\right) - u(t,x)M(t,x) \quad \text{(by simplification)}.
\end{aligned}$$

Thus, $p(t)$ is determined by the condition (3.53). We note that $p(t)$ depends on the existence of a nonzero function $u(t,x)$. Moreover, upon integration, we have

$$\int u(t,x)N(t,x)dx = V(t,x(t)) = \int p(s)ds + c$$
$$= \int \left[\frac{\partial}{\partial t}\left(\int u(t,x)N(t,x)dx \right) - u(t,x)M(t,x) \right] dt + c.$$

This implies that

$$\int u(t,x)N(t,x)dx - \int \left[\frac{\partial}{\partial t}\left(\int u(t,x)N(t,x)dx \right) - u(t,x)M(t,x) \right] dt = c. \tag{3.55}$$

The solution to (3.38) in the context of (3.50) is implicitly given by (3.55). From this (if possible), one can explicitly or implicitly solve for $x(t)$. This completes the solution representation procedure for the class of deterministic differential equations (3.38) in the context of (3.46) and (3.50).

(ii) We further examine the condition (3.53). This provides an additional insight for the selection of the function $u(t,x)$ in the problem-solving process. For this purpose, we assume that we can interchange the differentiation operation with integration. Furthermore, we assume that the functions $u(t,x), N(t,x)$, and $M(t,x)$ are continuously differentiable in both of the variables t and x. Under this conditions, we rewrite (3.53) as follows:

$$\frac{\partial}{\partial t}\left(\int u(t,x)N(t,x)dx \right) - u(t,x)M(t,x)$$
$$= \int \frac{\partial}{\partial t}(u(t,x)N(t,x))dx$$
$$- u(t,x)M(t,x) \quad \left(\text{by interchange of } \frac{\partial}{\partial t} \text{ with } \int \right)$$
$$= \int \frac{\partial}{\partial t}\left(u(t,x)N(t,x)\right) dx - \int \frac{\partial}{\partial x}\left(u(t,x)M(t,x)\right) dx$$
$$+ B(t) \quad \text{(by the partial antiderivative)},$$

where $B(t)$ is a constant of integration that depends on the initial point $x = c$. The above integral expression can be further rewritten as

$$\frac{\partial}{\partial t}\left(\int u(t,x)N(t,x)dx \right) - u(t,x)M(t,x)$$
$$= \int \left[\frac{\partial}{\partial t}(u(t,x)N(t,x)) - \frac{\partial}{\partial x}(u(t,x)M(t,x)) \right] dx + B(t)$$
$$= \int \left[N(t,x)\frac{\partial}{\partial t}u(t,x) + u(t,x)\frac{\partial}{\partial t}N(t,x) \right]$$

$$- \left[M(t,x)\frac{\partial}{\partial x}u(t,x) + u(t,x)\frac{\partial}{\partial x}M(t,x) \right] dx + B(t)$$

$$= \int \left[N(t,x)\frac{\partial}{\partial t}u(t,x) - M(t,x)\frac{\partial}{\partial x}u(t,x) \right]$$

$$+ u(t,x)\left[\frac{\partial}{\partial t}N(t,x) - \frac{\partial}{\partial x}M(t,x) \right] dx + B(t). \tag{3.56}$$

(iii) The feasibility of the condition (3.53) is guaranteed by the following conditions:

(a) If

$$\frac{\partial}{\partial x}\left(u(t,x)M(t,x)\right) = \frac{\partial}{\partial t}\left(u(t,x)N(t,x)\right) \tag{3.57}$$

for some $u(t,x)$, then (3.53) is valid. This is true in view of (3.56). In this case, $u(t,x)$ in (3.57) is termed the general integrating factor. The energy method (i) reduces to the generalized method of the integrating factor.

(b) If (3.57) is valid for $u(t,x) = 1$, then (3.57) reduces to

$$\frac{\partial}{\partial x}M(t,x) = \frac{\partial}{\partial t}N(t,x), \tag{3.58}$$

a special case of (3.57). In this case, $u(t,x) = 1$ in (3.53) is determined. The energy method (i) reduces to the method of the exact differential equation. This is considered to be a sufficient condition for the reduction of the differential equation (3.38) in its integrable form of (3.46). The differential equation is said to be exact if one can find an energy/Lyapunov function $V(t,x)$ so that

$$\frac{\partial}{\partial t}V(t,x)dt + \frac{\partial}{\partial x}V(t,x)dx = M(t,x)dt + N(t,x)dx. \tag{3.59}$$

By the method of undetermined coefficients, this implies that

$$\frac{\partial}{\partial t}V(t,x) = M(t,x) \quad \text{and} \quad \frac{\partial}{\partial x}V(t,x) = N(t,x). \tag{3.60}$$

Furthermore, if $M(t,x)$ and $N(t,x)$ are continuously differentiable, the condition (3.60) is also a necessary condition. This is due to the following

$$\frac{\partial}{\partial x}M(t,x) = \frac{\partial^2}{\partial x \partial t}V(t,x) = \frac{\partial^2}{\partial t \partial x}V(t,x)$$

$$= \frac{\partial}{\partial t}N(t,x) \quad \text{(by A. Clairaut's Theorem [85])}.$$

(c) If (3.57) is valid for $u(t,x) \equiv u(t)$, then (3.57) reduces to

$$\frac{\partial}{\partial x}\left(u(t)M(t,x)\right) = \frac{\partial}{\partial t}\left(u(t)N(t,x)\right). \tag{3.61}$$

This implies that

$$\frac{d}{dt}u(t) = -\frac{\frac{\partial}{\partial t}N(t,x) - \frac{\partial}{\partial x}M(t,x)}{N(t,x)}u(t) \tag{3.62}$$

is a special case of (3.57). In this case, $u(t,x) \equiv u(t)$ in (3.53) is determined by (3.62). The energy method (iii) reduces to the method of the integrating factor.

(d) If (3.57) is valid for $u(t,x) \equiv u(x)$, then (3.57) reduces to

$$\frac{\partial}{\partial x}(u(x)M(t,x)) = \frac{\partial}{\partial t}(u(x)N(t,x)). \tag{3.63}$$

This implies that

$$\frac{d}{dx}u(x) = -\frac{\frac{\partial}{\partial x}M(t,x) - \frac{\partial}{\partial t}N(t,x)}{M(t,x)}u(x) \tag{3.64}$$

is a special case of (3.57). In this case, $u(t,x) \equiv u(x)$ in (3.53) is determined by (3.64). The energy method (iii) reduces to the method of the integrating factor.

Example 3.4.1. Find a general solution to the given equation:

$$dx = -\frac{2t + x\cos(xt)}{t\cos(xt)}\,dt.$$

Solution procedure. We set $M(t,x) = 2t + x\cos(xt)$ and $N(t,x) = t\cos(xt)$. We seek functions p and V so that (3.46) is valid. By following the argument used in Observation 3.4.1(i), we define an energy function:

$$\begin{aligned}
V(t,x) &= \int u(t,x)t\cos(xt)dx \\
&= \int u(t,x)t\cos(xt)\,dx \quad \text{(by the integration-by-parts method)} \\
&= \int u\cos v dv \quad \text{(by the substitution method: } v = xt) \\
&= u\sin v - \int \sin v \frac{\partial}{\partial x}u\,dx \quad (\text{for } t \neq 0) \\
&= u(t,x)\sin(xt) - \int \sin(xt)\frac{\partial}{\partial x}u(t,x)\,dx \quad (\text{for } t \neq 0).
\end{aligned}$$

Now, we have

$$\frac{\partial}{\partial x}V(t,x) = \frac{\partial}{\partial x}\left(\int u(t,x)t\cos(xt)dx\right) = u(t,x)t\cos(xt),$$

$$\frac{\partial}{\partial t}V(t,x) = \frac{\partial}{\partial t}\left(\int u(t,x)t\cos(xt)dx\right) = \frac{\partial}{\partial t}(u\sin(xt)) - \frac{\partial}{\partial t}\int \sin(xt)\frac{\partial}{\partial x}u\,dx,$$

and the expressions in (3.53) becomes

$$\begin{aligned}
&\frac{\partial}{\partial t}\left(\int uN(t,x)dx\right) - uM(t,x) \\
&= \frac{\partial}{\partial t}(u\sin(xt)) - u(2t + x\cos(xt)) \\
&\quad - \frac{\partial}{\partial t}\int \sin(xt)\frac{\partial}{\partial x}u\,dx \quad \text{(by substitution)} \\
&= \sin(xt)\frac{\partial}{\partial t}u + u\frac{\partial}{\partial t}\sin(xt) - u(2t + x\cos(xt)) \\
&\quad - \frac{\partial}{\partial t}\int \sin(xt)\frac{\partial}{\partial x}u\,dx \quad \text{(by the product rule)} \\
&= \sin(xt)\frac{\partial}{\partial t}u + ux\cos(xt) - u(2t + x\cos(xt)) \\
&\quad - \frac{\partial}{\partial t}\int \sin(xt)\frac{\partial}{\partial x}u\,dx \quad \text{(by simplifying)} \\
&= \sin(xt)\frac{\partial}{\partial t}u - 2tu - \frac{\partial}{\partial t}\int \sin(xt)\frac{\partial}{\partial x}u\,dx \quad \text{(by simplifying).}
\end{aligned}$$

We choose u so that the condition (3.53) is satisfied, i.e. the above expression is a function of t only. This means that

$$\frac{d}{dt}u = \frac{\partial}{\partial x}u = 0.$$

That is to say, u is constant function. By using the method of Section 2.3, we have

$$u(t) = \mathrm{C},$$

where C is an arbitrary nonzero constant. In particular, we choose $\mathrm{C} = 1$. From this, the above expression reduces to

$$\frac{\partial}{\partial t}\left(\int uN(t,x)dx\right) - uM(t,x) = -2t.$$

This satisfies the conditions (3.53) and (3.54). This shows that the given differential equation belongs to the class of differential equations (3.38) defined by (3.54). Therefore, $p(t) = -2t$, from (3.49), the solution representation of the given problem is

$$m(t = V(t, x(t)) = \sin(xt) = \int -2tdt + \mathrm{C} = -t^2 + \mathrm{C}.$$

By solving the above equation for $x(t)$, we obtain

$$x(t) = \frac{1}{t}\sin^{-1}(-t^2 + \mathrm{C}) \quad \text{for } t \neq 0.$$

This completes the solution process of the given example.

Example 3.4.2. Find a general solution to the following differential equation:

$$dx = -\frac{2t\tan x + 2xt^2 + x - 2t}{\sec^2 x + t}\,dt.$$

Solution procedure. We set $M(t,x) = 2t\tan x + 2xt^2 + x - 2t$ and $N(t,x) = \sec^2 x + t$. By following the argument used in Example 3.4.1, we have

$$\begin{aligned}
V(t,x) &= \int u\left(\sec^2 x + t\right) dx \\
V(t,x) &= \int u\left(\sec^2 x + t\right) dx \quad \text{(by partial integration with respect to } x\text{)} \\
&= u(\tan x + tx) - \int_a^x (\tan x + tx)\,\frac{\partial}{\partial x}\,u\,dx \quad \text{(by integration by parts)}, \\
\frac{\partial}{\partial x}\,V(t,x) &= \frac{\partial}{\partial x}\left(\int_a^x u\left(\sec^2 x + t\right) dx\right) = u\,\frac{\partial}{\partial x}\,(\tan x + tx) = u(\sec^2 x + t), \\
\frac{\partial}{\partial t}\,V(t,x) &= \frac{\partial}{\partial t}\left(\int_a^x u\left(\sec^2 x + t\right) dx\right) \\
&= \frac{\partial}{\partial t}\left(u(\tan x + tx) - \int_a^x (\tan x + tx)\,\frac{\partial}{\partial x}\,u\,dx\right) \\
&= \frac{\partial}{\partial t}\,[u(\tan x + tx)] - \frac{\partial}{\partial t}\left[\int_a^x (\tan x + tx)\,\frac{\partial}{\partial x}\,u\,dx\right],
\end{aligned}$$

and the expressions (3.53) is:

$$\begin{aligned}
&\frac{\partial}{\partial t}\left(\int uN(t,x)dx\right) - uM(t,x) \\
&\quad = \frac{\partial}{\partial t}\,(u[\tan x + tx]) - u(2t\tan x + 2xt^2 + x - 2t) \\
&\qquad - \frac{\partial}{\partial t}\left[\int_a^x (\tan x + tx)\frac{\partial}{\partial x}udx\right] \quad \text{(by substitution)} \\
&\quad = (\tan x + tx)\frac{\partial}{\partial t}\,u + u\frac{\partial}{\partial t}(\tan x + tx) - u\left(2t\tan x + 2xt^2 + x - 2t\right) \\
&\qquad - \frac{\partial}{\partial t}\left[\int_a^x (\tan x + tx)\frac{\partial}{\partial x}udx\right] \quad \text{(by the product rule)} \\
&\quad = (\tan x + tx)\frac{\partial}{\partial t}\,u + ux - u\left(2t\tan x + 2xt^2 + x - 2t\right) \\
&\qquad - \frac{\partial}{\partial t}\left[\int_a^x (\tan x + tx)\frac{\partial}{\partial x}udx\right] \quad \text{(by simplifying)}
\end{aligned}$$

$$= (\tan x + tx)\frac{\partial}{\partial t}u - 2tu(\tan x + xt) - 2tu$$
$$- \frac{\partial}{\partial t}\left[\int_a^x (\tan x + tx)\frac{\partial}{\partial x}udx\right] \text{ (by simplifying)}$$
$$= (\tan x + tx)\left(\frac{\partial}{\partial t}u - 2tu\right) + 2tu$$
$$- \frac{\partial}{\partial t}\left[\int_a^x (\tan x + tx)\frac{\partial}{\partial x}u\,dx\right] \quad \text{(by regrouping).}$$

We choose u so that the condition (3.53) is satisfied, i.e. the above expression is a function of t only. This means that

$$\frac{\partial}{\partial t}u - 2tu = 0, \quad \frac{\partial}{\partial x}u = 0.$$

That is to say, we can assume that u is a function of t only. By using the method of Section 2.4, we have

$$u(t) = \exp\left[t^2\right]\text{C},$$

where C is an arbitrary nonzero constant. In particular, we choose $\text{C} = 1$. From this, the above expression reduces to

$$\frac{\partial}{\partial t}\left(\int uN(t,x)dx\right) - uM(t,x) = 2t\exp\left[t^2\right].$$

This establishes the validity of the condition (3.53). This shows that the given differential belongs to the class determined by (3.54). Therefore, $p(t) = 2t\exp\left[t^2\right]$, from (3.49), the solution representation of the given problem is

$$m(t = V(t,x(t)) = \exp\left[t^2\right][\tan x(t) + tx(t)] = \exp\left[t^2\right] + \text{C}.$$

Thus, the solution process is given by

$$\tan x(t) + tx(t) = 1 + \text{C}\exp\left[-t^2\right].$$

This completes the solution process of the given example.

Illustration 3.4.1 (conservation of energy [40, 85]). We recall the material from *Calculus III* concerning the concept of vector field: $\mathbf{F}(t,x) = P(t,x)\mathbf{i} + Q(t,x)\mathbf{j}$ in R^2. The differential equation (3.38) in the context of (3.50) can be considered as the description of the vector field, where $M(t,x) = -P(t,x)$ and $N(t,x) = Q(t,x)$. We assume that this is a force vector field, and it moves an object along a directed path C (curve in an R^2) described by $\mathbf{r}(t)$, $a \le t \le b$, from the initial point $A(\mathbf{r}(a))$ to the terminal point $B(\mathbf{r}(b))$ in R^2. There are different forms of energies: chemical, electrical, kinetic, nuclear, potential, thermal, etc. These energies are used to perform the task. In particular, we also recall the concepts of kinetic and potential energies. Due to the motion of the bodies or objects, the kinetic energy is defined by

$$K(t) = \frac{1}{2} m \left| \frac{d}{dt} \mathbf{r}(t) \right|^2$$

for t in $[a, b]$, where m is the mass;

$$v = \frac{d}{dt} \mathbf{r}(t), \quad a(t, x) = \frac{d^2}{dt^2} \mathbf{r}(t)$$

are the velocity and the acceleration of the object, respectively. The vector field generated by the motion of the body,

$$A(t, x) = A(\mathbf{r}(t)) = m \frac{d^2}{dt^2} \mathbf{r}(t),$$

is due to Newton's second law of motion. The potential energy of the object or body at a location (t, x) in R^2 is defined by $-V(t, x)$, and the force associated with this varying position (t, x) of the system is $F(t, x) = -\bigtriangledown V(t, x)$.

One of the motivations for the concept of the line integral of a vector field along the directed curve is that it represents the work done by the force for moving a body from one point to another along the curve. In the following, using the concept of the line integral, we compute the work done by the above-described forces associated with the concepts of kinetic and potential energy.

First, we find the work done by the motion of the body:

$$\begin{aligned}
W_k &= \int_C A(\mathbf{r}) \cdot d\mathbf{r} = \int_a^b A(\mathbf{r}(t)) \cdot d\mathbf{r}(t) \\
&= m \int_a^b \frac{d^2}{dt^2} \mathbf{r}(t) \cdot \frac{d}{dt} \mathbf{r}(t) dt \quad \text{(by substitution)} \\
&= m \int_a^b \frac{d}{dt} \mathbf{v}(t) \cdot \mathbf{v}(t)\, dt \quad \left[\text{by } \mathbf{v}(t) = \frac{d}{dt} \mathbf{r}(t)\right] \\
&= \frac{1}{2} m \int_a^b \frac{d}{dt} [\mathbf{v}(t) \cdot \mathbf{v}(t)]\, dt \quad \left[\text{by } \frac{d}{dt}(\mathbf{v} \cdot \mathbf{v}) = 2 \frac{d}{dt} \mathbf{v} \cdot \mathbf{v}\right] \\
&= \frac{1}{2} m \int_a^b \frac{d}{dt} \left[|\mathbf{v}(t)|^2\right] dt \quad [\text{by } (\mathbf{v} \cdot \mathbf{v}) = |\mathbf{v}|^2] \\
&= \frac{1}{2} m \left[|\mathbf{v}(t)|^2\right]_a^b \quad \text{(by the fundamental theorem of calculus [85])} \\
&= \frac{1}{2} m \left(|\mathbf{v}(b)|^2 - |\mathbf{v}(a)|^2\right) \quad \text{(from the notation).}
\end{aligned}$$

Therefore, the work done due to the motion of the body from the initial point $A(\mathbf{r}(a))$ to the terminal point $B(\mathbf{r}(b))$ is

$$\begin{aligned}
W_k &= \frac{1}{2}m|\mathbf{v}(b)|^2 - \frac{1}{2}m|\mathbf{v}(a)|^2 \\
&= \text{the kinetic energy at } B(\mathbf{r}(b)) - \text{the kinetic energy at } A(\mathbf{r}(a)) \\
&= K(b) - K(a) \\
&= \text{the change in kinetic energy at the terminal and the initial point.}
\end{aligned}$$

Now, we find the work done by the force $F(t,x) = -\nabla V(t,x)$, which is generated by the varying position of the body from the initial position to the terminal position. It is determined by

$$\begin{aligned}
W_p &= \int_C F(\mathbf{r}) \cdot d\mathbf{r} \quad [\text{by notation } F(t,x) = F(\mathbf{r}(t)) \text{ along the curve: } C] \\
&= -\int_C \nabla V(\mathbf{r}(t)) \cdot d\mathbf{r}(t) \quad [\text{by substitution/notation: } V(t,x) = V(\mathbf{r}(t))] \\
&= -\int_a^b \frac{d}{dt} V(\mathbf{r}(t))\, dt \quad \left[\text{by } \nabla V(\mathbf{r}) \cdot d\mathbf{r} = V_x \frac{d}{dt} x + V_y \frac{d}{dt} y\right] \\
&= -V(\mathbf{r}(t))|_a^b \quad (\text{by the fundamental theorem of calculus}) \\
&= -[V(\mathbf{r}(b)) - V(\mathbf{r}(a))] \quad (\text{from the notation}).
\end{aligned}$$

Therefore, the work done due to the application of the potential force of the body from the change in position from the initial point $A(\mathbf{r}(a))$ to the terminal point $B(\mathbf{r}(b))$ is

$$\begin{aligned}
W_p &= V(\mathbf{r}(a)) - V(\mathbf{r}(b)) \\
&= \text{the potential energy at } A(\mathbf{r}(a)) - \text{the potential energy at } B(\mathbf{r}(b)) \\
&= P(a) - P(b) \\
&= \text{the change in potential energy at the initial and the terminal point.}
\end{aligned}$$

We know that $W_k = W_p$. This implies that $K(b) - K(a) = P(a) - P(b)$. This shows that

$$P(a) + K(a) = P(b) + K(b).$$

Thus, for a given vector field $F(t,x) = -\nabla V(t,x)$, the sum of the potential energy and the kinetic energy is constant. If this is true for any two given points and any smooth curve, then it is called the law of conservation of mechanical energy. The vector field, $F(t,x) = -\nabla V(t,x)$ is called conservative. Now, given a vector field: $\mathbf{F}(t,x) = P(t,x)\mathbf{i} + Q(t,x)\mathbf{j}$ in R^2, under what conditions is this conservative? The answer to this question centers around the conditions that the differential equation (3.38) in the context of (3.50) is exact. The energy/Lyapunov function that provides an implicit/explicit solution to the exact differential equation is the potential energy function. This justifies our naming scheme — the "energy function method."

Observation 3.4.2. From (3.49), we note that $V(t,x)$ depends on an arbitrary constant of integration and an arbitrary choice of the function $p(t)$ in (3.45). In fact, $p(t)$ is determined by (3.53). In short, $V(t,x)$ is determined by $p(t)$, and $p(t)$ is determined by the rate coefficient in (3.38) and the condition (3.53).

3.4 Exercises

(1) Show that the given differential equation is exact. Then solve it by finding the energy function.

(a) $\left(2x^5t^2 + tx\right)dt + \left(\frac{10}{3}x^4t^3 + \frac{1}{2}t^2\right)dx = 0$
(b) $(\sin x \cos t)\,dt + (\sin t \cos x)\,dx = 0$
(c) $\left(x^2 \sec^2 t + \ln(1+t)\right)dt + (2x \tan t + \exp[2x])\,dx = 0,$
(d) $2xt\,dt + \left(x^2 + t^2\right)dx = 0$
(e) $(2t \exp[x] + 1)\,dt + t^2 \exp[x]\,dx = 0$
(f) $-2xy\,dx + \left(y^2 - x^2\right)dy = 0$

(2) Determine whether the given differential equation describes a conservative vector field in R^2.

(a) $(x - t\cot x)\,dt + \left(t + \frac{1}{2}t^2 \csc^2 x\right)dx = 0$
(b) $(2 + \sin x)\,dt + (5 + t\cos x)\,dx = 0$
(c) $\left(t^2 + 3x\right)dt + 3\left(t + x^2\right)dx = 0$
(d) $(\exp[t] + x\cos t)\,dt + (\exp[x] - x\sin t)\,dx = 0$

(3) Find the general solution representation of the differential equations in Exercises 1 and 2.

(4) Solve the following equations:

(a) $\left(4xx + 3x^2 - t\right)dt + t\,(t + 2x)\,dx = 0$
(b) $(2x \sinh t - 1)\,dt + \cosh t\,dx = 0$
(c) $\left(x^3 + xy^4\right)dx + 2y^4\,dy = 0$
(d) $\left(x^2 + t^2 - 1\right)dt + x\,(t - 2x)\,dx = 0$
(e) $(tx + 1)\,dt + t\,(t + 4x - 2)\,dx = 0$
(f) $x\,(t + x)\,dt + (t + 2x - 1)\,dx = 0$

3.5 Linear Nonhomogeneous Reduced Equations

In this section, we find another very broad class of differential equations (3.38). In fact, we make a special choice of the convenient function F in (3.42) as follows: $F(r,m) = \mu(t)m + p(t)$.

3.5.1 *General problem*

We consider a reduction problem for another broad class of nonlinear differential equations (3.38). This class of differential equations is reduced to the following first-order nonhomogeneous scalar differential equation with variable coefficients:

$$dm = [\mu(t)m + p(t)]\, dt, \tag{3.65}$$

where μ and p are continuous functions defined on R into R.

3.5.2 *Procedure for finding a general solution representation*

We present a procedure for a solution representation for a broader class of differential equations (3.38). We outline the method in the same spirit as Section 3.4. The procedure consists of: (LDE_1) determining a class of rate functions f in (3.38), (LDE_2) finding the energy/Lyapunov function, and (LDE_3) looking for the convenient choice of the functions μ and p. From this reduction procedure and using a general solution (3.65), a general solution expression of the differential equation (3.38) is determined in either explicit or implicit form.

Now, we briefly describe the following procedure to determine the solution representation of the nonlinear differential equation (3.38).

Step 1. We follow steps 1 and 2 of the procedure outlined in Subsection 3.3.2 to arrive at (3.40).

Step 2. By choosing the conceptual choice function as $F(t, V(t,x)) = \mu(t)V(t,x) + p(t)$, we repeat step 3 of the procedure in Subsection 3.3.2. Again, our goal is to find a broad class of differential equations (3.38) that is reducible to (3.65). For this purpose, we rewrite (3.43) as

$$\mu(t)V(t,x) + p(t) = \frac{\partial}{\partial t} V(t,x) + f(t,x)\frac{\partial}{\partial x} V(t,x). \tag{3.66}$$

This is a sufficient condition for the reducibility of (3.38) to the first-order nonhomogeneous scalar differential equation with variable coefficients (3.65).

In summary, (3.66) determines the class of differential equations (3.38) reducible to (3.65). Of course, the class depends on the conceptual choice of rate functions in the reducible differential equation (3.65).

Step 3. Now, we are ready to use step 4 of the procedure in Subsection 3.3.2 to find $V(t,x)$ as well as the convenient choice of the functions μ and p in (3.65).

From (3.66), (3.40), (3.41), and (3.65), we compute the differential of $m(t) = V(t, x(t))$ along with the solution process of (3.38):

$$\begin{aligned} dm(t) &= dV(t,x(t)) = LV(t,x(t))\, dt \\ &= \left[\frac{\partial}{\partial t} V(t,x(t)) + f(t,x(t))\,\frac{\partial}{\partial x} V(t,x(t))\right] dt \end{aligned}$$

$$= [\mu(t)V(t, x(t)) + p(t)]\, dt$$
$$= [\mu(t)m(t) + p(t)]\, dt. \tag{3.67}$$

By applying the method of Section 2.5, the general solution process of (3.67) is given by

$$V(t, x(t)) = m(t) = \exp\left[\int_a^t \mu(u)du\right] \mathrm{C} + \int_a^t \exp\left[\int_s^t \mu(u)du\right] p(s)\, ds, \tag{3.68}$$

where C is an arbitrary nonzero constant of integration. Thus, the energy function V is determined by the solution to (3.65).

Step 4. Under condition (3.66), the nonlinear differential equation (3.38) reduces to (3.65). The condition (3.66) determines the class of differential equations (3.38) depending on the conceptual coefficients of the first-order linear nonhomogeneous differential equation (3.65). This issue is addressed immediately, in the observation.

By employing the property of V and notations in step 1 of Procedure 3.2 and using (3.68), the general solution to (3.38) is represented by

$$x(t) = E(t, m(t))$$
$$= E\left(t, \exp\left[\int_a^t \mu(u)du\right] \mathrm{C} + \int_a^t \exp\left[\int_s^t \mu(u)du\right] p(s)\, ds\right), \tag{3.69}$$

where $m(t)$ is as determined in step 3, and $x(t)$ is a general solution process of the differential equation (3.38). The following flowchart describes the process of linear nonhomogeneous equation reduction.

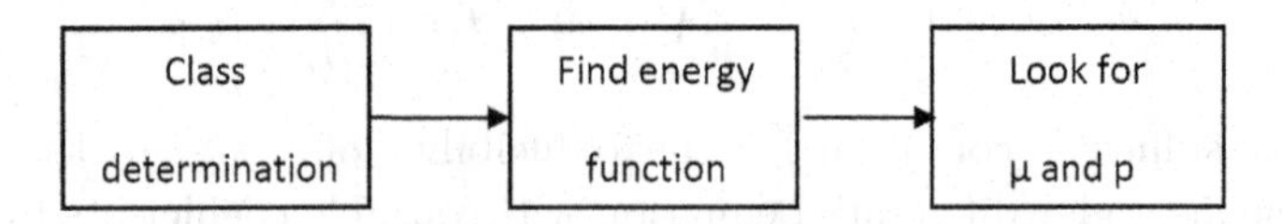

Fig. 3.3 Flowchart: Energy function method and nonhomogeneous equation.

Observation 3.5.1

(i) Step 4 of Procedure 3.4 shows that the coefficients of the reducible differential equation dependent on the class of differential equations (3.38), implicitly. This observation is analogous to Observation 3.4.1(i). In this case, we consider the representation similar to (3.50) with a certain modification:

$$f(t, x) = -\frac{M(t, x) - R(t, x)}{N(t, x)}, \tag{3.70}$$

where $M(t,x)$ and $N(t,x)$ are as defined in (3.50) and $R(t,x)$ is a smooth function. From (3.65) and (3.70), we have

$$\mu(t)V(t,x) + p(t) = \frac{\partial}{\partial t}V(t,x) - \frac{u(t,x)\left(M(t,x) - R(t,x)\right)}{u(t,x)N(t,x)}\frac{\partial}{\partial x}V(t,x), \tag{3.71}$$

where $\mu(t)$, $p(t)$, and $V(t,x)$ are unknown functions and $u(t,x) \neq 0$ is an arbitrary continuous function. Our goal is to find $\mu(t)$, $p(t)$, and $V(t,x)$. For this purpose, we imitate the procedure outlined in Observation 3.4.1(i), and we assume that the functions u, N, M, and R satisfy the conditions

$$V(t,x) = \int u(t,x)N(t,x)dx, \tag{3.72}$$

$$\frac{\partial}{\partial t}\left(\int u(t,x)N(t,x)\,dx\right) - u(t,x)M(t,x) \tag{3.73}$$

is a function of t only, and

$$\frac{R(t,x)}{N(t,x)}\frac{\partial}{\partial x}V(t,x) = \mu(t)V(t,x). \tag{3.74}$$

From (3.73), it follows that

$$\frac{\partial}{\partial x}\left(\frac{\partial}{\partial t}\left(\int u(t,x)N(t,x)\,dx\right) - u(t,x)M(t,x)\right) = 0. \tag{3.75}$$

The conditions (3.74) and (3.75) determine the class of functions $f(t,x)$ defined in (3.38). Of course, we note that this class of functions depends on the choice of $u(t,x)$.

Now we consider

$$\begin{aligned}
&\frac{\partial}{\partial t}V(t,x(t)) + f(t,x(t))\frac{\partial}{\partial x}V(t,x(t)) \\
&= \frac{\partial}{\partial t}V(t,x) - \frac{u(t,x)(M(t,x) - R(t,x))}{u(t,x)N(t,x)}\frac{\partial}{\partial x}V(t,x) \quad \text{[from (3.70)]} \\
&= \frac{\partial}{\partial t}\left(\int u(t,x)N(t,x)dx\right) - \frac{u(t,x)(M(t,x) - R(t,x))}{u(t,x)N(t,x)} \\
&\quad \times u(t,x)N(t,x) \quad \text{(by substitution)} \\
&= \frac{\partial}{\partial t}\left(\int u(t,x)N(t,x)dx\right) - u(t,x)M(t,x) \\
&\quad + u(t,x)R(t,x) \quad \text{(by simplification)} \\
&= \mu(t)V(t,x) + p(t).
\end{aligned}$$

Thus, $p(t)$ and $\mu(t)$ are determined by the conditions (3.73) and (3.74), respectively. We note that $p(t)$and $\mu(t)$ depend on the existence of a nonzero function $u(t,x)$.

From (3.68) and (3.72), we obtain

$$\int u(t,x)N(t,x)dx = V(t,x(t))$$
$$= \exp\left[\int_a^t \mu(u)du\right] \mathrm{C} + \int_a^t \exp\left[\int_s^t \mu(u)du\right] p(s)ds.$$

Hence,

$$\int u(t,x)N(t,x)dx = \exp\left[\int_a^t \mu(u)du\right] \mathrm{C} + \int_a^t \exp\left[\int_s^t \mu(u)du\right] p(s)ds. \tag{3.76}$$

In the context of (3.70), the solution to (3.38) is implicitly or explicitly given by (3.76). This completes our goal of finding the solution representation procedure for the class of deterministic differential equations (3.38) in the context of (3.70), (3.74), and (3.75).

(ii) The presented sufficient conditions (3.74) and (3.75) which determine the class reducible to linear differential equations seem not to be completely clear. This is due to the limited usage of the analytic background. Here, we provide definite sufficient conditions for a broader class of differential equations (3.38) that are not only reducible to linear differential equations but also arbitrary nonlinear solvable differential equations. To justify this, we reconsider (3.70) as

$$f(t,x) = \frac{H(t, \int^x N(t,s)ds) - M(t,x)}{N(t,x)}, \tag{3.77}$$

where $R(t,x)$ is replaced by $H(t, \int^x N(t,s)\, ds)$, and choosing $u(t,x) = 1$. Now, repeating the argument used in (i), one can conclude that the $V(t,x)$ is determined by (3.72), and the reduced differential equation is

$$dm = [H(t,m) + p(t)]\, dt. \tag{3.78}$$

(iii) If $H(t,m) = \mu(t)\, m$, then (3.77) reduces to (3.65) as a special case.

(iv) Moreover, if $N(t,x) = \frac{\partial}{\partial x} h(t,x)$, $M(t,x) = \frac{\partial}{\partial t} h(t,x)$, for a smooth real-valued function $h(t,x)$ defined on its domain, and $H(t,x) = P(t)h(t,x) + Q(t)(h(t,x))^n$. In this case, the reduced differential equation (3.78)

$$dm = [P(t)m + Q(t)m^n]\, dt. \tag{3.79}$$

Example 3.5.1. Find a general solution to

$$dx = -\frac{2t + x\cos(xt) + 2\sin(xt)}{t\cos(xt)}\, dt.$$

Solution procedure. We set $M(t,x) = 2t + x\cos(xt)$, $R(t,x) = 2\sin(xt)$, and $N(t,x) = t\cos(xt)$. We seek functions p, μ, and V so that (3.71) is satisfied. By following the argument used in Observation 3.5.1(i) and Example 3.4.1, we define an energy function,

$$\begin{aligned} V(t,x) &= \int u(t,x)t\cos(xt)\,dx \\ &= u\sin(xt) - \int \sin(xt)\frac{\partial}{\partial x}u\,dx \quad (\text{for } t \neq 0), \end{aligned}$$

and compute expressions in (3.67):

$$\begin{aligned} &\frac{\partial}{\partial t}V(t,x(t)) + f(t,x(t))\frac{\partial}{\partial x}V(t,x(t)) \\ &\quad = \frac{\partial}{\partial t}\left(\int u(t,x)N(t,x)dx\right) - u(t,x)M(t,x) - u(t,x)R(t,x) \\ &\quad = \frac{\partial}{\partial t}(u\sin(xt)) - u(2t + x\cos(xt)) - \frac{\partial}{\partial t}\int \sin(xt)\frac{\partial}{\partial x}u\,dx - 2u\sin(xt) \\ &\quad = \sin(xt)\frac{\partial}{\partial t}u + u\frac{\partial}{\partial t}\sin(xt) - u(2t + x\cos(xt)) \\ &\quad\quad - \frac{\partial}{\partial t}\int \sin(xt)\frac{\partial}{\partial x}u\,dx - 2u\sin(xt) \\ &\quad = \sin(xt)\frac{\partial}{\partial t}u - 2tu - \frac{\partial}{\partial t}\int \sin(xt)\frac{\partial}{\partial x}u\,dx - 2u\sin(xt). \end{aligned}$$

Imitating the argument used in Example 3.4.1, we choose u so that

$$u(t) = 1.$$

Hence,

$$\frac{\partial}{\partial t}V(t,x(t)) + f(t,x(t))\frac{\partial}{\partial x}V(t,x(t)) = -2\sin(xt) - 2t = -2V(t,x(t)) - 2t.$$

Here, $p(t) = -2t$ and $\mu(t) = -2$. Now, from (3.68), we get

$$\begin{aligned} V(t,x(t)) = m(t) &= \exp[-2t]\,\mathrm{C} - 2\int^t s\exp[-2(t-s)]\,ds \\ &= \exp[-2(t-a)]\,\mathrm{C} - (t-a) + \frac{1}{2}(1 - \exp[-2(t-a)]). \end{aligned}$$

By solving for $x(t)$ we fulfill our desired goal:

$$x(t) = \frac{1}{t}\sin^{-1}\left(\exp[-2(t-a)]\,\mathrm{c} - \left(t - a + \frac{1}{2}(1 - \exp[-2(t-a)])\right)\right), \quad \text{for } t \neq 0.$$

The following example illustrates the scope of this approach beyond the linear reducible differential equations.

Example 3.5.2. Find a general solution to the given equation

$$dx = -\frac{1}{2}\frac{(\tan x + xt)^3 + 2x}{\sec^2 x + t}\,dt.$$

Solution procedure. We set

$$M(t,x) = -\frac{1}{2}(\tan x + xt)^3 - x$$

and $N(t,x) = \sec^2 x + t$. We note that

$$\frac{\partial}{\partial x}M(t,x) = -\frac{3}{2}(\tan x + xt)^2\left(\sec^2 x + t\right) - 1,$$
$$\frac{\partial}{\partial t}N(t,x) = 1.$$

From this and Observation 3.4.1(iii), the given differential is neither exact nor reducible to being exact by the integrating factor. By following the argument used in Example 3.4.2, we arrive at

$$\begin{aligned}
V(t,x) &= \int u(\sec^2 x + t)dx \\
&= u(\tan x + tx) - \int_a^x (\tan x + tx)\frac{\partial}{\partial x}u\,dx \quad \text{(by integration by parts)},
\end{aligned}$$
$$\frac{\partial}{\partial x}V(t,x) = u\frac{\partial}{\partial x}(\tan x + tx) = u\left(\sec^2 x + t\right),$$
$$\frac{\partial}{\partial t}V(t,x) = \frac{\partial}{\partial t}[u(\tan x + tx)] - \frac{\partial}{\partial t}\left[\int_a^x (\tan x + tx)\frac{\partial}{\partial x}u\,dx\right].$$

The expression in (3.67)

$$\begin{aligned}
&\frac{\partial}{\partial t}V(t,x(t)) + f(t,x(t))\frac{\partial}{\partial x}V(t,x(t)) \\
&= \frac{\partial}{\partial t}\left(\int uN(t,x)dx\right) - uM(t,x) \\
&= \frac{\partial}{\partial t}[u(\tan x + tx)] - \frac{\partial}{\partial t}\left[\int_a^x (\tan x + tx)\frac{\partial}{\partial x}u\,dx\right] \\
&\quad - u\frac{1}{2}\left((\tan x + xt)^3 + 2x\right) \\
&= u\frac{\partial}{\partial t}(\tan x + tx) + (\tan x + tx)\frac{\partial}{\partial t}(u) \\
&\quad - \frac{\partial}{\partial t}\left[\int_a^x (\tan x + tx)\frac{\partial}{\partial x}u\,dx\right] - u\frac{1}{2}(\tan x + xt)^3 - ux.
\end{aligned}$$

For the choice of $u = 1$, the above expression reduces to

$$\begin{aligned}
&\frac{\partial}{\partial t} V(t, x(t)) + f(t, x(t)) \frac{\partial}{\partial x} V(t, x(t)) \\
&= \frac{\partial}{\partial t}(\tan x + tx) - \frac{1}{2} (\tan x + xt)^3 - x \\
&= x - \frac{1}{2} (\tan x + xt)^3 - x \\
&= -\frac{1}{2} (\tan x + xt)^3 \\
&= -\frac{1}{2} (V(t, x(t)))^3.
\end{aligned}$$

The reduced differential equation is

$$dm = -\frac{1}{2} m^3.$$

Its solution will be determined by the procedure described in Section 3.6. After finding the solution to this equation, one can mirror the procedure discussed in Observation 3.5.1(i). This completes the solution process of the given example.

3.5 Exercises

(1) Determine whether the given differential equation describes a conservative vector field in R^2.

(a) $(x - t\cot x)\, dt + \left(t + \frac{1}{2}t^2 \csc^2 x\right) dx = 0$
(b) $(2 + \sin x)\, dt + (5 + t\cos x)\, dx = 0$
(c) $\left(t^2 + 3x\right) dt + 3\left(t + x^2\right) dx = 0$
(d) $(\exp[t] + x\cos t)\, dt + (\exp[x] - x\sin t)\, dx = 0$

(2) Find the general solution to following differential equations:

(a) $(1 - xt)\, dt + t\,(y - t)\, dx = 0$
(b) $\left(2xt + 3t^2 x + 3x^2\right) dt + \left(t^2 + 2x\right) dx = 0$
(c) $\left(\ln x + 3x^2\right) dt + \left(\frac{t}{x} + 6tx\right) dx = 0$
(d) $x\,(1 + xt)\, dt + t\,(1 - xt)\, dx = 0$

(3) Find the general solution representation of the differential equations in Exercises 1 and 2.

(4) Find the general solution to the following differential equations:

(a) $2\,(\csc hx + t)\, dx + \left[\left(\ln\left(\tanh\left(\frac{x}{2}\right)\right) + xt\right)^3 + 2x\right] dt = 0$

(b) $4\left(\sec^2 x + t\right) dx + \left[(\tan x + xt)^5 + 4x\right] dt = 0$

(c) $4\,(\tanh x + t)\, dx + \left[(\ln(\cosh x) + tx)^5 + 4x\right] dt = 0$

(d) $2\left((1+\exp[-x])^{-1}+t\right)dx+\left[\ln\left(\frac{\exp[x]}{1+\exp[x]}+tx\right)^3+2x\right]dt=0$

(e) $2\left(\frac{d}{dx}h+t\right)dx+\left[(h(x)+tx)^3+2x\right]dt=0$

(f) $4\left(\frac{1}{\sqrt{1+x^2}}+t\right)dx+\left[\left(\ln\left(x+\sqrt{1+x^2}\right)+xt\right)^5+4x\right]dt=0$

Hint:

$$\int\frac{dx}{\sqrt{1+x^2}}=\ln\left(x+\sqrt{1+x^2}\right),$$

$$\int\frac{dx}{x\sqrt{1-x^2}}=-\ln\left(\frac{1+\sqrt{1-x^2}}{x}\right),$$

$$\int\operatorname{csc}hx\,dx=\ln\left(\tanh\left(\frac{x}{2}\right)\right),$$

$$\int\tanh x\,dx=\ln(\cosh x),$$

$$\int\frac{dx}{1+\exp[x]}=\ln\left(\frac{\exp[x]}{1+\exp[x]}\right).$$

(5) Find the general solution to the following differential equations:

(a) $4(g(x)+t)\,dx+\left[(G(x)+tx)^5+4x\right]dt=0$, where $G(x)=\int^x g(s)\,ds$

(b) $2\left(\frac{d}{dx}H(x)+b(t)\right)dx+\left[(H(x)+xb(t))^3+2x\frac{d}{dt}b(t)\right]dt=0$

3.6 Variable Separable Equations

In this section, we apply the energy function method developed in Section 3.4 to a particular class of differential equations (3.38). This class of differential equations is easily reducible to (3.45). The structure of this class of differential equations is characterized by the rate functions $f(t,x)$ in (3.38). In fact, the nonlinear rate function is decomposed into a product of two functions, one of which is a function of an independent variable, and the other a function of a dependent variable.

Definition 3.6.1. A function F defined on $J\times R$ into R is said to be a separable function in t and x variables if $F(t,x)=H(t)G(x)$. H and G are functions of t and x, respectively.

Definition 3.6.2. A differential equation (3.38) is said to be a separable differential equation if its rate function $f(t,x)$ is a separable function in t and x variables.

($H_{\mathbf{3.6}}$): In addition to hypothesis ($H_{\mathbf{3.3}}$), we assume that $f(t,x)$ is a separable function in t and x variables: $f(t,x)=a(t)b(x)$, a and b are continuous, and

$$G(x)=\int_c^x\frac{ds}{b(s)},$$

where G is invertible.

3.6.1 *General problem*

In the light of hypothesis ($\mathbf{H_{3.6}}$), (3.38) reduces to

$$dx = a(t)b(x)\,dt, \tag{3.80}$$

where the functions a and b satisfy conditions in hypothesis ($\mathbf{H_{3.6}}$).

3.6.2 *Procedure for finding a general solution*

The basic ideas [(3.40) and (3.41)] in step 2 of the procedure in Subsection 3.3.2 remain valid.

Step 1. In this case, it is obvious that we seek $V(t, x) \equiv V(x)$ in (3.46) and $p(t) = a(t)$. Thus, (3.46) reduces to

$$\begin{aligned} a(t) &= f(t,x)\frac{\partial}{\partial x}V(x) \\ &= a(t)b(x)\frac{\partial}{\partial x}V(x) \quad \text{(by substitution).} \end{aligned} \tag{3.81}$$

This implies that

$$b(x)\frac{\partial}{\partial x}V(x) = 1 \quad [\text{provided that } a(t) \neq 0]$$

and hence

$$\frac{\partial}{\partial x}V(x) = \frac{1}{b(x)}. \tag{3.82}$$

The partial derivative in (3.81) is indeed an ordinary derivative. Therefore, the differential equation (3.82) is an integrable differential equation, and it can be written as

$$dV(x) = \frac{1}{b(x)}\,dx. \tag{3.83}$$

The differential equation (3.83) is in an integrable form (Section 2.3). Hence,

$$V(x) = \int_c^x \frac{1}{b(s)}\,ds + C, \tag{3.84}$$

where C is an arbitrary constant of integration. Thus, $V(t,x) \equiv V(x)$ is determined. Moreover, from ($\mathbf{H_{3.6}}$), $V(x) = G(x)$.

Step 2. Now, from (3.84), we compute the differential of $V(t,x) \equiv G(x)$ along the given solution to (3.80) in the context of a separable differential equation. We get

$$\begin{aligned} \frac{\partial}{\partial x}V(t,x(t))\,dx(t) &\equiv dG(x(t)) \\ &= \frac{1}{b(x(t))}\,dx(t) \quad [\text{from (3.83)}] \end{aligned}$$

$$= \frac{1}{b(x(t))} b(x(t))a(t)\, dt \quad [\text{by } f(t,x) = b(x)a(t)]$$
$$= a(t)\, dt. \tag{3.85}$$

The differential equation (3.85) is in an integrable form. Therefore, by direct integration of both sides with respect to t, we have

$$G(x(t)) = \int_a^t a(s)\, ds + \mathrm{C}, \tag{3.86}$$

where a and C are arbitrary real numbers.

Step 3. Now, by applying step 5 of Section 3.4, we have

$$x(t) = G^{-1}\left(\int_a^t a(s)\, ds + \mathrm{C}\right), \tag{3.87}$$

where G^{-1} (if possible) stands for the inverse function of G ($G(x) \equiv V(t,x)$). Thus, the solution process in (3.87) is written as

$$x(t) = E\left(t, \left(\int_a^t a(s)\, ds + \mathrm{C}\right)\right) \equiv G^{-1}\left(\int_c^x a(s)\, ds + \mathrm{C}\right). \tag{3.88}$$

This is based on the notation defined in step 1 of the general procedure in Section 3.3.

This completes our procedure with respect to the class of separable deterministic differential equations (3.80) in the context of the energy function method.

Example 3.6.1. Let $x = [S]$ be a concentration of a reactant (substrate). It decomposes under the presence of an enzyme E. Let x_0 be the initial concentration of a substrate at the initial time $t = t_0$. Let $[E^*]$ be the total enzyme concentration in the steady state. From Illustration 3.2.2, we have

$$dx = -\frac{g(t)x}{k+x}\, dt, \quad x(t_0) = x_0,$$

where $g(t)$ is a continuous function and k is a positive constant. Find the solution to this IVP.

Solution procedure. We observe that the given differential equation is separable. Here, $a(t) = -g(t)$ and $b(x) = \frac{x}{k+x}$. We use the logic of the procedure in Subsection 3.6.2, and we have

$$-g(t) = -\frac{g(t)x}{k+x} \frac{\partial}{\partial x} G(x),$$
$$dG(x) = \frac{k+x}{x}\, dx.$$

This is an integrable differential equation, and its general solution is given by

$$V(t,x) \equiv G(x) = \int^{x} \frac{k+s}{s}\, ds + \mathrm{C}$$
$$= \int_{c}^{x} \left[\frac{k}{s}+1\right] ds + \mathrm{C} = k\ln(|x|) + x + \mathrm{C},$$

and hence

$$G(x) - G(x_0) = [k\ln(|x|) + x + C]|_{x_0}^{x} = k\ln\left(\left|\frac{x}{x_0}\right|\right) + (x - x_0).$$

On the other hand, from (3.86), we have

$$G(x(t)) - G(x_0) = -\int_{t_0}^{t} g(s)\, ds.$$

By comparing the above two expressions for $G(x(t)) - G(x_0)$, we obtain

$$k\ln\left(\left|\frac{x(t)}{x_0}\right|\right) + (x - x_0) = -\int_{t_0}^{t} g(s)\, ds.$$

Thus, the solution to the given IVP is given by the above implicit expression.

Illustration 3.6.1 (hydrolysis of esters [86, 99]). Esters are composed of a strong acid. The reaction mechanism of acid-catalyzed hydrolysis is considered to be described by

$$\mathrm{S} + \mathrm{HA} \rightleftarrows \mathrm{SH}^+ + \mathrm{A}^- \text{ (reaction rate: } k_1\text{“}\rightarrow\text{” and } k_{-1}\text{“}\leftarrow\text{”)} \text{ — fast reaction,}$$
$$\mathrm{SH}^+ + \mathrm{H_2O} \rightarrow \mathrm{P_1} + \mathrm{P_2} \text{ (reaction rate: } k_2\text{“}\rightarrow\text{”)} \text{ — slow reaction,}$$

where S stands for esters ($R'CO_2R$); HA, acid; SH^+, base ($R'C^+OOHR$); A^-, acid ($R'COOH$); P_1 and P_2, products (ROH and H^+). An acid acts as a catalyst. It is recognized by Bronsted that acid is capable of donating a proton to another molecule. Water is an amphoteric solvent. It has at least one replaceable hydrogen atom to form an intermediate molecule SH. Moreover, it participates in

$$\mathrm{HA} + \mathrm{SH} \rightleftarrows \mathrm{SH}_2^+ + \mathrm{A}^- \quad \text{(acid + base} \rightleftarrows \text{acid + base).}$$

Let us denote by a and b the initial concentration of ester (RCO_2R') and H^+ ions, respectively. Let $x = [RCO_2R']$ be the concentration reduction of the esters. During the reaction the strong acid will produce $x = [H^+]$.

(a) Show that $dx = k(a-x)(b+x)\, dt$, $x(0) = 0$.
(b) Find the solution to $x(t)$.
(c) Make a sketch of the solution process.

Solution procedure. We use the ideas of Illustrations 3.2.1 and 3.2.2 to derive the rate equations, and we arrive at the mathematical model of the dynamic process of the hydrolysis of ester:

$$\begin{aligned}\frac{d}{dt}[\mathrm{S}] &= -k_1[\mathrm{S}][\mathrm{HA}] + k_{-1}\left[\mathrm{SH}^+\right]\left[\mathrm{A}^-\right],\\ \frac{d}{dt}\left[\mathrm{SH}^+\right] &= k_1[\mathrm{S}][\mathrm{HA}] - k_{-1}\left[\mathrm{SH}^+\right]\left[\mathrm{A}^-\right] - k_2\left[\mathrm{SH}^+\right],\\ \frac{d}{dt}[\mathrm{HA}] &= -k_1[\mathrm{S}][\mathrm{HA}] + k_{-1}\left[\mathrm{SH}^+\right]\left[\mathrm{A}^-\right],\\ \frac{d}{dt}\left[\mathrm{A}^-\right] &= k_1[\mathrm{S}][\mathrm{HA}] - k_{-1}\left[\mathrm{SH}^+\right]\left[\mathrm{A}^-\right].\end{aligned}$$

Under the stationary state hypothesis (Illustration 3.2.1) and the strong acid condition, [86, 99] we have

$$\begin{gathered}k_1[\mathrm{S}][\mathrm{HA}] - k_{-1}\left[\mathrm{SH}^+\right]\left[\mathrm{A}^-\right] = 0,\\ \mathrm{HA} \rightleftarrows \mathrm{H}^+ + \mathrm{A}^- \text{ implies } k_1'[\mathrm{HA}] = k_{-1}'\left[\mathrm{H}^+\right]\left[\mathrm{A}^-\right].\end{gathered}$$

Hence,

$$\left[\mathrm{SH}^+\right] = \frac{k_1[\mathrm{S}][\mathrm{HA}]}{k_{-1}\left[\mathrm{A}^-\right]}, \quad \mathrm{K} = \frac{\left[\mathrm{H}^+\right]\left[\mathrm{A}^-\right]}{[\mathrm{HA}]}.$$

Thus,

$$\frac{d}{dt}\left[\mathrm{SH}^+\right] = -k_2\frac{k_1[\mathrm{S}][\mathrm{HA}]}{k_{-1}\left[\mathrm{A}^-\right]} = -\frac{k_2k_1[\mathrm{S}]\left[\mathrm{H}^+\right]}{k_{-1}\mathrm{K}} = -k[\mathrm{S}]\left[\mathrm{H}^+\right].$$

We know that the rate of product formation is equal to

$$-\frac{d}{dt}\left[\mathrm{SH}^+\right] = -\frac{d}{dt}[\mathrm{S}].$$

Hence,

$$dx = d(a - [\mathrm{S}]) = -\frac{d}{dt}[\mathrm{S}] = k[\mathrm{S}]\left[\mathrm{H}^+\right] = k(a-x)(b+x)\,dt, \quad x(0) = 0.$$

This validates (a).

To find the solution to the IVP in (a), we follow the procedure in Subsection 3.6.2 and the argument in Example 3.6.1 to get

$$\begin{aligned}k &= k(a-x)(b+x)\frac{\partial}{\partial x}G(x),\\ dG(x) &= \frac{1}{(a-x)(b+x)}\,dx.\end{aligned}$$

This is an integrable differential equation, and its general solution is given by

$$\begin{aligned}
V(t,x) \equiv G(x) &= \int^x \frac{1}{(a-s)(b+s)}\,ds + C \\
&= \frac{1}{a+b}\int_c^x \left[\frac{1}{a-s} + \frac{1}{b+s}\right] ds + C \\
&= \frac{1}{a+b}\left[\ln(b+x) - \ln(a-x)\right] + C \quad \text{(by partial fractions)} \\
&= \frac{1}{a+b}\ln\left(\frac{b+x}{a-x}\right) + C \quad \text{(by algebra and } a-x>0\text{)}.
\end{aligned}$$

Hence,

$$G(x) - G(x_0) = \frac{1}{a+b}\ln\left(\frac{a(b+x)}{b(a-x)}\right) = \ln\left(\frac{a(b+x)}{b(a-x)}\right)^{\frac{1}{a+b}}.$$

On the other hand, from (3.86), we have

$$G(x(t)) - G(x_0) = kt.$$

By comparing the above two expressions for $G(x) - G(x_0)$, we obtain

$$\ln\left(\frac{a(b+x(t))}{b(a-x(t))}\right)^{\frac{1}{a+b}} = kt.$$

From this and the fact that the exponential function with base e is the inverse of ln, we have

$$\left(\frac{a(b+x(t))}{b(a-x(t))}\right)^{\frac{1}{a+b}} = \exp[kt].$$

By algebraic simplification, we get

$$\frac{a(b+x(t))}{b(a-x(t))} = \exp\left[k(a+b)t\right]$$

and hence

$$a(b+x) = b(a-x)\exp[k(a+b)t].$$

Thus,

$$x(t)(a + b\exp[k(a+b)t]) = ab(\exp[k(a+b)t] - 1).$$

Therefore,

$$x(t) = \frac{ab(\exp[k(a+b)t]-1)}{a+b\exp[k(a+b)t]} = \frac{ab(1-\exp[-k(a+b)t])}{a\exp[-k(a+b)t]+b}$$
$$= \frac{ab(1-\exp[-k(a+b)t])}{\exp[-k(a+b)t]\,a+b}.$$

This is the solution to the given IVP (b).

To make the sketch, we note that: (i) $x(0) = 0$, and its horizontal asymptote is a, i.e. $x(t) \to a$ as $t \to \infty$; (ii) its domain and range are $[0, \infty)$ and $[0, a)$, respectively; and (iii) its sketch is a sigmoid curve.

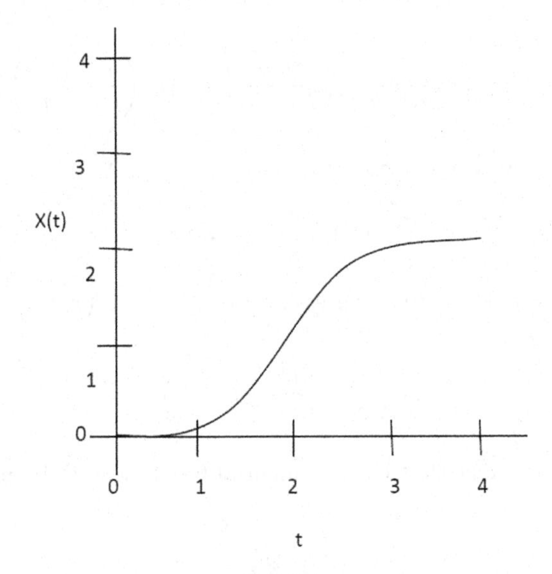

Fig. 3.4

Example 3.6.2. The following is given: $dN = g(t)N(k-N)dt$, $N(t_0) = N_0$, where k is a positive constant and $g(t)$ is a continuous function of t. Find the solution to this IVP. Make a sketch of the solution curve.

Solution procedure. By using the procedure in Subsection 3.6.2 and the arguments in Illustration 3.6.1, we obtain

$$g(t) = g(t)N(k-N)\frac{\partial}{\partial N}G(N),$$
$$dG(N) = \frac{1}{N(k-N)}\,dN.$$

This is an integrable differential equation, and its general solution is given by

$$\begin{aligned}
V(t,N) \equiv G(N) &= \int^{N} \frac{1}{s(k-s)}\,ds + C \\
&= \frac{1}{k}\int_{c}^{N}\left[\frac{1}{s}+\frac{1}{k-s}\right]ds + C \\
&= \frac{1}{k}\left[\ln(N) - \ln(|k-N|)\right] + C \quad \text{(by partial fractions)} \\
&= \frac{1}{k}\ln\left(\left|\frac{N}{k-N}\right|\right) + C \quad \text{(by algebra and } a-N>0\text{)}.
\end{aligned}$$

Hence,

$$G(N) - G(N_0) = \frac{1}{k}\ln\left(\left|\frac{N(k-N_0)}{N_0(k-N)}\right|\right).$$

On the other hand, from (3.86), we have

$$G(N(t)) - G(N_0) = \int_{t_0}^{t} g(s)\,ds.$$

By comparing the above two expressions for $G(N(t)) - G(N)$, we obtain

$$\ln\left[\left|\frac{N(t)(k-N_0)}{N_0(k-N(t))}\right|\right] = k\int_{t_0}^{t} g(s)\,ds,$$

which implies that

$$\left|\frac{N(t)(k-N_0)}{N_0(k-N(t))}\right| = \exp\left[k\int_{t_0}^{t} g(s)\,ds\right].$$

We solve this equation for $N \equiv N(t)$ and obtain a solution to the IVP:

$$\begin{aligned}
N(t) &= \frac{kN_0\exp\left[k\int_{t_0}^{t} g(s)\,ds\right]}{(k-N_0) + N_0\exp\left[k\int_{t_0}^{t} g(s)\,ds\right]} \\
&= \frac{kN_0}{N_0 + (k-N_0)\exp\left[-k\int_{t_0}^{t} g(s)\,ds\right]}.
\end{aligned}$$

To make the sketch, we note that: (i) $N(t_0) = N_0$, and its horizontal asymptote is k, i.e. $x(t) \to k$ as $t \to \infty$; (ii) for $N_0 > 0$, its domain and range are $[0,\infty)$ and $[N_0, k)$ (or $(k, N_0]$), respectively; and (iii) its sketch is a sigmoid curve.

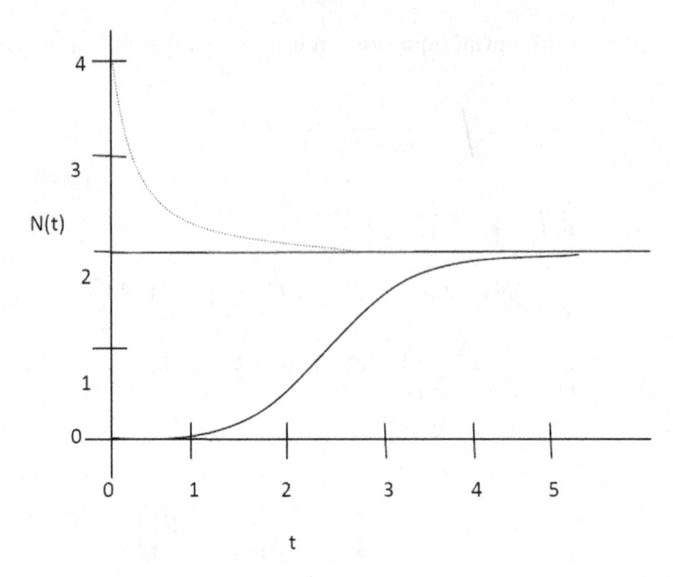

Fig. 3.5

Illustration 3.6.2 (water leakage problem: Torricelli's law [11, 24, 27]). A tank contains a liquid of uniform density ρ. It has been observed that the bottom of the tank has a hole with an area a. Let $h(t)$ and $V(t)$, respectively, be the height and volume of the liquid in the tank at a time t. Under normal physical conditions, the rate of leakage of the liquid (velocity) at a time t is given by the conservation-of-energy formula: $P(a) + K(a) = P(b) + K(b)$ (Illustration 3.4.1). Using the rectangular coordinate system at the center of the hole, we note that the motion of the liquid leakage is vertically downward. In this case, the initial position and initial velocity of the liquid element are h and 0, respectively, and the final position and the velocity are 0 and v, respectively. Hence, $P(a) = mgh$, $K(a) = 0$, $P(b) = 0$, and

$$K(b) = \frac{1}{2}v^2.$$

Therefore, from the conservation of energy, we have

$$P(a) + K(a) = P(b) + K(b),$$
$$mgh + 0 = 0 + \frac{1}{2}v^2 \quad \text{(by substitution)}.$$

We solve for v, and obtain

$$v = \sqrt{2mgh} = \sqrt{2\rho gh}.$$

Thus, the rate of outflow of the mass through the hole is $a\sqrt{2\rho g h}$. On the other hand, due to the leakage, the rate of change of volume of the liquid is given by

$$\frac{d}{dt}V(t).$$

Hence, the rate of outflow of the mass of the liquid is

$$\rho\frac{d}{dt}V(t).$$

Therefore,

$$\rho\frac{d}{dt}V(t) = -a\sqrt{2\rho g h}.$$

By algebraic simplification, this implies that

$$\frac{d}{dt}V(t) = -k\sqrt{h},$$

where

$$k = a\sqrt{\frac{2g}{\rho}}.$$

For water ($\rho = 1$), the above statement (mathematical) is called Torricelli's law for water draining.

Knowing the geometric nature of the tank, we can rewrite Torricelli's law in the form of a differential equation. To do this, we need to find an expression for

$$\frac{d}{dt}V(t)$$

as a function of h. Suppose that $A(h)$ stands for the horizontal cross-sectional area of the tank at height h above the hole. We use the method of finding the volume of a solid by cross-sections (calculus [85]), and we have

$$V(h) = \int_0^h A(y)\,dy,$$

which implies that

$$\begin{aligned}
\frac{d}{dt}V(h) &= \frac{d}{dt}\left[\int_0^h A(y)dy\right] \quad \text{(by differentiating both sides)}\\
&= \frac{d}{dh}\left[\int_0^h A(y)dy\right]\frac{dh}{dt} \quad \text{(by the chain rule)}\\
&= A(h)\,\frac{dh}{dt} \quad \text{(by the fundamental theorem of calculus)}.
\end{aligned}$$

By comparing this with

$$\frac{d}{dt}V(t) = -k\sqrt{h},$$

we have

$$A(h)\,\frac{dh}{dt} = -k\sqrt{h}.$$

This is an alternative form of Torricelli's law and is rewritten as

$$dh = -k\frac{\sqrt{h}}{A(h)}dt.$$

This is a variable separable differentiable equation. By following the method (Procedure 3.6) described in this section, the general solution to this differential equation is given by

$$V(t, h(t)) \equiv G(h(t)) = \int_0^h \left[\frac{A(s)}{\sqrt{s}}\right] ds = -kt,$$

where the initial time is $t = 0$.

Example 3.6.3. Suppose that a spherical shape tank with a diameter of 4 m has a circular hole with a radius of 1 cm. Let $g = 10\,\mathrm{m/s}^2$ and note that the tank is half-full of water.

(a) Show that the IVP associated with the example is

$$dh = -0.0001\pi\frac{\sqrt{20h}}{A(h)}dt, \quad h(0) = 2.$$

(b) Find the solution representation of the IVP in (a).
(c) How long will it take to drain the water completely?

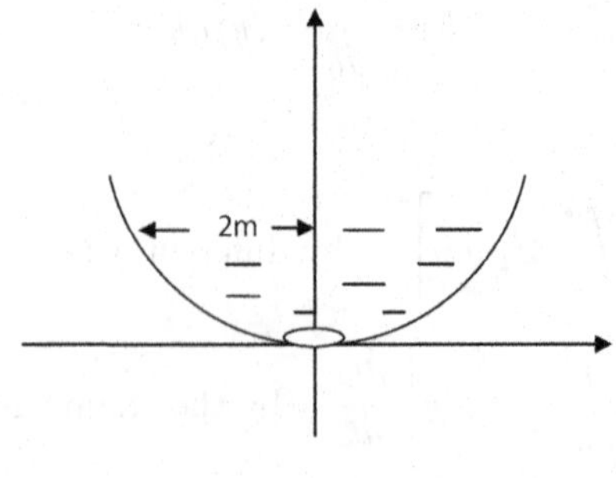

Fig. 3.6

Solution procedure. Let $h(t)$ be the height of the water in the tank. The area of the hole is 0.0001 square meters. From Illustration 3.6.2, the differential equation is given by

$$dh = -k\frac{\sqrt{h}}{A(h)}dt = -0.0001\pi\frac{\sqrt{20h}}{A(h)}dt, \quad h(0) = 2,$$

where $A(h) = \pi r^2 = \pi\left[4 - (2-h)^2\right] = \pi\left(4h - h^2\right)$. By following the procedure described in Illustration 3.6.1, we have

$$\begin{aligned}
-0.0001\sqrt{5}\,t = G(h(t)) &= \int_2^h \left[\frac{(4s - s^2)}{\sqrt{s}}\right] ds \\
&= \int_0^h \left(4s^{\frac{1}{2}} - s^{\frac{3}{2}}\right) ds = \left(\frac{8}{3}s^{\frac{3}{2}} - \frac{2}{5}s^{\frac{5}{2}}\right)\Big|_0^h \\
&= \left(\frac{8}{3}h(t)^{\frac{3}{2}} - \frac{2}{5}h(t)^{\frac{5}{2}}\right) - \left(\frac{16}{3}\sqrt{2} - \frac{8}{5}\sqrt{2}\right) \\
&= \left(\frac{8}{3}h(t)^{\frac{3}{2}} - \frac{2}{5}h(t)^{\frac{5}{2}}\right) - \frac{56}{15}\sqrt{2}.
\end{aligned}$$

The height of the water tank at any time is given by the above implicit expression of h. The tank will be empty when $h(t) = 0$. From this, the solution expression reduces to

$$-0.0004\sqrt{5}\,t = -\frac{56}{15}\sqrt{2}.$$

Now, we solve for t and get

$$t = \frac{\frac{56}{15}\sqrt{2}}{0.0004\sqrt{5}} = 10000\frac{14}{75}\sqrt{10}.$$

$$dh = -k\frac{\sqrt{h}}{A(h)}dt = -0.0001\pi\frac{\sqrt{20h}}{A(h)}dt, \quad h(0) = 2.$$

This is the length of time required to empty the tank.

Illustration 3.6.3. [30, 31, 99] Let A, B, and B' be chemical substances with initial concentrations $a = [A_0]$, $b = [B_0]$, and $b' = [B'_0]$, respectively. Let $[A]$, $[B]$, and $[B']$ be the concentrations of substances A, B, and B', respectively, at time t. Let us consider a general chemical reaction mechanism described by

Concurrent reaction: $A + B \to C$ (reaction rate: k) and $A + B' \to D$ (reaction rate: k').

Assume that the orders of reactions relative to A, B, and B', are $n = m$ for the first reaction and $n' = m' = n$ for the second reaction.

By following the procedure outlined in Illustrations 3.2.1 and 3.2.2, we arrive at

$$d[B] = k(a - [A])^n(b - [B])^n\,dt \quad \text{and} \quad d[B'] = k'(a - [A])^n\,(b' - [B'])^n\,dt.$$

The relationship of the concentrations between the substances B and B' can be established by eliminating the independent variable t (in this case, by taking the ratios of these rates):

$$d[B] = \frac{k}{k'} \frac{(b - [B])^n}{(b' - [B'])^n} \, d\,[B'].$$

This differential equation describes a growth rate of the reactant B with respect to the reactant B' as a kind of a new "chemical time," with B' playing the role of a "chemical clock." This is referred to as the relative growth rate in biological sciences [45, 86, 115, 133].

By setting $\frac{k}{k'} = \alpha, y = (b - [B])$, $z = b' - [B']$, and using the concept of the differential and its properties, the above differential equation reduces to

$$dy = \alpha \frac{y^n}{z^n} \, dz = \alpha \left(\frac{y}{z}\right)^n \, dz = f(z, y) \, dz.$$

We note this is a separable differential equation. By following the method described in this section, the general solution to this differential equation is given by

$$\begin{aligned} V(t, y(z)) \equiv G(y(z)) &= \int^y \frac{1}{s^n} \, ds = \frac{1}{1-n} y^{1-n} \\ &= \int \frac{1}{z^n} \, dz + C = \frac{1}{1-n} z^{1-n} + C, \end{aligned}$$

and hence

$$y(z) = \left(z^{1-n} + C\right)^{\frac{1}{1-n}},$$

where C is an arbitrary constant of integration.

This relation gives the solution representation of the differential equation in an explicit form.

Brief summary. The following flowchart summarizes the described steps for solving variable separable equations by using the energy function method:

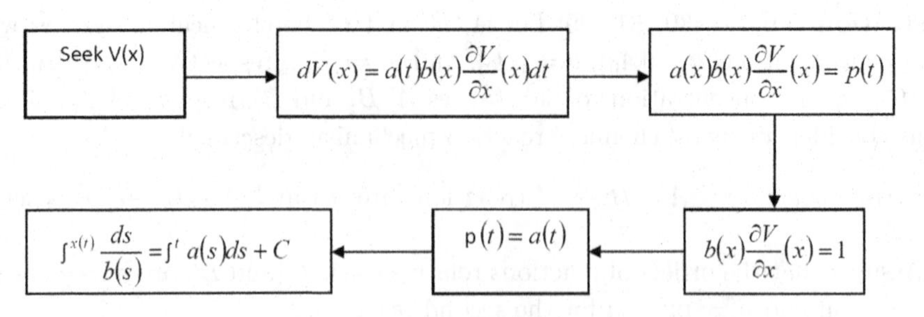

Fig. 3.7 Flowchart: Energy function method applied to separable equations.

3.6 Exercises

(1) Determine whether the given differential equation is separable. If so, solve it.

(a) $dx = (1 + t)a^{-x} dt$ (b) $dx = \sec t \sqrt{a + bx} \, dt$

(c) $dx = e^t \frac{\sqrt{a+bx}}{x} dt$ (d) $dx = \frac{\sqrt{(a^2-x^2)^3}}{x} dt$

(e) $dx = e^t \sqrt{a+bx}\, dt$ (f) $dx = \frac{\sqrt{(a^2-x^2)^3}}{x} dt$

(g) $dx = (1+t)3^{-x} dt$ (h) $dx = \sqrt{1+2x}\, dt$

(i) $dx = \frac{\sqrt{4+3x}}{x} dt$

(2) Determine whether the given differential equation is separable. If so, find the solution.

(a) $dx = -\frac{\ln a}{2}(1+t)^2 a^{-2x} dt$, $a > 0$

(b) $dx = \left[-\frac{\ln a}{2}(1+t)^2 a^{-2x} + 2a^{-x}\right] dt$, $x(t_0) = x_0$, $a > 0$

(c) $dx = -\frac{\ln a}{2}(1+t)^2 a^{-x}\, dt$, $x(t_0) = x_0$, $a > 0$

(d) $dx = \sec t \sqrt{a+bx}\, dt$, $b \neq 0$

(3) Determine whether the given differential equation is separable. If so, find the solution.

(a) $dx = e^t \frac{\sqrt{a+bx}}{x} dt$, $b \neq 0$ (b) $dx = -\frac{ae^{2t}}{2x^3} dt$, $x(t_0) = x_0$, $b \neq 0$

(c) $dx = \left[-\frac{\ln 3}{2(1+t)} 3^{-2x} + 23^{-x}\right] dt$ (d) $dx = \left[-\frac{\ln 3}{2(1+t)} 3^{-2x}\right] dt$, $x(0) = 1$

(4) Let us consider two tanks with an equal cross-sectional area A. The tanks are connected by a pipe in a such a way that the rate of flow of water from the first tank to the second tank is proportional to $(h - H)^{\frac{1}{2}}$, where h and H are the depths of the first and second tanks, respectively, and $h \geq H$. It is assumed that the tanks are empty at $t = 0$. The water is pumped into the first tank at a constant rate β. Let V_1 and V_2 be the volumes of water in the first and second tanks, respectively, at a time t. Thus, $V_1 = Ah$ and $V_2 = AH$. Show that [21]:

(a) $dh = [b - a(h-H)^{\frac{1}{2}}]dt$ and $dH = [a(h-H)^{\frac{1}{2}}]dt$

(b) $d(h+H) = b$

(c) $dH = a(bt - 2H)^{\frac{1}{2}} dt$

(d) $dz = \frac{b-2az}{2z} dt$ (hint: $z^2 = bt - 2H$)

(e) the relationship between H and t is

$$\ln\left(1 - \frac{2a}{b}\sqrt{bt-H}\right) + \frac{2a}{b}\sqrt{bt-H} + \frac{2a^2}{b} t = 0.$$

3.7 Homogeneous Equations

In this section, we present a subclass of differential equations (3.38) that is reducible to (3.65) or (3.45). This class of differential equations is referred to as homogeneous differential equations. Prior to introducing the definition of a homogeneous differential equation, we need a mathematical concept of the homogeneous function of degree n, with n being a real number.

Definition 3.7.1. Let $F(t, x)$ be defined on $J \times R$ into R. The function F is said to be homogeneous in (t, x) with degree n if for $t \neq 0$ and $x = tv$, $F(t, tv) = t^n F(1, v)$, for some real number n. In particular, for $n = 0$, F is simply called a homogeneous function.

Definition 3.7.2. A differential equation (3.38) is said to be a homogeneous differential equation if the rate function $f(t, x)$ in (3.38) is a homogeneous function of degree zero.

3.7.1 *General problem*

We apply the procedure in Subsection 3.5.2 to the homogeneous differential equation

$$dx = f(t, x)dt, \tag{3.89}$$

where f is as defined in (3.38), and $t \neq 0$. This procedure leads to the reduction of the differential equation (3.65).

($H_{3.7}$): In addition to hypothesis ($H_{3.3}$), we assume that the rate function $f(t, x)$ in (3.38) is a homogeneous function of degree 0. Moreover, it is assumed that $f(1, u) - u \neq 0$, and

$$G(u) = \int^{u} \frac{ds}{f(1, s) - s}$$

is defined and invertible.

3.7.2 *Procedure for finding a general solution*

A procedure for reducing a subclass of a class of differential equations (3.38) in the procedure in Subsection 3.3.2 is repeated. With this subclass of homogeneous differential equations, we associate a suitable natural energy/Lyapunov function in a unified way.

Step 1. By using the nature of the subclass of differential equations described by (3.89), we associate an unknown energy/Lyapunov function in step 1 of the procedure in Subsection 3.3.2:

$$V(t, x) = G\left(\frac{x}{t}\right), \tag{3.90}$$

where G is an unknown energy/Lyapunov function. The problem of seeking an energy function V is equivalent to the problem of seeking an unknown function G.

Step 2. The computation of the differential $dV(t, x)$ of V in (3.40) is achieved by using the substitution

$$v = \frac{x}{t}, \tag{3.91}$$

as follows:

$$\begin{aligned}\frac{\partial}{\partial t}V(t,x) &= \frac{\partial}{\partial t}G\left(\frac{x}{t}\right) \quad \text{(by definition)}\\ &= \frac{d}{dv}G(v)\frac{\partial}{\partial t}v \quad \text{(by the chain rule)}\\ &= \frac{d}{dv}G(v)\left(-\frac{x}{t^2}\right) \quad \left[\text{from (3.91), } \frac{\partial}{\partial t}v = \frac{\partial}{\partial t}\left(\frac{x}{t}\right) = -\frac{x}{t^2}\right] \qquad (3.92)\\ \frac{\partial}{\partial x}V(t,x) &= \frac{\partial}{\partial x}G\left(\frac{x}{t}\right) \quad \text{(by definition)}\\ &= \frac{d}{dv}G(v)\frac{\partial}{\partial x}v \quad \text{(by the chain rule)}\\ &= \frac{d}{dv}G(v)\left(\frac{1}{t}\right) \quad \left[\text{from (3.91), } \frac{\partial}{\partial x}v = \frac{\partial}{\partial x}\left(\frac{x}{t}\right) = \frac{1}{t}\right]. \qquad (3.93)\end{aligned}$$

Hence, from (3.40)–(3.41) and (3.89)–(3.93), we have

$$\begin{aligned}dV(t,x) &= dG\left(\frac{x}{t}\right) \quad \text{(by definition)}\\ &= LG\left(\frac{x}{t}\right)dt \quad \text{[from (3.89), (3.39), and (3.93)]}\\ &= LG(v)\,dt \quad \text{(by substitution and homogeneity)}, \qquad (3.94)\end{aligned}$$

where $x(t)$ is determined by (3.89), and L in (3.41) is a linear operator associated with (3.89)–(3.94). Moreover,

$$\begin{aligned}LV(t,x) &= \frac{d}{dv}G(v)\left(-\frac{x}{t^2}\right) + f(t,x)\frac{d}{dv}G(v)\left(\frac{1}{t}\right) \quad \text{[from (3.92) and (3.93)]}\\ &= \frac{d}{dv}G(v)\left(-\frac{tv}{t^2}\right) + f(t,tv)\frac{d}{dv}G(v)\left(\frac{1}{t}\right) \quad \text{[from (3.91)]}\\ &= \frac{d}{dv}G(v)\left(-\frac{tv}{t^2}\right) + f(t,tv)\frac{d}{dv}G(v)\left(\frac{1}{t}\right) \quad \text{(by homogeneity)}\\ &= \frac{d}{dv}G(v)\left(-\frac{v}{t}\right) + f(1,v)\frac{d}{dv}G(v)\left(\frac{1}{t}\right) \quad \text{(by regrouping)}\\ &= LG(v)dt \quad \text{(by notation)}, \qquad (3.95)\\ LG(v) &= \left[\frac{f(1,v)-v}{t}\right]\frac{d}{dv}G(v). \qquad (3.96)\end{aligned}$$

In this case, from (3.90)–(3.91) and (3.94)–(3.96), the expressions in (3.46) reduce to

$$p(t) = \left[\frac{f(1,v)-u}{t}\right]\frac{d}{dv}G(v). \qquad (3.97)$$

Step 3. Now, by using (3.97), we find a candidate for $G(v)$. We choose $p(t) = \frac{1}{t}$. From this and recalling Procedure 3.6, in particular (3.81), (3.97) becomes

$$\left[\frac{f(1,v)-v}{t}\right]\frac{d}{dv}G(v) = \frac{1}{t}.$$

By simplifying the above equation [in view of (3.82)] and using hypothesis ($H_{3.7}$), we obtain

$$\frac{d}{dv}G(v) = \frac{1}{f(1,v)-u}. \tag{3.98}$$

By applying the solution procedure of Section 2.4, we have

$$G(v) = \int_c^v \frac{ds}{f(1,s)-s} + \mathrm{C}, \tag{3.99}$$

where C is an arbitrary constant of integration.

Step 4. From hypothesis ($H_{3.7}$), G is invertible. Thus, by replacing $v = \frac{x}{t}$, (3.99) gives the solution representation of (3.89) in an implicit form:

$$x(t) = t\left(G^{-1}\left(\int_c^{\frac{x}{t}} \frac{ds}{(f(1,s)-s)} + \mathrm{C}\right)\right) = tG^{-1}(\ln t + \mathrm{C}). \tag{3.100}$$

This completes the procedure.

Illustration 3.7.1 (allometric law [45, 97, 115, 133]). In general, due to the complexity and interconnectivity of the system, it is not easy to find expressions about individual measurements, but it is easier to find a comparative relationship between pairs of measurements. Comparative physiologists are interested in knowing how to count and discount the effects of the differences of the total absolute size with regard to their desire to estimate the comparative development of an organ in a series of related species or groups. The main idea is to exhibit the existence of a basic relation that describes the relative growth processes in the diverse field in a systematic way. It has been well documented [45, 115, 133] that there exists a relation between the dimensions/weight y of an organ with the dimensions/weight x of the entire body (particular reference organ). We note that these organs may differ in nature, structure, and function. In order to derive the relationship between the variables x and y, one can assume (postulate) that:

CGM 1 (Malthusian growth). The growth is a process of the self-reproduction of an animate/inanimate substance. The rate of growth of an organism growing uniformly in all of its parts at a given time is directly proportional to its size at that time.

CGM 2 (density dependence). The growth rate decreases as the size (age) of the organism increases.

CGM 3 (environmental effects). The growth rate is affected by the external environmental changes (e.g. temperature and nutrition).

CGM 4 (uniform distribution of density dependence and environmental effects). The effects of the density dependence and environment on the growth of an organism under consideration are exactly equal to the effects of the density dependence and environment on the entire body. The general factor dependent on the age (size) and environment is the same for all parts of the body. In short, the degree of tolerance/ adaptability to the external environmental changes and age is independent of any part of the body or the entire body.

In summary: from CGM 1 we have

$$\frac{d}{dt}y = ky, \quad \frac{d}{dt}x = lx,$$

where k and l are, respectively, the specific rates of the organ and the body. From CGM 2, CGM 3, and CGM 4, let E stand for the general factor that is dependent on the environment and age (size). The mathematical model of the above-described organ–body dynamic process is

$$dy = kyE\,dt, \quad dx = lxE\,dt.$$

We can rewrite this (by using differentials):

$$\frac{dy}{ky} = E\,dt = \frac{dx}{lx}.$$

By algebraic simplifications, we have

$$dy = \frac{ky}{lx}\,dx = \frac{k}{l}\,\frac{y}{x}dx = \kappa\frac{y}{x}\,dx = f(x,y)\,dx.$$

This is a very simple homogeneous differential equation (it is also in a variable separable form). By following the procedure in Subsection 3.7.2, its general solution is given by

$$G(v) = \int_c^v \frac{ds}{s(\kappa - 1)} = (\kappa - 1)\ln v = \ln x + C,$$

where $v = \frac{y}{x}$, and hence

$$\begin{aligned} v = G^{-1}(\ln x + C) &= \exp\left[\frac{1}{\kappa - 1}\ln x + C\right] \\ &= \exp\left[\ln(x)^{(\kappa-1)}\right]\exp[C] \quad \text{(by properties of ln)} \\ &= [x]^{(\kappa-1)}C \quad (\text{exp is the inverse of ln and } C \equiv \exp[C]). \end{aligned}$$

Thus,

$$\begin{aligned} y &= x[x]^{(\kappa-1)}C \\ &= x^{\kappa}C \quad \text{(by the law of exponents).} \end{aligned}$$

This represents a pattern of growth of an organ relative to the entire body. It is termed allometry [115, 133]. The expression

$$dy = \kappa \frac{y}{x}\,dx$$

is called the allometric law.

Below, we will present a few examples to illustrate the procedure described in this section.

Example 3.7.1. The following is given:

$$dx = \frac{11x^4 - 8x^2t^2 - 16t^4}{8x^3t}\,dt.$$

(a) Verify that the given differential equation is homogeneous, and (b) find its general solution.

Solution procedure. Our goal is to verify that the given differential is homogeneous, and then find a general solution to it. First, we need to show that the rate function f, defined by

$$f(t,x) = \frac{11x^4 - 8x^2t^2 - 16t^4}{8x^3t},$$

is a homogeneous function of degree 0. For this purpose, we substitute $x = vt$ into the function, and we get

$$\begin{aligned} f(t,tv) &= \frac{11(vt)^4 - 8(vt)^2t^2 - 16t^4}{8(vt)^3t} \\ &= \frac{11v^4t^4 - 8v^2t^4 - 16t^4}{8v^3t^4} = \frac{11v^4 - 8v^2 - 16}{8v^3} = f(1,v). \end{aligned}$$

This shows that the rate function is a homogeneous function of degree 0. Therefore, the given differential equation is homogeneous. From the above, we have

$$\frac{f(1,v)-v}{t} = \frac{\frac{11v^4-8v^2-16}{8v^3} - v}{t} = \frac{11v^4 - 8v^2 - 16 - 8v^4}{8tv^3} = \frac{3v^4 - 8v^2 - 16}{8tv^3}.$$

By using the procedure in Subsection 3.7.2, we have

$$\frac{d}{dv}G(v) = \frac{1}{f(1,v)-v} = \frac{8v^3}{3v^4 - 8v^2 - 16}.$$

Hence

$$G(v) = \int_c^v \frac{8s^3}{3s^4 - 8s^2 - 16}\, ds + C.$$

The rest of the solution can be found by computing the above integral by the method of partial fractions and following the procedure in Subsection 3.7.2. The details are left to the reader.

Example 3.7.2. Solve the IVP

$$dx = \frac{1}{t}\left(x + \sqrt{t^2 - x^2}\right) dt,$$

where $x(t_0) = x_0$ for $t \geq t_0 > 0$.

Solution procedure. For this purpose, we follow the argument of Example 3.7.1. First, we need to find a general solution to it. We observe that for $x = tv$ and $t \neq 0$

$$\begin{aligned} f(t,x) &= \frac{1}{t}\left(x + \sqrt{t^2 - x^2}\right) = f(t, tv) \\ &= \frac{1}{t}\left(tv + \sqrt{t^2 - t^2v^2}\right) = \left(v + \sqrt{1 - v^2}\right) = f(1, v). \end{aligned}$$

This shows that f is a homogeneous function of degree 0. Therefore, by Definition 3.7.1, the given differential equation is homogeneous. In addition, we compute

$$\frac{f(1,v) - v}{t} = \frac{(v + \sqrt{1 - v^2}) - v}{t} = \frac{(v + \sqrt{1 - v^2}) - v}{t} = \frac{\sqrt{1 - v^2}}{t}.$$

From (3.49), (3.99) and step 4 of the procedure in Subsection 3.7.2, the solution to the given IVP is

$$G(v(t)) = \int_c^v \frac{1}{\sqrt{1 - s^2}}\, ds + C = \sin^{-1} v(t) + C,$$

$$G(v(t)) - G\left(v(t_0)\right) = \sin^{-1}\frac{x}{t} - \sin^{-1}\frac{x_0}{t_0} = \ln\left[\frac{t}{t_0}\right],$$

and hence

$$\sin^{-1}\frac{x}{t} = \sin^{-1}\frac{x_0}{t_0} + \ln\left[\frac{t}{t_0}\right].$$

Thus,

$$x(t) = t \sin\left(\sin^{-1}\frac{x_0}{t_0} + \ln\left[\frac{t}{t_0}\right]\right).$$

Example 3.7.3. The following is given:

$$dx = \left[\frac{x}{t}\right] - \frac{1}{2}\exp\left[-\frac{2x}{t}\right] dt,$$

where $x(t_0) = x_0$ for $t \geq t_0 > 0$.

(a) Verify that the given differential equation is homogeneous, and (b) find (i) the general solution, and (ii) the IVP of it.

Solution procedure. The goal is to find the solution to the given IVP. For this purpose, we follow the arguments used in Examples 3.7.1 and 3.7.2. Again, we observe that for $x = tv$ and $t \neq 0$

$$\begin{aligned} f(t,x) &= \frac{x}{t} - \frac{1}{2}\exp\left[-\frac{2x}{t}\right] = f(t,tv) \\ &= \frac{vt}{t} - \frac{1}{2}\exp\left[-\frac{2vt}{t}\right] = v - \frac{1}{2}\exp\left[-2v\right] = f(1,v). \end{aligned}$$

This shows that f is a homogeneous function of degree 0. Therefore, by Definition 3.7.1, the given differential equation is homogeneous. Moreover, we have

$$\frac{f(1,v)-v}{t} = \frac{2v - \exp[-2v] - 2v}{2t} = \frac{-\exp[-2v]}{2t}.$$

From (3.49), (3.99) and step 4 of the procedure in Subsection 3.7.2, the solution to the given IVP is

$$G(v(t)) = \int_c^v 2\exp\left[2s\right] ds + C = \exp\left[2v(t)\right] + C,$$

$$G(v(t)) - G\left(v(t_0)\right) = \exp\left[2\frac{x}{t}\right] - \exp\left[2\frac{x_0}{t_0}\right] = \ln\left[\frac{t}{t_0}\right],$$

and hence

$$\exp\left[2\frac{x}{t}\right] = \exp\left[2\frac{x_0}{t_0}\right] + \ln\left[\frac{t}{t_0}\right].$$

Thus,

$$x(t) = \frac{t}{2}\left(\ln\left(\exp\left[2\frac{x_0}{t_0}\right] + \ln\left[\frac{t}{t_0}\right]\right)\right).$$

Illustration 3.7.2 (dynamic of a swarm). Points P and Q are, respectively, the location of bird nests and the food source. Let us choose the origin of the rectangular coordinate system to be at the location $P(0,0)$. Assume that the coordinates of point Q are $(x_0, 0)$. In addition, assume that the swarm of birds left the location of the food source location and flew with a constant speed κ relative to the wind. The wind is blowing due south (relative to the ground) with constant speed λ. The birds maintain their desire to meet their offspring in the nests.

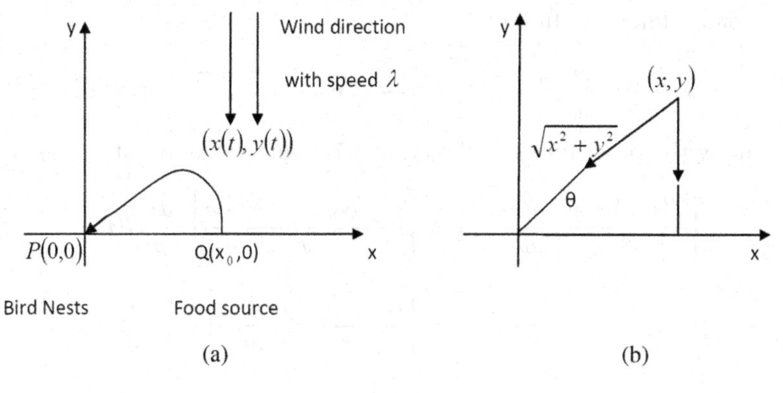

Fig. 3.8

Let $(x(t), y(t))$ (the projection of its actual position in 3D) be the instantaneous position of the swarm relative to the ground at a time t. The instantaneous velocity components of the swarm relative to the ground are

$$\frac{d}{dt}x, \quad \frac{d}{dt}y,$$

respectively. Moreover, they are given by

$$dx = -\kappa\cos\theta dt = \frac{-\kappa x}{\sqrt{x^2+y^2}}\,dt.$$

$$dy = (-\kappa\sin\theta - \lambda)\,dt = \left(\frac{-\kappa y}{\sqrt{x^2+y^2}} - \lambda\right)dt = \left(\frac{-\kappa y - \lambda\sqrt{x^2+y^2}}{\sqrt{x^2+y^2}}\right)dt.$$

By algebraic manipulations, we have

$$\frac{dx}{\kappa x} = -\frac{dt}{\sqrt{x^2+y^2}} = \frac{dy}{\kappa y + \lambda\sqrt{x^2+y^2}}.$$

Hence, we obtain

$$\begin{aligned} dy &= \frac{\kappa\, y + \lambda\sqrt{x^2+y^2}}{\kappa x}\,dx = \left(\frac{y}{x} + \frac{\lambda}{\kappa}\frac{\sqrt{x^2+y^2}}{x}\right)dx \\ &= \left(\frac{y}{x} + \alpha\frac{\sqrt{x^2+y^2}}{x}\right)dx = f(x,y)\,dx, \end{aligned}$$

where

$$\alpha = \frac{\lambda}{\kappa}, \quad f(x,y) = \frac{y}{x} + \alpha\frac{\sqrt{x^2+y^2}}{x}$$

is a homogenous function. In fact,

$$f(x, xv) = \frac{xv}{x} + \alpha\frac{\sqrt{x^2 + x^2v^2}}{x} = v + \alpha\sqrt{1 + v^2} = f(1, v).$$

Now, we apply the procedure in Subsection 3.7.2, and we arrive at

$$\begin{aligned}\left[\frac{f(1,v) - v}{x}\right]\frac{d}{dv}G(v) &= \left[\frac{v + \alpha\sqrt{1+v^2} - v}{x}\right]\frac{d}{dv}G(v)\\ &= \left[\frac{\alpha\sqrt{1+v^2}}{x}\right]\frac{d}{dv}G(v) = \frac{\alpha}{x}.\end{aligned}$$

By simplifying the equation, we obtain

$$\frac{d}{dv}G(v) = \frac{1}{\sqrt{1+v^2}}.$$

By applying the solution procedure in Subsection 2.3.2 and using trigonometric substitution, we obtain

$$G(v) = \int^v \frac{ds}{f(1,s) - s} = \ln\left(v + \sqrt{1+v^2}\right) = \alpha\ln(x) + C,$$

where C is an arbitrary constant of integration. To compute C, we note that

$$v(x_0) = \frac{y(x_0)}{x_0} = 0.$$

Hence, $G(v(x_0)) = 0 = \alpha\ln(x_0) + C$. Therefore, $C = -\alpha\ln(x_0)$. We substitute this value of C into the above equation, and we have

$$\ln\left(v + \sqrt{1+v^2}\right) = \alpha\ln(x) - \alpha\ln(x_0) = \alpha\ln\left(\frac{x}{x_0}\right) = \ln\left(\frac{x}{x_0}\right)^{\alpha}.$$

By taking the exponential of both sides and using the concept of the inverse, we have

$$\left(v + \sqrt{1+v^2}\right) = \left(\frac{x}{x_0}\right)^{\alpha}.$$

Our goal is to write v as a function of only x. For this purpose, we perform some algebraic manipulations:

$$\begin{aligned}\left(\frac{x}{x_0}\right)^{-\alpha} &= \frac{1}{v + \sqrt{1+v^2}}\\ &= \frac{v - \sqrt{1+v^2}}{\left(v + \sqrt{1+v^2}\right)\left(v - \sqrt{1+v^2}\right)} \quad \text{(by rationalizing the denominator)}\\ &= \frac{v - \sqrt{1+v^2}}{v^2 - (1+v^2)} = -\left(v - \sqrt{1+v^2}\right) \quad \text{(by simplifying).}\end{aligned}$$

Hence,

$$v - \sqrt{1+v^2} = -\left(\frac{x}{x_0}\right)^{-\alpha}.$$

Using this expression with the previous expression, we have

$$\begin{aligned}\left(\frac{x}{x_0}\right)^{\alpha} - \left(\frac{x}{x_0}\right)^{-\alpha} &= \left(v - \sqrt{1+v^2}\right) + \left(v + \sqrt{1+v^2}\right)\\ &= 2v\\ &= 2\frac{y(x)}{x} \quad \left(\text{by simplification and substitution: } v = \frac{y}{x}\right),\end{aligned}$$

which implies that

$$y(x) = \frac{1}{2}x\left[\left(\frac{x}{x_0}\right)^{\alpha} - \left(\frac{x}{x_0}\right)^{-\alpha}\right] = \frac{x_0}{2}\left[\left(\frac{x}{x_0}\right)^{\alpha+1} - \left(\frac{x}{x_0}\right)^{1-\alpha}\right].$$

This describes the projection of the path (2D) of the swarm's trajectory in 3D. In the engineering literature, this is also referred to as the phase plane trajectory of motion of the swarm [126]. Depending on the magnitude of α, one can draw a few conclusions about the reachability of the swarm to their nests. The details are left as an exercise.

Brief summary. The following flowchart summarizes the described steps for solving homogeneous equations by using the energy function method.

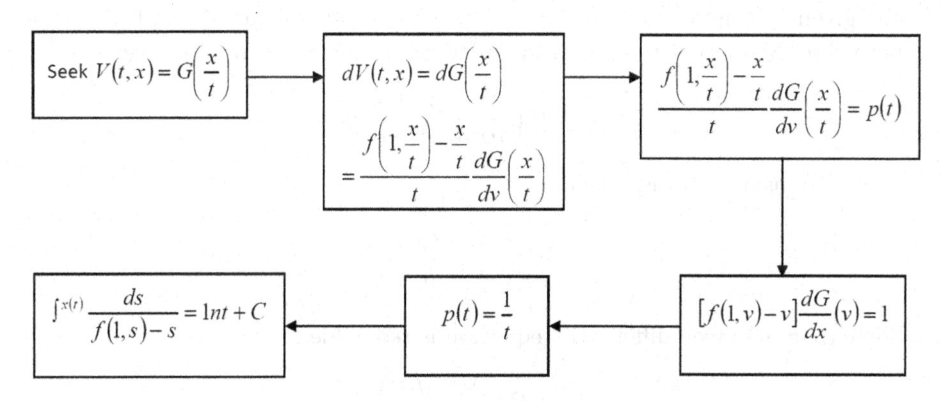

Fig. 3.9 Flowchart: Energy function method applied to homogeneous equations.

3.7 Exercises

(1) Let A, B, and d be any given real numbers. Decide which equations are homogeneous and which are not:

(a) $dx = \frac{t^2+2x^2}{2x\sqrt{t}}\,dt$ (b) $dx = \frac{t+x}{t-x}\,dt$

(c) $dx = \frac{x^2+d^2t^2}{x\sqrt{t}}\,dt$ (d) $dx = \frac{\sqrt{2dxt+x^2t^2}}{t}\,dt$

(e) $dx = \frac{8(1+B)x^4+(4B-2A)x^2t^2-2t^4}{8x^3t}\,dt$

(f) $dx = \left[\frac{x}{t} + \frac{t^3+xt^2}{(t-x)^3} + \frac{At}{(t-x)}\exp\left[\frac{x}{2t}\right] + B\frac{t+x}{t-x}\right]dt$

(g) $dx = \frac{(3+2B)x^4+2\left(B+Ad^2\right)x^2t^2-d^4t^4}{2x^3t}\,dt$, for any real number d

(h) $dx = \left[\frac{3x+(d+2A)t+B\sqrt{2dxt+x^2t^2}}{t}\right]dt$

(i) $dx = \left[\frac{x}{t} + \sin\frac{x}{t}\right]dt$

Hint:

$$\int \frac{ds}{\sqrt{2ds+s^2}} = \ln\left(s+d+\sqrt{2ds+s^2}\right),$$

$$\int \frac{(1-s)}{(1+s)}\,ds = 2\ln(1+s) - s.$$

(2) Let P be a continuous function. Also, assume that λ, μ, ν, and ψ are any given real numbers, and constants ν and λ are not both zero. Show that the following differential equation is reducible to homogeneous:

$$dx = P\left(\frac{\lambda x + \mu t}{\nu x + \psi t}\right)dt.$$

(3) Let P be a continuous function. Further assume that λ, μ, ν, ψ, ρ, and ϕ are any given real numbers with $e = \lambda\psi - \nu\mu \neq 0$, and constants ν and λ are not both zero. Show that the following differential equation is not homogeneous:

$$dx = P\left(\frac{\lambda x + \mu t + \rho}{\nu x + \psi t + \phi}\right)dt.$$

Hint: By using a transformation:

$$x = y + k, \quad dx = dy,$$
$$t = z + l, \quad dt = dz.$$

Show that the given differential equation is reducible to

$$dy = P\left(\frac{\lambda y + \mu z}{\nu y + \psi z}\right)dz,$$

where k and l are unknown real numbers that are determined by the following linear nonhomogeneous system of algebraic equations:

$$\lambda k + \mu l + \rho = 0,$$
$$\nu k + \psi k + \phi = 0.$$

3.8 Bernoulli Equations

In this section, we present another subclass of differential equations (3.38) that is reducible to (3.65). This class of differential equations is referred to as a class of Bernoulli differential equations. First, we define a Bernoulli differential equation.

Definition 3.8.1. A differential equation (3.38) is said to be a Bernoulli differential equation if the rate function $f(t,x)$ in (3.38) is in the following form: $f(t,x) = P(t)x + Q(t)x^n$, for any real number $n \neq 1$.

3.8.1 *General problem*

We follow the steps described in the procedure in Subsection 3.5.2 as it relates to the Bernoulli differential equation:

$$dx = [P(t)x + Q(t)x^n]\, dt, \tag{3.101}$$

where P and Q are continuous functions, and $n \neq 1$.

($\mathbf{H_{3.8}}$): It is assumed that P and Q are continuous functions. Moreover, (3.101) has a solution process. In addition, $P(t) \neq 0$ and $Q(t) \neq 0$.

3.8.2 *Procedure for finding a general solution*

The procedure for reducing a subclass of a class of differential equations (3.38) in Section 3.5 is repeated. With this subclass of Bernoulli-type differential equations, we associate a suitable natural energy/Lyapunov function in a unified way.

Step 1. We repeat steps 1 and 2 of the procedure in Subsection 3.3.2. By (3.101) and (3.66), this reduces to

$$\mu(t)V(t,x) + p(t) = \frac{\partial}{\partial t}V(t,x) + [P(t)x + Q(t)x^n]\frac{\partial}{\partial x}V(t,x), \tag{3.102}$$

Step 2. Again, by using (3.102), we find a candidate for $V(t,x)$. To do this, from (3.102) and the structure of rate functions in the Bernoulli differential equation (3.101): $f(t,x) = P(t)x + Q(t)x^n$ for $n \neq 1$. We have two different ways of making the choice of $V(t,x)$:

$$\begin{aligned} &\text{(i)} \quad P(t)x\frac{\partial}{\partial x}V(t,x) = \mu(t)V(t,x), \\ &\text{(ii)} \quad Q(t)x^n\frac{\partial}{\partial x}V(t,x) = p(t). \end{aligned} \tag{3.103}$$

Choice 1. We assume that $P(t)x \neq 0$ and choose $V(t,x)$, which satisfies (i) in (3.103), where $\mu(t) = \delta P(t)$ for an unknown nonzero constant $\delta(\delta \neq 0)$. The goal is to find $V(t,x)$, and the rate functions $\mu(t)$ and $p(t)$ of the reduced differential equation (3.65) [(3.102)]. We note that to find $\mu(t)$, we only need to find δ. First, we rewrite (i) as

$$\frac{\partial}{\partial x}V(t,x) = \frac{\mu(t)}{P(t)x}V(t,x) \quad \text{[by assumption } P(t)x \neq 0]$$
$$= \frac{\delta}{x}V(t,x) \quad \text{[from hypothesis (H}_{3.8}\text{)]}. \tag{3.104}$$

From this, it is clear that the quotient of

$$\frac{\partial}{\partial x}V(t,x)$$

with $V(t,x)$ is independent of t. Therefore, we can assume that the $V(t,x) \equiv V(x)$, i.e. $V(t,x)$, is independent of t.

By applying the method in Subsection 2.3.2, for each t in J, we have

$$V(t,x) \equiv V(x) = \exp\left[\delta \int_c^x \frac{ds}{s}\right] C$$
$$= \exp\left[\delta \ln\left(\frac{|x|}{|c|}\right)\right] C$$
$$\left[\text{from } \int_c^x \frac{ds}{s} = \ln(|x|) - \ln\left(|c| = \ln\left(\frac{|x|}{|c|}\right)\right)\right]$$
$$= \exp\left[\ln\left(\left(\frac{|x|}{|c|}\right)^{\delta}\right)\right] C \quad [\text{from } \delta \ln M = \ln(M)^{\delta}]$$
$$= x^{\delta} C \quad \text{(exp is the inverse of ln, and a generic constant)}, \tag{3.105}$$

where C is a nonzero arbitrary constant of integration. Otherwise, if $C = 0, V(x) \equiv 0$. This is not useful for finding the solution representation of (3.101).

We compute $\frac{\partial}{\partial x}V(t,x)$

$$\frac{d}{dx}V(x) = \delta x^{\delta-1}C. \tag{3.106}$$

From (3.103) and (3.106), we have

$$p(t) = Q(t)x^n \frac{\partial}{\partial x}V(t,x)$$
$$= Q(t)x^n \delta x^{\delta-1}C \quad \text{[by substitution from (3.106)]}$$
$$= \delta Q(t)x^{n+\delta-1}C \quad \text{(by simplification)}. \tag{3.107}$$

By using assumption ($H_{3.8}$) and separating the functions of t and x from (3.107), we have

$$x^{n+\delta-1} = \frac{p(t)}{\delta Q(t)C}. \tag{3.108}$$

We note that the right-hand side of (3.108) is a function of t only. Therefore, we choose $p(t)$ and δ so that

$$x^{n+\delta-1} = x^0 = 1, \tag{3.109}$$

$$p(t) = \delta Q(t)C. \tag{3.110}$$

From (3.109), we have

$$n + \delta - 1 = 0.$$

Hence,

$$\delta = 1 - n. \tag{3.111}$$

Moreover,

$$p(t) = (1-n)Q(t)C. \tag{3.112}$$

Using (3.105), (3.106), and (3.111), we now compute the right-hand side expression in (3.102):

$$\begin{aligned}
&\frac{d}{dt}V(x) + [P(t)x + Q(t)x^n]\frac{d}{dx}V(x)\\
&\quad = \delta[P(t)x + Q(t)x^n]x^{\delta-1}C \quad \text{(by substitution)}\\
&\quad = \delta\left[P(t)x^\delta + Q(t)x^{n+\delta-1}\right]C \quad \text{(by simplifying)}\\
&\quad = (1-n)\left[P(t)x^\delta + Q(t)\right]C \quad \text{[from (3.109) and (3.111)]}\\
&\quad = (1-n)P(t)x^\delta C + (1-n)Q(t)C \quad \text{(by simplifying)}.
\end{aligned} \tag{3.113}$$

From (3.105) and (3.113), (3.102) can be rewritten as

$$\begin{aligned}
\mu(t)V(t,x) + p(t) &= \frac{\partial}{\partial t}V(t,x)\\
&\quad + [P(t)x + Q(t)x^n]\frac{\partial}{\partial x}V(t,x) \quad \text{(by substitution)}\\
\mu(t)Cx^\delta + p(t) &= (1-n)P(t)Cx^\delta + (1-n)Q(t)C.
\end{aligned} \tag{3.114}$$

Now, from (3.114) and by using the method of undetermined coefficients, we compare the coefficients of x and conclude that

$$\mu(t)C = (1-n)P(t)C, \quad p(t) = (1-n)Q(t)C.$$

Hence,

$$\mu(t) = (1-n)P(t), \tag{3.115}$$

$$p(t) = (1-n)Q(t)C, \tag{3.116}$$

where C is a nonzero arbitrary constant of integration. In particular, one can pick $C = 1$. In this case, from (3.103), (3.112), (3.115), (3.116) and with $C = 1$, the reduced differential equation (3.65) is

$$dm = [(1-n)P(t)m + (1-n)Q(t)]\,dt. \tag{3.117}$$

From the above discussion, we note that the case of $\delta = 0$ is not feasible. In fact, the presented procedure does not does provide the existence of a desired energy function.

Choice 2. In this case, we assume that $Q(t)x \neq 0$, and we choose a $V(t,x)$ which satisfies (ii) in (3.103), where $p(t) = \gamma Q(t)$ for an unknown nonzero constant $\gamma(\gamma \neq 0)$. The goal is to find $V(t,x)$ and the rate functions $\mu(t)$ and $p(t)$ of the reduced differential equation (3.65) [(3.102)]. We note that to find $p(t)$, we need only to find γ. For this purpose, we consider (ii) in (3.103):

$$Q(t)x^n \frac{\partial}{\partial x} V(t,x) = p(t). \tag{3.118}$$

From (3.118) and the choice of $p(t)$, we solve for

$$\frac{\partial}{\partial x} V(t,x)$$

and get

$$\begin{aligned} \frac{\partial}{\partial x} V(t,x) &= \frac{p(t)}{Q(t)x^n} \\ &= \frac{\gamma Q(t)}{Q(t)x^n} \quad \text{(by substitution)} \\ &= \frac{\gamma}{x^n} \quad \text{(by simplification).} \end{aligned} \tag{3.119}$$

The right-hand side is independent of t, and therefore the left-hand side must also be independent of t. By integrating both sides of this expression with respect to x, we get

$$\begin{aligned} V(t,x) \equiv V(x) &= \frac{\gamma}{(1-n)x^{n-1}} + C \\ &= x^{n-1} + C \quad \text{(by the choice of } \gamma \text{ and the law of exponents),} \end{aligned} \tag{3.120}$$

where $\gamma = 1 - n$ and C is an arbitrary constant of integration. Again, we compute $\frac{\partial}{\partial x} V(t,x)$:

$$\frac{d}{dx} V(x) = \frac{d}{dx}\left(x^{1-n} + C\right) = (1-n)x^{-n}. \tag{3.121}$$

Similarly, by using (3.119), (3.120), and (3.121), we compute the right-hand side of the expression in (3.102) and we get the same expression as in (3.113) except that the appearance of the arbitrary constant C is additive instead of multiplicative.

$$\frac{d}{dt}V(x) + [P(t)x + Q(t)x^n]\frac{d}{dx}V(x) = (1-n)\left[P(t)x^{1-n} + Q(t)\right]. \tag{3.122}$$

From (3.120) and (3.122), (3.102) can be rewritten as

$$\mu(t)V(t,x) + p(t) = \frac{\partial}{\partial t}V(t,x) + [P(t)x + Q(t)x^n]\frac{\partial}{\partial x}V(t,x) \quad \text{(by substitution)},$$
$$\mu(t)x^{1-n} + p(t) + \mu(t)C = (1-n)P(t)x^{1-n} + (1-n)Q(t). \tag{3.123}$$

Now, from (3.103), (3.123), (3.121) and again by using the method of undetermined coefficients, we compare the coefficients of x's, and conclude that

$$\begin{aligned} \mu(t)V(t,x) &= P(t)x\frac{d}{dx}V(x) \\ \mu(t)(x^{1-n} + C) &= P(t)x(1-n)x^{-n} \quad \text{[from (3.118) and (3.121)]}, \\ \mu(t) &= (1-n)P(t), \quad C = 0 \quad \text{[in (3.120)]}, \end{aligned} \tag{3.124}$$
$$p(t) = (1-n)Q(t), \tag{3.125}$$

where C is an arbitrary constant of integration. Now, we pick $C = 0$.

In this case, from (3.120), (3.124), (3.125) and with $C = 0$, we have (3.117).

Step 3. By the condition (3.103), the nonlinear Bernoulli differential equation (3.101) reduces to (3.117). Hence, the condition (3.103) determines the class of functions f in (3.38) that is described in (3.101). Moreover, for this class of differential equations (3.101), its representation of the general solution is given by the inverse of function $V(x) = x^{1-n}$ [using (3.105) with $C = 1$]:

$$x(t) = E(t, m(t)), \tag{3.126}$$

where $m(t)$ is a general solution process of (3.117) (by applying the method of Section 2.5), and $x(t)$ is a general solution process of the differential equation (3.101), which belongs to the above-determined class. This completes the order reduction and solution representation process of (3.38) under the specified conditions on f in the Bernoulli differential equation (3.101).

Observation 3.8.1 We recall that the either one of the two relations in (3.103) plays a role in determining the candidate for the energy function. Moreover, from the procedure in Subsection 3.8.2, it is clear that the arguments used in finding the candidate in the procedure provide different energy functions, namely (i) $V_\delta(x) = x^{1-n}C$ and (ii) $V_\gamma(x) = x^{1-n} + C$, depending on the choices (i) and (ii) in (3.103), respectively, where C's are arbitrary constants of integration. We note that the arbitrary constant C in (i) is nonzero and in (ii) is any arbitrary constant C. It can be justified that no other combination of relations in (3.103) provides the feasibility of an energy function.

We will now present a few examples to illustrate the procedure described in this section.

Illustration 3.8.1. We are given $dx = [Px + Qx^n]\,dt$, where P and Q are arbitrary real numbers, and n is any real number that satisfies $n \neq 1$. Find a representation of the general solution to this differential equation.

Solution procedure. From steps 1 and 2 of the procedure detailing the solving of the Bernoulli type differential equation and choice 1, for any real number $n \neq 1$ and $\delta \neq 0$, the candidate for the energy function in (3.105) with $C = 1$ reduces to

$$V(x) = x^{\delta}.$$

In this case, from (3.111), (3.112), (3.115), and (3.116), $V(t,x)$ in (3.105) and the rate functions in (3.65) [(3.102)] are $V(t,x) = x^{1-n}$ $(\delta = 1-n)$, $\mu(t) = (1-n)P$, and $(1-n)Q = p(t)$.

By using the method of Section 2.5 and based on the given problem, the general solution to the reduced differential equation (3.117) is

$$V(x(t)) = m(t) = \Phi(t)c + (1-n)Q \int_a^t \Phi(t,s)\,ds,$$

where

$$\Phi(t) = \exp\left[(1-n)Pt\right] \quad \text{and} \quad V(x(t)) = (x(t))^{1-n}.$$

The general solution to the original given differential equation is

$$\begin{aligned} x(t) &= \left(\Phi(t)c + (1-n)Q \int \Phi(t,s)ds\right)^{-\frac{1}{(n-1)}} \\ &= \left(\Phi(t)c - \frac{Q}{P}\left(1 - \exp[(1-n)]\right)pt\right)^{-\frac{1}{(n-1)}}. \end{aligned}$$

Example 3.8.1 (economic growth: Solow model [130]). Let us consider the growth model of Solow under the Cobb–Douglas production function described in Example 3.2.3:

$$dr = [sr^a - nr]\,dt, \quad r(t_0) = r_0.$$

Find the expression for the capital per worker in the above model.

Solution procedure. We note that this is a Bernoulli-type differential equation. To find the expression for the capital per worker at any time t, we compare this equation with the equation in Illustration 3.8.1. Thus, we have $V(t,r) = r^{1-a}$ $(\delta = 1-a)$, $P = -n$, and $Q = s$. In this case, $\mu(t) = (1-a)(-n) = -n(1-a)$ and

$p(t) = (1-a)(s-0) = (1-a)s$. The reduced differential equation (3.117) is

$$dm = [-n(1-a)m + (1-a)s]\, dt.$$

By using the method of Section 2.5, the general solution to this differential equation is

$$V(r(t)) = m(t) = \Phi(t)C + (1-a)s \int_a^t \Phi(t,s)\, ds,$$

where

$$\Phi(t) = \exp\left[-(1-a)nt\right], \quad V(r(t)) = [r(t)]^{1-a}.$$

To find the solution to the given IVP, we need to compute the constant of integration C. For this purpose, we use $r(t_0) = r_0$, and we have

$$V(r_0) = [r_0]^{1-a} = \Phi(t_0)C + (1-a)s \int_a^{t_0} \Phi(t_0, s)\, ds.$$

We solve for C and obtain

$$\begin{aligned}
C &= \Phi^{-1}(t_0)\left[[r_0]^{1-a} - (1-a)s \int_a^{t_0} \Phi(t_0,s)\, ds\right] \\
&= \Phi^{-1}(t_0)[r_0]^{1-a} + \Phi^{-1}(t_0)\left[-(1-a)s \int_a^{t_0} \Phi(t_0,s)\, ds\right] \\
&= \Phi^{-1}(t_0)[r_0]^{1-a} + \Phi^{-1}(t_0)\Phi(t_0)\left[-(1-a)s \int_a^{t_0} \Phi^{-1}(s)\, ds\right] \\
&= \Phi^{-1}(t_0)[r_0]^{1-a} - (1-a)s \int_a^{t_0} \Phi^{-1}(s)\, ds.
\end{aligned}$$

We now substitute in the above equation

$$\begin{aligned}
V(r(t)) = [r(t)]^{1-a} &= \Phi(t)\left[C + (1-a)\int_a^t \Phi^{-1}(s)\, ds\right] \\
&= \Phi(t)\left[\Phi^{-1}(t_0)[r_0]^{1-a} - (1-a)s \int_a^{t_0} \Phi^{-1}(s)\, ds\right. \\
&\quad \left. +(1-a)s \int_a^t \Phi^{-1}(s) ds\right] \quad \text{(by substitution)} \\
&= \Phi(t)\left[\Phi^{-1}(t_0)[r_0]^{1-a} + (1-a)s \int_{t_0}^t \Phi^{-1}(s)\, ds\right] \quad \text{(by simplifying)} \\
&= \Phi(t,t_0)[r_0]^{1-a} + (1-a)s \int_{t_0}^t \Phi(t,s)\, ds \quad \text{(by properties of } \Phi) \\
&= \left[\Phi(t,t_0)\left([r_0]^{1-a} - \frac{s}{n}\right) + \frac{s}{n}\right] \quad \text{(by integration and simplification).}
\end{aligned}$$

Finally, the capital per worker $r(t)$ at any time t is given by

$$r(t) = \left(\Phi\left(t, t_0\right)\left(\left[r_0\right]^{1-a} - \frac{s}{n}\right) + \frac{s}{n}\right)^{\frac{1}{1-a}}.$$

Example 3.8.2. We are given $dx = \left[x \sin^2 t + x^2 \sin 2t\right] dt$ and $x(t_0) = x_0$. Find the general solution and the particular solution to the IVP.

Solution procedure. We observe that this is a Bernoulli-type differential equation with $n = 2$, $P(t) = \sin^2 t$, and $Q(t) = \sin 2t$. The candidate for the energy function in (3.105) with $C = 1$ is $V(x) = x^{\delta}$. In this case, from (3.111), (3.112), (3.115), and (3.116), $V(t,x)$ in (3.105) and the rate functions in (3.65) [(3.102)] are $V(t,x) = x^{-1}$ ($\delta = -1$), $\mu(t) = -\sin^2 t$, and $p(t) = -\sin 2t$. The reduced differential equation (3.117) is

$$dm = \left[-\sin^2 tm - \sin 2t\right] dt,$$

and its solution is

$$\begin{aligned} x^{-1}(t) = V(x(t)) &= m(t) \\ &= \Phi(t,a)c + \int_a^t \Phi(t,s)p(s)\, ds \\ &= \Phi(t,a)c - \int_a^t \Phi(t,r) \sin 2r\, dr, \end{aligned}$$

where

$$\Phi(t,a) = \exp\left[-\int_a^t \sin^2 u\, du\right] = \exp\left[\frac{1}{2}(t-a) - \frac{1}{4}\left(\sin 2t - \sin 2a\right)\right].$$

Thus, the general solution to the given differential equation is

$$x(t) = \left(\Phi(t,a)c - \int_a^t \Phi(t,r) \sin 2r\, dr\right)^{-1}.$$

The solution to the IVP is given for $t = t_0$, $x(t_0) = x_0$ and for $x_0 \neq 0$, and we have

$$\Phi^{-1}\left(t_0, a\right) x_0^{-1} = c - \int_a^{t_0} \Phi^{-1}(r,a) \sin 2r\, dr,$$

and hence

$$c = \Phi^{-1}\left(t_0, a\right) x_0^{-1} + \int_a^{t_0} \Phi^{-1}(r,a) \sin 2r\, dr.$$

By substituting this into the above expression of $x(t)$ and using the properties of the integral, we get our desired solution to the IVP:

$$x(t) = \left(\Phi\left(t, t_0\right) x_0^{-1} - \int_{t_0}^t \Phi(t,r) \sin 2r\, dr\right)^{-1}.$$

Example 3.8.3. We are given

$$dx = \left[4x + 5x^{\frac{2}{3}}\right] dt.$$

Find the general solution to the given differential equation.

Solution procedure. We observe that this is a Bernoulli-type differential equation with

$$n = \frac{2}{3}, \quad P(t) = 4, \quad Q(t) = 5.$$

The candidate for the energy function in (3.105) with $C = 1$ is $V(x) = x^{\delta}$. In this case, from (3.111), (3.112), (3.115), and (3.116), $V(t, x)$ in (3.105) and the rate functions in (3.65) [(3.102)] are

$$V(t,x) = x^{\frac{1}{3}} \quad \left(\delta = \frac{1}{3}\right),$$
$$\mu(t) = \frac{1}{3} 4 = \frac{4}{3},$$
$$p(t) = \frac{1}{3} 5 = \frac{5}{3}.$$

The reduced differential equation (3.117) is

$$dm = \left(\frac{4}{3}m + \frac{5}{3}\right) dt,$$

and its solution is given by

$$\begin{aligned} x^{\frac{1}{3}}(t) &= V(x(t)) = m(t) \\ &= \Phi(t)c + \frac{5}{3}\int \Phi(t,s)1\, ds \\ &= \Phi(t)c + \frac{5}{4}\left[\Phi(t) - 1\right], \end{aligned}$$

where

$$\Phi(t) = \exp\left[\frac{4}{3}t\right].$$

Therefore, the general solution to the given differential equation is

$$x(t) = \left(\exp\left[\frac{4}{3}t\right] c - \frac{5}{4}\left(1 - \exp\left[\frac{4}{3}t\right]\right)^3\right).$$

Example 3.8.4. We are given

$$dx = \left[1x + 4x^{\frac{4}{5}}\right] dt.$$

Find the general solution to the given differential equation.

Solution procedure. We observe that this is a Bernoulli-type differential equation with

$$n = \frac{4}{5}, \quad P(t) = 1, \quad Q(t) = 4.$$

The candidate for the energy function in (3.105) with $C = 1$ is $V(x) = x^{\delta}$. In this case, from (3.111), (3.112), (3.115), and (3.116), $V(t, x)$ in (3.105) and the rate functions in (3.65) [(3.102)] are

$$V(t,x) = x^{\frac{1}{5}} \quad \left(\delta = \frac{1}{5}\right),$$
$$\mu(t) = \frac{1}{5},$$
$$p(t) = \frac{1}{5}\,4 = \frac{4}{5}.$$

The reduced differential equation (3.117) is

$$dm = \left[\frac{1}{5}m + \frac{4}{5}\right] dt,$$

and its solution is

$$\begin{aligned} x^{\frac{1}{5}}(t) &= V(x(t)) = m(t) \\ &= \Phi(t)c + \frac{4}{5}\int \Phi(t,s)\, ds \\ &= \Phi(t)c + 4\,[\Phi(t) - 1], \end{aligned}$$

where

$$\Phi(t) = \exp\left[\frac{1}{5}t\right].$$

Thus, the general solution to the given differential equation is

$$x(t) = \left(\exp\left[\frac{1}{5}t\right](c+4) - 4\right)^5.$$

Example 3.8.5 [epidemic model (Illustration 3.2.4)]. The following IVP is given $dI = \alpha I\left(\overline{N} - I\right) dt$, $I(0) = 1$. Find: (i) the general solution and (ii) the particular solution to the IVP.

Solution procedure. We observe that this is a Bernoulli-type differential equation with $n = 2$, $P(t) = \alpha\,\overline{N}$, and $Q(t) = -\alpha$. The candidate for the energy function in (3.105) with $C = 1$ is $V(I) = I^{\delta}$. In this case, from (3.111), (3.112), (3.115), and (3.116), $V(t, I)$ in (3.105) and the rate functions in (3.65) [(3.102)] are $V(t, I) = I^{-1}$ $(\delta = -1)$, $\mu(t) = -\alpha\,\overline{N}$, and $p(t) = -(-\alpha) = \alpha$. The reduced differential equation (3.117) is

$$dm = \left[-\alpha\,\overline{N}\,m + \alpha\right] dt,$$

and its solution is

$$\begin{aligned} I^{-1}(t) = V(I(t)) &= m(t) \\ &= \Phi(t,a)c + \int_a^t \Phi(t,s)\alpha\, ds \\ &= \Phi(t,a)c + \frac{1}{\overline{N}}\,[1 - \Phi(t,a)], \end{aligned}$$

where

$$\Phi(t,a) = \exp\left[-\alpha\,\overline{N}(t-a)\right].$$

Thus, the general solution to the given differential equation is

$$I(t) = \left(\exp\left[-\alpha\,\overline{N}(t-a)\right]c + \frac{1}{\overline{N}}\left[1 - \exp\left[-\alpha\,\overline{N}(t-a)\right]\right]\right)^{-1}.$$

The solution to the IVP is given for $t = t_0$, $I(0) = I_0 = 1$, and for $1 \neq 0$,

$$\Phi^{-1}(t_0,a)\,I_0^{-1} = c + \alpha\int_a^{t_0}\Phi^{-1}(r,a)\,dr,$$

and hence

$$c = \Phi^{-1}(t_0,a)\,I_0^{-1} - \alpha\int_a^{t_0}\Phi^{-1}(r,a)\,dr.$$

By substituting this into the above expression of $I(t)$ and using the properties of the integral, we get

$$\begin{aligned} I(t) &= \left(\Phi(t,t_0)\,I_0^{-1} + \alpha\int_{t_0}^t \Phi(t,r)dr\right)^{-1} = \left(\frac{\Phi(t,t_0) + \alpha\,I_0\int_{t_0}^t \Phi(t,r)dr}{I_0}\right)^{-1} \\ &= \left(\frac{\exp\left[-\alpha\,\overline{N}(t-t_0)\right] + I_0\frac{1}{\overline{N}}\left[1 - \exp\left[-\alpha\,\overline{N}(t-t_0)\right]\right]}{I_0}\right)^{-1} \\ &= \left(\frac{\overline{N}\exp\left[-\alpha\,\overline{N}(t-t_0)\right] + I_0\left[1 - \exp\left[-\alpha\,\overline{N}(t-t_0)\right]\right]}{\overline{N}\,I_0}\right)^{-1} \\ &= \frac{\overline{N}\,I_0}{\overline{N}\exp\left[-\alpha\,\overline{N}(t-t_0)\right] + I_0\left[1 - \exp\left[-\alpha\,\overline{N}(t-t_0)\right]\right]} \\ &= \frac{\overline{N}\,I_0\exp\left[\alpha\,\overline{N}(t-t_0)\right]}{\overline{N} + I_0\left[\exp\left[\alpha\,\overline{N}(t-t_0)\right] - 1\right]}. \end{aligned}$$

This is the desired solution to the IVP.

Brief summary. The following flowchart summarizes the described steps for solving Bernoulli equations by using the energy function method.

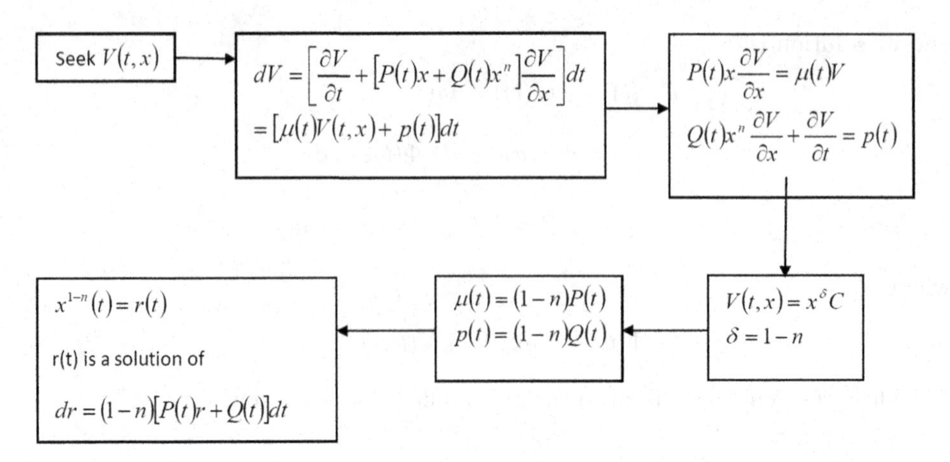

Fig. 3.10 Flowchart: Energy function method applied to Bernoulli equations.

3.8 Exercises

(1) Find the general solutions to the following differential equations:

(a) $dx = \left(4x + 5x^{\frac{1}{2}}\right) dt$ (b) $dx = \left(2x + 5x^{\frac{6}{7}}\right) dt$

(c) $dx = \left[\cos tx + \sin tx^2\right] dt$ (d) $dx = \left[\cos tx + \sin tx^2\right] dt$

(e) $dx = \left[\cos^2 tx + \sin 2tx^2\right] dt$ (f) $dx = \tan t \left(x + \cot tx^5\right) dt$

(g) $dx = \left[\frac{1}{\sqrt{1+t}}x + \sqrt{1+t}\,x^{\frac{1}{2}}\right] dt$ (h) $dx = \left[x \sec^2 t + x^3 \sec^2 t \tan t\right] dt$

(i) $dx = \left[\frac{x}{1+t} + \frac{x^5}{(1+t)^2}\right] dt$ (j) $dx = \left[x \cot t + \frac{1}{x^3(1+\cos^2 t)}\right] dt$

2. Find the solutions to the following initial value problems:

(a) $dx = \left(1x + 2x^{\frac{4}{5}}\right) dt$, $x(0) = 2$

(b) $dx = \left(4x + 5x^{\frac{2}{3}}\right) dt$, $x(0) = 3$

(c) $dx = \left[\cos^2 tx + \sin 2tx^3\right] dt$, $x(0) = 1$

(d) $dx = \left[\cos tx + \sin tx^2\right] dt$, $x(0) = 1$,

(e) $dx = \left[\sin^2 tx + \sin 2tx^3\right] dt$, $x(0) = 1$

(f) $dx = \left[\frac{x}{1+t} + \frac{x^5}{(1+t)^2}\right] dt$, $x(0) = 4$

3. Let $P(t)$ be an arbitrary continuous function. For any real number $n \neq 1$, find the general solution to the following differential equation:

$$dx = [P(t)x + Q(t)x^n]\, dt,$$

where

(a) $Q(t) = \frac{P(t)}{1+\int_a^t P(s)ds}$,

(b) $Q(t) = P(t)\left[\int_a^t P(s)ds\right]$

(c) $Q(t) = P(t)\left[\int_a^t P(s)ds\right]^2$.

4. Let $P(t)$ and $Q(t)$ be functions defined on R into R, and let $h(x)$ be a continuously differentiable function with $\frac{d}{dx}h(x) \neq 0$. For any $n \neq 1$, find the general solution to the following differential equation:

$$dx = \left[P(t)h(x) + Q(t)(h(x))^n\right] / \frac{dh}{dx}(x)\, dt.$$

5. Find the general solution to the following differential equation:

$$dx = \left[P(t)(a-x)(b+x) + \left(\frac{b+x}{a-x}\right)^{\frac{n-1}{a+b}}(a-x)(b+x)Q(t)\right] dt,$$

$x(t_0) = x_0$, where $P(t)$ and $Q(t)$ are as defined in Exercise 3.

3.9 Essentially Time-Invariant Equations

In this section, we present another subclass of differential equations (3.38) that is reducible to either (3.45) or (3.65). This class of differential equations is equivalent to time-invariant differential equations.

Definition 3.9.1. A differential equation (3.38) is said to be an essentially time-invariant differential equation if the rate function $f(t,x)$ in (3.38) has the following form: $f(t,x) = F(ax + bt + c)$.

3.9.1 *General problem*

The procedure in Subsection 3.5.2 is used to find a solution representation for an essentially time-invariant differential equation of the type

$$dx = F(ax + bt + c)\, dt, \tag{3.127}$$

where F is smooth enough to assure the existence of the solution process of (3.127) and a, b, and c are given arbitrary real numbers and $a \neq 0$. The described procedure leads to the reduction of the differential equation (3.65).

($\mathbf{H_{3.9}}$): In addition to hypothesis ($H_{3.3}$), we assume that $F(v)$ is continuous and satisfies the following condition:

$$G(x) = \int_c^x \frac{ds}{aF(s) + b}$$

is defined and invertible.

3.9.2 *Procedure for finding a general solution*

The procedure described in Sections 3.4 and 3.5 [for reducing a subclass of a class of differential equations (3.38) into (3.45) or (3.65)] is repeated. With this subclass of differential equations, we associate a suitable natural energy/Lyapunov function in a systematic way.

Step 1. By using the nature of the subclass of differential equations described in (3.127), we associate an unknown energy/Lyapunov function in step 1 of the procedure in Subsection 3.3.2:

$$V(t,x) = G(ax+bt+c), \tag{3.128}$$

where G is an unknown energy/Lyapunov function. The problem of seeking the energy function V is equivalent to the problem of seeking the unknown function G. a, b, and c are given arbitrary real numbers [as defined in (3.127)].

Step 2. The computation of the differential $dV(t,x)$ in (3.40) is achieved by using the substitution

$$v = ax+bt+c. \tag{3.129}$$

From (3.128) and (3.129), we have

$$\begin{aligned}
\frac{\partial}{\partial t}V(t,x(t)) &= \frac{\partial}{\partial t}G(ax+bt+c) \quad \text{(by the chain rule)}\\
&= \frac{d}{dv}G(v)\frac{\partial}{\partial t}v \quad \left[\text{from}(3.129),\ \frac{\partial}{\partial t}v = \frac{\partial}{\partial t}(ax+bt+c) = b\right]\\
&= \frac{d}{dv}G(v)b,
\end{aligned} \tag{3.130}$$

$$\begin{aligned}
\frac{\partial}{\partial x}V(t,x(t)) &= \frac{\partial}{\partial x}G(ax+bt+c) \quad \text{(by the chain rule)}\\
&= \frac{d}{dv}G(v)\frac{\partial}{\partial x}v \quad \left[\text{from (3.129)},\ \frac{\partial}{\partial x}v = \frac{\partial}{\partial x}(ax+bt+c) = a\right]\\
&= \frac{d}{dv}G(v)a.
\end{aligned} \tag{3.131}$$

Hence, from (3.40), (3.41), and (3.127)–(3.131), we have

$$\begin{aligned}
dV(t,x(t)) &= dG(ax(t)+bt+c) \quad [\text{from (3.40) and (3.128)}]\\
&= LG(ax+bt+c)\,dt \quad [\text{from (3.41)}],
\end{aligned} \tag{3.132}$$

where $x(t)$ is determined by (3.127), and L in (3.41) is a linear operator associated with (3.127) and (3.128). Moreover,

$$\begin{aligned}
LG(ax+bt+c) &= \frac{d}{dv}G(v)b + F(ax(t)+bt+c)\\
&\quad \times \frac{d}{dv}G(v)a \quad [\text{from (3.41), (3.130), and (3.131)}]\\
&= b\frac{d}{dv}G(v) + aF(v)\frac{d}{dv}G(v) \quad [\text{from (3.129)}].
\end{aligned}$$

Thus, from this and (3.128),

$$LG(v) = [aF(v) + b] \frac{d}{dv} G(v). \tag{3.133}$$

In this case, from (3.128), (3.129), (3.132), and (3.133), the expressions in (3.66) reduce to

$$\mu(t)G(v) + p(t) = [aF(v) + b] \frac{d}{dv} G(v). \tag{3.134}$$

Step 3. From (3.134), knowing that G, μ, and p are arbitrary functions, and a and b are given constants, we conclude that the left-hand side cannot be a function of time. This means that $p(t) \equiv$ constant function. Here, we choose $\mu(t) \equiv 0$. Thus, (3.134) reduces to

$$p = [aF(v) + b] \frac{d}{dv} G(v), \tag{3.135}$$

where $p(t) = p$ constant. In fact, $p = 1$. From (3.135), we obtain

$$\frac{d}{dv} G(v) = \frac{p}{aF(v) + b}, \tag{3.136}$$

provided that $aF(v(t)) + b \neq 0$.

Step 4. By applying the solution procedure in Section 2.3, we have

$$G(v) = \left[p \int_c^v \frac{ds}{aF(s) + b} + C \right]. \tag{3.137}$$

From (3.128), (3.132), and (3.134), the solution to the reduced differential equation $[dm(t) = pdt]$ is

$$m(t) = 1t + C \quad \text{(solution to the reduced equation)}$$

where

$$V(t, x(t)) = G(ax(t) + bt + c) \quad \text{[from (3.128)]},$$
$$V(t, x(t)) = m(t) \quad \text{[from (3.132)]}.$$

Thus,

$$G(ax(t) + bt + c) = t + C, \tag{3.138}$$

where C is an arbitrary constant of integration. If G invertible, then, by replacing $v = ax + bt + c$, (3.137) gives the solution to (3.127) in an implicit form:

$$x(t) = G^{-1}([t + C]) - bt - c. \tag{3.139}$$

Next, we present a few examples to highlight the procedure described in this section.

Example 3.9.1. Here, we are given

$$dx = \frac{1}{a}\left[\cos^2(ax + bt + c) - b\right] dt, \quad x(t_0) = x_0,$$

where a, b, and c are any given real numbers, and $a \neq 0$. Find: (i) the general solution and (ii) the solution of the given initial value problem.

Solution procedure. First, we note that this is a differential equation of the type (3.127) with

$$F(ax + bt + c) = \frac{1}{a}\left[\cos^2(ax + bt + c) - b\right].$$

Following the steps of the procedure in Subsection 3.8.2, we compute

$$aF(v) + b = \cos^2 v - b + b = \cos^2 v.$$

From this and (3.137), the general solution to the given differential equation is given by

$$G(v) = \left[\int_c^v \frac{ds}{\cos^2 v}\right] + C = \tan v + C$$

with $p = 1$, and

$$V(t, x(t)) = G(ax(t) + bt + c) = t + C.$$

Thus,

$$\tan(ax(t) + bt + c) = t + C,$$
$$x(t) = \frac{1}{a}\left[\tan^{-1}(t + C) - bt - c\right].$$

This is the general solution to the given differential equation. To find the solution to the IVP, we need to find the value of the constant of integration C. To do this, we use the initial data in the following manner:

$$\tan(ax(t_0) + bt_0 + c) = t_0 + C.$$

We compute C and substitute in the above expression of $x(t)$, and obtain

$$C = \tan(ax_0 + bt_0 + c) - t_0,$$
$$\begin{aligned} x(t) &= \frac{1}{a}\left[\tan^{-1}(t + \tan(ax_0 + bt_0 + c) - t_0) - bt - c\right] \\ &= \frac{1}{a}\left[\tan^{-1}(t - t_0 + \tan(ax_0 + bt_0 + c)) - bt - c\right]. \end{aligned}$$

This is the solution to the given IVP.

Example 3.9.2. The following is given:

$$dx = \frac{1}{1 + (ax + bt + c)^2}\, dt, \quad x(t_0) = x_0.$$

For $a \neq 0 \neq b$ and

$$1 > -\frac{1}{b},$$

find (i) the general solution and (ii) the solution to the given initial value problem.

Solution procedure. By following the procedure described in Subsection 3.8.2 for solving deterministic differential equations, we arrive at an energy-like function,

$$G(v) = \int^v \frac{ds}{aF(s) + b},$$

where

$$aF(v) + b = a\frac{1}{1 + v^2} + b = \frac{a + b\left(1 + v^2\right)}{1 + v^2},$$

for $v = ax + bt + c$, $\nu = 1$, and $C = 0$. Hence,

$$G(v) = \int^v \frac{ds}{aF(s) + b} = \int^v \frac{s^2 + 1}{bs^2 + a + b}\, ds = \int^v \left(\frac{1}{b} - \frac{\frac{a}{b}}{bs^2 + a + b}\right) ds$$
$$= \frac{1}{b}\, v - \frac{a}{b^2} \int^v \frac{1}{s^2 + \tau^2}\, ds = \frac{1}{b}\, v - \frac{a}{\tau b^2} \tan \frac{v}{\tau}, \quad \tau^2 = 1 + \frac{a}{b} > 0,$$

for $b \neq 0$.

In this case, the reducible differential equation (3.65) is $dm = dt$, and hence

$$G(v(t)) = \frac{1}{b}\, v(t) - \frac{a}{\tau b^2} \tan \frac{v(t)}{\tau} = t + C,$$

which implies that

$$\frac{1}{b}\left[ax(t) + bt + c\right] - \frac{a}{\tau b^2} \tan \frac{ax(t) + bt + c}{\tau} = t_0 + C,$$

where C is an arbitrary constant of integration. The implicit general solution representation of the given differential is given by the above expression. This completes the solution process of (i). To solve the IVP (ii), we utilize the initial data and solve of C as

$$\frac{1}{b}\left[ax_0 + bt_0 + c\right] - \frac{a}{\tau b^2} \tan \frac{ax_0 + bt_0 + c}{\tau} = t_0 + C,$$

and hence

$$\frac{1}{b}\left[ax_0 + bt_0 + c\right] - \frac{a}{\tau b^2}\tan\left[\frac{ax_0 + bt_0 + c}{\tau}\right] - t_0 = C.$$

We substitute the expression of C, and after simplification the solution $x(t)$ is given by

$$\tan\frac{ax(t) + bt + c}{\tau} = \tan\frac{ax_0 + bt_0 + c}{\tau} + \frac{\tau b}{a} \times \left[a(x(t) - x_0) + b(t - t_0) - b(t + t_0)\right].$$

Brief summary. The following flowchart summarizes the described steps for solving essentially time-invariant equations by using the energy function method.

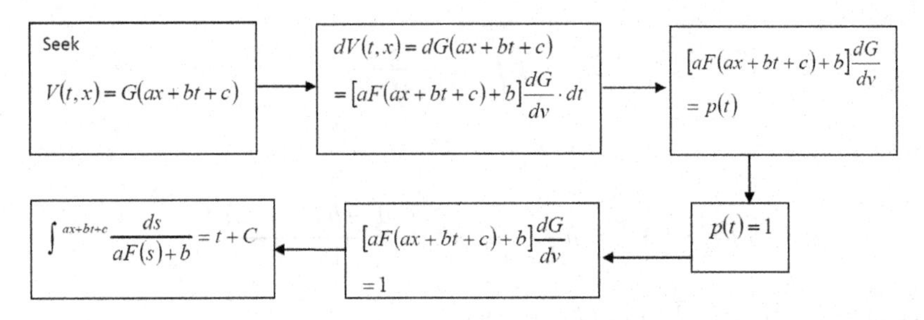

Fig. 3.11 Flowchart: Energy function method applied to essentially time-invariant equations.

3.9 Exercises

(1) Find the general solutions to the following differential equations:
  (a) $dx = \sqrt{3}\,(x + 2t + 1)\,dt$
  (b) $dx = \left[5(x + 2t + 1)^{\frac{2}{3}} + 2\right] dt$
  (c) $dx = \left[4(x + 2t + 1) + 5(x + 2t + 1)^{\frac{2}{3}} + 2\right] dt$
  (d) $dx = \sqrt{3}\left[x + 2t + 1 + (x + 2t + 1)^{\frac{2}{3}}\right] dt$
  (e) $dx = \left[(3x + 2t + 1))^{\frac{1}{3}} + 2\right] dt$

(2) Find the solution to the following initial value problems:
  (a) $dx = \sqrt{3}\,(2x + t - 1)\,dt$, $x(0) = 1$
  (b) $dx = \left[\sin 2t(x + t + 1)^{-2} + 1\right] dt$, $x(0) = 1$
  (c) $dx = \left[5(2x + t - 1)^{\frac{2}{3}} + 1\right] dt$, $x(0) = 1$
  (d) $dx = \left(1(x - 3t + 5) + 2(x - 3t + 5)^{\frac{4}{5}} - 3\right) dt$, $x(0) = 1$
  (e) $dx = \left[\cos^2 t(x + t + 1) + \sin 2t(x + t + 1)^{-2} + 1\right] dt$, $x(0) = 1$

3.10 Notes and Comments

A refined approach to deterministic modeling of dynamic processes is outlined in Section 3.2. Illustration 3.2.1 (chain reactions: Rice–Herzfeld mechanism [30, 31, 86, 99, 132]), Illustration 3.2.2 (enzymatic reaction process), [38, 47, 86, 125, 132] and Example 3.2.1 [86, 99, 132] are based on the fundamental ideas in chemical kinetics. The outlines of Illustration 3.2.3 (ecological process [16, 55, 77, 94, 104, 129]) are formulated by using ecological research. Example 3.2.2 (population genetics [20, 34, 49, 118, 119, 121]) further illustrates the modeling aspect of population genetics study. Illustration 3.2.4 (epidemiological process [5, 55, 136]) is based on epidemiological modeling. A single composite-commodity model by Robert M. Solow [130] and its special cases [28, 130, 131] are discussed in Illustration 3.2.5 and Example 3.2.3. Exercise 3 (unimolecular reaction [99, 132]) and Exercises 4 and 5 (economic growth theory [28, 130, 131]) further challenge students' minds. A very general outline of the 21st century open-ended approach, namely the energy or Lyapunov-like function method, is presented in Section 3.3. The authors hope that this will provide a source for undergraduate research that the mathematical community is lacking as compared to the other sciences.

By employing the knowledge gained in Chapter 2, the feasibility of the general procedure described in Section 3.3 is further justified in Sections 3.4 and 3.5. This is achieved by classifying nonlinear scalar differential equations that are reducible to integrable and linear scalar differential equations. Finally, Sections 3.6 and 3.7 show that variable separable differential equations (Section 3.6) and homogeneous differential equations (Section 3.7) belong to the class of integrable reduced differential equations. The Bernoulli-type differential equations in Section 3.8 and the time-invariant differential equations in Section 3.9 belong to the class of linear reduced differential equations. The energy or Lyapunov-like function method (Section 3.3) and its special cases (Sections 3.4–3.9) are adapted from Ref. 50. The label "energy method" is justified by Illustration 3.4.1 (conservation of energy [40, 85]), a classical two-dimensional dynamical vector field. Example 3.5.2 exhibits the role and scope of the energy/Lyapunov function method of solving scalar differential equations. This example cannot be solved by using any approaches given in current textbooks. Many applied illustrations and examples are detailed to illustrate the mathematical problem-solving approach. Some examples are Illustration 3.6.1 (hydrolysis of esters [86, 99]), Illustration 3.6.2 (water leakage problem: Torricelli's law [11, 24, 27]), Illustration 3.6.3 (concurrent reaction [30, 31, 99]), Illustration 3.7.1 (allometric law [45, 97, 115, 133]), Illustration 3.7.2 (dynamics of swarm), Illustration 3.8.1 (economic growth: Solow model [28, 130]), and Example 3.8.5 (Illustration 3.2.4). In addition, several other useful mathematical examples and observations (for example, Observation 3.5.1) are discussed to maximize the reader's curiosity and to, hopefully, minimize any frustrations.

3.10 Notes and Comments

A [illegible] approach [illegible] modeling [illegible] processes [illegible] found in Section 3.2. Illustration [illegible] reaction [illegible] (20, 31, 90, 99, 152); Illustration [illegible] process (38, [illegible]) and Section 3.2.1 (30, 90, [illegible]) [illegible] about [illegible] chemical [illegible] The [illegible] Illustration [illegible] [illegible] 16, [illegible] 34, [illegible] 129) [illegible] by [illegible] [illegible] population genetics [28, 34, 39, 78, 118, 129] [illegible] reader [illegible] genetic study [illegible] is based on [illegible] [illegible] Robert [illegible] [illegible] [illegible] example [illegible] [illegible] [illegible] [illegible] second [illegible] of [illegible] model [illegible] the [illegible] Section [illegible] [illegible] that the [illegible] [illegible] [illegible] the other [illegible] section [illegible].

[illegible] gained [illegible] [illegible] [illegible] [illegible] [illegible] [illegible] [illegible] different [illegible] [illegible] [illegible] [illegible] equations [illegible] [illegible] [illegible] [illegible] [illegible] [illegible] Section [illegible] [illegible] Illustration 3.11 [illegible] [illegible] [illegible] [illegible] equations [illegible] [illegible] [illegible] approaches [illegible] [illegible] [illegible] Illustration [illegible] Illustration 3.3 [illegible] [illegible] [illegible] [31, [illegible]] [illegible] Illustration 7.2 [illegible] [illegible] [28, 150] [illegible] In addition, several other [illegible] [illegible] [illegible] we discussed [illegible] [illegible] [illegible] [illegible].

Chapter 4

First-Order Systems of Linear Differential Equations

4.1 Introduction

The mathematical modeling procedures for solving first-order linear system differential equations and their theoretical analysis are the premise of this chapter. In addition, efforts are made to contrast the scalar versus the system of differential equations. The format and presentation of this chapter is similar to that of Chapter 2. In Section 4.2, several dynamic processes in the biological, chemical, engineering, medical, physical, and social sciences are discussed to illustrate the modeling procedures. Section 4.3 begins with first-order systems of differential equations with constant coefficients. The eigenvalue type of method for solving scalar homogeneous differential equations is extended to solve systems of homogeneous differential equations with constant coefficients. Again, the method for solving systems of linear homogeneous differential equations reduces to the problem of solving linear systems of algebraic equations with constant coefficients. By introducing the concept of the fundamental matrix, its role and scope for finding the IVP is addressed. The general step-by-step procedures for finding the general solution and the solution to IVPs are systematically outlined. Finding the fundamental matrix solution is presented in Section 4.4. The byproduct of this is utilized to solve the IVPs. In addition, several applied and mechanical examples as well as illustrations are given to demonstrate the procedures and related math. In Section 4.5, the procedures for seeking solutions to linear systems of homogeneous differential equations are extended to a general system of homogeneous differential equations with constant coefficients. The general step-by-step procedures for finding the general solution and the solution to IVPs are logically outlined. The limitation on the technique of finding a closed-form solution even with constant coefficients is emphasized. Numerous applied and mechanical examples with illustrations are given to explain the procedures. Section 4.6 deals with systems of first-order nonhomogeneous differential equations. The method of variation of constant parameters is used to find both the general solution and the solution to the IVPs. Section 4.7 offers further insights, and it is motivated by computational insight of the previous sections of this chapter as well as Section 2.6.

This is needed for studying systems of linear differential equations with time-varying coefficient matrix functions. Again, this section provides a foundation for answering the remaining questions posed at the beginning of the section and can be used for studying the higher-order differential equations in Chapter 5.

4.2 Mathematical Modeling

The material in this section is a natural extension of Sections 2.2 and 3.2. In general, dynamic processes in the biological, chemical, engineering, medical, physical, and social sciences are highly complex, with several interacting and interconnected subcomponents. In fact, for a better understanding of multispecies dynamic processes, a single-state-variable description of the multispecies dynamic system is inadequate. It is natural to except the development in a multivariable-state description. Again, the development is based on a theoretical experimental setup, the fundamental laws in science and engineering, and the knowledge-based information about dynamic processes.

Multistate Deterministic Modeling Procedure 4.2.1. In this section, we extend the mathematical modeling procedure of Chapter 2 to multistate-interacting dynamic processes. Again, we use the description of multispecies processes in science and engineering which consist of d components of the dynamic system $[x_1, x_2, \ldots, x_l, \ldots, x_d]^T \in R^d$, as a "$d$-dimensional state vector," such as the "concentration vector" of d chemical reactants in a chemical reaction process, the "size/biomass vector of d species" in a biological process, or the "price vector" of d commodities/services in a business and sociological process. It is measured as vector quantity.

Let $x(t)$ be a d-dimensional state vector of a system at a time t. The state of the system is observed over an interval of $[t, t + \Delta t]$, where Δt is a small increment in t. Without loss of generality, we assume that Δt is positive. For the sake of simplicity, we assume that the process is operating under a controlled environment. We observe the state vectors,

$$x(t_0) = x(t), x(t_1), x(t_2), \ldots, x(t_k), \ldots, x(t_n) = x(t + \Delta t),$$

at

$$t_0 = t, t_1 = t + \tau, t_2 = t + 2\tau, \ldots, t_k = t + k\tau, \ldots, t_n = t + \Delta t = t + n\tau,$$

over the interval $[t, t + \Delta t]$, where n belongs to $\{1, 2, 3, \ldots\}$ and

$$\tau = \frac{\Delta t}{n}.$$

These observations are made under the following conditions:

MDM 1. The observations of the state vectors of the system are made at $t_1, t_2, \ldots, t_k, \ldots, t_n$ under the controlled environment.

MDM 2. The dynamics of the system of interacting components follow its own laws (if any). It is assumed that it is evolving in the closed environment without any uncertainties. The state of the system is observed on every time subinterval of length τ.

MDM 3. The dynamic state of multicomponent processes is measured by the intra-component (within the component itself) and intercomponent (cross-interactions between the components) interactions.

MDM 4. For each $l \in \{1, 2, \ldots, d\}$ and each $k = 1, 2, \ldots, n$, the lth component $x_l(t_k)$ of the state vector $x(t_k)$ at the time t_k is either increased or decreased, or there is no change in the presence of the gth and rth components of the dynamic process for any $g, r \in \{1, 2, \ldots, d\}$. From the description in Section 2.2, $\triangle\alpha_{ll}(t_k)$, $\triangle\alpha_{gl}(t_k)$, and $\triangle\beta_{lr}(t_k)[\triangle\beta_{ll}(t_k) = 0]$ stand for experimental or knowledge-based observed microscopic or local increments at t_k to the lth component influenced by itself, by the lth to the gth component and the rth to the lth component of the state of the system, respectively, over the subinterval of length τ. Moreover, we further note that for a pair of pairs (g, l) and (l, r), the nature of increments depends on the particular qualities of the lth, gth, and rth components of the multicomponent process, and the influence of the cross-interactions between components. We define $d \times d$ matrices $G(t_k) = \triangle\alpha_{gl}(t_k)$ and $R(t_k) = (\triangle\beta_{lr}(t_k))_{d\times d}$, whose entries represent increments to the components of the d-dimensional multicomponent state dynamic process.

Following the modeling description in Section 2.2, we assume that all elements $\triangle\alpha_{gl}(t_k)$ and $\triangle\beta_{lr}(t_k)$ of the matrices $G(t_k)$ and $R(t_k)$ defined in MDM 4 are based on the dynamic law(s) of the processes MDM 2 and MDM 3. These increments may

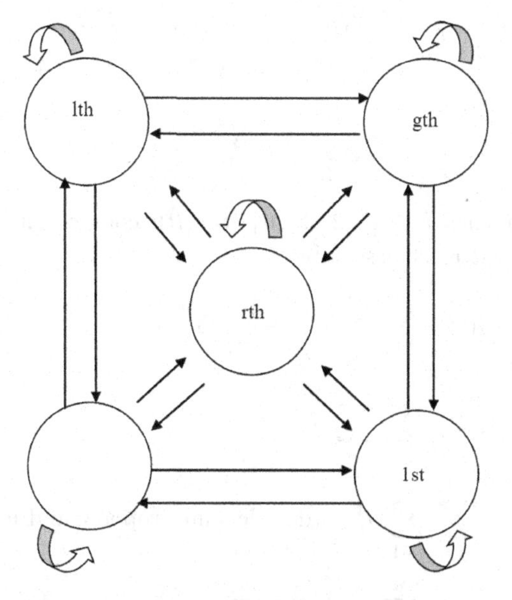

Fig. 4.1 Multicomponent interaction process.

depend on the initial state vector of the system $x(t)$ at a time t and the length of the subinterval τ.

In summary, for each $l \in \{1, 2, \ldots, d\}$, an initial state, and $2n - 1$ experimental or knowledge-based observed state increments $G_{gl}^k = \triangle\alpha_{gl}(t_k)$ and $R_{lr}^k = \triangle\beta_{lr}(t_k)$ for any $g, r \in \{1, 2, \ldots, d\}$ and $k \in \{1, 2, \ldots, n\}$, we have

$$\begin{aligned}
x_l(t_0) &= x_l(t), \\
x_l(t_1) - x_l(t_0) &= G_{1l}^1 + \cdots + G_{gl}^1 + \cdots + G_{dl}^1 \\
&\quad + R_{l1}^1 + \cdots + R_{ll-1}^1 + R_{ll+1}^1 + \cdots + R_{ld}^1 \\
&= \sum_{j=1}^{d} G_{jl}^1 + \sum_{j \neq l}^{d} R_{lj}^1, \\
x_l(t_2) - x_l(t_1) &= G_{1l}^2 + \cdots + G_{gl}^2 + \cdots + G_{dl}^2 \\
&\quad + R_{l1}^2 + \cdots + R_{ll-1}^2 + R_{ll+1}^2 + \cdots + R_{ld}^2 \\
&= \sum_{j=1}^{d} G_{jl}^2 + \sum_{j \neq l}^{d} R_{lj}^2, \\
&\cdots\cdots\cdots\cdots\cdots\cdots\cdots\cdots\cdots\cdots, \qquad (4.1) \\
x_l(t_k) - x_l(t_{k-1}) &= G_{1l}^k + \cdots + G_{gl}^k + \cdots + G_{dl}^k \\
&\quad + R_{l1}^k + \cdots + R_{ll-1}^k + R_{ll+1}^k + \cdots + R_{ld}^k \\
&= \sum_{j=1}^{d} G_{jl}^k + \sum_{j \neq l}^{d} R_{lj}^k, \\
&\cdots\cdots\cdots\cdots\cdots\cdots\cdots\cdots\cdots\cdots, \\
x_l(t_n) - x_l(t_{n-1}) &= G_{1l}^n + \cdots + G_{gl}^n + \cdots + G_{dl}^n \\
&\quad + R_{l1}^n + \cdots + R_{ll-1}^n + R_{ll+1}^n + \cdots + R_{ld}^n \\
&= \sum_{j=1}^{d} G_{jl}^n + \sum_{j \neq l}^{d} R_{lj}^n.
\end{aligned}$$

From (4.1), for each $l \in \{1, 2, \ldots, d\}$, the lth component of the state of the process at $t_n = t + \triangle t$ is expressed by

$$\begin{aligned}
x_l(t + \triangle t) &= x_l(t) + \sum_{i=1}^{n} \sum_{s=1}^{d} G_{sl}^i + \sum_{i=1}^{n} \sum_{j \neq l}^{d} R_{lj}^i \\
&= x_l(t) + \sum_{j=1}^{d} \sum_{i=1}^{n} G_{jl}^i \\
&\quad + \sum_{j \neq l}^{d} \sum_{i=1}^{n} R_{lj}^i \quad \text{(double sum property and index change)} \\
&= x_l(t) + \sum_{i=1}^{n} G_{ll}^i + \sum_{j \neq l}^{d} \sum_{i=1}^{n} (G_{jl}^i + R_{lj}^i) \quad \text{(by regrouping)}, \qquad (4.2)
\end{aligned}$$

where

$$\sum_{j=1}^{d}\sum_{i=1}^{n} G^i_{jl}, \quad \sum_{j\neq l}^{d}\sum_{i=1}^{n} R^i_{lj}$$

are referred to as global aggregate increments to the lth component from itself and from all other remaining components of the dynamic process, respectively, over the interval $[t, t+\triangle t]$ of length $\triangle t$. The increments

$$\sum_{i=1}^{n} G^i_{gl} \quad \text{and} \quad \sum_{i=1}^{n} R^i_{lr}$$

are called local aggregate increments by the lth to the lth and gth components, and by the rth to the lth component of the process, respectively, over the interval $[t, t+\triangle t]$ of length $\triangle t$. These local aggregate increments are denoted by

$$\triangle^n_{gl}(l) = \sum_{i=1}^{n} G^i_{gl},$$

$$\triangle^n_{lr}(r) = \sum_{i=1}^{n} R^i_{lr}.$$

From the expression (4.2), the aggregate change of the state of the lth component $x_l(t+\triangle t) - x_l(t)$ of the multicomponent system under n observations over the given time interval $[t, t+\triangle t]$ of length $\triangle t$ is given by

$$\begin{aligned} x_l(t+\triangle t) - x_l(t) &= \sum_{i=1}^{n} G^i_{ll} + \sum_{r\neq l}^{d}\sum_{i=1}^{n}(G^i_{rl} + R^i_{lr}) \\ &= \left[\sum_{i=1}^{n} \triangle\alpha_{ll}(t_i) + \sum_{r\neq l}^{d}\sum_{i=1}^{n} \triangle\alpha_{rl}(t_i) + \sum_{k=1}^{n} \triangle\beta_{lr}(t_i)\right] \\ &= \left[S^{ll}_n(\alpha) + \sum_{r\neq l}^{d}(S^{rl}_n(\alpha) + S^{lr}_n(\beta))\right]\frac{\triangle t}{\tau}, \end{aligned} \tag{4.3}$$

where $G^k_{gl} = \triangle\alpha_{gl}(t_k)$, $R^k_{lr} = \triangle\beta_{lr}(t_k)$,

$$S^{rl}_n(\alpha) = \frac{1}{n}\left[\sum_{k=1}^{n} \triangle\alpha_{rl}(t_k)\right],$$

$$S^{lr}_n(\beta) = \frac{1}{n}\left[\sum_{k=1}^{n} \triangle\beta_{lr}(t_k)\right],$$

for $k = 1, 2, \ldots, n$ and $l, r = 1, 2, \ldots, d$.

$S^{rl}_n(\alpha)$ and $S^{lr}_n(\beta)$ are the local sample averages of the observed aggregate increment data. Under the assumption of small τ, large n with $\tau n = \triangle t$, and the nature of the system, we have

$$\lim_{\tau\to 0^+}\left[\frac{S_n^{rl}(\alpha)}{\tau}\right] = \lambda_{rl} \quad \text{and} \quad \lim_{\tau\to 0^+}\left[\frac{S_n^{lr}(\beta)}{\tau}\right] = \chi_{lr}. \tag{4.4}$$

From MDM 3 and (4.4), (4.3) reduces to

$$\begin{aligned} x_l(t+\Delta t) - x_l(t) &= (\text{aggregate increment due to the } l\text{th component}) \\ &\quad + (\text{aggregate increment due to the } r\text{th component for } r \neq l) \\ &\approx \left[\lambda_{ll} + \sum_{r\neq l}^{d}(\lambda_{rl} + \chi_{lr})\right]\Delta t. \end{aligned} \tag{4.5}$$

Thus, the aggregate change $x_l(t+\Delta t) - x_l(t)$ of the lth component state of the system in (4.5) under the controlled environmental conditions over the given interval $[t, t+\Delta t]$ of length Δt is approximated by the instantaneous rates due to the presence of the components.

Now, we define the $d \times d$ multicomponent instantaneous rate matrix $C = (c_{lr})_{d\times d}$ of the system as follows:

$$c_{lr} = \begin{cases} \chi_{lr}, & \text{for } r \neq l, \\ \lambda_{ll} + \sum_{r\neq l}^{d} \lambda_{rl}, & \text{for } r = l, \end{cases} \tag{4.6}$$

where λ_{ll}, λ_{gl}, and χ_{lr} are microscopic or local rates of change of the lth state component in the presence of itself, the gth and rth components due to the influence of the lth component itself, the gth component, and the influence of the rth component of the process on the lth component, respectively, over an interval of length Δt. From MDM 4 and (4.6), (4.5) reduces to

$$\begin{aligned} x(t+\Delta t) - x(t) &= (\text{total amount of the positive increment}) \\ &\quad - (\text{total amount of the negative increment}) \\ &\approx C\bar{1}\Delta t, \end{aligned} \tag{4.7}$$

where $\bar{1} = [1, 1, \ldots, 1]^T$, $x(t)$, $C = (c_{lr})_{d\times d}$, and c_{lr} are as defined in (4.6).

Thus, the overall approximate aggregate change $x(t+\Delta t) - x(t)$ of state of the system in (4.7) under the controlled environment over the given interval $[t, t+\Delta t]$ of length Δt is given by $C\bar{1}\Delta t$. If Δt is very small, then its differential is $dt = \Delta t$. From (4.7), the differential of x is defined by

$$dx = C\bar{1}\, dt. \tag{4.8}$$

The expression (4.8) is called as a system differential equation. It is usually denoted by

$$\frac{dx}{dt} = C\bar{1}, \tag{4.9}$$

where $C\bar{1}$ is called the instantaneous rate of change of state of the system, and C may depend on time t and the state x of the system.

Illustration 4.2.1 (multieconomy processes [3, 28]). The single-component process mathematical model (Illustration 2.2.1) is extended to multicomponent dynamic processes in economics, management and information sciences. We simply modify the description of Illustration 2.2.1 by replacing a "state" with a "state vector." Here, the state vector $x(t)$ of the system stands for a specific rate of the price or value vector of a multidimensional entity. E is a collection of multidimensional assets, information, markets, products, services per unit item or size vector $\overline{1} = [1, \ldots, 1, \ldots, 1]_{1\times d}$ per unit time at $t \in J = [a, b]$, $a, b \in R$ and $J \subseteq R$. The rate or specific rate of the price or value of the entity is observed over an interval of $[t, t + \Delta t]$, where Δt is a small increment in t. Its mathematical description can be recasted by following the development of Multistate Modeling Procedure 4.2.2 as

$$x(t + \Delta t) - x(t) \approx C\overline{1}\Delta t,$$

which implies that

$$dx = C\overline{1}\, dt.$$

We note that if $x(t)$ is the specific rate of the price or value vector at a time t, then the coefficient matrix C is called a measure of the instantaneous specific rate matrix of the multiple-component price or value of the entity at a time t over an interval of small length $\Delta t = dt$; if $x(t)$ is the price or value vector at a time t, the matrix C is called a measure of the instantaneous rate of growth or decay of the price or value vector of the entity at a time t over an interval of small length $\Delta t = dt$.

Illustration 4.2.2 (multispecies competitive–cooperative processes [46, 73, 116, 126]). Illustration 2.2.2 is extended to several multicomponent decay/growth dynamic processes in the biological, chemical, compartmental, pharmacological, physical, and social sciences. Here, a state vector of a system $x(t)$ stands for either the specific rate of an amount (size) or an amount (size) of an entity, where it is a member of a collection (substance, population, charge, ionizing beam of particles, etc.) of a multicomponent system at t in $J = [a, b]$, $a, b \in R$ and $J \subseteq R$. The specific rate of an amount (size) or an amount (size) of the entity is observed over an interval of $[t, t + \Delta t]$, where Δt is a small increment in t. Again, here, the size may be a volume or weight or diameter or number, etc. Without loss of generality, we assume that Δt is positive. By following the above-described modeling procedure, we have

$$x(t + \Delta t) - x(t) \approx C\overline{1}\Delta t$$

and hence

$$dx = C\overline{1}\, dt.$$

As before (Illustration 4.2.1), we note that if $x(t)$ is the specific rate of a multidimensional entity vector at a time t, then the coefficient matrix C is called a measure of the instantaneous competitive–cooperative specific rate (per capita competitive–cooperative rate) of the multiple-component amount (size) vector of the entity at a time t; $x(t+\triangle t) - x(t)$ is the change of the per capita rate or rate of the multiamount (size) of the multidimensional entity at a time t over an interval of small length $\triangle t = dt$; if $x(t)$ is the amount or size of an entity at a time t, the matrix C is called the measure of the instantaneous rate of the amount or size of the of an entity at a time t.

Illustration 4.2.3 (Multi-component Diffusion Processes [119]). This illustration generalizes Illustration 2.2.4 to the multi-component diffusion processes in biological, biochemical, biophysical, chemical, compartmental, epidemiological, pharmacological, physical, physiological, social and systems under limited resources/capacity/environmental conditions/controlled-mechanism. As before, a state vector $x(t)$ of the system can be defined as in Illustration 4.2.2. The details are left to the reader.

Observation 4.2.1

(i) We further emphasize that based on experimental observations, information, and basic scientific laws/principles in the biological, chemical, engineering, medical, physical, and social sciences, we can infer that, in general, the instantaneous microscopic or local rates depend on both the initial time t and the initial state $x(t) \equiv x$. As a result of this, the coefficient matrix (C) defined in (4.6) need not be an absolute constant. It may depend on both the initial time t and the initial state $x(t) \equiv x$ of the system, as long as their dependence on t and x is very smooth. As a result of this, (4.7) reduces to

$$x(t+\triangle t) - x(t) \approx C(t,x)\triangle t, \tag{4.10}$$

where $C(t,x)$ is referred to as the average/expected/mean specific rate of the state of the system on the interval of length $\triangle t$. Moreover, the differential equation (4.8) reduces to

$$dx = C(t,x)dt. \tag{4.11}$$

(ii) Furthermore, the mathematical model of the multiagent dynamic process is described by a system of deterministic differential equations. Moreover, the differential equations corresponding to (4.11) reduce to

$$dx = C(t,x)x\,dt. \tag{4.12}$$

In particular,

$$dx = [C(t)x + c(t)]\,dt, \tag{4.13}$$

where C is a $d \times d$ multicomponent specific state rate matrix function and $c(t)$ is a d-dimensional input rate function.

The following specific illustrations and examples justify the scope and the significance of Observation 4.2.1(ii) in the modeling of dynamic processes in the sciences.

The illustrations and examples are in the framework of the above-presented mathematical model building process of this section.

Illustration 4.2.4 (electrical circuit network [35, 40, 126]). First, let us augment the discussion of the RC circuit (Exercise 6, Section 2.2, Electrical Circuit Problem). Besides the basic elements in the circuit analysis, namely the register and capacitor, we introduce the inductor. We recall the roles of these elements. Moreover, in the electrical network, the magnitudes of voltage drops and electric current are interconnected, and follow laws that are known as Kirchhoff's laws:

ECN 1. The presence of the register in the circuit tends to resist the flow of the electric current (Ohm's law: $RI = E$).

ECN 2. The capacitor stores the electrons. The effects of this storage are to block slowly varying current and pass rapidly varying current with high magnitude (Coulomb's law: $\frac{d}{dt}E = \frac{1}{C}I$).

ECN 3. The presence of the inductor in the circuit tends to impede a change in current magnitude. The slower the change, the lesser the obstruction to the flow of the current ($L\frac{d}{dt}I = E$).

ECN 4 (Kirchhoff's current law). The sum of the currents flowing through any point in the network is zero.

ECN 5 (Kirchhoff's voltage law). The sum of the instantaneous voltage drops when measured across each element in a specified direction is zero.

C, E, R, and I are as defined in Exercise 6, Section 2.2. L stands for inductance, and it is measured in henrys. In the following, let us consider an RL circuit (Figure 4.2) as the prototype model in an electrical network.

Step 1 (Kirchhoff's current law). In this circuit, there are three independent variables, I_1, I_2, and I_3. It is assumed that when the switch is closed, all currents

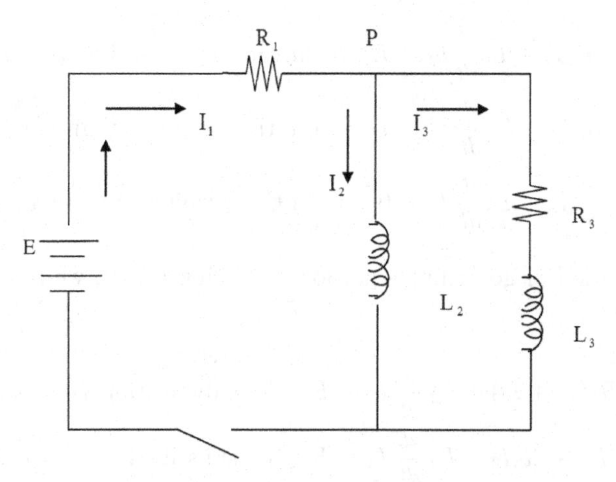

Fig. 4.2 RL circuit.

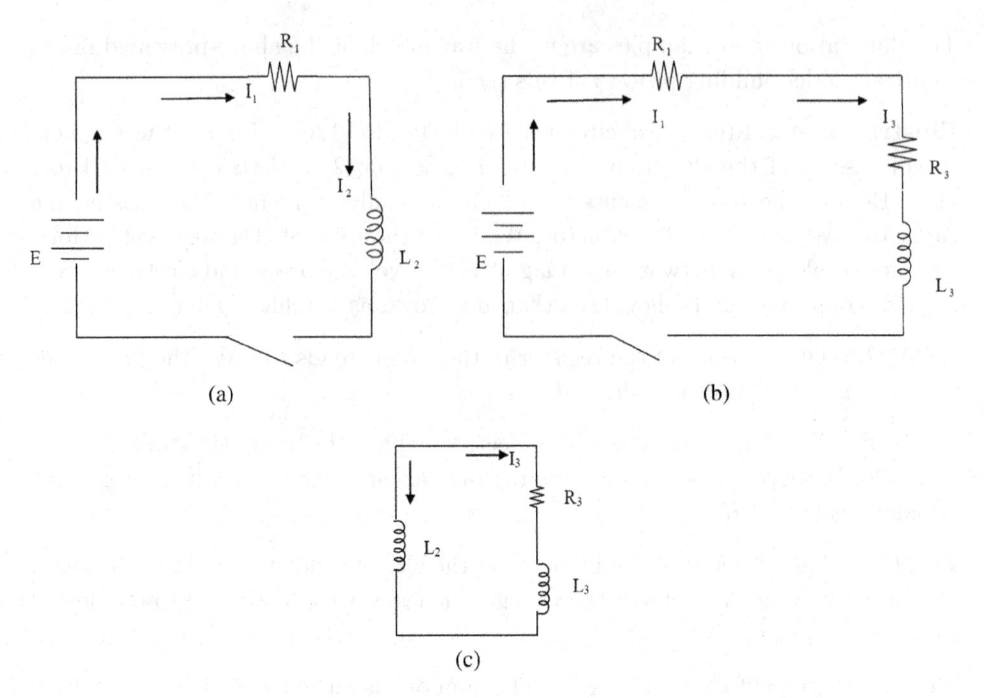

Fig. 4.3 RL subcircuit.

and charges are zero. From Kirchhoff's current law, we have

$$I_1 - I_2 - I_3 = 0 \quad \text{(at a point } P \text{ in the circuit).} \tag{4.14}$$

Step 2 (Kirchhoff's voltage law). The given circuit can be decomposed into three subcircuits, as described in Figures 4.3(a)–4.3(c). From the application of Kirchhoff's voltage law to the three subcircuits, we get

$$R_1 I_1 + L_2 \frac{d}{dt} I_2 = E \quad \text{[from the circuit in Figure 4.3(a)],} \tag{4.15}$$

$$R_1 I_1 + R_3 I_3 + L_3 \frac{d}{dt} I_3 = E \quad \text{[from the circuit in Figure 4.3(b)],} \tag{4.16}$$

$$R_3 I_3 - L_2 \frac{d}{dt} I_3 + L_3 \frac{d}{dt} I_3 = 0 \quad \text{[from the circuit in Figure 4.3(c)].} \tag{4.17}$$

Step 3 (Formation of governing equations). Using (4.14), we rewrite (4.15) and (4.16) as follows:

$$R_1(I_2 + I_3) + L_2 \frac{d}{dt} I_2 = E \quad \text{(by substitution } I_1 = I_2 + I_3\text{),}$$

$$R_1(I_2 + I_3) + R_3 I_3 + L_3 \frac{d}{dt} I_3 = E \quad \text{(by substitution } I_1 = I_2 + I_3\text{).} \tag{4.18}$$

From (4.18) with algebraic simplifications and notations, we get

$$dI_2 = \left(-\frac{R_1}{L_2} I_2 - \frac{R_1}{L_2} I_3 + \frac{E}{L_2}\right) dt,$$

$$dI_3 = \left(-\frac{R_1}{L_3} I_2 - \frac{R_1 + R_3}{L_3} I_3 + \frac{E}{L_3}\right) dt.$$

Its matrix representation is

$$dI = CI dt + p\, dt, \tag{4.19}$$

where

$$I = \left[I_2\ I_3\right]^T, \quad C = \begin{bmatrix} -\dfrac{R_1}{L_2} & -\dfrac{R_1}{L_2} \\ -\dfrac{R_1}{L_3} & -\dfrac{R_1 + R_3}{L_3} \end{bmatrix}, \quad p = \left[\frac{E}{L_2}\ \frac{E}{L_3}\right]^T. \tag{4.20}$$

This completes the derivation of the deterministic mathematical model for the given electrical network.

Illustration 4.2.5 (cellular dynamics [29, 57, 117, 119]). The biological catalysts are different from the catalysts in the chemistry. The biological cell is regarded as a chemical system of interacting molecules. In the catalysis of cellular processes, the concentrations of reactants of both the small molecules/metabolites and the macromolecules (DNA, RNA, and protein) at an instant in a cell play a significant role. It is natural to study the properties of a biological cell by studying the specific molecular species constituting the cell. The maintenance of the cell depends on the chemical reactions taking place within the cell as well as the material entering and leaving the cell from the environment.

Let us consider a prototype model of a cellular dynamics governed by a two-species monomolecular chemical reaction in the cell influenced by the intra- and extracellular environments of the cell. Let $[A_1]$ and $[A_2]$ be the concentrations of two chemical substances of A_1 and A_2, respectively. The reaction is described by

$$\rightleftarrows A_1 \rightleftarrows A_2 \rightleftarrows \text{ (with rates } \rightleftarrows : k_{1e}, k_{21}, k_{02}, k_{2e}, k_{12},\ k_{01}).$$

Step 1. In this step, we need to find $\frac{d}{dt}[A_1]$. The rate of change of $\frac{d}{dt}[A_1]$ is determined by

$$\rightarrow A_1\,(k_{1e}), A_2 \rightarrow A_1\,(k_{21}) \quad \text{and} \quad A_1 \rightarrow A_2\,(k_{12}), A_1 \rightarrow (k_{10}).$$

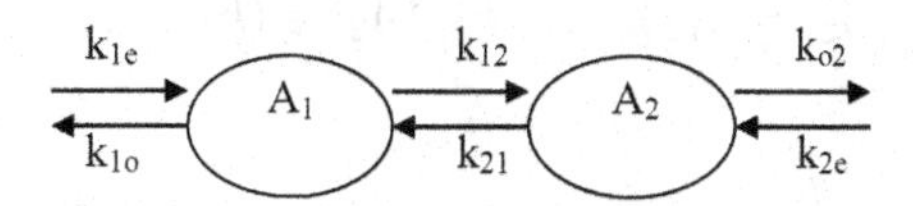

Fig. 4.4 Multimolecular open chemical reaction.

A_1 is formed by two processes: $\to A_1(k_{1e})$ and $A_2 \to A_1(k_{21})$. Therefore, the rate of formation of A_1 is

$$\frac{d}{dt}[A_1] = k_{21}[A_2] + k_{1e}.$$

A_1 is decomposed by two processes: $A_1 \to A_2(k_{12})$ and $A_1 \to (k_{10})$. Thus, the rate of decomposition is given by

$$-\frac{d}{dt}[A_1] = k_{12}[A_1] + k_{10}.$$

Therefore, the net change of A_1 is

$$\frac{d}{dt}[A_1] = -k_{12}[A_1] + k_{21}[A_2] + (k_{1e} - k_{10}).$$

Step 2. In this step, we need to find $\frac{d}{dt}[A_2]$. The rate of change of $\frac{d}{dt}[A_2]$ is determined by

$$\to A_2(k_{2e}), A_1 \to A_2(k_{12}) \quad \text{and} \quad A_2 \to A_1(k_{21}), A_2 \to (k_{02}).$$

A_2 is formed by two processes: $\to A_2(k_{2e})$ and $A_1 \to A_2(k_{12})$. Therefore, the rate of formation of A_2 is

$$\frac{d}{dt}[A_2] = k_{12}[A_1] + k_{2e}.$$

A_2 is decomposed by these two processes: $A_2 \to A_1(k_{12})$ and $A_2 \to (k_{02})$. Thus, the rate of decomposition is given by

$$-\frac{d}{dt}[A_2] = k_{21}[A_1] + k_{02}.$$

Therefore, the net change of A_2 is

$$\frac{d}{dt}[A_2] = k_{12}[A_1] - k_{21}[A_2] + (k_{2e} - k_{02}).$$

The mathematical description of a two-species monomolecular chemical reaction system is summarized by

$$\frac{d}{dt}[A_1] = -k_{12}[A_1] + k_{21}[A_2] + (k_{1e} - k_{10}),$$
$$\frac{d}{dt}[A_2] = k_{12}[A_1] - k_{21}[A_2] + (k_{2e} - k_{02}).$$

Thus,

$$dc = (Ac + p)dt, \tag{4.21}$$

where

$$c = \left[[A_1] \quad [A_2] \right]^T, \quad p = \left[(k_{1e} - k_{10}) \; (k_{2e} - k_{20}) \right]^T,$$

$$A = \begin{bmatrix} -k_{12} & k_{21} \\ k_{12} & -k_{21} \end{bmatrix}.$$

Illustration 4.2.6 (multispecies ecosystems [90, 94]). Let us consider an n species ecosystem. Let $N = [N_1, N_2, \ldots, N_l, \ldots, N_n]^T$ be the state of an n-species population vector whose components represent the size of a single species in the community. The possible types of interactions between any two species of the n-species community ecosystem are commensalism, amensalism, symbiosis/mutualism, competition, and general predator–prey which includes plant–herbivore, parasite–host, etc.

For each $l = 1, \ldots, l, \ldots, n$, and by following the description given by Lotka, the population change process for the lth species in the community during the time interval of length $\triangle t$ is described by

$$\begin{aligned} \triangle N_l &= \text{(net gain or loss)} \\ &= \text{(change in the absence of interactions)} \\ &\quad + \text{(change due to interactions with others)} \\ &= \left(e_l + a_{ll} N_l + \sum_{j \neq 1}^{n} \alpha_{lj} N_j \right) \triangle t, \end{aligned}$$

which implies that

$$dN_l = \left(e_l + a_{ll} N_l + \sum_{j \neq 1}^{n} \alpha_{lj} N_j \right) dt.$$

Thus,

$$dN = (AN + e)\, dt, \tag{4.22}$$

where $A = (\alpha_{lj})_{n \times n}$ community matrix. The α_{lj} characterizes the influence of the jth species on the lth species in the community. The sign of α_{lj} determines the nature of the interactions between the (l, j)th pair of species, such as the effects of the jth species on the lth as positive, neutral, or negative ($\alpha_{lj} : +, 0, -$), depending on whether the population of the lth species is increased, unaffected, or decreased by the presence of the jth species. Thus, $(+,+)$, $(-,-)$, $(+,0)$, and $(+,-)$ denotes mutualism/symbiosis, competition, commensalism, and predator–prey, respectively. Furthermore, $\alpha_{ll} < 0$ means that the lth species is density-dependent, i.e. as the size of the lth species increases, its rate decreases, and vice versa.

Observation 4.2.2

(i) We note that most of the multispecies mathematical models are highly nonlinear and nonstationary, such as (Lotka–Volterra-type model)

$$dN_l = \left(e_l + \sum_{j=1}^{n} a_{lj} N_j \right) N_l \, dt. \tag{4.23}$$

Of course, the closed-form solutions to systems like (4.23) are not feasible. However, if the system is very close to its equilibrium state, $N^* = [N_1^*, N_2^*, \ldots, N_l^*, \ldots, N_n^*]^T$. The equilibrium states of the system are given by setting $dN = 0$. This implies that the population is unchanged. It shows that the population has reached its steady-state/equilibrium state. The equilibrium states are determined by solving the following system of algebraic equations associated with rate functions:

$$\left(e_l + \sum_{j=1}^{n} a_{lj} N_j^* \right) N_l^* = 0, \quad \text{for } l = 1, 2, \ldots, n. \tag{4.24}$$

Using one of the solutions N^* to these systems of algebraic equations, we can find the deviation of an unknown state N of the system with N^* by employing the transformation

$$x = N - N^*, \tag{4.25}$$

and we obtain

$$\begin{aligned} dx_l &= d\left(N_l - N_l^*\right) = dN_l \\ &= \left(e_l + \sum_{j=1}^{n} a_{lj} N_j \right) N_l \, dt \quad \text{[by substitution (4.23)]} \\ &= \left(e_l + \sum_{j=1}^{n} a_{lj} \left(N_j - N_j^* + N_j^*\right) \right) \left(N_l - N_l^* + N_l^*\right) \quad \text{(using a−a=0)} \\ &= \left(e_l + \sum_{j=1}^{n} a_{lj} \left(x_j + N_j^*\right) \right) \left(x_l + N_l^*\right) \quad \text{[from (4.25)]} \\ &= e_l \left(x_l + N_l^*\right) + \left[\sum_{j=1}^{n} a_{lj} x_j + \sum_{j=1}^{n} a_{lj} N_j^* \right] \left(x_l + N_l^*\right) \quad \text{(by multiplying)} \end{aligned}$$

$$= \left[\left(e_l + \sum_{j=1}^{n} a_{lj} N_j^* \right) N_l^* + \left(e_l + \sum_{j=1}^{n} a_{lj} N_j^* \right) x_l \right.$$

$$\left. + \sum_{j=1}^{n} N_l^* a_{lj} x_j + \sum_{j=1}^{n} a_{lj} x_j x_l \right] dt.$$

From the above expression and assuming that the population in the community is very near to the equilibrium state, we have the steady-state approximation (by neglecting the terms of order greater than 1, i.e. the term $\sum_{j=1}^{n} a_{lj} x x_l$ of the state dynamic)

$$dx_l = \left[\sum_{j=1}^{n} N_l^* a_{lj} x_j + \left(e_l + \sum_{j=1}^{n} a_{lj} N_j^* \right) x_l \right] dt,$$

which can be rewritten as

$$dx = A\left(N^*\right) x\, dt, \tag{4.26}$$

where the $n \times n$ matrix $A\left(N^*\right)$, whose elements are defined by

$$a_{lj}\left(N^*\right) = \begin{cases} e_l + 2a_{ll} N_l^* + \sum\limits_{j \neq l}^{n} a_{lj} N_j^*, & \text{for } l = j, \\ N_l^* a_{lj}, & \text{for } j \neq l. \end{cases}$$

This mathematically-described process is referred to as the linearization process of the nonlinear system (4.23). The system in (4.26) is called a linear approximation of the nonlinear system (4.23). This is the standard procedure that is used to obtain a linear system corresponding to the nonlinear system with respect to the given equilibrium state N^*. By using the linear approximation system (4.26), one can find the closed-form solution and can draw some inferences about the nonlinear system. The inferences are reasonable if the nonlinear system is operating in a very close neighborhood of the equilibrium state N^*. Otherwise, one needs to use the nonlinear methods to analyze nonlinear systems of type (4.23). A couple of well-known nonlinear techniques will be highlighted in Chapter 6.

(ii) The above-described linearized process can be carried out by finding the Jacobian matrix (1.37) (Observation 1.5.3) with regard to the rate vector associated with the system (4.23). For example, in the case of (4.23), the vector is $f(t, N) = [f_1(t, N), \ldots, f_l(t, N), \ldots, f_n(t, N)]^T$, where $f_l(t, N)$ is defined by

$$f_l(t, N) = \left(e_l + \sum_{j=1}^{n} a_{lj} N_j \right) N_l = e_l N_l + a_{ll} N_l^2 + \sum_{j \neq l}^{n} a_{lj} N_j N_l,$$

and its gradient vector is

$$\frac{\partial}{\partial N}f_l(t,N) = \left[\frac{\partial}{\partial N_1}f_l(t,N),\ldots,\frac{\partial}{\partial N_l}f_l(t,N),\ldots,\frac{\partial}{\partial N_j}f_l(t,N),\ldots,\frac{\partial}{\partial N_n}f_l(t,N)\right]$$

$$= \left[a_{l1}N_l,\ldots,\left(e_l + 2a_{ll}N_l + \sum_{j\neq l}^{n} a_{lj}N_j\right),\ldots,a_{lj}N_l,\ldots,a_{ln}N_l\right].$$

We note that this gradient vector is the lth row in the Jacobian matrix of $f(t,N)$. Hence the Jacobian matrix associated with $f(t,N)$ is

$$\frac{\partial}{\partial N}f(t,N)$$

$$= \begin{bmatrix} \left(e_1 + 2a_{11}N_1 + \sum_{j\neq 1}^{n} a_{1j}N_j\right) & \cdots & a_{1n}N_1 \\ \cdots & \cdots & \cdots \\ a_{n1}N_n & \cdots & \left(e_n + 2a_{nn}N_n + \sum_{j\neq n}^{n} a_{nj}N_j\right) \end{bmatrix}.$$

The value of this Jacobian matrix function at N^* is

$$\frac{\partial}{\partial N}f(t,N^*).$$

This is exactly equal to the matrix $A(N^*)$ in (4.26), i.e.

$$\frac{\partial}{\partial N}f(t,N^*) = A(N^*).$$

If the rate vector function is continuously differentiable, then this is the most standard method of linearizing the nonlinear system of differential equations.

Example 4.2.1. We consider a classical Lotka–Volterra model [55, 94, 104] with a one-predator, one-prey system. Let N_1 and N_2 be the size of the prey and the predator population, respectively. By imitating the Lotka-type description outlined in Illustration 4.2.6, we have

$$\begin{aligned}\Delta N_1 &= (\text{net gain or loss}) \\ &= (\text{change in the absence of interactions}) \\ &\quad + (\text{change due to interactions with others})\end{aligned}$$

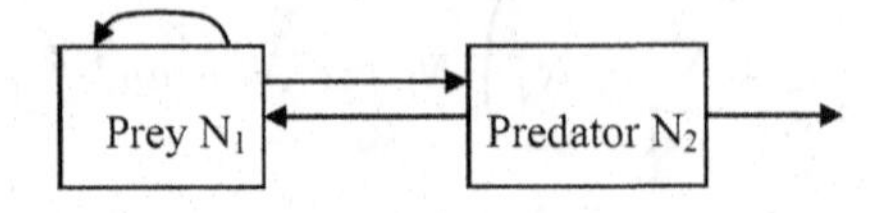

Fig. 4.5 Prey–predator interactions.

$$= (\text{net change in size formed due to } N_1) + (\text{change due to the destruction by } N_2) = (\alpha_0 N_1 - \alpha_1 N_1 N_2)\,\triangle t \quad \text{(Illustration 4.2.6).}$$

Similarly,

$$\begin{aligned}\triangle N_2 &= (\text{net gain or loss}) \\ &= (\text{change in the absence of interactions}) \\ &\quad + (\text{change due to interactions with others}) \\ &= (\text{death due to the starvation of } N_2) \\ &\quad + (\text{change due to the formation from } N_1) \\ &= (-\beta_0 N_2 + \beta_1 N_1 N_2)\,\triangle t \\ &\qquad \text{(Illustration 4.2.6 and number of contacts),}\end{aligned}$$

where α_0, α_1, β_0, and β_1 are positive numbers.

Thus,

$$\begin{aligned} dN_1 &= (\alpha_0 N_1 - \alpha_1 N_1 N_2)\,dt, \\ dN_2 &= (-\beta_0 N_2 + \beta_1 N_1 N_2)\,dt. \end{aligned}$$

For this system, applying Observation 4.2.2, we obtain the algebraic equations analogous to system (4.24):

$$\begin{cases} \alpha_0 N_1^* - \alpha_1 N_1^* N_2^* = 0, \\ -\beta_0 N_2^* + \beta_1 N_1^* N_2^* = 0, \end{cases} \quad \text{which implies that} \begin{cases} \alpha_0 - \alpha_1 N_2^* = 0, \\ -\beta_0 + \beta_1 N_1^* = 0. \end{cases}$$

The nonzero solution to the above algebraic system is

$$N^* = \left[N_1^*\ N_2^*\right]^T = \left[\frac{\beta_0}{\beta_1}\ \frac{\alpha_0}{\alpha_1}\right]^T.$$

$$dN = CNdt, \tag{4.27}$$

where $N = \left[N_1\ N_2\right]^T$ and

$$C = \begin{bmatrix} \alpha_0 - \alpha_1 N_2^* & -\alpha_1 N_1^* \\ \beta_1 N_2^* & -\beta_0 + \beta_1 N_1^* \end{bmatrix} = \begin{bmatrix} 0 & -\dfrac{\alpha_1 \beta_0}{\beta_1} \\ \dfrac{\beta_1 \alpha_0}{\alpha_1} & 0 \end{bmatrix}.$$

The solution of this mathematical model will be presented in Section 4.3.

Illustration 4.2.7 (multicompartment system [46, 58, 111, 119]). In the last 70 years, the concept of a compartment has been extensively studied in the scientific literature. A substance with a given size (mass/volume) is called a compartment. This term is also defined as a state characterized by spatial localization and the nature of the substance. For example, the intra- and extracellular glucose content may be viewed as two distinct compartments. Furthermore, two identifiable different substances that occupy the same volume (physical) may also be called compartments. For example, free iodine and serum-bound iodine in the blood may be called two different compartments. In short, the idea of a compartment covers a broad class of processes in the biological, chemical, medical, and social sciences [46, 58].

A system is a set of interacting components/parts that are viewed as a whole. A multicompartment system consists of two or more than two interconnected compartments:

MCS 1. The measurable qualities of a system are called the states of the system.

MCS 2. The compartment system is static in the absence of any external disturbance.

MCS 3. The compartment is characterized by its steady-state flux of matter into and/or out of it. Each flux has its own channel.

MCS 4. It is assumed that a substance can enter a given compartment from an exterior at a steady rate and vice versa.

MCS 5. In general, it is assumed that the net inflow to the compartment is equal to the net outflow of the material.

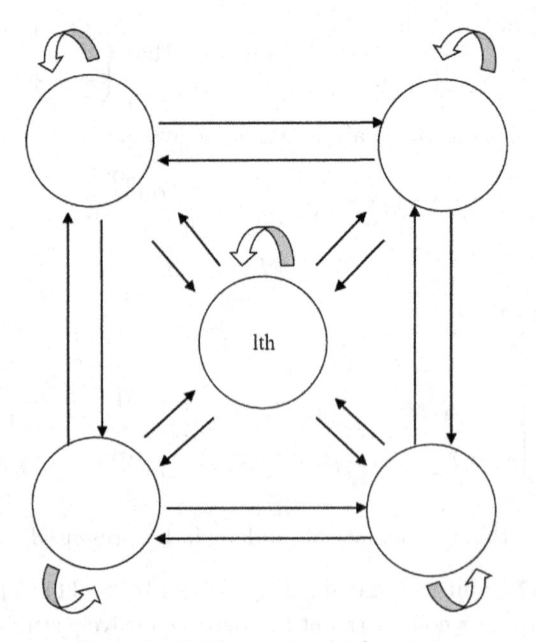

Fig. 4.6 Multicompartmental systems.

MCS 6. The transfer of the material from one region to another may be an actual transport of the substance, or due to the causes, such as, diffusion, osmosis, and pressure gradient.

Let $q = [q_1, q_2, \ldots, q_1, \ldots, q_n]^T$ be a state of a system of n compartments. q_l represents the amount of the observable substance (size/mass) (MCS 1) in the lth compartment. Let e_l and α_{li} be the rates of material that enters the lth compartment from the environment (MCS 4) and the ith compartment (MCS 3), respectively, and α_{0l} and α_{jl} be the rates of the amount of the substance transferred from the lth compartment to the environment (MCS 4) and the jth compartment (MCS 3), respectively. α's are all nonnegative. For each $l = 1, 2, \ldots, n$, the change of the material process in the lth compartment during the time interval of length $\triangle t$ is described by

$$\begin{aligned}\triangle q_l &= \text{(net gain or loss)}\\ &= \text{(change in the absence of interactions)}\\ &\quad + \text{(change due the interactions with others)}\\ &= (e_l - \alpha_{0l} q_l)\triangle t + \left(\sum_{j\neq l}^{n} \alpha_{lj} q_j - \sum_{j\neq l}^{n} \alpha_{jl} q_l\right)\triangle t\\ &= \left(e_l - \left(\alpha_{0l} + \sum_{j\neq l}^{n} \alpha_{jl}\right) q_l + \sum_{j\neq l}^{n} \alpha_{lj} q_j\right)\triangle t.\end{aligned}$$

This implies that

$$dq_l = \left(e_l - \left(\alpha_{0l} + \sum_{j\neq l}^{n} \alpha_{jl}\right) q_l + \sum_{j\neq l}^{n} \alpha_{lj} q_j\right) dt.$$

Hence,

$$dq = (Qq + e)\,dt, \tag{4.28}$$

where $e = [e_1, e_2, \ldots, e_l, \ldots, e_n]^T$, $Q = (q_{lj})_{n\times n}$, and

$$q_{lj} = \begin{cases} -\left(\alpha_{0l} + \sum\limits_{j\neq l}^{n} \alpha_{jl}\right), & \text{for } j = l,\\ \alpha_{lj}, & \text{for } j \neq l.\end{cases}$$

Here,

$$|q_{ll}| = \alpha_{0l} + \sum_{j\neq l}^{n} \alpha_{jl}$$

is called the total outflow rate from the lth compartment. This is the kinetic form introduced in Ref. 134 into the expression of a drug from a subcutaneous depot.

The measurement of the amount of the substance is not, in general, feasible. The observations of compartmental concentrations are most commonly encountered.

Therefore, it is natural to convert (4.28) in the form of a concentration framework. For this purpose, we define q_l by

$$c_l = \frac{q_l}{V_l}, \tag{4.29}$$

where V_l is the volume and it is assumed to be the constant parameter for the lth compartment. With this transformation, (4.28) reduces to

$$dc = (Ac + a)\,dt, \tag{4.30}$$

where

$$a = \left[\frac{e_1}{V_1}, \frac{e_2}{V_2}, \ldots, \frac{e_l}{V_l}, \ldots, \frac{e_n}{V_n}\right]^T,$$

$A = (a_{lj})_{n\times n}$, and

$$a_{lj} = \begin{cases} -\frac{1}{V_l}\left(\alpha_{0l} + \sum_{j\neq l}^{n} \alpha_{jl}\right), & \text{for } j = l, \\ \dfrac{\alpha_{lj} V_j}{V_l}, & \text{for } j \neq l. \end{cases}$$

We note that the above transformation leads to a coefficient matrix in (4.30) without biophysical significance [unless V_l is regarded as a volume such as the fractional clearance rates in (4.28)]. Hereafter, we will be using (4.30) in our discussion. This completes the compartmental modeling process.

Example 4.2.2 (pharmacokinetics [46, 119, 134]). For clinical, diagnostic, therapeutic, or testing purposes, the physicians or physiologists frequently administer some drug in the human or animal body. The quantity and quality effects depend on the concentration of the drug in certain tissues/organs and/or its circulation in the blood. The knowledge about the time and the concentration of the substance in the various crucial parts of the body plays a significant role, such as: (i) finding the best way to dose and to administer the drug; (ii) how to modify the drug or dose to have the desired therapeutical effects.

In general, the substance is more frequently introduced into the digestive channels "per os" and, in rare situations, "per rectum." Most of the medicines are administered by two somewhat different types of injection: (a) somewhat indirectly — subcutaneous, intramuscular, or peritoneal injection; (b) directly — intervascular injection (usually in the intravenous form), such as: (i) "prompt" injection, i.e. a one-time single dose; (ii) "intermittent" injection — a smaller portion over a longer period; and (iii) "drop/continuous" injection (Dauer injection). Under these drug administration processes, the drug typically goes to the: (1) the blood, (2) tissues, and (3) kidneys (lungs, bile, etc.). The drug transfer from any one of the above locations to any of these specified locations or any other location follows Fick's law of the diffusion process. By following the procedures described in Illustrations 2.5.3

and 2.5.4 and given the nature of the cell membrane, we derive the substance distribution kinetics system [Teorell's pharmacokinetics (TPK)] due to Ref. 134, under the assumptions that:

TPK 1. The substance transfer process obeys Fick's law (osmotic pressure gradient concentration gradient).

TPK 2. The electric potential gradient which is assumed to be negligible, (i.e. electrically charged particles (the ions/colloid particles), the membrane effects, and permeability of the membrane are absent).

TPK 3. A partial amount of the drug is inactivated in the tissue. This is due to the chemical reaction that takes place in the tissue, such as oxidation and coupling with other substances. In the following, we consider one of the somewhat indirect substance administration processes, such as subcutaneous, intramuscular, or peritoneal injection. This generates another location of the source of the substance as an input to the blood component, as stated above. In this case, the substance transfer flow pathways are described by the following directed substance transfer diagram:

$$S \rightleftarrows B \rightleftarrows T \rightarrow \text{(with rates: } k_{10}, \mathrm{k}_{01}, k_{21}, k_{12}, k_{20}, \text{ and } k_{31})$$

$$\downarrow$$

$$K$$

where, S, B, K, and T stand for subcutaneous depot (source S), blood, kidney, and tissue, respectively. Let us denote by a_0, a_1, and a_2 the amounts in the number of gram molecules (or gram or any unit) of the substance in the subcutaneous depot, the blood circulation, and tissues, respectively. In addition, let V_0, V_1, and V_2 be the volumes of the depot, the blood, and the tissues, respectively, and c_0, c_1, and c_2 be the concentrations of the substance in S, B, and T, respectively. Here,

$$k_{10} = \frac{\mu_1}{V_0}, \quad k_{21} = \frac{\mu_2}{V_1}, \quad k_{12} = \frac{\mu_2}{V_2}, \quad k_{31} = \frac{\mu_3}{V_1}$$

and

$$k_{20} = \frac{\mu_0}{V_2}$$

are rate coefficients; μ_1, μ_2, μ_3, and μ_0 are permeability coefficients. The substance transfer is described by Fick's law in the diffusion process.

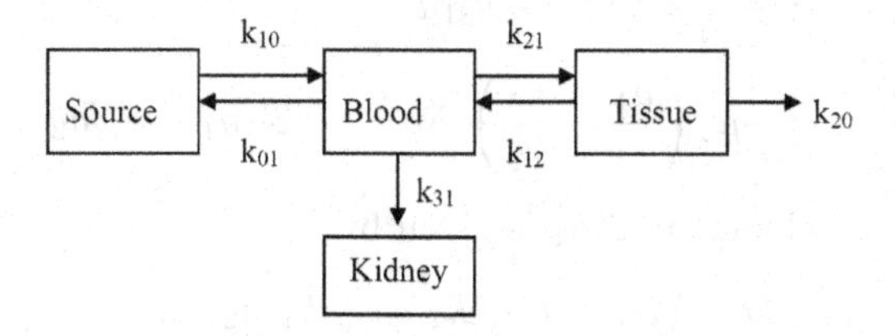

Fig. 4.7 Substance transfer flow.

Step 1. In this step, we need to find $\frac{d}{dt}a_0$. The rate of change of $\frac{d}{dt}a_0$ is determined by $S \rightleftarrows B$ only. The change in the transfer of the amount in the subcutaneous depot to the blood over an interval of time $\triangle t$ due to the resorption process is

$$\begin{aligned}&\text{Change in amount}\\&\quad= \text{diffusion coefficient} \times \text{concentration difference}\\&\qquad \times \text{surface area} \times \triangle t\\&\quad= \text{permeability coefficient} \times \text{concentration difference}\\&\qquad \times \text{surface area} \times \triangle t,\\&\quad \triangle a_0 = \mu_1 (c_1 - c_0) \triangle t = \mu_1 \left(\frac{a_1}{V_1} - \frac{a_0}{V_0}\right) dt,\end{aligned}$$

and

$$\frac{1}{V_1} < \frac{1}{V_0}, \quad \frac{1}{V_0}$$

is very large. Therefore,

$$\triangle a_0 = -\frac{\mu_1}{V_0} a_0 \triangle t = -k_{10} a_0 \triangle t,$$

and hence

$$da_0 = -k_{10} a_0 \, dt, \quad a(0) = m_0.$$

The solution process of this differential equation is given by the procedure described in Section 2.4:

$$a_0(t) = m_0 \exp\left[-k_{10} t\right],$$

where m_0 is the total initial amount of the substance that is administered to the subject.

Step 2. In this step, we need to find $\frac{d}{dt}a_1$. The rate of change of $\frac{d}{dt}a_1$ is determined by $S \rightleftarrows B \rightleftarrows T$ and $B \to K$. By following the argument used in step 1, the change in the transfer of the amount in the blood determined by the transfer of the amounts from the depot to the blood, from the blood to the kidney (elimination), and between the blood and the tissue (take-up) over an interval of time $\triangle t$ is

$$\begin{aligned}\triangle_{sb} a_1 &= k_{10} a_0 \triangle t \quad \text{(expression of } \triangle a_0 \text{ in step 1)},\\ \triangle_{bk} a_1 &= -\frac{\mu_3}{V_1} a_1 \triangle t = -k_{31} a_1 \triangle t,\\ \triangle_{bt} a_1 &= -\mu_2 \left(\frac{a_1}{V_1} - \frac{a_2}{V_2}\right) \triangle t = -\left(k_{21} a_1 - k_{12} a_2\right) \triangle t,\end{aligned}$$

respectively. The net change in $\triangle a_1$ is given by

$$\begin{aligned}\triangle a_1 &= k_{10} a_0 \triangle t - k_{31} a_1 \triangle t - (k_{21} a_1 - k_{12} a_2) \triangle t\\ &= \left[-\left(k_{21} + k_{31}\right) a_1 + k_{12} a_2 + k_{10} a_0\right] \triangle t \quad \text{(by regrouping)}.\end{aligned}$$

Thus, we infer that a_1 satisfies the differential equation

$$da_1 = \left[-\left(k_{21}+k_{31}\right)a_1 + k_{12}a_2 + k_{10}a_0\right]dt.$$

Moreover, from step 1, we have

$$\begin{aligned} da_1 &= \left[-\left(k_{21}+k_{31}\right)a_1 + k_{12}a_2 + k_{10}a_0\right]dt \\ &= \left[-\left(k_{21}+k_{31}\right)a_1 + k_{12}a_2 + k_{10}m_0 \exp\left[-k_{10}t\right]\right]dt. \end{aligned}$$

Step 3. In this step we need to find $\frac{d}{dt}a_2$. The rate of change of $\frac{d}{dt}a_2$ is determined by $B \rightleftarrows T \rightarrow$. By following the arguments used steps 1 and 2, the change in the transfer of the amount in the tissues is given by: the transfer between blood and tissue and the amount due to the inactivation in the tissue (TPK 3) over an interval of time $\triangle t$ as

$$\begin{aligned} \triangle_{\mathrm{tb}}a_2 &= \left(k_{21}a_1 - k_{12}a_2\right)\triangle t, \\ \triangle_{\mathrm{to}}a_2 &= -\frac{\mu_0}{V_2}a_2\triangle t = -k_{20}a_2\triangle t, \end{aligned}$$

respectively. Thus, the net change in $\triangle a_2$ is given by

$$\begin{aligned} \triangle a_2 &= \left(k_{21}a_1 - k_{12}a_2\right)\triangle t - k_{20}a_2\triangle t \\ &= \left[\left(k_{21}\right)a_1 - \left(k_{12}+k_{20}\right)a_2\right]\triangle t. \end{aligned}$$

From this we conclude that a_2:

$$da_2 = \left[\left(k_{21}\right)a_1 - \left(k_{12}+k_{20}\right)a_2\right]dt.$$

The mathematical model of the substance kinetics of two-compartment systems consisting of the blood circulation and the tissues is summarized by

$$\begin{aligned} da_1 &= \left[-\left(k_{21}+k_{31}\right)a_1 + k_{12}a_2 + k_{10}m_0 \exp\left[-k_{10}t\right]\right]dt, \\ da_2 &= \left[\left(k_{21}a_1 - \left(k_{12}+k_{20}\right)a_2\right]dt, \end{aligned}$$

which can be rewritten as

$$da = \left(Ka + k(t)\right)dt, \quad a_0 = [0,0]^T, \tag{4.31}$$

where

$$K = \begin{bmatrix} -\left(k_{21}+k_{31}\right) & k_{12} \\ k_{21} & -\left(k_{12}+k_{20}\right) \end{bmatrix}$$

and $k(t) = \left[k_{10}m_0 \exp\left[-k_{10}t, 0\right]\right]^T$.

Illustration 4.2.8 (group dynamics [44, 127]). There are an innumerable number of verbal propositions concerning human group dynamics under diverse conditions, namely cultural, economic, educational, geographic, historical, etc. In Ref. 44, Homans formulated and discussed certain verbal propositions concerning the behavior of human groups. In particular, he considered the analysis of groups in terms of the interrelations between specific variables such as activity, interaction, and sentiment. In the following, we present three basic verbal propositions proposed by Homans. They are as follows:

HGD 1. "If the frequency of interactions between two or more persons or groups increases, the degree of their liking for one another will increase, and vice versa."

HGD 2. "... if the scheme of activities is changed, the scheme of interactions will, in general, change also, and vice versa."

HGD 3. "... persons who feel sentiments of liking for one another will express those sentiments in activities over and over the activities of the external system, and these activities may strengthen the sentiments of liking."

Based on these three basic propositions, Simon [127] formulated and studied the mathematical model of group interactions. Following Simon, we denote by a, i, f, and e the number of activities, the number of interactions, the level of friendship, and the exogenous input that characterizes externally imposed activities, respectively. From HGD 1 and HGD 2, we have

$$i = \alpha_0 f + \beta_0 a,$$

where α_0 and β_0 are some positive constants. Now, let $\triangle a$ and $\triangle f$ be changes in the number of activities and the level of friendship in the group, respectively, over a time interval of length $\triangle t$. $\triangle a$ and $\triangle f$ are determined by

$$\begin{aligned}
\triangle a &= (\text{net gain or loss}) \\
&= (\text{change in the absence of interactions}) \\
&\quad + (\text{change due to the level of friendship}) \\
&= \gamma_1 (f - \beta_1 a) \triangle t + \gamma_2 (e - a) \triangle t \quad (\text{from HGD 3 and HGD 1}) \\
&= [-(\gamma_1 \beta_1 + \gamma_2) a + \gamma_1 f + \gamma_2 e] \triangle t \quad (\text{by rearrangement}), \\
\triangle f &= (\text{net gain or loss}) \\
&= (\text{fraction of the difference between the activity} \\
&\qquad \text{and some fraction of friendship}) \\
&= \alpha_1 (i - \beta_1 f) \triangle t \quad (\text{from HGD 1}) \\
&= \alpha_1 (\alpha_0 f + \beta_0 a - \beta_1 f) \triangle t \quad (\text{by substitution}), \\
&= [(\alpha_0 \alpha_1 - \beta_1) f + \beta_0 \alpha_1 a] \triangle t \quad (\text{by simplification})
\end{aligned}$$

where α_1, β_1, γ_1, and γ_2 are positive constants, and α_0 and β_0 are as defined before. From the above discussion, we have

$$\begin{aligned}
da &= [-\alpha_{11} a + \alpha_{12} f + \gamma_2 e]\, dt \\
df &= [\alpha_{21} a - \alpha_{22} f]\, dt.
\end{aligned}$$

Thus,

$$dx = (Ax + p)\,dt, \tag{4.32}$$

where $\alpha_{11} = \gamma_1\beta_1 + \gamma_2$, $\alpha_{12} = \gamma_1$, $\alpha_{21} = \beta_0\alpha_1$ positive numbers, $\alpha_{22} = \alpha_0\alpha_1 - \beta_1$, and $x = [a, f]^T, A = (\alpha_{ij})_{2\times 2}$ is a constant matrix and $p = [\gamma_2 e, 0]^T$ is an input to the group dynamic.

This completes the modeling process of group/individual social interaction dynamics.

Example 4.2.3 (Kendall's mathematical marriage model [48]). In 1949, D. G. Kendall presented a simple deterministic mathematical model under the following assumptions:

KMM 1. The growth of a single species is directly proportional to the size of the population.

KMM 2. The unmarried male and unmarried female birth and death rates are identical.

KMM 3. The birth and death rates of married and unmarried persons are directly proportional to the number of couples in the community.

KMM 4. The rate of change of marriages of single persons is described by a rate function K.

KMM 5. In particular, K in KMM 4 is $K(F, M) = \rho \min\{F, M\}$, where F and M are the number of unmarried females and of unmarried males, respectively.

KMM 6. Initially the number of unmarried males is greater than that of unmarried females.

Let C be the number of married couples in a community at a time t. In addition, let β_1 and β_2 be the birth rates of the female and the male population in the community, and δ_1 and δ_2 be the death rates of unmarried females and unmarried males, respectively. Now, by following Illustration 4.2.8, the change of unmarried singles and married couples is given by

$$\begin{aligned}
\triangle F &= \text{(net gain or loss in the unmarried female population)}\\
&= \text{(change by death)} + \text{(change by birth)}\\
&\quad + \text{(change due to the death of the married male)}\\
&\quad + \text{(change due to marriage)}\\
&= [-\delta_1 F + \beta_1 C + \delta_2 C - K(F, M)]\,\triangle t \quad \text{(from KMM 3 and KMM 4)},\\
\triangle M &= \text{(net gain or loss in the unmarried male population)}\\
&= \text{(change by death)} + \text{(change by birth)}\\
&\quad + \text{(change due to the death of the married female)}
\end{aligned}$$

$$\begin{aligned}
&\quad + \text{(change due to marriage)}\\
&= [-\delta_2 M + \beta_2 C + \delta_1 C - K(F, M)]\triangle t \quad \text{(from KMM 3 and KMM 4)},\\
\triangle C &= \text{(net gain or loss in the unmarried male population)}\\
&= \text{(change due to the death of the married couple)}\\
&\quad + \text{(change due to marriage)}\\
&= [-(\delta_1 + \delta_2) C + K(F, M)] \triangle t \quad \text{(from KMM 3 and KMM 4)},
\end{aligned}$$

where $\beta_1, \beta_2, \delta_1$, and δ_2 are positive constants. From the above discussion, we have the following nonlinear system of differential equations:

$$\begin{aligned}
dF &= [-\delta_1 F + \beta_1 C + \delta_2 C - K(F, M)]\, dt,\\
dM &= [-\delta_2 M + \beta_2 C + \delta_1 C - K(F, M)]\, dt,\\
dC &= [-(\delta_1 + \delta_2) C + K(F, M)]\, dt.
\end{aligned}$$

Moreover,

$$\begin{aligned}
dF &= [-\delta_1 F + \Lambda_f(F, M) - K(F, M)]\, dt,\\
dM &= [-\delta_2 M + \Lambda_m(F, M) - K(F, M)]\, dt,\\
dC &= [-(\delta_1 + \delta_2)C + K(F, M)]\, dt,
\end{aligned}$$

where $\Lambda_f(F, M) = \beta_1 C + \delta_2 C$ and $\Lambda_m(F, M) = \beta_2 C + \delta_1 C$ birth rates due to the married couples. From this, assumption KMM 2, setting $\delta_1 = \delta_2 = \delta$ and $\beta_1 = \beta_2 = \beta$, we have

$$\begin{aligned}
dF &= [-\delta F + \beta C + \delta C - K(F, M)]\, dt,\\
dM &= [-\delta M + \beta C + \delta C - K(F, M)]\, dt,\\
dC &= [-(2\delta)C + K(F, M)]\, dt.
\end{aligned}$$

In this case, $\Lambda_f(F, M) = \Lambda_m(F, M) = \beta C + \delta C$. Now, by subtracting the second equation from the first, we have

$$\begin{aligned}
d(F - M) = dF - dM &= [-\delta_1(F - M) + (\beta_1 + \delta_2) - (\beta_2 + \delta_1)]\, C dt\\
&= -\delta\,(F - M)\, dt \quad \text{(by KMM 2)}.
\end{aligned}$$

The solution to this equation (by the application of the procedure for solving this type of differential equation in Section 2.4) is

$$F(t) - M(t) = (F(0) - M(0)) \exp[-\delta t]\,.$$

From this, we infer that any initial excess population of males or females disappears in the course of time. From this observation and KMM 6, we have

$F(t) < M(t)$. From KMM 5, $K(F(t), M(t)) = \rho \min\{F(t), M(t)\} = \rho F(t)$. Moreover, in this case, the above system of differential equations reduces to

$$dF = [-(\delta+\rho)F + (\beta+\delta)C]\,dt,$$
$$dM = [-\delta M + (\beta+\delta)C - \rho F]\,dt,$$
$$dC = [-2\delta C + \rho F]\,dt.$$

We note that the first and third equations in the above are independent of M (decoupled with M). Therefore, one can solve this system by solving the system of the first two coupled differential equations,

$$dF = [-(\delta+\rho)F + (\beta+\delta)C]\,dt,$$
$$dC = [-2\delta C + \rho F]\,dt,$$

i.e.

$$dx = Axdt, \quad x(0) = x_0, \tag{4.33}$$

where $x = [F, C]^T$,

$$A = \begin{bmatrix} -(\delta+\rho) & (\beta+\delta) \\ \rho & -2\delta \end{bmatrix},$$

$x_0 = [F(0), C(0)]^T$.

Then, using this solution and applying the procedure in Subsection 2.5.4, one can solve the following differential equation:

$$dM = [-\delta M + (\beta+\delta)\,C - \rho F]\,dt.$$

Illustration 4.2.9 (US government system [92]). More than 225 years ago, the founding fathers of the United States of America very wisely and cleverly created a unique national governance document, namely the "Constitution of the United States of America." The preamble to the US Constitution [92] begins with: "We the People of the United States, in Order to form a more perfect Union, establish justice, insure domestic tranquility, provide for the common defense, promote the general Welfare, and secure the Blessings of Liberty to ourselves and our Posterity, do ordain and establish this Constitution for the United States of America." It sets the specific goal, with inclusion, openness, universal tone and directions, coupled with the fundamental rights and responsibilities for the common good for the benefit of all human beings great or small, rich or poor, or healthy or unhealthy.

In fact, this document is the description of an open competitive–cooperative dynamic government system with a definite goal as outlined in the preamble. The US government system consists of three branches (components), namely the legislative branch, the executive branch, and the judiciary branch. An environment for this system consists of intranational and extranational environments. The verbal description of the US government system is given in the US Constitution [92]. We employ a few fundamental ideas that are used to develop a dynamic model of the US government system.

Article I (Legislative Branch [92])

Section 1. "All legislative Powers herein granted shall be vested in a Congress of the United States, which shall consist of a Senate and House of Representatives."

Section 7. "All Bills for raising Revenue shall originate in the House of Representatives; but the Senate may propose or concur with Amendments as on other Bills."

Section 8. "The Congress shall have Power to lay and collect Taxes, Duties, Imports and Excises, to pay the Debts and provide for the common Defense and general Welfare of the United States; ...."

"To make all Laws which shall be necessary and proper for carrying into Execution the foregoing Powers, and all other Powers vested by this Constitution in the Government of the United States, or in any Department or Officer thereof."

Article II (Executive Branch [92])

Section 1. "The executive Power shall be vested in a President of the United States of America. He shall hold his Office during the Term of four Years, and, together with the Vice-president, chosen for the same Term, be elected, as follow."

Section 2. "The President shall be Commander in Chief of Army and Navy of the United States, and ...."

"He shall have Power, by and with the Advice and Consent of the Senate, to make Treaties, provided two-thirds of the Senators present concur; he shall nominate, and by and with the Advice and Consent of the Senate, shall appoint Ambassadors, other public Ministers and Consuls, Judges of the Supreme Court, and all other Officers of the United States, whose Appointments are not herein otherwise provided for, and which shall be established by Law: but the Congress may by Law vest the Appointment of such inferior officers, as they think proper, in the President alone, in the Courts of Law, or in the heads of Departments."

"The President shall have power to fill up all Vacancies that may happen during the recess of the Senate, by granting Commissions which shall expire at the End of their next Session."

Section 3. "He shall time to time give to the Congress Information of the State of the Union, and recommend to their Consideration such Measures as he shall judge necessary and expedient; he may, on extraordinary occasions, convene both Houses, or either of them, and in Case of Disagreement between them, with respect to the time of Adjournment, he may adjourn them to such time as she/he thinks proper; he shall receive Ambassadors and other public Ministers; he shall Care that the Laws be faithfully executed, and shall Commission all the Officers of the United States."

Article III (Judiciary Branch [92])

Section 1. "The judicial Power of the United States shall be vested in one supreme Court, and in such inferior Courts as the Congress may from time to time ordain and establish...."

Section 2. "The judicial power shall extend to all Cases, in Law and Equity, arising under this Constitution, the Laws of the United States, and Treaties made, or which shall be made, under their Authority;— to all Cases affecting Ambassadors, other public Ministers and Consuls;— to all Cases of admiralty and maritime Jurisdiction: — to Controversies to which the United Sates shall be a Party;— to Controversies between two or more States:— between a State and Citizens of another State;— between Citizens of different States;— between Citizens of the same State claiming Lands under Grants of different States, and between a State, or the Citizens thereof, and foreign States, Citizens or Subjects...."

Article V (Open system)
"The Congress, whenever two thirds of both Houses shall deem it necessary, shall amendments to this Constitution, or on the Application of the Legislatures of two thirds of the several States, shall call a Convention for proposing Amendments, which, in either Case, shall be valid to all Intents and Purposes, as Part of this Constitution, when ratified by the Legislatures of three fourths of the several States, or by Conventions in three fourths thereof, as the one or the other Mode of Ratification may be proposed by the Congress; Provided that no Amendment which may be made prior to the Year One thousand eight hundred and eight shall in any Manner affect the first and fourth clauses in the Ninth Section of the first Article; and that no State, without its Consent, shall be deprived of its equal Suffrage in the Senate."

Bill of Rights (Intranational environment)

Amendment I. "Congress shall make no law respecting an establishment of religion, or prohibiting the free exercise thereof; or abridging the freedom of speech, or of the press; or the right of the people peaceably to assemble, and to petition the Government for a redress of grievances."

Amendment IV. "The right of people to be secure in their persons, houses, papers, and effects, against unreasonable searches and seizures, shall not be violated, and no warrants shall issue, but upon probable cause, supported by Oath or affirmation, and particularly describing the place to be searched, and the persons or things to be seized."

The measure of the success of the government depends upon the handling of short-term and long-term issues affecting the current and future daily life of the ordinary citizens. Based on the Constitution and knowing the checks and balances between the three components of the government, we present the mathematical model of the US government system. The state dynamic of the government system is characterized by actions. Based on the duties and power of the decision-making

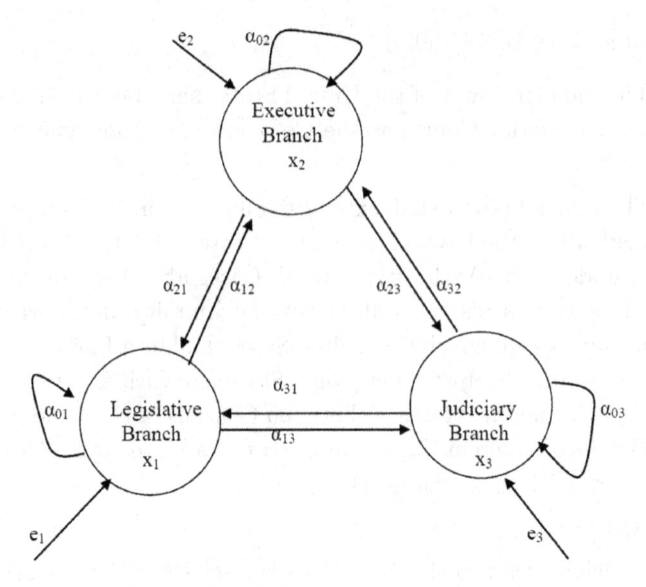

Fig. 4.8 United States government dynamic.

process as outlined in the US Constitution, let x_1, x_2, and x_3 be the number of decisions/actions (bills) taken by the legislative, executive, and judiciary branches of the government, respectively. In addition, let Δx_1, Δx_2, and Δx_3 be net changes in the number of decisions/actions of the three branches of the government over a time interval of length Δt. By following the argument used in Illustration 4.2.7 and by employing the US constitutional rights and responsibilities of these components of the government, for $l = 1, 2, 3$, we have

$$\begin{aligned}\Delta x_l &= \text{(change in the absence of interactions)}\\ &\quad + \text{(change in the presence of interactions)}\\ &= \left[e_l - \alpha_{0l}x_l - \sum_{j\neq l}^{3}\alpha_{lj}x_l + \sum_{j\neq l}^{3}\alpha_{jl}x_j\right]\Delta t\\ &= \left[e_l - \left(\alpha_{0l} + \sum_{j\neq l}^{3}\alpha_{lj}\right)x_l + \sum_{j\neq l}^{3}\alpha_{jl}x_j\right]\Delta t.\end{aligned}$$

Thus,

$$dx_l = \left[e_l - \left(\alpha_{0l} + \sum_{j\neq l}^{3}\alpha_{lj}\right)x_l + \sum_{j\neq l}^{3}\alpha_{jl}x_j\right]dt,$$

and moreover

$$dx = (Ax + e)\,dt, \tag{4.34}$$

where $e = [e_1, e_2, e_3]^T$, $A = (a_{jl})_{3\times 3}$, and

$$a_{lj} = \begin{cases} -\left(\alpha_{0l} + \sum\limits_{j\neq l}^{3} \alpha_{lj}\right), & \text{for } j = l, \\ \alpha_{jl}, & \text{for } j \neq l. \end{cases}$$

Here,

$$|a_{ll}| = \alpha_{0l} + \sum_{j\neq l}^{3} \alpha_{lj}$$

is called an aggregate decision/action outflow rate from the lth branch of the government to other branches including the environment. α_{ll}'s are inactionable and actionable actions/decisions in the respective branches of the government. As per the preamble to the Constitution, it is assumed and expected that all of the elected and appointed government officials are working toward the goal stated in the preamble. Thus, it is natural to assume that all parameters in the system (4.34) are at least nonnegative. The more natural interpretation of the system parameters is: α_{0l}'s represent the fractions of the expected decisions/actions about the issues that are of interest to the citizens (Amendments I and IV), and are under the review of the lth branch of the US government. Moreover, for $j \neq l$, α_{lj} characterizes the rate of flow of the possible expected number of actionable decision/actions sent to the lth branch by the jth branch of the US government. We note that the rate α_{31} is negligible except for the role in the Presidential Impeachment trial (Article I, Section 3) and Advice and Consent of the Senate for the appointment of Federal Court Judges (Article II, Section 2).

4.2 Exercises

(1) ***Reversible bacterial mutation*** [119]. In a bacterial cell culture process, it has been observed that the variant of the initial form appears. The process of formation of a variant is referred to as mutations. The variant form of an initial form of bacteria is called a mutant. It is also known that certain variant cells are converted back into the original form by a process known as back-mutation/reversion. Let A and B be two types of bacterial cells that are genetically connected. B is formed by the mutation process from an initial cell type A. At a time t, let N_1 and N_2 be the number of cells of types A and B, respectively. The dynamic interactions of these cells are described by: $A \rightleftarrows B$; the rates k_{11} and k_{22} are fractional growth rates of A and B, respectively; k_{12} is the fractional mutation rate of the initial form of cell A to cell B, and k_{21} is

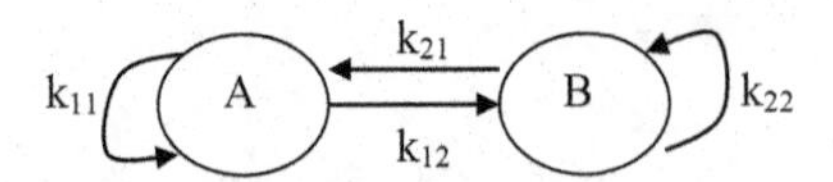

Fig. 4.9 Mutation dynamic process.

the fractional back-mutation rate of variant cell B to the original cell. k_{ij}'s for all $1 \leq i, j \leq 2$ are considered to be positive.

Derive the deterministic mathematical description of the mutation dynamic process.

(2) ***Pharmacokinetics — "prompt" injection*** [134]. Assume that all the conditions of Example 4.2.2 are satisfied.

(a) Derive Teorell's mathematical model for the drug distributions of intravenously administered "prompt" injection.
(b) Compare the mathematical models in (a) with the models described in Example 4.2.2.

(3) ***Kendall's two-sex model*** [48, 106]. In 1949, D. G. Kendall presented a two-sex deterministic model under the following assumptions: (i) the growth of a single species is directly proportional to the size of the population; (ii) let F and M be the number of females and the number of males at a time t, and it is assumed that

$$dF = \left[-\mu F + \frac{1}{2}\Lambda(F, M)\right] dt,$$

$$dM = \left[-\mu M + \frac{1}{2}\Lambda(F, M)\right] dt,$$

where $\Delta(F, M)$ represents an arbitrary interactions between the female and the male population. Rewrite Kendall's systems of differential equations, if:

(a) $\Lambda(F, M) = \rho FM$,
(b) $\Lambda(F, M) = 2\rho\sqrt{FM}$,
(c) $\Lambda(F, M) = 2\rho(F + M)$;
(d) Give the interpretation for Kendall's interaction $\Lambda(F, M)$ in (a)–(c).

(4) ***Goodman's two-sex model*** [36, 106]. Assume that all the conditions of Exercise 3 in Section 4.2 are satisfied. Under Goodman's interaction function, $\Lambda(F, M) = 2\rho \min\{F, M\}$, show that Kendall's deterministic two-sex model reduces to $dF = (-\mu F + \rho M)\, dt$ and $dM = (-\mu + \rho)\, M\, dt$.

(5) In Kendall's deterministic marriage dynamic model (Example 4.2.3), the conditions KMM 2 and KMM 6 are dropped, and the condition KMM 5 is replaced by KMM 5*:

$$K(F, M) = \frac{1}{2}\rho(M + F).$$

Show that under these assumptions Kendall's deterministic marriage dynamic model described in Example 4.2.3 is reduced to the following system of differential equations:

$$dF = [-\delta_1 F + \beta_1 C + \delta_2 C - K(F, M)]\, dt,$$
$$dM = [-\delta_2 M + \beta_2 C + \delta_1 C - K(F, M)]\, dt,$$
$$dC = [-(\delta_1 + \delta_2) C + K(F, M)]\, dt.$$

Hence,

$$dF = \left[-\left(\delta_1 + \frac{1}{2}\rho\right) F - \frac{1}{2}\rho M + (\beta_1 + \delta_2) C\right] dt,$$
$$dM = \left[-\frac{1}{2}\rho F - \left(\delta_2 + \frac{1}{2}\rho\right) M + (\beta_2 + \delta_1) C\right] dt,$$
$$dC = \left[\frac{1}{2}\rho F + \frac{1}{2}\rho M - (\delta_1 + \delta_2) C\right] dt.$$

(6) ***Generalized compartmental model.*** Recalling that dynamic modeling was based on the fact that the stop-clock is perfect. In general, it is subject to random fluctuations. Let F be a cumulative distribution function (cdf) of a random variable T (time). For $x \in R^n$ be the state of the multidimensional compartmental system. Show that the mathematical model (4.30) reduces to

$$dc = (Ac + a)\, dF,$$

where A, c, and a are as defined in (4.30). Moreover, give the interpretation of this model. (*Hint*: An equivalent representation is

$$c(t) = c_0 + \int_{-\infty}^{t} (Ac(s) + a)\, dF(s).$$

Under further assumptions on F, show that

$$c(t) = c_0 + \int_{t_0}^{t} (Ac(s) + a)\, dF(s).$$

(7) ***Generalized Kendall's two-sex model.*** Under the assumptions of Exercise 6 in Section 4.2, reformulate Kendall's mathematical marriage model (Example 4.2.3).

4.3 Linear Homogeneous Systems

In this section, we extend the eigenvalue type method described in Section 2.4 to find a solution to the first-order linear homogeneous system of differential equations with constant coefficients.

4.3.1 *General problem*

Let us consider a first-order linear homogeneous deterministic system of differential equations of type

$$dx = Ax\,dt, \tag{4.35}$$

where x is an n-dimensional column vector (or $n \times 1$ matrix); $A = (a_{ij})_{n\times n}$ is an $n \times n$ constant matrix whose entries a_{ij}'s are real numbers ($a_{ij} \in R$).

The goal is to present a procedure for finding a general solution to (4.35). We are also interested in solving an IVP associated with this system. In addition, we will lay down a few basic ideas and algorithms (theorems) that can be easily applied to: (a) time-varying systems, (b) nonhomogeneous systems, and (c) nonlinear systems of differential equations. For this purpose, we first need to define the concept of a solution process of (4.35).

Definition 4.3.1. Let $J = [a, b]$, $J \subseteq R$ be an interval $a, b \in R$. A solution to the system of differential equations (4.35) is an n-dimensional differentiable column vector (or $n \times 1$ matrix) function x defined on J into R^n such that it satisfies the system (4.35) on J. In short, if we substitute $x(t) = (x_i(t))_{n\times 1}$ and $dx(t) = (dx_i(t))_{n\times 1}$ into the system (4.35), then the system of differential equations remains valid for all t in J.

Example 4.3.1. We are given $dx_1 = (x_1 + 2x_2)\,dt$ and $dx_2 = (4x_1 + 3x_2)\,dt$. Show that $x_1(t) = e^{-t}\begin{bmatrix}1 & -1\end{bmatrix}^T$ and $x_2(t) = e^{5t}\begin{bmatrix}1 & 2\end{bmatrix}^T$ are solutions to the given system of differential equations, where T stands for the transpose of a matrix.

Solution procedure. By using the concept of the product of matrices, the matrix representation of the given differential equations is $dx = Ax\,dt$, where

$$A = \begin{bmatrix}1 & 2\\ 4 & 3\end{bmatrix}, \quad x = \begin{bmatrix}x_1\\ x_2\end{bmatrix}$$

are 2×2 and 2×1 matrices, respectively.

We apply deterministic differential calculus [85] and Definition 1.5.3 to $x_1(t) = e^{-t}\begin{bmatrix}1 & -1\end{bmatrix}^T$, and compute the following differential:

$$\begin{aligned} dx_1(t) &= d\left(e^{-t}\begin{bmatrix}1 & -1\end{bmatrix}^T\right) \quad \text{(given expression)}\\ &= -e^{-t}\begin{bmatrix}1 & -1\end{bmatrix}^T dt \quad \text{[Theorem 1.5.1(A2) and the chain rule]}\\ &= \begin{bmatrix}-e^{-t} & e^{-t}\end{bmatrix}^T dt \quad \text{(scalar multiplication to matrix).}\end{aligned}$$

We also compute $Ax_1(t)\,dt$ as

$$Ax_1(t)\,dt = \begin{bmatrix}1 & 2\\ 4 & 3\end{bmatrix}\begin{bmatrix}e^{-t}\\ -e^{-t}\end{bmatrix} dt \quad \text{(substitution and Definition 1.3.12)}$$

$$= \begin{bmatrix} e^{-t} & -2e^{-t} \\ 4e^{-t} & -3e^{-t} \end{bmatrix} dt \quad \text{(Definition 1.2.11)}$$

$$= \begin{bmatrix} -e^{-t} \\ e^{-t} \end{bmatrix} dt \quad \text{(simplification)}$$

$$= \begin{bmatrix} -e^{-t} & e^{-t} \end{bmatrix}^T dt \quad \text{(Definition 1.3.12)}$$

$$= dx_1(t) \quad \text{(by substitution)}.$$

This shows that the given process $x_1(t) = e^{-t}\begin{bmatrix} 1 & -1 \end{bmatrix}^T$ satisfies the given system of differential equations. Therefore, it is a solution to the given system of differential equations. Similarly, by applying the above argument to $x_2(t) = e^{5t}\begin{bmatrix} 1 & 2 \end{bmatrix}^T$, we have

$$dx_2(t) = d\left(e^{5t}\begin{bmatrix} 1 & 2 \end{bmatrix}^T\right) = 5e^{5t}\begin{bmatrix} 1 & 2 \end{bmatrix}^T dt,$$

$$Ax_2(t)dt = \begin{bmatrix} 1 & 2 \\ 4 & 3 \end{bmatrix}\begin{bmatrix} e^{5t} \\ 2e^{5t} \end{bmatrix} dt = \begin{bmatrix} e^{5t} + 4e^{5t} \\ 4e^{5t} + 6e^{5t} \end{bmatrix} dt = \begin{bmatrix} 5e^{5t} \\ 10e^{5t} \end{bmatrix} dt = 5\begin{bmatrix} e^{5t} \\ 2e^{5t} \end{bmatrix} dt.$$

By comparing the expressions of $dx_2(t)$ and $Ax_2(t)dt$ in the context of definition of the transpose of a matrix (Definition 1.3.12), we conclude that $x_2(t) = e^{5t}\begin{bmatrix} 1 & 2 \end{bmatrix}^T$ is a solution process of the given system of differential equations.

4.3.2 *Procedure for finding a general solution*

We extend the eigenvalue method introduced in Section 2.4 to systems of linear homogeneous differential equations with constant coefficients. We present a four-step procedure for determining a solution to (4.35). It is as follows:

Step 1. Let us seek a solution to (4.35) of the form

$$x(t) = \exp[\lambda t]\, c, \tag{4.36}$$

where $c = (c_1, c_2, \ldots, c_n)^T$ is an n-dimensional unknown constant column vector (or an $n \times 1$ matrix) and λ is an unknown scalar quantity (a real or complex number).

We note that if $c = 0$, i.e. $c_1 = c_2 = \cdots = c_n = 0$, then $x(t)$ in (4.36) is a trivial solution [zero solution, i.e. zero function $x(t) \equiv 0$ on J] to (4.35).

We seek nontrivial solutions (nonzero solutions) of (4.35). For this purpose, we need to compute λ and c. This is achieved as follows. We differentiate x in (4.36) with respect to t. After substituting $x(t)$ and $dx\,(t)$ in (4.35), we obtain

$$\lambda \exp[\lambda t]\, c = A \exp[\lambda t]\, c = \exp[\lambda t]\, Ac.$$

From this and the fact that $\exp[\lambda t] \neq 0$ on J, we have $\lambda c = Ac$, and hence

$$(A - \lambda I)\, c = 0, \tag{4.37}$$

where I is an $n \times n$ identity matrix.

Step 2. Now, the problem of finding a solution to the system of differential equations (4.35) reduces to the problem of solving the linear homogeneous system of algebraic equations (4.37) with an n-dimensional unknown nonzero constant column vector (or an $n \times 1$ matrix) c and a scalar quantity λ (a real or complex number).

From the study of the linear homogeneous system of algebraic equations of Sections 1.3 and 1.4, for the existence of the nonzero solution vector c, it is necessary and sufficient [Observation 1.4.3(b)] that

$$\det(A - \lambda I) = 0, \tag{4.38}$$

where "det" stands for the determinant of the matrix $A - \lambda I$. Equation (4.38) is referred to as the characteristic equation of the matrix A. Moreover, it is an nth-degree polynomial equation in λ. Its leading coefficient is always 1. Hence, from the fundamental theorem of algebra, the characteristic equation (4.38) always has n roots $\lambda_1, \ldots, \lambda_k, \ldots, \lambda_n$ (real or complex number) which are not necessarily all distinct. We further remark that these roots are uniquely determined by the coefficient matrix A in (4.35). In short, these roots depend on only the entries of the coefficient matrix A.

Step 3. For each characteristic root $\lambda_k \equiv \lambda_k(A)$ determined in step 2, there exists at least one nonzero vector $c_k \equiv c_k(A)$. This is computed by substituting $\lambda = \lambda_k$ into the system (4.37), and then solving the system of n linear homogeneous algebraic equations

$$(A - \lambda_k I)\, c = 0, \tag{4.39}$$

for each $k = 1, 2, \ldots, n$. Of course, the system (4.39) has infinitely many non-zero solutions. Associated with each k, $k = 1, 2, \ldots, n$, we choose a nonzero solution to (4.39), and denote it by $c_k \equiv c_k(A)$. We note that the choice of c_k is not unique. It is based on the choice of an individual problem-solver.

Traditionally, for $\lambda_k \neq 0$, $\frac{1}{\lambda_k}$ is called an eigenvalue of the matrix A. But, usually, λ_k is called an eigenvalue of the matrix A, and c_k is the corresponding eigenvector of the matrix A.

Step 4. In step 3, we found n eigenvectors $c_1, \ldots, c_k, \ldots, c_n$ of the coefficient matrix A in (4.35) corresponding to n roots $\lambda_1, \ldots, \lambda_k, \ldots, \lambda_n$ of (4.38) in step 2. From (4.36), we have n solutions,

$$x_1(t) = \exp[\lambda_1 t]\, c_1, \ldots, x_k(t) = \exp[\lambda_k t]\, c_k, \ldots, x_n(t) = \exp[\lambda_n t]\, c_n,$$

to the system of differential equations (4.35). This completes the procedure for finding solutions to (4.35).

Example 4.3.2. We are given $dx(t) = Ax\, dt$, where

$$A = \begin{bmatrix} 5 & -4 \\ 3 & -2 \end{bmatrix}.$$

Find solutions to the given system of differential equations.

Solution process. This is the system of two ($n = 2$) first-order linear differential equations. To find solutions to this system of differential equations, we imitate the procedure in Subsection 4.3.2. We seek a solution of the form described in step 1, i.e. $x(t) = \exp[\lambda t]\, c$. The primary goal is to find an unknown nonzero two-dimensional vector $c = [c_1 \ c_2]^T$ and a scalar number λ. For this purpose, we use (4.37), and we have

$$0 = (A - \lambda I)\, c = \left(\begin{bmatrix} 5 & -4 \\ 3 & -2 \end{bmatrix} - \lambda \begin{bmatrix} 1 & 0 \\ 0 & 1 \end{bmatrix} \right) c \quad \text{(by substitution)}$$

$$= \begin{bmatrix} 5 - \lambda & -4 \\ 3 & -2 - \lambda \end{bmatrix} c. \quad \text{[by Theorem 1.3.1(S2)].}$$

This can be rewritten as

$$\begin{bmatrix} 5 - \lambda & -4 \\ 3 & -2 - \lambda \end{bmatrix} c = \begin{bmatrix} (5 - \lambda)\, c_1 - 4c_2 \\ 3c_1 - (2 + \lambda)\, c_2 \end{bmatrix} = 0 \quad \text{(by Definition 1.3.11).}$$

Thus,

$$(5 - \lambda)\, c_1 - 4c_2 = 0,$$

$$3c_1 - (2 + \lambda)\, c_2 = 0.$$

The characteristic equation (4.38) of the matrix A in the context of the given example is

$$0 = \det(A - \lambda I) = -(5 - \lambda)(2 + \lambda) + 12 \quad \text{(step 1 in Example 1.4.7)}$$

$$= \lambda^2 - 3\lambda + 2 \quad \text{(by simplifying)}$$

$$= (\lambda - 2)(\lambda - 1) \quad \text{(step 2 in Example 1.4.7).}$$

The eigenvalues are $\lambda_1 = 1$ and $\lambda_2 = 2$. This completes step 2 of the procedure in Subsection 4.3.2.

By using the argument of step 3 of the procedure in Subsection 4.3.2, we find eigenvectors corresponding to the eigenvalues $\lambda_1 = 1$ and $\lambda_2 = 2$. For this purpose, we need to solve the systems of linear homogeneous algebraic equations (4.39) corresponding to these eigenvalues. First, we consider $(A - \lambda_1 I)\, c = (A - 1I)\, c = 0$, which can be written as

$$4c_1 - 4c_2 = 0,$$

$$3c_1 - 3c_2 = 0.$$

One can imitate the procedure outlined in Example 1.4.9 to find the solution set. However, in this case ($n = 2$), we can directly solve one of the equations in the above system, in particular $4c_1 - 4c_2 = 0$. The solution set

$$S = \left\{ c : c = \begin{bmatrix} c_1 & c_1 \end{bmatrix}^T \right\} = \operatorname{Span}\left(\left\{ \begin{bmatrix} 1 & 1 \end{bmatrix}^T \right\} \right).$$

Any one of the members of the span can be chosen as an eigenvector of the matrix A corresponding to the eigenvalue $\lambda_1 = 1$. Here, we choose $c_1(A) = \begin{bmatrix} 1 & 1 \end{bmatrix}^T$. We repeat the above discussion with regard to the eigenvalue $\lambda_2 = 2$, and we have $(A - 2I)\,c = 0$, and hence

$$3c_1 - 4c_2 = 0,$$
$$3c_1 - 4c_2 = 0.$$

Again, its solution set is

$$S = \left\{ c : c = \begin{bmatrix} \frac{4}{3} c_2 & c_2 \end{bmatrix}^T \right\} = \text{Span}\left(\left\{ \begin{bmatrix} \frac{4}{3} & 1 \end{bmatrix}^T \right\} \right).$$

As before, we choose an eigenvector of the matrix A corresponding to the eigenvalue $\lambda_2 = 2$. Here, we choose $c_2(A) = \begin{bmatrix} 4 & 3 \end{bmatrix}^T$. From step 4, we found two solutions,

$$x_1(t) = \begin{bmatrix} e^t \\ e^t \end{bmatrix}, \quad x_2(t) = \begin{bmatrix} 4e^{2t} \\ 3e^{2t} \end{bmatrix},$$

corresponding to the two eigenvalues $\lambda_1 = 1$ and $\lambda_2 = 2$ and two eigenvectors, $c_1(A) = \begin{bmatrix} 1 & 1 \end{bmatrix}^T$ and $c_2(A) = \begin{bmatrix} 4 & 3 \end{bmatrix}^T$. This completes the solution procedure.

Example 4.3.3. We are given $dx_1 = x_2\,dt$. Find a solution to this system of differential equations.

Solution procedure. The given differential equation does not look like a system of differential equations. However, after careful observation, we have the following representation. Using the product of matrices, the matrix representation of the given differential equation is $dx\,(t) = Ax\,dt$, where

$$A = \begin{bmatrix} 0 & 1 \\ 0 & 0 \end{bmatrix}, \quad x = \begin{bmatrix} x_1 \\ x_2 \end{bmatrix}$$

are 2×2 and 2×1 matrices, respectively.

Now, applying the procedure in Subsection 4.3.2 and using the argument in Example 4.3.2, we have

$$\begin{bmatrix} -\lambda & 1 \\ 0 & -\lambda \end{bmatrix} c = \begin{bmatrix} -\lambda c_1 + c_2 \\ -\lambda c_2 \end{bmatrix} = 0,$$

which implies that

$$-\lambda c_1 + c_2 = 0,$$
$$-\lambda c_2 = 0.$$

The characteristic equation (4.38) of the matrix A in the context of the given example is

$$0 = \det(A - \lambda I) = (-\lambda)(-\lambda) = \lambda^2.$$

The eigenvalues are $\lambda_1 = 0$ and $\lambda_2 = 0$. This completes step 2 of the procedure in Subsection 4.3.2. We note that $\lambda_1 = 0$ is the double (repeated) root of the characteristic equation of A. The eigenvector corresponding to the eigenvalue $\lambda_1 = 0$ or $\lambda_2 = 0$ is given by $(A - \lambda_1 I)\,c = (A - 0I)\,c = Ac = 0$. This can be rewritten as

$$\begin{aligned} 0c_1 + c_2 &= 0, \\ 0c_1 - 0c_2 &= 0. \end{aligned}$$

From this system of algebraic equations, we conclude that the first equation ($0c_1 + c_2 = 0$) is more significant. From this equation, we infer that $c_2 = 0$ and c_1 is an arbitrary number. Thus, the solution set of this system is

$$S = \left\{c : c = \begin{bmatrix} c_1 & 0 \end{bmatrix}^T\right\} = \text{Span}\left(\left\{\begin{bmatrix} 1 & 0 \end{bmatrix}^T\right\}\right).$$

Any one of the members of the span can be chosen as an eigenvector of the matrix A corresponding to the eigenvalue $\lambda_1 = 0$ or $\lambda_2 = 0$. Here, we choose $c_1(A) = \begin{bmatrix} 1 & 0 \end{bmatrix}^T$. From step 4, we found one solution,

$$x_1(t) = e^{0t}\begin{bmatrix} 1 \\ 0 \end{bmatrix} = \begin{bmatrix} 1 \\ 0 \end{bmatrix} \quad (\text{by } e^{0t} = e^0 = 1),$$

corresponding to the eigenvalue $\lambda_1 = 0$ and the eigenvector $c_1(A) = \begin{bmatrix} 1 & 0 \end{bmatrix}^T$. This completes the solution procedure.

Observation 4.3.1. From the characteristic polynomial $\det(A - \lambda I)$ in (4.38), we note that for $AI = A$, $\det(A - \lambda I) = \det(A - A) = 0$. This observation is in fact the Cayley–Hamilton theorem, which states: "Every square matrix satisfies its own characteristic equation" [4].

Example 4.3.4. Show that A in Example 4.3.2 satisfies the Cayley–Hamilton theorem.

Solution process. Let us verify the validity of the Cayley–Hamilton theorem. In this case, $\det(A - \lambda I) = \lambda^2 - 3\lambda + 2 = p(\lambda)$. Now, we compute $p(A)$ as

$$\begin{aligned} p(A) &= A^2 - 3A + 2I \\ &= \begin{bmatrix} 5 & -4 \\ 3 & -2 \end{bmatrix}\begin{bmatrix} 5 & -4 \\ 3 & -2 \end{bmatrix} - 3\begin{bmatrix} 5 & -4 \\ 3 & -2 \end{bmatrix} + 2\begin{bmatrix} 1 & 0 \\ 0 & 1 \end{bmatrix} \quad (\text{by substitution}) \end{aligned}$$

$$= \begin{bmatrix} 25-12 & -20+8 \\ 15-6 & -12+4 \end{bmatrix} + \begin{bmatrix} -15 & 12 \\ -9 & 6 \end{bmatrix} + \begin{bmatrix} 2 & 0 \\ 0 & 2 \end{bmatrix}$$

(Definitions 1.3.9 and 1.3.11)

$$= \begin{bmatrix} 25-12-15+2 & -20+8+12 \\ 15-6-9 & -12+4+6+2 \end{bmatrix} \quad \text{[Theorem 1.3.1(A2)]}$$

$$= \begin{bmatrix} 0 & 0 \\ 0 & 0 \end{bmatrix} \quad \text{(by simplification).}$$

This validates the Cayley–Hamilton theorem.

Theorem 4.3.1 (fundamental property). *Let $x_1(t), x_2(t), \ldots, x_k(t), \ldots, x_m(t)$ be m solutions to* (4.35) *on J. Then,*

$$\begin{aligned} x(t) &= a_1x_1(t) + a_2x_2(t) + \cdots + a_kx_k(t) + \cdots + a_mx_m(t) \\ &= \sum_{k=1}^{m} a_kx_k(t) \end{aligned} \tag{4.40}$$

is also a solution to (4.35) *on J, where $a_1, a_2, \ldots, a_k, \ldots, a_m$ are arbitrary constants (real or complex numbers).*

Proof. Let us find the differential of $x(t)$ with respect to t both sides of (4.40). Now, using the rules of the derivative, the concept of the solution of the system of differential equations, and the matrix algebra, we obtain

$$\begin{aligned} dx(t) &= d\left[a_1x_1(t) + a_2x_2(t) + \cdots + a_kx_k(t) + \cdots + a_mx_m(t)\right] \quad \text{(definition)} \\ &= a_1dx_1(t) + a_2dx_2(t) + \cdots + a_kdx_k(t) + \cdots + a_mdx_m(t) \\ &\quad \text{(Observation 1.5.2)} \\ &= a_1Ax_1(t)dt + a_2Ax_2(t)dt + \cdots + a_kAx_k(t)dt + \cdots + a_mAx_m(t)dt \\ &\quad \text{(by solution)} \\ &= A\left[a_1x_1(t) + a_2x_2(t) + \cdots + a_kx_k(t) + \cdots + a_mx_m(t)\right]dt \\ &\quad \text{(Theorem 1.3.1)} \\ &= Ax(t)dt \quad \text{[from (4.40)],} \end{aligned}$$

which shows that $x(t)$ is a solution to (4.35). This completes the proof of the theorem. □

Example 4.3.5. Let $x_1(t)$ and $x_2(t)$ be solutions to the system of differential equations in Example 4.3.2. Show that

$$x(t) = a_1 x_1(t) + a_2 x_2(t) = a_1 \begin{bmatrix} e^t \\ e^t \end{bmatrix} + a_2 \begin{bmatrix} 4e^{2t} \\ 3e^{2t} \end{bmatrix} = \begin{bmatrix} a_1 e^t + 4a_2 e^{2t} \\ a_1 e^t + 3a_2 e^{2t} \end{bmatrix}$$

is the solution process of the given systems of differential equations, where a_1 and a_2 are arbitrary real numbers.

Solution procedure. We imitate the argument used in the proof of Theorem 4.3.1. We consider

$$\begin{aligned}
dx(t) &= d\left(a_1 x_1(t) + a_2 x_2(t)\right) = d\left(\begin{bmatrix} e^t \\ e^t \end{bmatrix} a_1 + \begin{bmatrix} 4e^{2t} \\ 3e^{2t} \end{bmatrix} a_2\right) \quad \text{(definition)} \\
&= a_1 dx_1(t) + a_2 dx_2(t) = a_1 d\left(\begin{bmatrix} e^t \\ e^t \end{bmatrix}\right) + a_2\left(\begin{bmatrix} 4e^{2t} \\ 3e^{2t} \end{bmatrix}\right) \quad \text{(Observation 1.5.2)} \\
&= a_1 dx_1(t) + a_2 dx_2(t) = a_1 \begin{bmatrix} e^t dt \\ e^t dt \end{bmatrix} + a_2 \begin{bmatrix} 8e^{2t} dt \\ 6e^{2t} dt \end{bmatrix} \quad \text{(usual calculus)} \\
&= a_1 dx_1(t) + a_2 dx_2(t) = a_1 \begin{bmatrix} e^t \\ e^t \end{bmatrix} dt + a_2 \begin{bmatrix} 8e^{2t} \\ 6e^{2t} \end{bmatrix} dt \quad \text{(Definition 1.3.9)} \\
&= \begin{bmatrix} a_1 e^t + 8a_2 e^{2t} \\ a_1 e^t + 6a_2 e^{2t} \end{bmatrix} dt \quad \text{[Theorem 1.3.1(S2)].}
\end{aligned}$$

Now, we compute $Ax(t)dt$ as follows:

$$\begin{aligned}
Ax(t)dt &= \begin{bmatrix} 5 & -4 \\ 3 & -2 \end{bmatrix} \begin{bmatrix} a_1 e^t + 4a_2 e^{2t} \\ a_1 e^t + 3a_2 e^{2t} \end{bmatrix} dt \quad \text{(substitution)} \\
&= \begin{bmatrix} 5\left(a_1 e^t + 4a_2 e^{2t}\right) - 4\left(a_1 e^t + 3a_2 e^{2t}\right) \\ 3\left(a_1 e^t + 4a_2 e^{2t}\right) - 2\left(a_1 e^t + 3a_2 e^{2t}\right) \end{bmatrix} dt \quad \text{(Definition 1.3.11)} \\
&= \begin{bmatrix} a_1 e^t + 8a_2 e^{2t} \\ a_1 e^t + 6a_2 e^{2t} \end{bmatrix} dt \quad \text{(simplification).}
\end{aligned}$$

By comparing these calculations, we conclude that $dx(t) = Ax(t)\,dt$. From Definition 4.3.1, we infer that the linear combination $x(t)$ of the given solutions $x_1(t)$ and $x_2(t)$ is also the solution to the given system of differential equations.

Observation 4.3.2. The right-hand side expression in (4.40) is called a linear combination of m solutions $x_1(t), x_2(t), \ldots, x_k(t), \ldots, x_m(t)$ of (4.35). Theorem 4.3.1 suggests that a span (Observation 1.4.5) of a set of solutions to (4.35) is a vector space (Definition 1.4.4). A set of solutions $\{x_1(t), x_2(t), \ldots, x_k(t), \ldots, x_m(t)\}$ is said to be a complete set of solutions if any solution to (4.35) can be found by the linear combination of the members of this set. It can also be called the maximal linearly independent set (basis — Definition 1.4.6) of solutions to (4.35).

4.3.3 *Initial value problem*

Let us formulate an IVP relative to (4.35) for the purpose of solving applied problems. The complete set of solutions is used to solve any given IVP. We consider

$$dx = Ax\,dt, \quad x(t_0) = x_0. \tag{4.41}$$

Here x and A are as defined in (4.35); t_0 and x_0 are in J and R^n, respectively. (t_0, x_0) is called an initial data or initial condition. In short, x_0 is the value of the solution function at $t = t_0$ (the initial/given time), and it is an initial/given state vector of the system. The problem of finding a solution to (4.41) is referred to as the initial value problem (IVP). Its solution is represented by $x(t) = x(t, t_0, x_0)$ for $t \geq t_0$ and t in J.

Definition 4.3.2. A solution to the IVP (4.41) is a continuously differentiable n-dimensional column vector (or $n \times 1$ matrix) function x defined on $[t_0, b) \subseteq J$, which satisfies both the given system of differential equations (4.41) and any given initial condition (t_0, x_0). In short, (i) $x(t)$ is a solution to (4.35) (Definition 4.3.1) defined on $[t_0, b) \subseteq J$, and (ii) $x(t_0) = x_0$.

Example 4.3.6. We are given $dx_1 = x_2\,dt$, $x(t_0) = x_0 \in R^2$. Show that for $t \geq t_0$, $x(t, t_0, x_0) = [x_{10} + (t - t_0)\,x_{20}, x_{20}]^T$ is the solution to the given IVP, where T stands for the transpose of a matrix.

Solution procedure. By using the argument in the solution process of Example 4.3.3, we arrive at $dx(t) = Ax\,dt$, where A and x are 2×2 and 2×1 matrices, respectively, and are as described in Example 4.3.3. From Definition 4.3.2, we need to show that: (i) $x(t, t_0, x_0) = [x_{10} + (t - t_0)\,x_{20}, x_{20}]^T$ satisfies the given differential equation for $t \geq t_0$, and (ii) $x(t_0, t_0, x_0) = x_0$. We compute $dx(t, t_0, x_0)$ as follows:

$$\begin{aligned} dx(t, t_0, x_0) &= d\left[x_{10} + (t - t_0)\,x_{20}, x_{20}\right]^T \quad \text{(notation)} \\ &= \left[d\left(x_{10} + (t - t_0)\,x_{20}\right), dx_{20}\right]^T \quad \text{(Observation 1.5.2)} \\ &= [x_{20}, 0]^T\,dt \quad \text{(differential calculus)}. \end{aligned}$$

On the other hand, we compute $Ax(t, t_0, x_0)\,dt$ as

$$\begin{aligned} Ax(t, t_0, x_0)\,dt &= \begin{bmatrix} 0 & 1 \\ 0 & 0 \end{bmatrix} \begin{bmatrix} x_{10} + (t - t_0)\,x_{20} \\ x_{20} \end{bmatrix} dt \quad \text{(substitution)} \\ &= \begin{bmatrix} x_{20} \\ 0 \end{bmatrix} dt \quad \text{(Definition 1.3.11)} \end{aligned}$$

$$= \begin{bmatrix} x_{20} & 0 \end{bmatrix}^T dt \quad \text{(Definition 1.3.12)}$$
$$= dx(t, t_0, x_0) \quad \text{(from the above expression).}$$

From the above discussion, we conclude that $x\,(t, t_0, x_0)$ satisfies the given system of differential equations. Finally, we must prove that $x(t_0, t_0, x_0) = x_0$. We observe that

$$x(t_0, t_0, x_0) = [x_{10} + (t_0 - t_0)x_{20}, x_{20}]^T = [x_{10}, x_{20}]^T = x_0.$$

This completes the solution process of the example.

4.3.4 *Procedure for solving the IVP*

Here is the procedure for finding a solution to the IVP (4.41).

Step 1. It is safe to assume that m solutions $x_1(t), x_2(t), \ldots, x_k(t), \ldots, x_m(t)$ defined in Theorem 4.3.1 were determined by the four-step procedure in Subsection 4.3.2 for finding solutions to (4.35). Therefore, from Theorem 4.3.1, we conclude that a linear combination of m solutions $x_1(t), x_2(t), \ldots, x_k(t), \ldots, x_m(t)$ of (4.35),

$$x(t) = a_1 x_1(t) + a_2 x_2(t) + \cdots + a_k x_k(t) + \cdots + a_m x_m(t), \tag{4.42}$$

is also a solution to (4.35).

Step 2. By using the information in step 1, we try to solve the IVP (4.41). For this purpose, we simply need to verify that a solution to (4.35) defined in (4.42) satisfies the given initial conditions, i.e.

$$x(t_0) = a_1 x_1(t_0) + a_2 x_2(t_0) + \cdots + a_k x_k(t_0) + \cdots + a_m x_m(t_0) = x_0. \tag{4.43}$$

This means that for any given initial data (t_0, x_0) in (4.41), the system of nonhomogeneous algebraic equations (4.43) must have a solution. We recall that $a_1, a_2, \ldots, a_k, \ldots, a_m$ are arbitrary constants. The system of linear algebraic equations (4.43) consists of n equations with m unknown variables. We recall that this system of linear nonhomogeneous algebraic equations has a unique solution, provided that: (i) $m = n$ and (ii) the determinant of the coefficient matrix of the system (4.43) is different from zero, i.e. $\det(x_{ik}\,(t_0)) \neq 0$, where $x_{ik}\,(t_0)$ is the ith component of the kth solution, for $k = 1, 2, \ldots, n$. See Observation 1.4.2.

Step 3. Moreover, under the conditions in step 2 and substituting the solutions to (4.43) (the values of $a_1, a_2, \ldots, a_k, \ldots, a_n$) into (4.42), the solution to the IVP (4.41) is determined.

Observation 4.3.3

(i) We also note that the given initial data (t_0, x_0) are arbitrary. Therefore, the mathematical statement "$\det(x_{ik}(t_0)) \neq 0$" must be valid for any given/initial time t_0 and x_0 in (4.41). We further note that, from Theorem 4.3.1, the solution representation of (4.35) can be rewritten as

$$x(t) = \Phi(t)a, \tag{4.44}$$

where Φ is an $n \times n$ matrix function defined by $\Phi(t) = (x_{ik}(t))_{n\times n}$, whose columns are the solutions $x_1(t), x_2(t), \ldots, x_k(t), \ldots, x_n(t)$ to (4.35); an n-dimensional unknown constant column vector ($n \times 1$ matrix)

$$a = [a_1, a_2, \ldots, a_k, \ldots, a_n]^T$$

is composed of arbitrary coefficients of the solutions $x_k(t)$ in (4.42) for $k = 1, 2, \ldots, n$.

(ii) The solution represented in (4.44) is called a general solution to (4.35). We remark that $\Phi(t)$ defined in (4.44) depends on the eigenvalues and corresponding eigenvectors of the coefficient matrix A. However, $\Phi(t)$ is not uniquely determined by the eigenvectors corresponding to the eigenvalues of the matrix A. This procedure for finding the general solution is summarized in Figure 4.10.

(iii) We further remark that the observations in (i) and (ii) are also applicable to the IVP for the system of linear differential equations with a time-varying system (the matrix A is a function of the independent variable t in J).

(iv) In the case of the system (4.41) (time-invariant, i.e. A is a constant matrix), the validity of $\det(\Phi(t_0)) \neq 0$ is equivalent to the fact that $\det(\Phi(0)) \neq 0$ $[0 \neq \det(\Phi(t_0)) = \exp[\lambda_1 t_0] \exp[\lambda_2 t_0] \cdots \exp[\lambda_n t_0] \det(\Phi(0)]$ by Theorem 1.4.1(D4)). This is in view of the representation of the solution (4.36) and $\exp[\lambda_k t_0] \neq 0$ for any t_0 in R. In the light of this, it is enough to verify that the $n\times n$ matrix $\Phi(0) = [c_1, \ldots, c_k, \ldots, c_n]$, associated with n eigenvectors as the columns of the matrix, is invertible (Theorem 1.4.4). This implies that

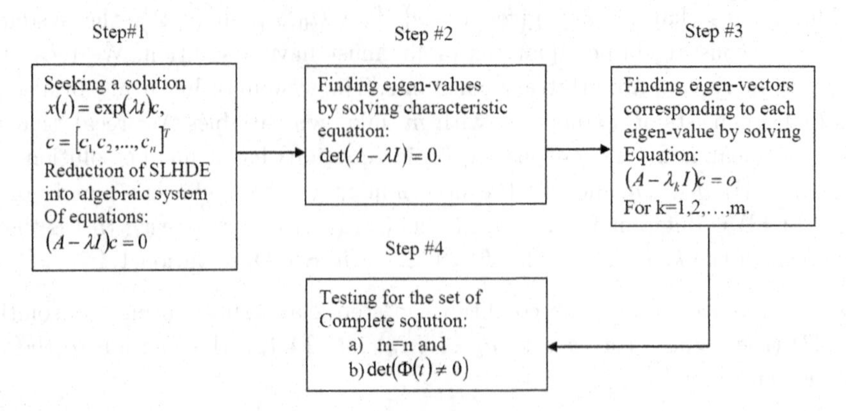

Fig. 4.10 Flowchart for finding a general solution.

$\Phi(t_0)$ has inverse $\Phi^{-1}(t_0)$ (Observation 1.4.3). Hence, from (4.43) and (4.44) (Theorem 1.3.4), we have

$$a = \Phi^{-1}(t_0)x_0. \tag{4.45}$$

(v) The solution to the IVP (4.41) determined in step 3 can be represented by substituting the expression (4.45) into (4.44). Hence, we have

$$\begin{aligned} x(t) &= \Phi(t)\Phi^{-1}(t_0)x_0 \\ &= \Phi(t, t_0)\, x_0, \end{aligned} \tag{4.46}$$

where $\Phi(t, t_0) = \Phi(t)\Phi^{-1}(t_0)$. Thus, the solution $x(t, t_0, x_0) = x(t)$ to the IVP (4.41) is represented by

$$x(t, t_0, x_0) = x(t) = \Phi(t, t_0)x_0.$$

This solution is referred to as a particular solution to (4.35). The procedure for finding the solution to the IVP (4.41) is summarized in Figure 4.11.

(vi) From (iv), we further remark that $\Phi(0) = [c_1, \ldots, c_k, \ldots, c_n]$ is not uniquely determined. It depends on a particular choice of eigenvector corresponding to the eigenvalue of the matrix A. Hence, $\Phi(0)$ is not unique.

Example 4.3.7. We are given $dx_1 = x_2dt$, $x(t_0) = x_0$. (a) Show that solutions $x_1(t) = \begin{bmatrix}1 & 0\end{bmatrix}^T$ and $x_2(t) = \begin{bmatrix}t & 1\end{bmatrix}^T$ form a complete set of solutions to the given system of differential equations, where T stands for the transpose of a matrix. (b) Solve the IVP.

Solution procedure. From Example 4.3.3, we note that $x_1(t) = [1 \quad 0]^T$ is the solution process of the given system of differential equations. We only need to show that $x_2(t) = [t \quad 1]^T$ satisfies the given system of differential equations. For this purpose, we apply the deterministic differential calculus (Observation 1.5.2) to $x_2(t) = [t \quad 1]^T$, and we compute the differential

$$\begin{aligned} dx_2(t) &= d([t \quad 1]^T) \quad \text{(by the given expression)} \\ &= [1 \quad 0]^T\, dt \quad \text{[Observation 1.5.2 and the chain rule]}. \end{aligned}$$

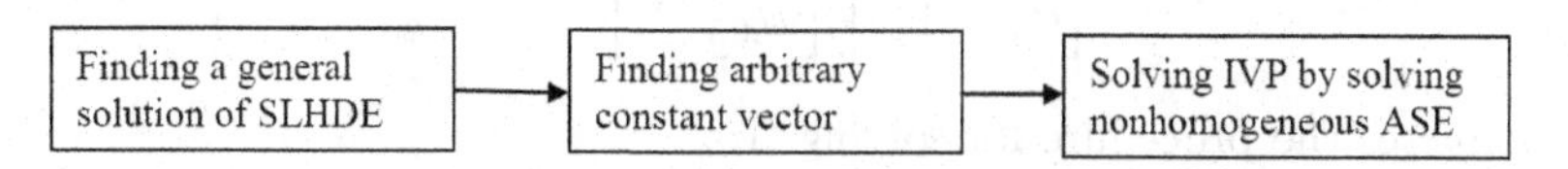

Fig. 4.11 Flowchart for finding a particular solution.

We also compute $Ax_2(t)dt$:

$$\begin{aligned} Ax_2(t)dt &= \begin{bmatrix} 0 & 1 \\ 0 & 0 \end{bmatrix} \begin{bmatrix} t \\ 1 \end{bmatrix} dt \quad \text{(substitution and Definition 1.3.12)} \\ &= \begin{bmatrix} 1 \\ 0 \end{bmatrix} dt \quad \text{(Definition 1.3.11)} \\ &= \begin{bmatrix} 1 & 0 \end{bmatrix}^T dt \quad \text{(Definition 1.3.12)} \\ &= dx_2(t) \quad \text{(substitution).} \end{aligned}$$

The above-presented discussion shows that the given process $x_2(t) = [t \quad 1]^T$ satisfies the given system of differential equations. Therefore, it is a solution to the given system of differential equations. Here, $n = 2$. We need to verify that the set consisting of these two solutions is linearly independent (basis). For this purpose, from Observation 4.3.3, we have

$$\Phi(t) = \begin{bmatrix} 1 & t \\ 0 & 1 \end{bmatrix},$$

and must compute its determinant. The determinant is $\det(\Phi(t)) = 1 \neq 0$ for any $t \in J$. Therefore, by the application of Theorem 1.4.6, we conclude that the set of solutions $\{x_1(t), x_2(t)\}$ is linearly independent. From Observation 4.3.2, we conclude that the set $\{x_1(t), x_2(t)\}$ is the complete set of solutions to the given system.

To solve (b), we need to follow step 3 of the procedure in Subsection 4.3.4, and find $\Phi^{-1}(t_0)$. By imitating the arguments used in Examples 1.4.3, 1.4.5, and 1.4.8, we have

$$\Phi^{-1}(t_0) = \begin{bmatrix} 1 & -t_0 \\ 0 & 1 \end{bmatrix}.$$

This together with formula (4.45) gives us

$$a = \Phi^{-1}(t_0)x_0 = \begin{bmatrix} 1 & -t_0 \\ 0 & 1 \end{bmatrix} x_0.$$

From (4.46), the solution to the given IVP is as follows:

$$\begin{aligned} x(t, t_0, x_0) = x(t) = \Phi(t, t_0)x_0 &= \begin{bmatrix} 1 & t \\ 0 & 1 \end{bmatrix} \begin{bmatrix} 1 & -t_0 \\ 0 & 1 \end{bmatrix} x_0 \\ &= \begin{bmatrix} 1 & t - t_0 \\ 0 & 1 \end{bmatrix} \begin{bmatrix} x_{10} \\ x_{20} \end{bmatrix} = \begin{bmatrix} x_{10} + (t - t_0)\, x_{20} \\ x_{20} \end{bmatrix}. \end{aligned}$$

This completes the procedure for solving (b).

Observation 4.3.4

(i) From the definition of the differential of a matrix function (Observation 1.5.2) and the matrix algebra (Theorem 1.3.1), we observe that

$$\begin{aligned} d\Phi(t) &= d\,(x_{ik}(t))_{n\times n} = d\,[x_1(t)\,x_2(t)\ \dots, x_k(t)\dots x_n(t)] \quad \text{[from (4.44)]} \\ &= (dx_{ik}(t))_{n\times n} = [dx_1(t)\,dx_2(t), \dots, dx_k(t)\dots dx_n(t)] \\ &\qquad \text{(Observation 1.5.2)} \\ &= [Ax_1(t)\,Ax_2(t)\ \dots, Ax_k(t)\ \dots\ Ax_n(t)]\,dt \quad \text{[by solution]} \\ &= A\,[x_1(t)\,x_2(t)\ \dots, x_k(t)\ \dots\ x_n(t)]\,dt \quad \text{[by (1.8)]} \\ &= A\Phi(t)dt \quad \text{(by definition of } \Phi). \end{aligned} \tag{4.47}$$

This shows that the $\Phi(t)$ defined in (4.44) is also a solution to (4.35). If $\det(\Phi(t_0)) \neq 0$, then Φ is called a general fundamental matrix solution to (4.35) at $t = t_0$. Moreover, $\Phi(t, t_0)$ is a fundamental matrix solution to (4.35). $\Phi(t, t_0)$ is a particular fundamental solution to (4.35). From (4.46), we have $\Phi(t, t_0) = \Phi(t)\Phi^{-1}(t_0)$ and $\Phi(t_0, t_0) = I$. $\Phi(t, t_0)$ is also referred to as the state transition matrix or the normalized fundamental matrix solution to (4.41) at $t = t_0$. Hence, $x(t, t_0, x_0) = x(t) = \Phi(t, t_0)x_0$ is called a particular solution to (4.35).

(ii) We recall that $\Phi(t)$ defined in (4.44) is not uniquely determined. Therefore, at this stage, it is not clear regarding the uniqueness of the solution $x(t, t_0, x_0) = x(t) = \Phi(t, t_0)x_0$ to (4.41). This will be clarified at a later stage of the chapter.

Example 4.3.8. We are given $dx_1 = (x_1 + 2x_2)dt$ and $dx_1 = (4x_1 + 3x_2)dt$. Show that (a) $x_1(t) = e^{-t}[1 \quad -1]^T$ and $x_2(t) = e^{5t}[1 \quad 2]^T$ form a complete set of solutions, and (b) $[x_1(t)\ x_2(t)]^T$ is the general fundamental matrix solution to the given system of differential equations at $t = t_0$, where T stands for the transpose of a matrix.

Solution procedure. From the solution process of Example 4.3.1, we conclude that the vector functions $x_1(t)$ and $x_2(t)$ are the solution process of the given system of differential equations that is represented by $dx(t) = Ax\,dt$, where, A and x are 2×2 and 2×1 matrices, respectively, described in Example 4.3.1. Now, by following the discussion in Example 4.3.7, we set $\Phi(t)$:

$$[x_1(t) \quad x_2(t)]^T = \Phi(t) = \begin{bmatrix} e^{-t} & e^{5t} \\ -e^{-t} & 2e^{5t} \end{bmatrix}, \quad \text{with } \Phi(t_0) = \begin{bmatrix} e^{-t_0} & e^{5t_0} \\ -e^{-t_0} & 2e^{5t_0} \end{bmatrix}.$$

We compute $\det(\Phi(t_0))$ as follows:

$$\begin{aligned} \det(\Phi(t_0)) &= e^{-t_0}(2e^{5t_0}) - (-e^{-t_0})e^{5t_0} \quad \text{(by Definition 1.4.2)} \\ &= 3e^{4t_0} \quad \text{(by simplification)} \\ &\neq 0 \quad \text{(the property of exponential functions).} \end{aligned}$$

Therefore, by Observation 4.3.3(iv), we conclude that the column vector functions of $\Phi(t)$ are linearly independent. Moreover, for $n = 2$, these solutions form a complete set of solutions to the given system of differential equations. This completes the proof of (a). To prove (b), we compute $A\Phi(t)dt$ and $d\Phi(t)$ as

$$A\Phi(t)dt = \begin{bmatrix} 1 & 2 \\ 4 & 3 \end{bmatrix} \begin{bmatrix} e^{-t} & e^{5t} \\ -e^{-t} & 2e^{5t} \end{bmatrix} dt \quad \text{(by substitution)}$$

$$= \begin{bmatrix} e^{-t} - 2e^{-t} & e^{5t} + 4e^{5t} \\ 4e^{-t} - 3e^{-t} & 4e^{5t} + 6e^{5t} \end{bmatrix} dt \quad \text{(by Definition 1.3.11)}$$

$$= \begin{bmatrix} -e^{-t} & 5e^{5t} \\ e^{-t} & 10e^{5t} \end{bmatrix} dt \quad \text{(by simplification),}$$

$$d\Phi(t) = d\left(\begin{bmatrix} e^{-t} & e^{5t} \\ -e^{-t} & 2e^{5t} \end{bmatrix}\right) \quad \text{(by substitution)}$$

$$= \begin{bmatrix} d(e^{-t}) & d(e^{5t}) \\ d(-e^{-t}) & d(2e^{5t}) \end{bmatrix} \quad \text{(by Observation 1.5.2)}$$

$$= \begin{bmatrix} -e^{-t} & 5e^{5t} \\ e^{-t} & 10e^{5t} \end{bmatrix} dt \quad \text{(deterministic differential and Definition 1.3.9).}$$

By comparing the right-hand side expressions of $A\Phi(t)dt$ and $d\Phi(t)$, from part (a) of the problem, and applying Definition 4.3.1, we conclude that $\Phi(t)$ is a general fundamental matrix solution to the given system of differential equations at $t = t_0$ (Observation 4.3.4).

Prior to the practical/computational procedures for finding the n eigenvalue and corresponding n eigenvectors associated with (4.37), we present a calculus-dependent test for verifying the validity of $\det(\Phi(t_0)) \neq 0$, where $\Phi(t) = (x_{ik}(t))_{n\times n} = [x_1(t), x_2(t), \ldots, x_k(t), \ldots, x_n(t)]$, whose columns are the solutions $x_1(t), x_2(t), \ldots, x_k(t), \ldots, x_n(t)$ to (4.35). This test is easy to verify and computationally attractive. It is a conceptual test. It does not require the knowledge of the fundamental matrix solution. Moreover, it is applicable to both time-invariant and time-varying linear systems of differential equations.

Theorem 4.3.2 (Abel–Jacobi–Liouville [15, 42]). *Let*

$$x_1(t), x_2(t), \ldots, x_k(t), \ldots, x_n(t)$$

be any n solutions to (4.35) *on J. Let*

$$\Phi(t) = [x_1(t), x_2(t), \ldots, x_k(t), \ldots, x_n(t)] = [\Phi_1^T(t), \ldots, \Phi_k^T(t), \ldots, \Phi_n^T(t)]^T$$

be the $n \times n$ matrix function defined on J, where

$$\Phi_i(t) = [x_{i1}(t), x_{i2}(t), \ldots, x_{ij}(t), \ldots, x_{in}(t)]$$

is the ith row of $\Phi(t)$. *Let*

$$\mathrm{tr}(A) = \sum_{i=1}^{n} a_{ii} \quad (\textit{Trace of matrix } A).$$

Then

$$d\det(\Phi(t)) = \mathrm{tr}(A)\det(\Phi(t))dt \tag{4.48}$$

and

$$\exp\left[-\int_{t_0}^{t}\mathrm{tr}(A)ds\right]\det(\Phi(t)) = c \quad (\textit{constant}). \tag{4.49}$$

Proof. From Observation 4.3.4, we conclude that $\Phi(t)$ is the matrix solution to (4.35), i.e.

$$d\Phi(t) = (dx_{ik}(t))_{n\times n} = A\Phi(t)dt = \left(\sum_{j=1}^{n} a_{ij}x_{jk}(t)\right)_{n\times n} dt. \tag{4.50}$$

We apply Theorem 1.5.3 to $\det(\Phi(t))$, and obtain

$$d\det(\Phi(t)) = \sum_{i=1}^{n} W(\Phi_1(t), \ldots, d\Phi_i(t), \ldots, \Phi_n(t)). \tag{4.51}$$

For each i, $1 \leq i \leq n$, we first rewrite the ith term in (4.51) as follows:

$$\begin{aligned} & W(\Phi_1(t), \ldots, d\Phi_i(t), \ldots, \Phi_n(t)) \\ &= \begin{vmatrix} x_{11} & \cdots & x_{1k} & \cdots & x_{1n} \\ \cdots & \cdots & \cdots & \cdots & \cdots \\ dx_{i1} & \cdots & dx_{ik} & \cdots & dx_{in} \\ \cdots & \cdots & \cdots & \cdots & \cdots \\ x_{n1} & \cdots & x_{nk} & \cdots & x_{nn} \end{vmatrix} \\ &= \begin{vmatrix} x_{11} & \cdots & x_{1k} & \cdots & x_{1n} \\ \cdots & \cdots & \cdots & \cdots & \cdots \\ \sum_{j=1}^{n} a_{ij}x_{j1}dt & \cdots & \sum_{j=1}^{n} a_{ij}x_{jk}dt & \cdots & \sum_{j=1}^{n} a_{ij}x_{jn}dt \\ \cdots & \cdots & \cdots & \cdots & \cdots \\ x_{n1} & \cdots & x_{nk} & \cdots & x_{nn} \end{vmatrix}. \end{aligned}$$

By repeated application of the property of the determinant [Theorem 1.4.1(D6)], we have

$$W(\Phi_1(t), \ldots, d\Phi_i(t), \ldots, \Phi_n(t)) = \begin{vmatrix} x_{11} & \cdots & x_{1k} & \cdots & x_{1n} \\ \cdots & \cdots & \cdots & \cdots & \cdots \\ a_{ii}x_{i1}dt & \cdots & a_{ii}x_{ik}dt & \cdots & a_{ii}x_{in}dt \\ \cdots & \cdots & \cdots & \cdots & \cdots \\ x_{n1} & \cdots & x_{nk} & \cdots & x_{nn} \end{vmatrix}.$$

This Together with Theorem 1.4.1(D4) gives us

$$W(\Phi_1(t), \ldots, d\Phi_i(t), \ldots, \Phi_n(t)) = a_{ii} \det(\Phi(t))dt,$$

for each i, $1 \leq i \leq n$. We substitute this expression into (4.51), and obtain

$$\begin{aligned} d\det(\Phi(t)) &= \sum_{i=1}^{n} a_{ii} \det(\Phi(t))dt \\ &= \operatorname{tr}(A) \det(\Phi(t))dt. \end{aligned}$$

This establishes the validity of the linear homogeneous scalar differential equation (4.48).

A general solution to the differential equation (4.48) (Section 2.4) is

$$\det(\Phi(t)) = \exp\left[\int_{t_0}^{t} \operatorname{tr}(A)ds\right] c.$$

By using the algebraic properties of the exponential function, we can solve for c and obtain

$$\exp\left[-\int_{t_0}^{t} \operatorname{tr}(A)ds\right] \det(\Phi(t)) = c.$$

This establishes the validity of (4.49). Moreover, from $t = t_0$, if $\det(\Phi(t_0))$ is known, then $\det(\Phi(t_0)) = c$. Thus, c is determined by the initial data $(t_0, \det(\Phi_0))$. Therefore, the particular solution to (4.48) can be determined for any given value of $\Phi(t)$ at $t = t_0$. □

Example 4.3.9. Let us consider Example 4.3.8. Show that

$$\exp\left[\int_{t_0}^{t} \operatorname{tr}(A)ds\right] \det(\Phi(t_0)) = \det(\Phi(t)) = 3e^{4t}.$$

Solution procedure. By following the argument used in the solution process of Example 4.3.8, we have $\det(\Phi(t)) = 3e^{4t}$. From this, we compute

$$d\det(\Phi(t)) = d(3e^{4t}) = 12e^{4t} = 4\det(\Phi(t)).$$

We note that $\operatorname{tr}(A) = 4$. From these statements we conclude that $d\det(\Phi(t))$ satisfies the differential equation (4.48). Thus, by employing the argument used in the proof of Theorem 4.3.2, we have

$$\det(\Phi(t)) = \det(\Phi_0)\exp\left[\int_{t_0}^{t} \operatorname{tr}(A)ds\right] = \det(\Phi_0)e^{4(t-t_0)}.$$

From Example 4.3.8, we have $\det(\Phi_0) = 3e^{4t_0}$, and hence

$$\det(\Phi(t)) = 3e^{4t_0}e^{4(t-t_0)} = 3e^{4t}.$$

Observation 4.3.5

(i) From (4.49), we can immediately conclude that $\det(\Phi(t)) \neq 0$ on J if and only if $\det(\Phi(t_0)) = c \neq 0$. Therefore, $\Phi(t)$ is a fundamental matrix solution to (4.35) if and only if $\det(\Phi(t_0)) = c \neq 0$.
(ii) If $\det(\Phi(t^*)) = 0$ for some $t = t^*$ in J, then $\det(\Phi(t)) = 0$ for all t in J. This means that the set of solutions is linearly dependent (Definition 1.4.5).
(iii) Any set $\{x_1(t), x_2(t), \ldots, x_k(t), \ldots, x_n(t)\}$ of n nonzero solutions to (4.35) is a complete set of solutions to (4.35) (basis for the solution space and Observation 4.3.2) if $\det(\Phi(t^*)) \neq 0$ for some t^* in J. This idea can be used to construct the complete set of solutions to (4.35). For example, a choice of initial states for n solutions can be made from the set $\{e_1, e_2, \ldots, e_j, \ldots, e_n\}$, where $e_j = (\delta_{ij})_{n\times 1}$, $\delta_{ij} = 0$ for $i \neq j$, and $\delta_{ij} = 1$ for $i = j$ (Observation 1.4.7).
(iv) We further remark that $\det(\Phi(t))$ is called the Wronskian of the set $\{x_1(t), x_2(t), \ldots, x_k(t), \ldots, x_n(t)\}$ of solutions to (4.35).
(v) $\det(\Phi(t)) \neq 0$ on J if and only if $\Phi(t)$ invertible on J (Theorem 1.4.4). At a later stage, we find the inverse of $\Phi(t)$, i.e. $\Phi^{-1}(t)$ on J.

Example 4.3.10. Let us consider Example 4.3.9. Find $\Phi^{-1}(t)$, where $\Phi(t)$ is the fundamental matrix solution to the given system of differential equations.

Solution procedure. From Example 4.3.9 and Observation 4.3.5(v), we conclude that $\Phi(t)$,

$$\Phi(t) = \begin{bmatrix} e^{-t} & e^{5t} \\ -e^{-t} & 2e^{5t} \end{bmatrix},$$

as defined in Example 4.3.8, is invertible. Now, we imitate the argument of finding the inverse of this 2×2 matrix as outlined in Examples 1.4.3 and 1.4.8, and we have

$$\Phi^{-1}(t) = \frac{\operatorname{adj}\Phi(t)}{\det(\Phi(t))} = \frac{\begin{bmatrix} 2e^{5t} & -e^{5t} \\ e^{-t} & e^{-t} \end{bmatrix}}{3e^{4t}} = \begin{bmatrix} \frac{2}{3}e^{t} & -\frac{1}{3}e^{t} \\ \frac{1}{3}e^{-5t} & \frac{1}{3}e^{-5t} \end{bmatrix}.$$

This completes the procedure for finding the inverse of $\Phi(t)$.

Example 4.3.11. Let us consider $dx = Ax\,dt$, $x(t_0) = x_0$, where

$$A = \begin{bmatrix} 5 & -4 \\ 3 & -2 \end{bmatrix}.$$

Solve the IVP.

Solution procedure. Here, $n = 2$. Step 1 of the procedure in Subsection 4.3.4 requires us to apply the four-step procedure in Subsection 4.3.2 to find solutions to the given system of differential equations. This was achieved in Example 4.3.2, and we have

$$x_1(t) = \begin{bmatrix} e^t \\ e^t \end{bmatrix}, \quad x_2(t) = \begin{bmatrix} 4e^{2t} \\ 3e^{2t} \end{bmatrix}.$$

From Example 4.3.5, we know that $x(t) = a_1x_1(t) + a_2x_2(t)$ is the solution process of the given system for any arbitrary constants a_1 and a_2. This completes step 2 of the procedure in Subsection 4.3.4. From Step 3, Observation 4.3.3(v), and by imitating the reasoning used in Example 4.3.10, we obtain $\det(\Phi(t)) = -e^{3t}$ and algebraic simplification yields

$$\Phi^{-1}(t_0)x_0 = \begin{bmatrix} -3e^{-t_0} & 4e^{-t_0} \\ e^{-2t_0} & -e^{-2t_0} \end{bmatrix} \begin{bmatrix} x_{10} \\ x_{20} \end{bmatrix} = \begin{bmatrix} 4x_{20}e^{-t_0} - 3x_{10}e^{-t_0} \\ x_{10}e^{-2t_0} - x_{20}e^{-2t_0} \end{bmatrix}.$$

Again, using (4.46), the solution to the given IVP is

$$\begin{aligned} x(t, t_0, x_0) = \Phi(t)\Phi^{-1}(t_0)x_0 &= \begin{bmatrix} e^t & 4e^{2t} \\ e^t & 3e^{2t} \end{bmatrix} \begin{bmatrix} (4x_{20} - 3x_{10})e^{-t_0} \\ (x_{10} - x_{20})e^{-2t_0} \end{bmatrix} \\ &= \frac{1}{3} \begin{bmatrix} e^{(t-t_0)}(4x_{20} - 3x_{10}) + e^{2(t-t_0)}4(x_{10} - x_{20}) \\ e^{-(t-t_0)}(4x_{20} - 3x_{10}) + e^{2(t-t_0)}3(x_{10} - x_{20}) \end{bmatrix}. \end{aligned}$$

This completes the procedure for solving the IVP.

The following result provides an analytic method for determining an inverse of a fundamental matrix solution to (4.35). Moreover, this result will be used subsequently.

Theorem 4.3.3. *Let Φ be a fundamental matrix solution to (4.35). Then, Φ is invertible, and its inverse Φ^{-1} satisfies the following matrix differential equation:*

$$d(\Phi^{-1}(t)) = -\Phi^{-1}(t)A\,dt. \tag{4.52}$$

Moreover,

$$\begin{aligned} d((\Phi^{-1}(t))) &= -A^T(\Phi^{-1})^T(t)\,dt \\ d((\Phi^T(t))^{-1}) &= -A^T(\Phi^T(t))^{-1}dt. \end{aligned} \tag{4.53}$$

Proof. Let Φ be a fundamental matrix solution to (4.35). Therefore, from Observation 4.3.5, it is invertible on J, and its inverse Φ^{-1} exists on J. This, together with Theorem 1.5.1(I1), gives us

$$\begin{aligned} d(\Phi^{-1}(t)) &= -\Phi^{-1}(t)\, d\Phi(t)\Phi^{-1}(t) \quad \text{[Theorem 1.5.1(I1)]} \\ &= -\Phi^{-1}(t)A\Phi(t)\, dt\Phi^{-1}(t) \quad \text{[substitution (4.47)]} \\ &= -\Phi^{-1}(t)A\, dt \quad \text{(Observation 1.3.5)}. \end{aligned}$$

This completes the proof of the expressions (4.52). The proof of the expressions (4.53) follow from the properties of the transpose and inverse of matrices (Theorems 1.3.2 and 1.3.3). This completes the proof of the theorem. □

Observation 4.3.6. From (4.53), we note that $(\Phi^{-1})^T(t)$ is a fundamental matrix solution to a system of linear differential equations:

$$dy = -A^T y\, dt, \quad y(t_0) = x_0. \tag{4.54}$$

This statement can also be justified by imitating the entire preceding procedure for finding solutions to the system (4.41) relative to the system (4.54). The system (4.54) is called the adjoint to the system (4.41). The system (4.54) is equivalent to the system

$$dy = -yA\, dt, \quad y(t_0) = x_0^T. \tag{4.55}$$

We note that the y in (4.55) is a row vector, and the y in (4.54) is a column vector. In the light of this notational understanding, $\Phi^{-1}(t)$ is a fundamental matrix solution to (4.55). This can be further justified from (4.52).

Example 4.3.12. Let us consider Example 4.3.10. Show that $\Phi^{-1}(t)$ is the fundamental matrix solution to $dx = -yA\, dt$, where A is as defined in Example 4.3.2.

Solution procedure. We remark that $\Phi^{-1}(t)$ is an invertible matrix, and its inverse is found in Example 4.3.10. Therefore, we only need to show that $\Phi^{-1}(t)$ satisfies (4.52). Now, we repeat the argument used in Example 4.3.8 to compute $-\Phi^{-1}(t)A\, dt$ and $d\Phi^{-1}(t)$, and we obtain

$$\begin{aligned} -\Phi^{-1}(t)A\, dt &= -\begin{bmatrix} \frac{2}{3}e^t & -\frac{1}{3}e^t \\ \frac{1}{3}e^{-5t} & \frac{1}{3}e^{-5t} \end{bmatrix} \begin{bmatrix} 1 & 2 \\ 4 & 3 \end{bmatrix} dt \\ &= -\begin{bmatrix} \frac{2}{3}e^t - \frac{4}{3}e^t & \frac{4}{3}e^t - e^t \\ \frac{1}{3}e^{-5t} + \frac{4}{3}e^{-5t} & \frac{2}{3}e^{-5t} + e^{-5t} \end{bmatrix} dt \end{aligned}$$

$$= \begin{bmatrix} \frac{2}{3}e^{t} & -\frac{1}{3}e^{t} \\ -\frac{5}{3}e^{-5t} & -\frac{5}{3}e^{-5t} \end{bmatrix} dt,$$

$$d\Phi^{-1}(t) = \begin{bmatrix} d\left(\frac{2}{3}e^{t}\right) & d\left(-\frac{1}{3}e^{t}\right) \\ d\left(\frac{1}{3}e^{-5t}\right) & d\left(\frac{1}{3}e^{-5t}\right) \end{bmatrix} = \begin{bmatrix} \frac{2}{3}e^{t} & -\frac{1}{3}e^{t} \\ -\frac{5}{3}e^{-5t} & -\frac{5}{3}e^{-5t} \end{bmatrix} dt.$$

By comparing the right-hand side expressions of $-\Phi^{-1}(t)A\,dt$ and $d\Phi^{-1}(t)$, and applying Definition 4.3.1, we conclude that $\Phi^{-1}(t)$ is general fundamental matrix solution to the given system of differential equations in the example. This completes the solution procedure.

4.3 Exercises

(1) Prove or disprove that the given functions are solutions to the corresponding system of differential equations:

(a) $dx = \begin{bmatrix} 1 & 1 \\ 4 & 1 \end{bmatrix} x dt$, $x_1(t) = e^{-t}\begin{bmatrix} -1 \\ 2 \end{bmatrix}$ and $x_2(t) = e^{3t}\begin{bmatrix} 1 \\ 2 \end{bmatrix}$

(b) $dx = \begin{bmatrix} 3 & 0 & 1 \\ 9 & -1 & 2 \\ -9 & 4 & -1 \end{bmatrix} x dt$, $x_1(t) = e^{3t}\begin{bmatrix} 4 \\ 9 \\ 0 \end{bmatrix}$,

$$x_2(t) = e^{-t}\begin{bmatrix} \cos t \\ \sin t + 2\cos t \\ -4\cos t - \sin t \end{bmatrix},\ x_3(t) = e^{-t}\begin{bmatrix} \sin t \\ 2\sin t - \cos t \\ \cos t - 4\sin t \end{bmatrix}$$

(c) $dx = \begin{bmatrix} 1 & 0 \\ -1 & 2 \end{bmatrix} x dt$, $x_1(t) = e^{t}\begin{bmatrix} 1 \\ 1 \end{bmatrix}$ and $x_2(t) = e^{2t}\begin{bmatrix} 0 \\ 1 \end{bmatrix}$

(d) $dx = \begin{bmatrix} 1 & 1 \\ 0 & 1 \end{bmatrix} x dt$, $x_1(t) = e^{t}\begin{bmatrix} 1 \\ 0 \end{bmatrix}$ and $x_2(t) = e^{t}\begin{bmatrix} t \\ 1 \end{bmatrix}$

(e) $dx = \begin{bmatrix} 1 & 0 \\ 1 & 1 \end{bmatrix} x dt$, $x_1(t) = e^{t}\begin{bmatrix} 0 \\ 1 \end{bmatrix}$ and $x_2(t) = e^{t}\begin{bmatrix} 1 \\ t \end{bmatrix}$

(f) $dx = \begin{bmatrix} -1 & 2 & 2 \\ 2 & 2 & 2 \\ -3 & -6 & -6 \end{bmatrix} x dt$, $x_1(t) = \begin{bmatrix} 0 \\ -1 \\ 1 \end{bmatrix}$, $x_2(t) = e^{-2t}\begin{bmatrix} -2 \\ 1 \\ 0 \end{bmatrix}$,

and $x_3(t) = e^{-3t}\begin{bmatrix} -2 \\ 1 \\ 0 \end{bmatrix}$

(g) $dx = \begin{bmatrix} 1 & 0 \\ 1 & 1 \end{bmatrix} x dt$, $x_1(t) = e^{t}\begin{bmatrix} 0 \\ 1 \end{bmatrix}$ and $x_2(t) = e^{t}\begin{bmatrix} 1 \\ t \end{bmatrix}$

(2) Find the Wronskian of the given functions in Exercise 1.
(3) By using the results of Exercise 2, prove or disprove that the set of functions in Exercise 1 is a linearly independent set of solutions to the system of differential equations.
(4) Which set of solutions in Exercise 1 forms the fundamental matrix solutions of the corresponding system of differential equations? Please show your reasons for your answer.
(5) By using the Abel–Jacobi–Liouville formula, prove or disprove the answer to the question in Exercise 4.
(6) Find the general solution to the differential equations in Exercise 1 (if possible).
(7) Find the solutions to initial value problems (if possible) with regard to systems in Exercise 1(a)–(e) with the following initial conditions: (i) $(0, x_0) = (0, [1 \quad 2]^T)$ and (ii) $(1, x_0) = (1, [2 \quad 1]^T)$.
(8) Under the assumption that stop-clock time is a random variable with a cumulative distribution function (cdf) F, develop a general solution procedure for the differential equation of the Riemann–Stieltjes type

$$dx = Ax\, dF, \quad x(t_0) = x_0,$$

where A is an $n \times n$ constant matrix with real entries and for further assumptions are as is Exercise 6, Section 4.2. Moreover, solve the given IVP. (**Hint**: Let $x(t) = \exp\left[\lambda \int^t dF(s)\right] c$ be a solution.)
(9) Solve the IVP associated with Exercise 7 of Section 4.2.

4.4 Procedure for Finding the Fundamental Matrix Solution

In the following, we present the procedure for finding the fundamental matrix solution to (4.35). The problem of finding the fundamental solution to (4.35) is equivalent to the problem of finding n linearly independent eigenvectors of the coefficient matrix A in (4.35) corresponding to the n eigenvalues determined by (4.37). There are three cases. They depend on the nature of the eigenvalues of (4.37), namely: (1) distinct real numbers, (2) distinct complex numbers, and (3) repeated, real or complex numbers. The procedures for finding n linearly independent eigenvectors of the coefficient matrix A and the corresponding n linearly independent solutions to (4.35) are outlined in the following discussion.

Distinct Real Eigenvalues 4.4.1. *Let $\lambda_1, \ldots, \lambda_k, \ldots, \lambda_n$ be the real and distinct roots of the characteristic equation* (4.38). *In addition, let $c_1, \ldots, c_k, \ldots, c_n$ be the eigenvectors corresponding to these eigenvalues determined by solving the system of algebraic equations* (4.39). *Then the set of solutions $\{x_1(t), x_2(t), \ldots, x_n(t)\}$ is the complete set of solutions* (4.35).

Proof. In this case, it can be verified that the Wronskian of the corresponding set $\{x_1(t), x_2(t), \ldots, x_k(t), \ldots, x_n(t)\}$ of solutions to (4.35) is different from zero.

This means that $\det(\Phi(t)) \neq 0$ on J. In particular, from Observation 4.3.3(iv), $\det(\Phi(0)) \neq 0$, where the $n \times n$ matrix $\Phi(0) = [c_1, \ldots, c_k, \ldots, c_n]$ is associated with n eigenvectors as the columns of the matrix. Thus, $[c_1, \ldots, c_k, \ldots, c_n]$ is invertible. Hence, the set $\{c_1, \ldots, c_k, \ldots, c_n\}$ is a basis for R^n.

Alternatively, one can use the Principle of Mathematical Induction 1.2.3 argument to show that n eigenvectors $c_1, \ldots, c_k, \ldots, c_n$ are linearly independent (Definition 1.4.5). In fact, $n = 1$, $\{c_1\}$ is clearly linearly independent because $c_1 \neq 0$, and hence $a_1c_1 = 0$ implies that the scalar $a_1 = 0$. We assume that $\{c_1, \ldots, c_k, \ldots, c_{k-1}\}$ is the set of linearly independent vectors, then we show that $\{c_1, \ldots, c_k, \ldots, c_k\}$ is also the set of linearly independent vectors (induction hypothesis). For this purpose, we consider

$$a_1c_1 + a_2c_2 + \cdots + a_{k-1}c_{k-1} + a_kc_k = 0, \tag{4.56}$$

where $a_1, a_2, \ldots, a_k$ are some scalars. We need to show that $a_1 = a_2 = \cdots = a_k = 0$. We multiply both sides by the matrix A to (4.56), and we have

$$\begin{aligned} 0 &= A0 \\ &= A\left[a_1c_1 + a_2c_2 + \cdots + a_jc_j + \cdots + a_kc_k\right] \quad \text{[Definition 1.3.7 and (1.8)]} \\ &= a_1Ac_1 + a_2Ac_2 + \cdots + a_jAc_j + \cdots + a_kAc_k \quad \text{(Theorem 1.3.1)} \\ &= a_1\lambda_1c_1 + a_2\lambda_2c_2 + \cdots + a_{k-1}\lambda_{k-1}c_{k-1} + a_k\lambda_kc_k. \\ &\quad \text{(by the eigenvalue concept).} \end{aligned}$$

By subtracting λ_k times (4.56) from the last part of the above equation, we get

$$a_1(\lambda_1 - \lambda_k)c_1 + a_2(\lambda_2 - \lambda_k)c_2 + \cdots + a_{k-1}(\lambda_{k-1} - \lambda_k)c_{k-1} = 0.$$

From this, and the set of linearly independent vectors $\{c_1, \ldots, c_k, \ldots, c_{k-1}\}$, we conclude that

$$a_1(\lambda_1 - \lambda_k) = a_2(\lambda_2 - \lambda_k) = \cdots = a_{k-1}(\lambda_{k-1} - \lambda_k) = 0.$$

From this and knowing the distinct eigenvalues of matrix A, $\lambda_1 - \lambda_k \neq 0$, $(\lambda_2 - \lambda_k) \neq 0, \ldots, (\lambda_{k-1} - \lambda_k))$ imply that the scalar quantities $a_1 = a_2 = \cdots = a_{k-1} = 0$. From this and (4.56), we have $a_kc_k = 0$ with eigenvector $c_k \neq 0$. Therefore, $a_k = 0$. Thus, we have shown that $a_j = 0$ for every j, $j = 1, 2, \ldots, k$. Hence, by the application of principle of mathematical induction 1.2.3, we conclude that the n eigenvectors $c_1, \ldots, c_k, \ldots, c_n$ of A are linearly independent in R^n. Moreover, they form a basis for R^n. This establishes the validity of $\det(\Phi(0)) \neq 0$ (Theorem 1.4.6).

In summary, from (4.44) and (4.46), the general solution to (4.35) and the solution to the IVP (4.41) are represented by $x(t) = \Phi(t)a$ and $x(t, t_0, x_0) = \Phi(t, t_0)x_0$, respectively. In the case of real and distinct eigenvalues of the coefficient matrix A, the computational procedure for finding the solution is completed. □

Illustration 4.4.1. We consider: $dx = Ax\,dt$, $x(t_0) = x_0$, where

$$A = \begin{bmatrix} a_{11} & a_{12} \\ a_{21} & a_{22} \end{bmatrix},$$

and a_{ij}'s are any given real numbers satisfying the condition $d(A) = (\mathrm{tr}\,(A))^2 - 4\det(A) > 0$; $\mathrm{tr}\,(A)$ and $\det(A)$ are as defined in Theorem 4.3.2 and Definition 1.4.2, respectively. Find: (a) a general solution, and (b) the solution to the given IVP.

Solution procedure. To solve an IVP (steps 1–3 of the procedure in Subsection 4.3.4), we need to find a general solution to the given system of differential equations (steps 1–4 of the procedure in Subsection 4.3.2 and Observation 4.3.3). Step 1 of the procedure in Subsection 4.3.4 is the four-step procedure in Subsection 4.3.2. We apply the four-step procedure in Subsection 4.3.2 to this example. We seek a solution of the form $x(t) = \exp[\lambda t]c$. The primary goal is to find an unknown nonzero two-dimensional vector $c = [c_1\ c_2]^T$ and a scalar number λ. For this purpose, we use (4.37), and we have

$$\begin{aligned} 0 = (A - \lambda I)c &= \left(\begin{bmatrix} a_{11} & a_{12} \\ a_{21} & a_{22} \end{bmatrix} - \lambda \begin{bmatrix} 1 & 0 \\ 0 & 1 \end{bmatrix}\right) c \quad \text{(by substitution)} \\ &= \begin{bmatrix} a_{11} - \lambda & a_{12} \\ a_{21} & a_{22} - \lambda \end{bmatrix} c \quad \text{[by Theorem 1.3.1(S2)]}. \end{aligned}$$

This can be rewritten as

$$\begin{bmatrix} a_{11} - \lambda & a_{12} \\ a_{21} & a_{22} - \lambda \end{bmatrix} c = \begin{bmatrix} (a_{11} - \lambda)\,c_1 + a_{12}c_2 \\ a_{21}c_1 + (a_{22} - \lambda)\,c_2 \end{bmatrix} = 0 \quad \text{(Definition 1.3.11)},$$

which implies that

$$\begin{aligned} (a_{11} - \lambda)\,c_1 + a_{12}c_2 &= 0, \\ a_{21}c_1 + (a_{22} - \lambda)\,c_2 &= 0. \end{aligned}$$

The characteristic equation (4.38) of the matrix A in the context of the given example is

$$\begin{aligned} 0 &= \det(A - \lambda I) \\ &= (a_{11} - \lambda)(a_{22} - \lambda) - a_{21}a_{12} \quad \text{(by step 1 in Example 1.4.7)} \\ &= \lambda^2 - (a_{11} + a_{22})\lambda + a_{11}a_{22} - a_{21}a_{12} \quad \text{(simplifying)} \\ &= \lambda^2 - \mathrm{tr}\,(A)\lambda + \det(A) \quad [\text{Definition 1.4.2 and } \mathrm{tr}\,(A)] \\ &= (\lambda - \lambda_1)(\lambda - \lambda_2) \quad \text{(by the quadratic formula)}, \end{aligned}$$

where

$$\lambda_1 = \frac{\operatorname{tr}(A) + \sqrt{(\operatorname{tr}(A))^2 - 4\det(A)}}{2},$$

$$\lambda_2 = \frac{\operatorname{tr}(A) - \sqrt{(\operatorname{tr}(A))^2 - 4\det(A)}}{2}$$

are the roots of the quadratic polynomial equation $\lambda^2 - \operatorname{tr}(A)\lambda + \det(A) = 0$. They are also the eigenvalues of the coefficient matrix in the illustration. From the assumption $(\operatorname{tr}(A))^2 - 4\det(A) > 0$, we conclude that these are distinct real roots of $0 = \det(A - \lambda I)$. We need to find eigenvectors corresponding to these eigenvalues λ_1 and λ_2. For this purpose, we need to solve a system of linear homogeneous algebraic equations corresponding to (4.39). For $k = 1, 2$, we consider $(A - \lambda_k I)\,c = 0$, which can be written as

$$\begin{cases}(a_{11} - \lambda_1)c_1 + a_{12}c_2 = 0,\\ a_{21}c_1 + (a_{22} - \lambda_1)c_2 = 0,\end{cases} \qquad \begin{cases}(a_{11} - \lambda_2)c_1 + a_{12}c_2 = 0,\\ a_{21}c_1 + (a_{22} - \lambda_2)\,c_2 = 0.\end{cases}$$

One can imitate the procedure outlined in Examples 1.4.7 and 1.4.9 to find the solution set.

Case 1: From $\det(A - \lambda I) = 0 = (a_{11} - \lambda)(a_{22} - \lambda) - a_{21}a_{12}$, we note that at least a_{12} or $a_{21} = 0$ if and only if the eigenvalues λ_1 and λ_2 are diagonal elements of the matrix. In this case, the corresponding solution spaces are either

$$\begin{aligned}
\text{(i)}\quad & S_1 = \{c : c = [c_1\ 0]^T\} = \operatorname{Span}(\{[1\ 0]^T\}),\\
& S_2 = \{c : c = [0\ c_2]^T\} = \operatorname{Span}(\{[0\ 1]^T\}),\\
\text{(ii)}\quad & S_1 = \{c : c = [c_1(a_{11} - a_{22})\ a_{21}c_1]^T\} = \operatorname{Span}(\{[a_{11} - a_{22}\ a_{21}]^T\}),\\
& S_2 = \{c : c = [0\ c_2]^T\} = \operatorname{Span}(\{[0\ 1]^T\}),\\
& \text{or}\\
\text{(iii)}\quad & S_1 = \{c : c = [c_1\ 0]^T\} = \operatorname{Span}(\{[1\ 0]^T\}),\\
& S_2 = \{c : c = [a_{12}c_1\ (a_{22} - a_{11})]^T\} = \operatorname{Span}(\{[a_{12}\ a_{22} - a_{11}]^T\}).
\end{aligned}$$

Hence, the choice of eigenvectors and of solutions are:

(i) for $a_{21} = 0 = a_{12}$ and $\lambda_1 = a_{11}$,

$$(a_{11} - a_{11})c_1 = 0, \quad (a_{22} - a_{11})c_2 = 0,$$

$$x_1(t) = e^{a_{11}t}\begin{bmatrix}1\\0\end{bmatrix};$$

for $a_{21} = 0 = a_{12}$ and $\lambda_2 = a_{22}$,

$$(a_{11} - a_{22})c_1 = 0, \quad (a_{22} - a_{22})c_2 = 0,$$

$$x_2(t) = e^{a_{22}t} \begin{bmatrix} 0 \\ 1 \end{bmatrix};$$

(ii) for $a_{21} \neq 0$, $a_{12} = 0$, and $\lambda_1 = a_{11}$,

$$(a_{11} - a_{11})c_1 = 0, \quad a_{21}c_1 + (a_{22} - a_{11})c_2 = 0,$$

$$x_1(t) = e^{a_{11}t} \begin{bmatrix} a_{11} - a_{22} \\ a_{21} \end{bmatrix};$$

for $a_{21} \neq 0$, $a_{12} = 0$, and $\lambda_2 = a_{22}$,

$$(a_{11} - a_{22})c_1 = 0, \quad a_{21}c_1 + (a_{22} - a_{22})c_2 = 0,$$

$$x_2(t) = e^{a_{22}t} \begin{bmatrix} 0 \\ 1 \end{bmatrix};$$

or

(iii) for $a_{12} \neq 0$, $a_{21} = 0$, and $\lambda_1 = a_{11}$,

$$(a_{11} - a_{11})c_1 + a_{12}c_2 = 0, \quad (a_{22} - a_{11})c_2 = 0,$$

$$x_1(t) = e^{a_{11}t} \begin{bmatrix} 1 \\ 0 \end{bmatrix};$$

for $a_{21} \neq 0$, $a_{12} = 0$, and $\lambda_1 = a_{22}$,

$$(a_{11} - a_{22})c_1 + a_{12}c_2 = 0 \quad \text{and} \quad (a_{22} - a_{22})c_2 = 0,$$

$$x_2(t) = e^{a_{22}t} \begin{bmatrix} a_{12} \\ a_{22} - a_{11} \end{bmatrix}.$$

Case 2: The negation of the statement in case 1 is that the off-diagonal elements are nonzero if and only if neither of the eigenvalues λ_1 and λ_2 is a diagonal element of the matrix. In this case, we can directly solve one of the equations in each of the above two systems. In particular, we choose $(a_{11} - \lambda_1)\, c_1 + a_{12}c_2 = 0$, $a_{12} \neq 0$, and $a_{21}c_1 + (a_{22} - \lambda_2)\, c_2 = 0$, $a_{21} \neq 0$.

The solution set

$$S_1 = \left\{ c : c = \begin{bmatrix} c_1 & -\dfrac{a_{11} - \lambda_1}{a_{12}} c_1 \end{bmatrix}^T \right\} = \text{Span}(\{[a_{12}\ \lambda_1 - a_{11}]^T\}).$$

Similarly,

$$S_2 = \left\{ c : c = \begin{bmatrix} -\dfrac{a_{22} - \lambda_2}{a_{21}} c_2 & c_2 \end{bmatrix}^T \right\} = \text{Span}(\{[\lambda_2 - a_{22}\ a_{21}]^T\}).$$

The members of the span can be chosen as the eigenvectors of the matrix A. The solutions to the given system corresponding to the eigenvalues λ_1 and λ_2 are

$$x_1(t) = e^{\lambda_1 t}\begin{bmatrix} a_{12} \\ \lambda_1 - a_{11}1 \end{bmatrix}, \quad x_2(t) = e^{\lambda_2 t}\begin{bmatrix} \lambda_2 - a_{22} \\ a_{21} \end{bmatrix}.$$

The roots of the characteristic polynomials are distinct and real. The solution sets corresponding to these cases are linearly independent. Therefore, the fundamental matrices formed by these solutions are

Case 1:

$$\text{(i)} \quad \Phi(t) = \begin{bmatrix} e^{a_{11}t} & 0 \\ 0 & e^{a_{22}t} \end{bmatrix},$$

$$\text{(ii)} \quad \Phi(t) = \begin{bmatrix} (a_{11} - a_{22})\, e^{a_{11}t} & 0 \\ a_{21}e^{a_{11}t} & e^{a_{22}t} \end{bmatrix},$$

$$\text{(iii)} \quad \Phi(t) = \begin{bmatrix} e^{a_{11}t} & a_{12}e^{a_{22}t} \\ 0 & (a_{22} - a_{11})\, e^{a_{22}t} \end{bmatrix};$$

Case 2:

$$\Phi(t) = \begin{bmatrix} a_{12}e^{\lambda_1 t} & (\lambda_2 - a_{22})\, e^{\lambda_2 t} \\ (\lambda_1 - a_{11})\, e^{\lambda_1 t} & a_{21}e^{\lambda_2 t} \end{bmatrix}.$$

In addition, using (4.44), the general solutions are represented by:

Case 1:

$$\text{(i)} \quad x(t) = \Phi(t)a = \begin{bmatrix} e^{a_{11}t} & 0 \\ 0 & e^{a_{22}t} \end{bmatrix} a,$$

$$\text{(ii)} \quad x(t) = \Phi(t)a = \begin{bmatrix} (a_{11} - a_{22})\, e^{a_{11}t} & 0 \\ a_{21}e^{a_{11}t} & e^{a_{22}t} \end{bmatrix} a,$$

$$\text{(iii)} \quad x(t) = \Phi(t)a = \begin{bmatrix} e^{a_{11}t} & a_{12}e^{a_{22}t} \\ 0 & (a_{22} - a_{11})\, e^{a_{22}t} \end{bmatrix} a;$$

Case 2:

$$x(t) = \Phi(t)a = \begin{bmatrix} a_{12}e^{\lambda_1 t} & (\lambda_2 - a_{22})\, e^{\lambda_2 t} \\ (\lambda_1 - a_{11})\, e^{\lambda_1 t} & a_{21}e^{\lambda_2 t} \end{bmatrix} a,$$

where $a = \begin{bmatrix} a_1 & a_2 \end{bmatrix}^T$ is an arbitrary two-dimensional constant vector.

Now, the solution to the IVP is determined by imitating the steps of the procedure in Subsection 4.3.4:

Case 1:

$$\text{(i)}\quad x(t) = \Phi(t,t_0)x_0 = \begin{bmatrix} e^{a_{11}(t-t_0)} & 0 \\ 0 & e^{a_{22}(t-t_0)} \end{bmatrix} x_0,$$

$$\text{(ii)}\quad x(t) = \Phi(t,t_0)x_0 = \begin{bmatrix} e^{a_{11}(t-t_0)} & 0 \\ \dfrac{a_{21}}{a_{11}-a_{22}}\left(e^{a_{11}(t-t_0)} - e^{a_{22}(t-t_0)}\right) & e^{a_{22}(t-t_0)} \end{bmatrix} x_0,$$

$$\text{(iii)}\quad x(t) = \Phi(t,t_0)x_0 = \begin{bmatrix} e^{a_{11}(t-t_0)} & \dfrac{a_{12}}{a_{22}-a_{11}}\left(e^{a_{22}(t-t_0)} - e^{a_{11}(t-t_0)}\right) \\ 0 & e^{a_{22}(t-t_0)} \end{bmatrix} x_0;$$

Case 2:

$$x(t) = \Phi(t,t_0)x_0 = \frac{2}{d(A) + (a_{22}-a_{11})\sqrt{d(A)}} \begin{bmatrix} \phi_{11}(t-t_0) & \phi_{12}(t-t_0) \\ \phi_{21}(t-t_0) & \phi_{22}(t-t_0) \end{bmatrix} x_0,$$

where

$$\phi_{11}(t-t_0) = a_{21}a_{12}e^{\lambda_1(t-t_0)} + (\lambda_2 - a_{22})(a_{11} - \lambda_1)e^{\lambda_2(t-t_0)},$$

$$\phi_{12}(t-t_0) = a_{12}(a_{22} - \lambda_2)\left(e^{\lambda_1(t-t_0)} - e^{\lambda_2(t-t_0)}\right),$$

$$\phi_{21}(t-t_0) = a_{21}(a_{11} - \lambda_1)\left(e^{\lambda_2(t-t_0)} - e^{\lambda_1(t-t_0)}\right),$$

$$\phi_{22}(t-t_0) = (\lambda_1 - a_{11})(a_{22} - \lambda_2)e^{\lambda_1(t-t_0)} + a_{21}a_{12}e^{\lambda_2(t-t_0)}.$$

In the context of case 2, we note that

$$\begin{aligned} \det 2(\Phi(0)) &= 2\left[a_{21}a_{12} + (\lambda_1 - a_{11})(a_{22} - \lambda_2)\right] \\ &= d(A) + (a_{22} - a_{11})\sqrt{d(A)} \neq 0, \end{aligned}$$

since $d(A) = (a_{11} - a_{22})^2 + 4a_{21}a_{12}$. This completes the solution procedure.

Example 4.4.1 (Kendall's mathematical marriage model [48]). We consider Kendall's deterministic system of differential equations in the two-sex dynamic model derived in Example 4.2.3,

$$dx = Ax\,dt, \quad x(0) = x_0,$$

$$dM = [-\delta M + (\beta + \delta)C - \rho F]\,dt, \quad M(0) = M_0,$$

where $x = [F, C]^T$,

$$A = \begin{bmatrix} -(\delta + \rho) & (\beta + \delta) \\ \rho & -2\delta \end{bmatrix}, \quad x_0 = [F(0), C(0)]^T.$$

Solve the IVP.

Solution procedure. We repeat the procedure described in Illustration 4.4.1. First, we note that the characteristic equation associated with the first system of homogeneous differential equations has real and distinct roots. For this purpose, we consider

$$\det(A - \lambda I) = 0 = \lambda^2 + (3\delta + \rho)\lambda - \rho(\beta + \delta) + 2\delta(\rho + \delta).$$

We have

$$\lambda = \frac{-(3\delta + \rho) \pm \sqrt{(\delta + \rho)^2 + 4\rho\beta}}{2}.$$

Here, the characteristic roots are real and distinct. Therefore, the procedure described in Illustration 4.4.1 (case 2) is applicable. Thus, the solution to the system of linear homogeneous differential equations is given by

$$x(t) = \Phi(t)x_0 = \frac{2}{d(A) + (a_{22} - a_{11})\sqrt{d(A)}} \begin{bmatrix} \phi_{11}(t - t_0) & \phi_{12}(t - t_0) \\ \phi_{21}(t - t_0) & \phi_{22}(t - t_0) \end{bmatrix} x_0,$$

where

$$\phi_{11}(t - t_0) = a_{21}a_{12}e^{\lambda_1(t-t_0)} + (\lambda_2 - a_{22})(a_{11} - \lambda_1)e^{\lambda_2(t-t_0)},$$

$$\phi_{12}(t - t_0) = a_{12}(a_{22} - \lambda_2)\left(e^{\lambda_1(t-t_0)} - e^{\lambda_2(t-t_0)}\right),$$

$$\phi_{21}(t - t_0) = a_{21}(a_{11} - \lambda_1)\left(e^{\lambda_2(t-t_0)} - e^{\lambda_1(t-t_0)}\right),$$

$$\phi_{22}(t - t_0) = (\lambda_1 - a_{11})(a_{22} - \lambda_2)e^{\lambda_1(t-t_0)} + a_{21}a_{12}e^{\lambda_2(t-t_0)},$$

$$a_{11} = -(\delta + \rho), \quad a_{12} = (\beta + \delta), \quad a_{21} = \rho, \quad a_{22} = -2\delta,$$

$$\lambda_1 = \frac{-(3\,\delta + \rho) + \sqrt{(\,\delta + \rho)^2 + 4\rho\beta}}{2}, \quad \lambda_2 = \frac{-(3\,\delta + \rho) - \sqrt{(\,\delta + \rho)^2 + 4\rho\beta}}{2},$$

$$d(A) = (a_{11} - a_{22})^2 + 4a_{21}a_{12} = (\delta + \rho)^2 + 4\rho\beta,$$

This completes the solution procedure of the first system of differential equations.

Now, we substitute the components of this solution into the second scalar differential equation:

$$dM = [-\delta M + (\beta + \delta)\,C(t) - \rho F(t)]\,dt, \quad M(0) = M_0.$$

This is the scalar nonhomogeneous differential equation. We apply the procedure in Subsection 2.5.4 to solve this IVP. Its solution is

$$M(t) = M_0 \exp[-\delta t] + \int_0^t \exp[-\delta(t - s)]\,[(\beta + \delta)\,C(s) - \rho F(s)]\,ds.$$

Illustration 4.4.2 (Dynamics of food passage in ruminants [8]). The ruminant animal species, such as the camel, cow, deer, goat, ox, and sheep, have a complicated stomach. The food they eat is processed through stages in the stomach. By following

Blaxter, Graham, and Wainman (BGW) [8], we develop the mathematical model of this dynamic process.

BGW 1. The freshly eaten but not chewed food is processed into a storage compartment referred to as the rumen.

BGW 2. At a later time, the unchewed food is chewed. Then the chewed food passes through the omasum into the abomasum, where it is further processed.

BGW 3. From the abomasum, it slowly enters the intestines, and then is discharged in the form of feces.

BGW 4. The change in the passage of food is directly proportional to the product of the size of the food in that part of the stomach and the time interval of length $\triangle t$.

BGW 5. The total amount of food V that enters the duodenum at a time t with the amount of feces $W = W(t)$. The duodenum receives exactly the same amount that leaves the abomasum. In summary, the food processing scheme is described by the flowchart

$$R \to A \to I \to F, \quad \text{with rates } k_1 \text{ and } k_2,$$

where k_1 and k_2 are positive constants, $k_1 \neq k_2$, and are called the specific rates of digestion.

Let U, C, and V be the amounts of unchewed and chewed food in the rumen, abomasum, and duodenum, respectively, at a time t. Initially $(t = 0)$, let $U(0) = U_0$, $C(0) = 0 = V(0)$ be the amount of food in the respective parts of the stomach. The change of unchewed food and of chewed food over a time interval of length $\triangle t$ are given by

$$\begin{aligned}
\triangle U &= \text{(net loss in the unchewed food in the rumen)} \\
&= \text{(change due to the passage from rumen to abomasum)} \\
&= -k_1 U \triangle t \quad \text{(from BGW 2 and BGW 4)}, \\
\triangle C &= \text{(net gain or loss in the food)} \\
&= \text{(change due to arrival from the rumen)} \\
&\quad + \text{(change due to the intestines)} \\
&= [k_1 U - k_2 C] \triangle t \quad \text{(from BGW 2, BGW 3, and BGW 4)}, \\
\triangle V &= \text{(net gain of food)} \\
&= k_2 C \triangle t \quad \text{(from BGW 5)}.
\end{aligned}$$

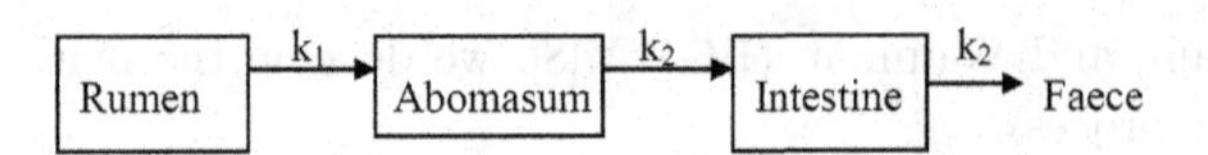

Fig. 4.12 Dynamic of food passage in ruminants.

From the above discussion, we have the linear system of differential equations

$$\begin{aligned} dU &= -k_1 U\,dt, & U(0) &= U_0, \\ dC &= [k_1 U - k_2 C]\,dt, & C(0) &= 0, \\ dV &= k_2 C\,dt, & V(0) &= 0, \end{aligned}$$

which can be rewritten as

$$dx = Mx\,dt, \quad x(0) = x_0,$$
$$dV = k_2 C\,dt, \quad V(0) = 0,$$

where $x = [U, C]^T$,

$$M = \begin{bmatrix} -k_1 & 0 \\ k_1 & -k_2 \end{bmatrix},$$

$x_0 = [U(0), C(0)]^T$.

By applying the procedure of Illustration 4.4.1 [case 1: (ii)], the solution to the IVP corresponding to the system of two coupled linear homogeneous differential equations is given by

$$\begin{aligned} x(t) = \Phi(t,0)x_0 &= \begin{bmatrix} e^{a_{11}t} & 0 \\ \dfrac{a_{21}}{a_{11} - a_{22}} \left(e^{a_{11}t} - e^{a_{22}t}\right) & e^{a_{22}t} \end{bmatrix} x_0 \\ &= \left[U_0 e^{-k_1 t}, \frac{k_1 U_0}{k_2 - k_1} \left(e^{-k_1 t} - e^{-k_2 t}\right) \right]^T \end{aligned}$$

where $a_{11} = -k_1$, $a_{12} = 0$, $a_{21} = k_1$, and $a_{22} = -k_2$. Moreover,

$$\begin{aligned} V(t) = k_2 \int_0^t C(s)\,ds &= \frac{k_2 k_1 U_0}{k_2 - k_1} \left[\left(\frac{e^{-k_1 s}}{-k_1} - \frac{e^{-k_2 s}}{-k_2} \right) \right]_0^t \\ &= U_0 - \frac{U_0}{k_2 - k_1} \left(k_2 e^{-k_1 t} - k_1 e^{-k_2 t}\right). \end{aligned}$$

We note that the mathematical model of the food-passing system of ruminant animals is considered to be an example of three-compartment systems, namely the rumen, abomasum, and duodenum. This is said to be an open (leaky) compartment system. Moreover, the compartment "duodenum" can be considered as a "sink" [46].

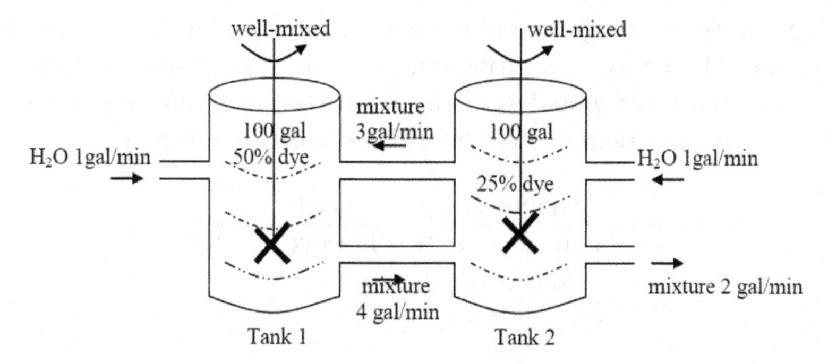

Fig. 4.13 Mixing problem.

Example 4.4.2 (mixing problem [55]). Let us consider two interconnected tanks with an inlet to each tank and an outlet to the second tank. Initially, assume that the two tanks contains 100 gal of a mixture of dye and water. For our purposes, 50% of the mixture in tank 1 and 25% of the mixture in tank 2 contain dye. Pure water flows into the first tank at a fixed rate of 1 gal/min, and into the second tank at the same rate. The mixtures of the two tanks are exchanged at a rate of 4 gal/min from tank 1 to tank 2, and 3 gal/min from tank 2 to tank 1. The well-mixed mixture is drawn from tank 2 at a rate of 2 gal/min. (a) Derive a pair of differential equations for the concentration of the dye in each tank; (b) find the concentration expressions in each tank.

Solution process. We need to use the following procedure to a set of differential equations.

Step 1. To solve this type of problem, we first need to find a net flow of liquid (dye and water) for each tank. This will provide information about the change in the volume of the liquid (if any) in each tank. For this purpose, we find

$$\begin{aligned}&(\text{net flow rate of tank 1})\\&\quad= (\text{net inflow rate of tank 1}) + (\text{net outflow rate of tank 1})\\&\quad= (1\,\text{gal water and 3 gal mixture}) - (4\,\text{gal mixture}) = 0.\end{aligned}$$

This shows that the volume of the liquid in tank 1 is unchanged. Similarly,

$$\begin{aligned}&(\text{net flow rate of tank 2})\\&\quad= (\text{net inflow rate of tank 2}) + (\text{net outflow rate of tank 2})\\&\quad= (1\,\text{gal water and}\ 4\,\text{gal mixture}) - (3 + 2\,\text{gal mixture}) = 0.\end{aligned}$$

This shows that the volume of the liquid in tank 2 is unchanged.

Step 2. Now, we are ready to find the concentration of the dye in each tank at a time t. For this purpose, we utilize the volume of the liquid in each tank as determined in step 1. Let q_1 and q_2 be the volume of the dye in each tank at a time t. Hence, the concentrations of the dye in these tanks are given by

$$c_1 = \frac{\text{volume of dye in tank 1}}{\text{volume of mixture in tank 1}} = \frac{q_1}{100},$$
$$c_2 = \frac{\text{volume of dye in tank 2}}{\text{volume of mixture in tank 2}} = \frac{q_2}{100}.$$

Step 3. Now, we compute the change in the volume of the dye in each tank over the interval of length $\triangle t$:

$$\begin{aligned}
\triangle q_1 &= (\text{net gain or loss in volume of dye in tank 1}) \\
&= (\text{net inflow of dye in tank 1}) \\
&\quad + (\text{net outflow of dye in tank 1}) \\
&= \frac{q_2}{100} \times 3\triangle t - \frac{q_1}{100} \times 4\triangle t \\
&= \left(-\frac{4}{100} q_1 + \frac{3}{100} q_2\right) \triangle t, \\
\triangle q_2 &= (\text{net gain or loss in volume of dye in tank 2}) \\
&= (\text{net inflow of dye in tank 2}) \\
&\quad + (\text{net outflow of dye in tank 2}) \\
&= \frac{q_1}{100} \times 4\triangle t - \frac{q_2}{100} \times 3\triangle t - \frac{q_2}{100} \times 2\triangle t \\
&= \left(\frac{4}{100} q_1 - \frac{5}{100} q_2\right) \triangle t.
\end{aligned}$$

From the above discussion, we have the linear system of differential equations

$$\begin{aligned}
dq_1 &= \left(-\frac{4}{100} q_1 + \frac{3}{100} q_2\right) dt, \quad q_1(0) = 50, \\
dq_2 &= \left(\frac{4}{100} q_1 - \frac{5}{100} q_2\right) dt, \quad q_2(0) = 25.
\end{aligned}$$

Step 4. The state of the mixing dye dynamic is the number of volumes of the dye in each tank. To express this system of differential equations in terms of concentrations as the state of the system, we need to use the transformations $q_1 = 100c_1$ and $q_2 = 100c_2$ if and only if

$$c_1 = \frac{q_1}{100}, \quad c_2 = \frac{q_2}{100}.$$

By using this transformation, the above-described system of differential equations reduces to

$$100dc_1 = \left(-4\frac{q_1}{100} + 3\frac{q_2}{100}\right)dt = (-4c_1 + 3c_2)dt, \quad c_1(0) = \frac{1}{2},$$

$$100dc_2 = \left(4\frac{q_1}{100} - 5\frac{q_2}{100}\right)dt = (4c_1 - 5c_2)dt, \quad c_1(0) = \frac{1}{4}.$$

Hence,

$$dc_1 = \left(-\frac{4}{100}c_1 + \frac{3}{100}c_2\right)dt, \quad c_1(0) = \frac{1}{2},$$

$$dc_2 = \left(\frac{4}{100}c_1 - \frac{5}{100}c_2\right)dt, \quad c_1(0) = \frac{1}{4},$$

which can be rewritten as

$$dc = Ac\,dt, \quad c(0) = c_0,$$

where $c = [c_1, c_2]^T$,

$$A = \begin{bmatrix} -\dfrac{4}{100} & \dfrac{3}{100} \\ \dfrac{4}{100} & -\dfrac{5}{100} \end{bmatrix},$$

$c_0 = [0.5, 0.25]^T$.

This is the desired system of differential equations. This completes the derivation of part (a). To complete the solution of part (b), we need to solve the above-described IVP in (a). For this purpose, we imitate the procedure of Illustration 4.4.1, and we obtain

$$\det(A - \lambda I) = 0 = \lambda^2 + 0.09\lambda + 0.0008,$$

$$\lambda = \frac{-0.09 \pm \sqrt{0.0081 - 0.0032)}}{2} = \frac{-0.09 \pm 0.07}{2}.$$

Thus, $\lambda_1 = -0.01$ and $\lambda_2 = -0.08$. Here, the characteristic roots are real and distinct. Therefore, the procedure described in Illustration 4.4.1 (case 2) is applicable. Thus, the solution to the given IVP is

$$c(t) = \Phi(t,0)c_0 = \frac{2}{d(A) + (a_{22} - a_{11})\sqrt{d(A)}} \begin{bmatrix} \phi_{11}(t) & \phi_{12}(t) \\ \phi_{21}(t) & \phi_{22}(t) \end{bmatrix} x_0,$$

where $a_{11} = -0.04$, $a_{12} = 0.03$, $a_{21} = 0.04$, and $a_{22} = -0.05$;

$$\phi_{11}(t) = 0.0012e^{-0.01t} + 0.0009e^{-0.08t}, \quad \phi_{12}(t) = 0.0009\left(e^{-0.01t} - e^{-0.08t}\right),$$
$$\phi_{21}(t) = -0.0012\left(e^{-0.08t} - e^{-0.01t}\right), \quad \phi_{22}(t) = 0.0009e^{-01t} + 0.0012e^{-0.8t},$$
$$d(A) = 0.0001 + 0.0048 = 0.0049.$$

Thus, we have our solution.

Example 4.4.3 (electric network [55]). Let us consider the following circuit network. Here, it is given that $C = 3\,\mathrm{F}$, $L = 1\,\mathrm{H}$, and $R = 0.25\,\Omega$. Find the general solution associated with the circuit network.

Solution procedure. We need to utilize the notations, definitions, and the general procedure outlined in Illustration 4.2.4. We also need to modify the RL circuit in Figures 4.2, 4.3(a), 4.3(b), and 4.3(c) by replacing R_1, R_3, and L_2 with L, R, and C, respectively, and deleting E and L_3. This creates the LRC circuit with exactly the same circuit structure as the RL circuit. A further detailed figure regarding this LRC circuit is left as an exercise to the reader.

Step 1. From step 1 of Illustration 4.2.4 in the context of the above-described circuit modifications and corresponding modified figures of the circuit, we have $I_1 + I_2 - I_3 = 0$ (at P in the LRC modified circuit).

Step 2. As in step 2 of Illustration 4.2.4 in the context of the above-described circuit modifications and corresponding modified LRC Figures 4.3(b) and 4.3(c) of the LRC circuit, we obtain

$$L\frac{d}{dt}I_1 + RI_3 = 0 \quad \text{[modified LRC Figure 4.3(b)]},$$
$$RI_3 + E_2 = 0 \quad \text{[modified LRC Figure 4.3(c)]}.$$

From step 1 and by using ENC 3 ($\frac{d}{dt}E_2 = \frac{I_2}{C}$) and by differentiating the second equation with respect to t, we have

$$R\frac{d}{dt}I_3 + \frac{d}{dt}E_2 = 0 = R\frac{d}{dt}I_3 + \frac{I_2}{C} = R\frac{d}{dt}I_3 + \frac{I_3 - I_1}{C}.$$

From this discussion after algebraic simplifications, we arrive at the following system of differential equations:

$$dI = AI\,dt, \quad \text{where } I = \begin{bmatrix} I_1 & I_3 \end{bmatrix}^T, \quad A = \begin{bmatrix} 0 & -\dfrac{R}{L} \\ \dfrac{1}{CR} & -\dfrac{1}{CR} \end{bmatrix}.$$

Step 3. We utilize the same procedure in Subsection 4.3.2 for finding the general solution to the above system. The characteristic equation of the matrix

$$\lambda^2 - \mathrm{tr}\,(A)\lambda + \det\,(A) = 0 = \lambda^2 + \frac{\lambda}{CR} + \frac{1}{CL}.$$

Thus, its eigenvalues are

$$\lambda = \frac{-\frac{1}{CR} \pm \sqrt{\frac{L-4CR^2}{C^2R^2L}}}{2} = \frac{-1 \pm \sqrt{\frac{L-4CR^2}{L}}}{2CR}.$$

For given $C = 3$, $L = 1$, and $R = 0.25$,

$$\lambda = \frac{-1 \pm \sqrt{1 - 3/4}}{3/2} = \frac{-1 \pm \sqrt{1/4}}{3/2} = \frac{-1 \pm 1/2}{3/2},$$

and $\lambda_1 = -\frac{1}{3}$, $\lambda_2 = -1$. The eigenvectors corresponding to these eigenvalues are $c_1 = \begin{bmatrix} 3 & 4 \end{bmatrix}^T$ and $c_2 = \begin{bmatrix} 1 & 4 \end{bmatrix}^T$, respectively. The general solution is given by $I(t) = \Phi(t)a$, where

$$\Phi(t) = \begin{bmatrix} 3e^{-\frac{1}{3}t} & e^{-t} \\ 4e^{-\frac{1}{3}t} & 4e^{-t} \end{bmatrix}.$$

This completes the solution process of the example.

Example 4.4.4. We are given $dx = Ax\,dt$, $x(t_0) = x_0$, where

$$A = \begin{bmatrix} 4 & -3 \\ 2 & -1 \end{bmatrix}.$$

Find: (a) a general solution, and (b) the solution to the given IVP.

Solution procedure. In this case, $d(A) = (4+1)^2 + 4(-6) = 1$. The coefficient matrix satisfies the condition in case 2 of Illustration 4.4.1. By repeating the argument of Illustration 4.4.1, we arrive at

$$\begin{bmatrix} 4-\lambda & -3 \\ 2 & -1-\lambda \end{bmatrix} c = \begin{bmatrix} (4-\lambda)\,c_1 - 3c_2 \\ 2c_1 - (1+\lambda)\,c_2 \end{bmatrix} = 0 \quad \text{(Definition 1.3.11)},$$

which implies that

$$\begin{aligned} (4-\lambda)\,c_1 - 3c_2 &= 0, \\ 2c_1 + -(1+\lambda)\,c_2 &= 0. \end{aligned}$$

The characteristic equation (4.38) of the matrix A in the context of the given example is

$$\begin{aligned} 0 &= \det(A - \lambda I) \\ &= (4-\lambda)(-1-\lambda) + 6 \quad \text{(step 1 in Example 1.4.7)} \\ &= \lambda^2 - 3\lambda + 2 \quad \text{(by simplifying)} \\ &= \lambda^2 - \operatorname{tr}(A)\lambda + \det(A) \quad [\text{Definition 1.4.2 and } \operatorname{tr}(A)] \\ &= (\lambda - 1)(\lambda - 2) \quad \text{(by the quadratic formula).} \end{aligned}$$

The eigenvectors corresponding to the eigenvalues $\lambda_1 = 2$ and $\lambda_2 = 1$ are given by $(a_{11} - \lambda_1)\, c_1 + a_{12}c_2 = 2c_1 - 3c_2 = 0$ and $a_{21}c_1 + (a_{22} - \lambda_2)\, c_2 = 2c_1 - 2c_2 = 0$, respectively. Thus, solution sets are:

$$S_1 = \left\{ c : c = \begin{bmatrix} c_1 & -\dfrac{a_{11} - \lambda_1}{a_{12}} c_1 \end{bmatrix}^T \right\} = \operatorname{Span}\left(\left\{ \begin{bmatrix} -3 & -2 \end{bmatrix}^T \right\} \right),$$

$$S_2 = \left\{ c : c = \begin{bmatrix} -\dfrac{(a_{22} - \lambda_2)}{a_{21}} c_1 & c_1 \end{bmatrix}^T \right\} = \operatorname{Span}\left(\left\{ \begin{bmatrix} 2 & 2 \end{bmatrix}^T \right\} \right).$$

The members of the span can be chosen as the eigenvectors of the matrix A, and the solutions to the given system corresponding to the eigenvalues λ_1 and λ_2 are

$$x_1(t) = e^{2t} \begin{bmatrix} -3 \\ -2 \end{bmatrix}, \quad x_2(t) = e^t \begin{bmatrix} 2 \\ 2 \end{bmatrix}.$$

The roots of the characteristic polynomials are distinct and real. The solution sets corresponding to this case are linearly independent. Applying case 2 of Illustration 4.4.1, the fundamental matrix solution, the general solution, and the solution to the IVP are

$$\Phi(t) = \begin{bmatrix} -3e^{2t} & 2e^t \\ -2e^{2t} & 2e^t \end{bmatrix},$$

$$x(t) = \Phi(t)\, a = \begin{bmatrix} -3e^{2t} & 2e^t \\ -2e^{2t} & 2e^t \end{bmatrix} a,$$

$$x(t) = \Phi(t)x_0 = -\frac{2}{4} \begin{bmatrix} \phi_{11}(t - t_0) & \phi_{12}(t - t_0) \\ \phi_{21}(t - t_0) & \phi_{22}(t - t_0) \end{bmatrix} x_0,$$

respectively, where

$$\phi_{11}(t - t_0) = -6e^{2(t - t_0)} + 4e^{(t - t_0)},$$

$$\phi_{12}(t - t_0) = -6\left(e^{(t - t_0)} - e^{2(t - t_0)} \right),$$

$$\phi_{21}(t-t_0) = 4\left(e^{2(t-t_0)} - e^{(t-t_0)}\right),$$

$$\phi_{22}(t-t_0) = 4e^{2(t-t_0)} - 6e^{(t-t_0)}.$$

In the context of case 2 of Illustration 4.4.1, we note that $d(A) + (a_{22} - a_{11})\sqrt{d(A)} = (1-5) \neq 0$, since $d(A) = (a_{11} - a_{22})^2 + 4a_{21}a_{12} = 25 - 24 = 1$. This completes the solution procedure.

Distinct Complex Eigenvalues 4.4.2. *Let $\lambda_1, \ldots, \lambda_k, \ldots, \lambda_n$ be the distinct complex roots of* (4.38)*. Then there is a complete set of solutions.*

We also note that real roots can be considered as complex roots due to the definition of the complex numbers. We recall the elementary definition of the complex number $z = a + ib$, where a and b are real numbers and $i = \sqrt{-1}$. Furthermore, a is called a real part of z, and it is denoted by Re $z = a$; b is called an imaginary part of z, and it is denoted by Im $z = b$. Of course, if $b = 0$, then $z = a =$ Re z. This justifies that every real number can be considered to be a complex number. Here, we focus on a complex number in which Im $z = b \neq 0$.

The discussion about a complex root $\lambda = \lambda_k$ of (4.38) for some k, $k = 1, 2, \ldots, n$, with $\operatorname{Re}\lambda_k = \lambda_k$ ($\operatorname{Im}\lambda_k = 0$) is exactly the same as for the above (distinct real roots). However, we need to shed light on a complex root $\lambda = \lambda_k$ with $\operatorname{Im}\lambda_k \neq 0$ for some k, $k = 1, 2, \ldots, n$. In this case, we first, recall the following well-known facts:

DDC 1. It is known that "complex roots of the characteristic equation (4.38) with real coefficients always occur in complex conjugate pairs." Thus, the total number of complex-valued roots of the characteristic equation (4.38) always occurs in even numbers, like $2, 4, 6, \ldots, 2k$, with $2k \leq n$. If $\lambda = \lambda_k$ is a complex root with $\operatorname{Im}\lambda_k \neq 0$, then its complex conjugate $\overline{\lambda} = \overline{\lambda}_k$ with $\operatorname{Im}\overline{\lambda}_k = -\operatorname{Im}\lambda_k \neq 0$ is also the complex root of the characteristic equation (4.38) with real coefficients.

DDC 2. Eigenvectors corresponding to the complex roots determined by (4.39) are ordinarily complex-valued eigenvectors. They also occur in complex conjugate pairs. (The coefficients of the matrix A in (4.35) are real numbers.) If $c = c_k$ is an eigenvector corresponding to the complex eigenvalue $\lambda = \lambda_k$ with $\operatorname{Im}\lambda_k \neq 0$, then its complex conjugate $\overline{c} = \overline{c}_k$ vector is also an eigenvector corresponding to the complex conjugate eigenvalue $\overline{\lambda} = \overline{\lambda}_k$.

DDC 3. In view of the facts of DDC 1 and DDC 2, the solutions corresponding to these complex eigenvalues and corresponding eigenvectors will also be complex-valued functions. Our goal is to find pairs of linearly independent real-valued solutions corresponding to each pair of complex conjugate eigenvalues and corresponding complex conjugate eigenvectors of the matrix A in (4.35). Thus, for complex conjugate pairs of eigenvalues $\lambda_k = \alpha_k + i\beta_k$, $\overline{\lambda}_k = \alpha_k - i\beta_k$ with $\operatorname{Im}\lambda_k = \beta_k \neq 0$ and corresponding complex conjugate eigenvectors $c_k = d_k + ie_k$, $\overline{c}_k = d_k - ie_k$, we write the corresponding solution representation as in (4.36) as:

$$\begin{aligned} x_k(t) &= \exp\left[(\alpha_k + i\beta_k)\, t\right] c_k = \exp\left[\alpha_k t\right] \exp\left[i\beta_k t\right] (d_k + ie_k) \\ &\qquad \text{(law of exponents)} \\ &= \exp\left[\alpha_k t\right] (\cos\beta_k t + i \sin\beta_k t)\, (d_k + ie_k) \quad \text{(Euler's formula)} \\ &= \exp\left[\alpha_k t\right] \left[(d_k \cos\beta_k t - e_k \sin\beta_k t) \right. \\ &\qquad \left. + i\,(e_k \cos\beta_k t + d_k \sin\beta_k t)\right] \quad \text{(product of complex numbers)}. \end{aligned} \tag{4.57}$$

Similarly,

$$\begin{aligned} \overline{x}_k(t) &= \exp\left[(\alpha_k - i\beta_k)\, t\right] \overline{c}_k \\ &= \exp\left[\alpha_k t\right] \left[(d_k \cos\beta_k t - e_k \sin\beta_k t) - i\,(e_k \cos\beta_k t + d_k \sin\beta_k t)\right]. \end{aligned} \tag{4.58}$$

From the sum and the difference of the complex conjugate pair of numbers or functions, we have

$$\operatorname{Re} x_k(t) = \frac{1}{2}\, x_k(t) + \frac{1}{2}\, \overline{x}_k(t) \quad \text{and} \quad \operatorname{Im} x_k(t) = \frac{1}{2}\, x_k(t) - \frac{1}{2}\, \overline{x}_k(t).$$

From this, (4.57), and (4.58), we get

$$\operatorname{Re} x_k(t) = \exp\left[\alpha_k t\right] (d_k \cos\beta_k t - e_k \sin\beta_k t)\,, \tag{4.59}$$

$$\operatorname{Im} x_k(t) = \exp\left[\alpha_k t\right] (e_k \cos\beta_k t + d_k \sin\beta_k t)\,. \tag{4.60}$$

From the application of Theorem 4.3.1 to $\operatorname{Re} x_k(t)$ and $\operatorname{Im} x_k(t)$, it follows that $\operatorname{Re} x_k(t)$ and $\operatorname{Im} x_k(t)$ are solutions to (4.35). Furthermore, these are real-valued functions. This provides the procedure for finding the complete set of solutions to (4.35) corresponding to distinct complex eigenvalues of A.

Illustration 4.4.3. We consider $dx = Ax\, dt$, $x(t_0) = x_0$, where

$$A = \begin{bmatrix} a_{11} & a_{12} \\ a_{21} & a_{22} \end{bmatrix},$$

and a_{ij}'s are any given real numbers satisfying the condition $d(A) = (\operatorname{tr}(A))^2 - 4\det(A) < 0$; $\operatorname{tr}(A)$ and $\det(A)$ are as defined in Theorem 4.3.2 and Definition 1.4.2, respectively. Find: (a) a general solution, and (b) the solution to the given IVP.

Solution procedure. Again, we note that $n = 2$. We imitate the arguments used in Illustration 4.4.1 and the procedure in Subsection 4.3.4 to find the solution to the IVP. As a result of this, we have

$$0 = \det(A - \lambda I) = \lambda^2 - \operatorname{tr}(A)\lambda + \det(A) = (\lambda - \lambda_1)(\lambda - \lambda_2)\,,$$

where λ_1 and λ_2 are as defined in Illustration 4.4.1, and are roots of the quadratic polynomial with real coefficients. From the assumption $(\operatorname{tr}(A))^2 - 4\det(A) < 0$ and from the above discussion (DDC 1), we conclude that λ_1 and λ_2 are complex conjugate pair roots of $0 = \det(A - \lambda I)$ (for instance, $\lambda_2 = \overline{\lambda}$), and they are distinct. We need to find eigenvectors corresponding to these eigenvalues λ_1 and λ_2. Again, from our discussion (DDC 1 and DDC 3), it is enough to find the eigenvector corresponding to either of the complex eigenvalues, say, λ_1 or its complex conjugate, λ_2. Without loss of generality, we choose $\lambda_1 = \alpha + i\beta$ with $\operatorname{Im}\lambda_1 = \beta \neq 0$. With this, we imitate the argument used in Illustration 4.4.1 in the context of case 2 and the discussion in DDC 3, and obtain eigenvectors and solutions corresponding to the eigenvalue $\lambda_1 = \alpha + i\beta$:

$$\begin{aligned}
x_1(t) &= e^{\lambda_1 t}\begin{bmatrix} a_{12} \\ \lambda_1 - a_{11}\end{bmatrix} = \exp\left[(\alpha + i\beta)\, t\right]\begin{bmatrix} a_{12} \\ \alpha + i\beta - a_{11}\end{bmatrix} \\
&= e^{\alpha t}\left(\cos\beta t + i\sin\beta t\right)\begin{bmatrix} a_{12} \\ (\alpha - a_{11}) + i\beta\end{bmatrix} \quad \text{(Euler's formula)} \\
&= e^{\alpha t}\begin{bmatrix} a_{12}\left(\cos\beta t + i\sin\beta t\right) \\ \left(\cos\beta t + i\sin\beta t\right)\left((\alpha - a_{11}) + i\beta\right)\end{bmatrix} \quad \text{(algebraic simplification)} \\
&= e^{\alpha t}\begin{bmatrix} a_{12}\cos\beta t + i a_{12}\sin\beta t \\ \left((\alpha - a_{11})\cos\beta t - \beta\sin\beta t\right) + i\left(\beta\cos\beta t + (\alpha - a_{11})\sin\beta t\right)\end{bmatrix} \\
&= e^{\alpha t}\left[\begin{bmatrix} a_{12}\cos\beta t \\ (\alpha - a_{11})\cos\beta t - \beta\sin\beta t\end{bmatrix} + i\begin{bmatrix} a_{12}\sin\beta t \\ \beta\cos\beta t + (\alpha - a_{11})\sin\beta t\end{bmatrix}\right].
\end{aligned}$$

From the discussion in DDC 3, the real and imaginary parts of the solution are linearly independent solutions to the given problem. Therefore, the matrix formed by these solutions,

$$\Phi(t) = e^{\alpha t}\begin{bmatrix} a_{12}\cos\beta t & a_{12}\sin\beta t \\ (\alpha - a_{11})\cos\beta t - \beta\sin\beta t & \beta\cos\beta t + (\alpha - a_{11})\sin\beta t\end{bmatrix},$$

is a fundamental matrix solution (Observations 4.3.2) with $\det(\Phi(t)) = \beta a_{12} e^{2\alpha t}$. By following the arguments used in Examples 1.4.5 and 1.4.8, we have

$$\Phi^{-1}(t) = \begin{bmatrix} \dfrac{e^{-\alpha t}}{\beta a_{12}}\left[(\alpha - a_{11})\sin\beta t + \beta\cos\beta t\right] & -\dfrac{1}{\beta}e^{-\alpha t}\sin\beta t \\ -\dfrac{e^{-\alpha t}}{\beta a_{12}}\left[(\alpha - a_{11})\cos\beta t - \beta\sin\beta t\right] & \dfrac{1}{\beta}e^{-\alpha t}\cos\beta t\end{bmatrix}.$$

Moreover, its general solution is given by (4.44), i.e.

$$\begin{aligned}
x(t) &= \Phi(t)a \\
&= e^{\alpha t}\begin{bmatrix} a_{12}\cos\beta t & a_{12}\sin\beta t \\ (\alpha - a_{11})\cos\beta t - \beta\sin\beta t & \beta\cos\beta t + (\alpha - a_{11})\sin\beta t\end{bmatrix} a,
\end{aligned}$$

where $a = \begin{bmatrix} a_1 & a_2 \end{bmatrix}^T$ is an arbitrary two-dimensional constant vector. Now, the solution to the IVP determined by the procedure in Subsection 4.3.4 and Observation 4.3.3 is

$$x(t) = \Phi(t, t_0)x_0 = \Phi(t)\Phi^{-1}(t_0)x_0 = \begin{bmatrix} \phi_{11}(t-t_0) & \phi_{12}(t-t_0) \\ \phi_{21}(t-t_0) & \phi_{22}(t-t_0) \end{bmatrix} x_0,$$

where

$$\phi_{11}(t-t_0) = \frac{1}{\beta} e^{\alpha(t-t_0)} \left[(a_{11}-\alpha)\sin\beta(t-t_0) + \beta\cos\beta(t-t_0)\right],$$

$$\phi_{12}(t-t_0) = \frac{a_{12}}{\beta} e^{\alpha(t-t_0)} \sin\beta(t-t_0),$$

$$\phi_{21}(t-t_0) = -\frac{(\alpha-a_{11})^2+\beta^2}{\beta a_{12}} e^{\alpha(t-t_0)} \sin\beta(t-t_0),$$

$$\phi_{22}(t-t_0) = \frac{1}{\beta} e^{\alpha(t-t_0)} \left[\beta\cos\beta(t-t_0) + (\alpha-a_{11})\sin\beta(t-t_0)\right].$$

Here, we have used the following trigonometric identities:

$$\sin(A-B) = \sin A\cos B - \cos A\sin B,$$

$$\cos(A-B) = \cos A\cos B + \sin A\sin B.$$

This completes the solution process of the given illustration.

Example 4.4.5 (predator–prey model [55, 104]). We consider a linearized version of the classical model of Lotka and Volterra with the one-predator one-prey system described in Example 4.2.1:

$$dN = CN\,dt, \quad N(0) = N_0,$$

where $N = \begin{bmatrix} N_1 & N_2 \end{bmatrix}^T$ and

$$C = \begin{bmatrix} 0 & -\dfrac{\alpha_1\beta_0}{\beta_1} \\ \dfrac{\beta_1\alpha_0}{\alpha_1} & 0 \end{bmatrix}.$$

Solve the IVP.

Solution procedure. The procedure mirrors that of Illustration 4.4.3. First, we note that the characteristic equation associated with the matrix C in the linearized predator–prey mathematical model is $\det(C-\lambda I) = 0 = \lambda^2 + \alpha_0\beta_0$. We have $\lambda = \pm\sqrt{\alpha_0\beta_0}\,i$. Here, the characteristic roots are distinct complex numbers. Therefore, the procedure described in Illustration 4.4.3 is applicable. Thus, the solution to the system of linear homogeneous differential equations is given by

$$x(t) = \Phi(t, t_0)x_0 = \Phi(t)\Phi^{-1}(t_0)x_0 = \begin{bmatrix} \phi_{11}(t-t_0) & \phi_{12}(t-t_0) \\ \phi_{21}(t-t_0) & \phi_{22}(t-t_0) \end{bmatrix} x_0,$$

where

$$\phi_{11}(t-t_0) = \cos\beta(t-t_0),$$

$$\phi_{12}(t-t_0) = -\frac{\alpha_1}{\beta_1}\sqrt{\frac{\beta_0}{\alpha_0}}\sin\beta(t-t_0),$$

$$\phi_{21}(t-t_0) = \frac{\beta_1}{\alpha_1}\sqrt{\frac{\alpha_0}{\beta_0}}\sin\beta(t-t_0),$$

$$\phi_{22}(t-t_0) = \cos\beta(t-t_0) \quad \text{and} \quad \beta = \sqrt{\alpha_0\beta_0}.$$

The solution procedure is completed.

Example 4.4.6 (electric network [55]). Let us consider the circuit network outlined in Example 4.4.3 with $C = 2\,\text{F}$, $L = 10\,\text{H}$, and $R = \frac{5}{2}\,\Omega$. Find the general solution associated with the circuit network.

Solution procedure. Utilizing all the notations, definitions, and the procedures outlined in Illustration 4.4.3 and Example 4.4.3, we have the following system of differential equations:

$$dI = AI\,dt, \qquad \text{where } I = \begin{bmatrix} I_1 & I_2 \end{bmatrix}^T, \; A = \begin{bmatrix} 0 & -\dfrac{R}{L} \\ \dfrac{1}{CR} & -\dfrac{1}{CR} \end{bmatrix}.$$

For given $C = 2$, $L = 10$, and $R = 2.5$, the eigenvalues with regard to the coefficient matrix are

$$\lambda = \frac{-1 \pm \sqrt{-4}}{10}, \; \lambda_1 = -0.1 + 0.2i, \; \lambda_2 = -0.1 - 0.2i$$

From the procedure outlined in Illustration 4.4.3, the eigenvectors, the fundamental matrix solution, and the general solution to the given system of differential equations are

$$c_1 = \begin{bmatrix} -0.25 & -0.1 + 0.2\,i \end{bmatrix}^T,$$

$$\Phi(t) = \begin{bmatrix} -0.25e^{-0.1t}\cos 0.2t & -0.25e^{-0.1t}\sin 0.2t \\ e^{-0.1t}\left[-0.1\cos 0.2t - 0.2\sin\beta t\right] & e^{-0.1t}\left[-0.1\sin 0.2t + 0.2\cos 0.2t\right] \end{bmatrix},$$

$$x(t) = \begin{bmatrix} -0.25e^{-0.1t}\cos 0.2t & -0.25e^{-0.1t}\sin 0.2t \\ e^{-0.1t}\left[-0.1\cos 0.2t - 0.2\sin\beta t\right] & e^{-0.1t}\left[0.1\sin 0.2t + 0.2\cos 0.2t\right] \end{bmatrix} a,$$

respectively, where $a = [a_1 \quad a_2]^T$ is an arbitrary two-dimensional constant vector. This completes the solution process of the example.

Example 4.4.7. We are given $dx = Ax\,dt$, $x(t_0) = x_0$, where

$$A = \begin{bmatrix} 2 & 1 \\ -4 & 2 \end{bmatrix}.$$

Find the solution to the IVP.

Solution procedure. In this case, $d(A) = (2+2)^2 - 4(8) = -16$. By repeating the argument used in Illustration 4.4.3, we arrive at

$$\begin{aligned} 0 &= \det(A - \lambda I) = (2-\lambda)(2-\lambda) + 4 \\ &= \lambda^2 - 4\lambda + 8 = (\lambda - 2 - 2i)(\lambda - 2 + 2i). \end{aligned}$$

Here, $n = 2$ and the eigenvalues of the coefficient matrix are complex conjugate numbers: $\lambda_1 = 2+2i$ and $\lambda_2 = 2-2i$. Here also, we work with $\lambda_1 = 2+2i = \alpha + i\beta$. This implies that $\alpha = 2$ and $\beta = 2$. Again, we imitate the argument used in Illustration 4.4.3, and obtain the following linearly independent real solutions:

$$\begin{aligned} x_1(t) &= e^{\lambda_1 t}\begin{bmatrix} a_{12} \\ \lambda_1 - a_{11} \end{bmatrix} = \exp\left[(2+i2)\,t\right]\begin{bmatrix} 1 \\ 2+i2-2 \end{bmatrix} \quad (a_{12} = 1 \text{ and } a_{11} = 2) \\ &= e^{2t}(\cos 2t + i\sin 2t)\begin{bmatrix} 1 \\ (2-2)+i2 \end{bmatrix} \quad \text{(Euler's formula)} \\ &= e^{2t}(\cos 2t + i\sin 2t)\begin{bmatrix} 1 \\ 2i \end{bmatrix} \quad \text{(simplification)} \\ &= e^{2t}\begin{bmatrix} \cos 2t \\ -2\sin 2t \end{bmatrix} + ie^{2t}\begin{bmatrix} \sin 2t \\ 2\cos 2t \end{bmatrix} \quad \text{(simplification)}. \end{aligned}$$

From the discussion in DDC 3, the real and imaginary parts of the solution are linearly independent solutions to the given system. Therefore, the fundamental matrix formed by these solutions is

$$\Phi(t) = \begin{bmatrix} e^{2t}\cos 2t & e^{2t}\sin 2t \\ -2e^{2t}\sin 2t & 2e^{2t}\cos 2t \end{bmatrix}$$

with $\det(\Phi(t)) = 2e^{4t}$. By following the arguments used in Examples 1.4.5 and 1.4.8, we have

$$\Phi^{-1}(t) = \begin{bmatrix} e^{-2t}\cos 2t & -\dfrac{1}{2}e^{-2t}\sin 2t \\ e^{-2t}\sin 2t & \dfrac{1}{2}e^{-2t}\cos 2t \end{bmatrix}.$$

Moreover, its general solution is

$$x(t) = \Phi(t)\,a = \begin{bmatrix} e^{2t}\cos 2t & e^{2t}\sin 2t \\ -2e^{2t}\sin 2t & 2e^{2t}\cos 2t \end{bmatrix} a,$$

where $a = \begin{bmatrix} a_1 & a_2 \end{bmatrix}^T$ is an arbitrary two-dimensional constant vector. Now, by following the procedure in Subsection 4.3.4, the solution to the IVP is:

$$x(t) = \Phi(t)\Phi^{-1}(t_0)x_0 = \begin{bmatrix} e^{2(t-t_0)} \cos 2(t-t_0) & \frac{1}{2} e^{2(t-t_0)} \sin 2(t-t_0) \\ -2e^{2(t-t_0)} \sin 2(t-t_0) & e^{2(t-t_0)} \cos 2(t-t_0) \end{bmatrix} x_0.$$

Example 4.4.8. We are given $dx = Ax\,dt$, $x(t_0) = x_0$, where

$$A = \begin{bmatrix} 0 & 1 \\ -1 & 0 \end{bmatrix}.$$

Find the solution to the IVP.

Solution procedure. In this case, $d(A) = (0+0)^2 - 4(1) = -4$. By repeating the argument used in Illustration 4.4.3, we have

$$0 = \det(A - \lambda I) = (-\lambda)(-\lambda) + 1 = \lambda^2 + 1 = (\lambda - i)(\lambda + i).$$

Here, $n = 2$ and the eigenvalues of the coefficient matrix are complex conjugate numbers $\lambda_1 = 0 + i$ and $\lambda_2 = 0 - i$. Here also, we work with $\lambda_1 = 0 + i = \alpha + i\beta$, $\alpha = 0$ and $\beta = 1$. We imitate the solution procedure of Example 4.4.7, and obtain

$$\begin{aligned} x_1(t) &= e^{\lambda_1 t} \begin{bmatrix} a_{12} \\ \lambda_1 - a_{11} \end{bmatrix} = \exp[(0+i)t] \begin{bmatrix} 1 \\ i \end{bmatrix} \quad (a_{12} = 1 \text{ and } a_{11} = 0) \\ &= e^{0t}(\cos t + i \sin t) \begin{bmatrix} 1 \\ i \end{bmatrix} = (\cos t + i \sin t) \begin{bmatrix} 1 \\ i \end{bmatrix} \quad \text{(simplification)} \\ &= \begin{bmatrix} \cos t \\ -\sin t \end{bmatrix} + i \begin{bmatrix} \sin t \\ \cos t \end{bmatrix} \quad \text{(simplification).} \end{aligned}$$

Furthermore, the fundamental matrix solution, its determinant, inverse, the general solution, and the solution to the IVP are

$$\Phi(t) = \begin{bmatrix} \cos t & \sin t \\ -\sin t & \cos t \end{bmatrix}, \quad \det(\Phi(t)) = 1,$$

$$\Phi^{-1}(t) = \begin{bmatrix} \cos t & -\sin t \\ \sin t & \cos t \end{bmatrix}, \quad x(t) = \Phi(t)a = \begin{bmatrix} \cos t & \sin t \\ -\sin t & \cos t \end{bmatrix} a,$$

$$x(t) = \Phi(t)\Phi^{-1}(t_0)x_0 = \begin{bmatrix} \cos(t-t_0) & \sin(t-t_0) \\ -\sin(t-t_0) & \cos(t-t_0) \end{bmatrix} x_0,$$

respectively.

Multiple Eigenvalues 4.4.3. *Let $\lambda_1, \ldots, \lambda_k, \ldots, \lambda_n$ be roots of* (4.38). *We assume that there are m distinct roots and $m < n$. This means that there is at least one root of the characteristic equation* (4.38) *of multiplicity $p > 1$. Determine a complete set of solutions to* (4.35). *The outline of the procedure is as follows:*

DME 1. We know that for each eigenvalue λ, there is always at least one nonzero solution to (4.35). As a result of this, there are at least m linearly independent eigenvectors corresponding to m distinct eigenvalues of A; for the details, see the case of distinct eigenvalues.

DME 2. To find a general solution to (4.35), we need a complete set of solutions to (4.35). For this purpose, we need to find the $n - m$ remaining linearly independent eigenvectors of A. The method for finding these remaining linearly independent eigenvectors of A is the subject of the following discussion.

DME 3. An eigenvalue of multiplicity p is said to be complete if it has p linearly independent eigenvectors. If all eigenvalues of the matrix A are complete, then it follows that A has its n linearly independent eigenvectors $c_1, \ldots, c_k, \ldots, c_n$ associated with the eigenvalues $\lambda_1, \ldots, \lambda_k, \ldots, \lambda_n$ (each repeated with its multiplicity). In this case, a general solution to (4.35) is given by the procedures outlined with regard to distinct real as well as complex roots.

DME 4. In the repeated root case, the problem is: What if at least one eigenvalue of multiplicity $p > 1$ is not complete? In this case, there is a deficiency in the number of linearly independent eigenvectors. An eigenvalue $\lambda = \lambda_i$ of multiplicity $p > 1$ is said to be defective if it is not complete. Assume that λ_i has only q linearly independent eigenvectors, $c_i(1), \ldots, c_i(l), \ldots, c_i(q)$, and of course $q < p$. These q eigenvectors, $c_i(1), \ldots, c_i(l), \ldots, c_i(q)$, are referred as the ordinary eigenvectors associated with the eigenvalue $\lambda = \lambda_i$. In this case, $p-q$ eigenvectors associated with this eigenvalue $\lambda = \lambda_i$ are not recovered. This number, $p - q$, is the deficiency in the number of linearly independent eigenvectors corresponding to the eigenvalue $\lambda = \lambda_i$. An algorithm for finding $p - q$ defective eigenvectors and corresponding solutions with respect to the defective eigenvalue $\lambda = \lambda_i$ is presented. Prior to this, we need to introduce two main concepts regarding a generalized eigenvector.

Definition 4.4.1. Let λ be any eigenvalue of the matrix A. A nonzero vector c is said to be a generalized eigenvector or root vector of the matrix A associated with the given eigenvalue $\lambda = \lambda_i$ if there exists a smallest positive integer r such that $(A - \lambda I)^r c = 0$ and $(A - \lambda I)^{r-1} c \neq 0$. Moreover, c is referred to as a rank $\mathbf{r}$ generalized eigenvector, and it is denoted by $c_i(r)$. In particular, if $r = 1$, then the generalized eigenvector $c_i(1) = c_i$ is an eigenvector associated with the eigenvalue $\lambda = \lambda_i$.

Definition 4.4.2. Let c be an ordinary eigenvector corresponding to the eigenvalue $\lambda = \lambda_i$ of the matrix. For each l, $1 \leq l \leq q$, $c_i(l)$ is an ordinary eigenvector corresponding to the eigenvalue λ_i of the matrix A. A length $\boldsymbol{k}$ chain of generalized eigenvectors with respect to the ordinary eigenvector $c_i(l)$ is a set $\{c_i(l), c_i(l,k-1), \ldots, c_i(l,2), c_i(l,1)\}$ of k generalized eigenvectors such that

$$\begin{aligned} (A-\lambda_i I)\, c_i(l,1) &= c_i(l,2), \\ (A-\lambda_i I)\, c_i(l,2) &= c_i(l,3), \\ &\cdots\cdots \\ (A-\lambda_i I)\, c_i(l,k-1) &= c_i(l). \end{aligned} \tag{4.61}$$

DME 5. With reference to the defective eigenvalue $\lambda = \lambda_i$ of the multiplicity p and $p-q$ defective solutions corresponding to λ_i in DME 4, we note that the length of the longest chain is at most $r = 1+(p-q)$. An algorithm for finding the defective solutions is given below.

Algorithm for finding defective eigenvectors and solutions. *For each l, $1 \le l \le q$, let $c_i(l)$ be an ordinary eigenvector corresponding to the eigenvalue $\lambda = \lambda_i$ of the matrix A. Let $\{c_i(l), c_i(l,k-1), \ldots, c_i(l,2), c_i(l,1)\}$ be a length k chain of generalized eigenvectors with respect to the ordinary eigenvector $c_i(l)$. Under this stipulation, we seek a solution to (4.35) of the form*

$$x(t) = \exp[\lambda t]\left(c_i(l)\frac{t^{k-1}}{(k-1)!} + \cdots + tc_i(l,2) + c_i(l,1)\right), \tag{4.62}$$

where $\lambda = \lambda_i$ and $c_i(l,1)$ is a nonzero solution to

$$(A-\lambda I)^k c = 0, \quad \text{for } 1 \le k \le 1+(p-q), \tag{4.63}$$

where $c_i(l,k-1), \ldots, c_i(l,2), c_i(l,1)$ are unknown nonzero vectors to be determined.

Solution procedure. We differentiate both sides of (4.62) with respect to t, and we have

$$\begin{aligned} dx(t) &= \lambda \exp[\lambda t]\left(c_i(l)\frac{t^{k-1}}{(k-1)!} + \cdots + tc_i(l,2) + c_i(l,1)\right) dt \\ &\quad + \exp[\lambda t]\left(\frac{t^{k-2}}{(k-2)!}c_i(l) + \cdots + tc_i(l,3) + c_i(l,2)\right) dt \quad \text{(product rule)} \\ &= \exp[\lambda t]\left(\frac{t^{k-1}}{(k-1)!}\lambda c_i(l) + (\lambda c_i(l,k-1) + c_i(l))\frac{t^{k-2}}{(k-2)!}\right. \\ &\quad \left. + \cdots + t(\lambda c_i(l,2) + c_i(l,3)) + (\lambda c_i(l,1) + c_i(l,2))\right) dt \quad \text{(regrouping)}. \end{aligned}$$

Under the assumption that the expression (4.62) is a solution to (4.35), we have

$$dx(t) = Ax(t)dt = A\exp[\lambda t]\left(\frac{t^{k-1}}{(k-1)!}c_i(l) + \cdots + tc_i(l,2) + c_i(l,1)\right)dt$$

$$= \exp[\lambda t]\left(\frac{t^{k-1}}{(k-1)!}Ac_i(l) + \cdots + tAc_i(l,2) + Ac_i(l,1)\right)dt$$

(Theorem 1.3.1).

Now, by comparing the coefficients of the terms $t^{k-1}, t^{k-1}, \ldots, t, t^0$ on both sides, we have the algorithm

$$\begin{aligned} Ac_i(l,1) &= \lambda c_i(l,1) + c_i(l,2),\\ Ac_i(l,2) &= \lambda c_i(l,2) + c_i(l,3),\\ &\cdots\cdots\cdots\\ Ac_i(l,k-1) &= \lambda c_i(l,k-1) + c_i(l),\\ Ac_i(l) &= \lambda c_i(l), \end{aligned} \tag{4.64}$$

which implies that

$$\begin{aligned} (A-\lambda I)c_i(l,1) &= c_i(l,2),\\ (A-\lambda I)c_i(l,2) &= c_i(l,3),\\ &\cdots\cdots\cdots\\ (A-\lambda I)c_i(l,k-1) &= c_i(l),\\ (A-\lambda I)c_i(l) &= 0. \end{aligned} \tag{4.65}$$

This provides a computational algorithm for finding the unknown coefficient vectors $c_i(l,k-1), \ldots, c_i(l,2), c_i(l,1)$ of $x(t)$ defined in (4.62). In fact, this is the method of undetermined coefficients. This algorithm is computationally attractive. One needs to successively multiply by $(A-\lambda I)$ to find the coefficients of lower exponents of t in terms of the polynomial factor in (4.62). In particular, for $q = 1 = l$, $r = p = k$ (c_i is an ordinary eigenvector associated with $\lambda = \lambda_i$). In this case, $c_i(1,p-1), \ldots, c_i(1,2), c_i(1,1)$ are generalized eigenvectors associated with λ_i. The set $\{c_i, c_i(1,p-1), \ldots, c_i(1,2), c_i(1,1)\}$ is referred to as a length p chain of generalized eigenvectors based on the eigenvector c_i. The corresponding p linearly independent solutions are

$$\begin{aligned} x_i^1 &= c_i\exp[\lambda_i t],\\ x_i^2 &= [tc_i + c_i(1,p-1)]\exp[\lambda_i t],\\ x_i^3 &= \left[\frac{1}{2!}t^2c_i(1,3) + tc_i(1,2) + c_i(1,1)\right]\exp[\lambda_i t],\\ &\cdots\cdots\cdots\\ x_i^p &= \left[\frac{1}{(p-1)!}t^{p-1}c_i(1,1) + \cdots + tc_i(1,2) + c_i(1,1)\right]\exp[\lambda_i t]. \end{aligned} \tag{4.66}$$

DME 6. It is known that "a collection of all vectors of any chain of generalized eigenvectors is a set of linearly independent vectors." (a) We repeat the procedure outlined in the step DME 5 with respect to each different ordinary eigenvector corresponding to the same eigenvalue, $\lambda = \lambda_i$ (if necessary). (b) We further repeat the process described in DME 4 and DME 5 with respect to each defective eigenvalue of the matrix A. It is known that "the union of any two chains of generalized eigenvectors with respect to linearly independent eigenvectors is a set of linearly independent vectors" (whether the two base eigenvectors are associated with the same eigenvalue or with different eigenvalues). The collection of all generalized eigenvectors of the matrix forms a set of linearly independent vectors. Using this complete set of generalized eigenvectors, a generalized fundamental matrix solution to the system (4.35) can be formed. This completes the procedure concerning the repeated root case of (4.38).

Illustration 4.4.4. Let us consider $dx = Ax\,dt$, $x(t_0) = x_0$, where

$$A = \begin{bmatrix} a_{11} & a_{12} \\ a_{21} & a_{22} \end{bmatrix},$$

and a_{ij}'s are any given real numbers satisfying the condition $d(A) = (\mathrm{tr}(A))^2 - 4\det(A) = 0$, where $\mathrm{tr}(A)$ and $\det(A)$ are as defined in Illustration 4.4.1. Find: (a) a general solution, and (b) the solution to the given IVP.

Solution procedure. Again, we note that $n = 2$. We mimic the arguments used in Illustration 4.4.1 and the procedure in Subsection 4.3.4 to find the solution to the IVP. As a result of this, we have

$$0 = \det(A - \lambda I) = \lambda^2 - \mathrm{tr}(A)\lambda + \det(A) = (\lambda - \lambda_1)(\lambda - \lambda_2),$$

where λ_1 and λ_2 are as described in Illustration 4.4.1, and are roots of the quadratic polynomial with real coefficients. From the assumption $d(A) = (\mathrm{tr}(A))^2 - 4\det(A) = 0$, we conclude that λ_1 and λ_2 are real and equal roots of $0 = \det(A - \lambda I)$. In fact, its repeated root is $\lambda_1 = \frac{\mathrm{tr}(A)}{2}$. Thus, the quadratic polynomial equation associated with the matrix A has a root of multiplicity 2. We need to find an eigenvector corresponding to this eigenvalue λ_1. Again, we imitate the argument used in Illustration 4.4.1 to obtain the eigenvector and a solution corresponding to the eigenvalue $\lambda_1 = \frac{\mathrm{tr}(A)}{2}$. Let c_1 be an eigenvector corresponding to λ_1, and its corresponding solution is $x_1(t) = e^{\lambda_1 t}c_1$. The details are exactly similar to those given in Illustration 4.4.1. To find the complete set of eigenvectors, we need to follow the procedure described in Multiple Eigenvalues 4.4.3.

First, we recall the procedure for finding the complete set of solutions concerning the multiplicity of eigenvalues. In this illustration, $\lambda_1 = \frac{\mathrm{tr}(A)}{2}$ is of multiplicity 2, $n = 2$, $p = 2$, $m = 1$. Therefore, $n - m = 1$ remaining linearly independent eigenvectors need to be determined. $p - q = 2 - 1 = 1$ is the deficiency. To find the remaining linearly independent solution, we need to imitate the algorithm for finding the defective eigenvector and corresponding solution. Using (4.62) (DME 5), we seek a solution as follows:

$$x(t) = e^{\lambda_1 t}\left(c_1 t + c_1(1,1)\right),$$

where $c_1(1,1)$ is the generalized eigenvector (Definition 4.4.1) corresponding to the ordinary eigenvector (definition in DME 4) c_1. By following the above-presented algorithm (DME 5), we obtain

$$\begin{aligned} dx(t) &= \left[\lambda_1 e^{\lambda_1 t}\left(c_1 t + c_1(1,1)\right) + e^{\lambda_1 t} c_1\right] dt = Ax(t)\,dt \\ &= A\left(e^{\lambda_1 t}\left(c_1 t + c_1(1,1)\right)\right) dt, \end{aligned}$$

which implies (by the method of undetermined coefficients) that

$$\begin{aligned} (A - \lambda_1 I)\, c_1(1,1) &= c_1, \\ (A - \lambda_1 I)\, c_1 &= 0. \end{aligned}$$

By knowing c_1, we can solve the first system of linear nonhomogeneous algebraic equations. Hence, a generalized eigenvector $c_1(1,1)$ is chosen as one of the nonzero solutions to this algebraic system. Thus, the corresponding solution is

$$x_2(t) = e^{\lambda_1 t}\left(c_1 t + c_1(1,1)\right).$$

In this case, the general solution to the given system is $x(t) = \Phi(t)a$, where

$$\Phi(t) = \begin{bmatrix} x_1(t) & x_2(t) \end{bmatrix} = \begin{bmatrix} e^{\lambda_1 t} c_1 & e^{\lambda_1 t}\left(c_1 t + c_1(1,1)\right) \end{bmatrix}.$$

Finally, the solution to the IVP can be computed by following the arguments used in Illustrations 4.4.1 and 4.4.2.

Example 4.4.9. We are given $dx = Ax$, $x(t_0) = x_0$, where

$$A = \begin{bmatrix} 0 & 1 \\ 0 & 0 \end{bmatrix}.$$

Solve the IVP.

Solution procedure. We follow the argument and the procedure described in Illustration 4.4.4. From Example 4.3.3, we already have the eigenvalues of the given matrix and a corresponding solution. In this example, $\lambda_1 = 0$ and

$$x_1(t) = e^{0t}\begin{bmatrix} 1 \\ 0 \end{bmatrix} = \begin{bmatrix} 1 \\ 0 \end{bmatrix}.$$

Again, recalling Example 4.3.3 and Illustration 4.4.4, $n = 2$, $p = 2$, $m = 1$. Therefore, $n - m = 1$ remaining linearly eigenvectors need to be determined. Here, 1 is the deficiency. To find the remaining linearly independent solution, we need to imitate the algorithm described in Multiple Eigenvalues 4.4.3 (DME 5) for finding the defective eigenvector and corresponding solution. $\lambda_1 = 0$ with its ordinary eigenvector $c_1 = \begin{bmatrix} 1 & 0 \end{bmatrix}^T$. We need to find $c_1(1,1)$, a solution to the following system of equations:

$$(A - \lambda_1 I)\, c_1(1,1) = c_1,$$
$$(A - \lambda_1 I)\, c_1 = 0.$$

By knowing c_1, we can solve the first system of linear nonhomogeneous algebraic equations, and we obtain

$$(A - \lambda_1 I)\, c_1(1,1)) = \begin{bmatrix} 0 & 1 \\ 0 & 0 \end{bmatrix} c_1(1,1) = \begin{bmatrix} 1 \\ 0 \end{bmatrix},$$

with the notation

$$c_1(1,1) = \begin{bmatrix} c_{11}(1,1) \\ c_{12}(1,1) \end{bmatrix},$$

which implies that

$$0c_{11}(1,1) + c_{12}(1,1) = 1, \quad 0c_{11}(1,1) + 0c_{12}(1,1) = 0.$$

By inspection, from the first equation, we have $c_{12}(1,1) = 1$ and $c_{11}(1,1)$ is an arbitrary real number. We choose $c_{11}(1,1) = 0$. Hence, the generalized eigenvector

$$c_1(1,1) = \begin{bmatrix} 0 \\ 1 \end{bmatrix}.$$

Now, by imitating the argument used in Illustration 4.4.4, the second solution is given by

$$x_2(t) = e^{0t}\left(c_1 t + c_1(1,1)\right) = t \begin{bmatrix} 1 \\ 0 \end{bmatrix} + \begin{bmatrix} 0 \\ 1 \end{bmatrix} = \begin{bmatrix} t \\ 0 \end{bmatrix} + \begin{bmatrix} 0 \\ 1 \end{bmatrix} = \begin{bmatrix} t \\ 1 \end{bmatrix}.$$

By repeating the earlier procedure, for instance Example 4.3.7, the general solution and the solution to the IVP are

$$x(t) = \Phi(t)a = \begin{bmatrix} 1 & t \\ 0 & 1 \end{bmatrix} a,$$

$$x(t, t_0, x_0) = \begin{bmatrix} 1 & t - t_0 \\ 0 & 1 \end{bmatrix} \begin{bmatrix} x_{10} \\ x_{20} \end{bmatrix} = \begin{bmatrix} x_{10} + (t - t_0)x_{20} \\ x_{20} \end{bmatrix}.$$

This completes the solution process of the example.

Example 4.4.10. Let us consider

$$A = \begin{bmatrix} -2 & -9 & 0 \\ 1 & 4 & 0 \\ 1 & 3 & 1 \end{bmatrix}.$$

Find (a) the multiplicity of each eigenvalue of the given matrix A, and (b) eigenvectors corresponding to eigenvalues; (c) if the eigenvalue of the multiplicity in (a) is not complete, then find (i) the ordinary eigenvectors, (ii) the deficiencies corresponding to the eigenvalues, (iii) the rank r of the generalized eigenvector corresponding to each eigenvalue, (iv) a length k chain of generalized eigenvectors with respect to each ordinary eigenvector, and (v) a general fundamental solution to $dx = Ax\,dt$.

Solution procedure. First, we compute

$$\det(A - \lambda I) = \begin{vmatrix} -2-\lambda & -9 & 0 \\ 1 & 4-\lambda & 0 \\ 1 & 3 & 1-\lambda \end{vmatrix}$$

$$= (1-\lambda)\left[(-2-\lambda)(4-\lambda)+9\right] = (1-\lambda)^3.$$

The solution to the corresponding equation (4.38) is $\lambda_1 = \lambda_2 = \lambda_3 = 1$. These are three eigenvalues of the given matrix A. They are identical. Therefore, $\lambda_1 = 1$ is called an eigenvalue of multiplicity 3.

To find the eigenvectors corresponding to these three identical eigenvalues $\lambda_1 = \lambda_2 = \lambda_3 = 1$, we utilize (4.39) with respect to the example, and solve the corresponding linear system of homogeneous algebraic equations as follows: for $\lambda_1 = 1$,

$$(A - \lambda I)c = \begin{bmatrix} -2-\lambda & -9 & 0 \\ 1 & 4-\lambda & 0 \\ 1 & 3 & 1-\lambda \end{bmatrix} \begin{bmatrix} c_1 \\ c_2 \\ c_3 \end{bmatrix} = \begin{bmatrix} -3 & -9 & 0 \\ 1 & 3 & 0 \\ 1 & 3 & 0 \end{bmatrix} \begin{bmatrix} c_1 \\ c_2 \\ c_3 \end{bmatrix} = \begin{bmatrix} 0 \\ 0 \\ 0 \end{bmatrix}.$$

This system represents the linear system of homogeneous algebraic equations

$$-3c_1 - 9c_2 = 0,$$

$$c_1 + 3c_2 = 0,$$

$$c_1 + 3c_2 = 0.$$

This system of linear homogeneous algebraic equations has infinitely many solutions. They are described by two arbitrary parameters:

$$c = \begin{bmatrix} -3c_2 & c_2 & c_3 \end{bmatrix}^T = c_2 \begin{bmatrix} -3 & 1 & 0 \end{bmatrix}^T + c_3 \begin{bmatrix} 0 & 0 & 1 \end{bmatrix}^T,$$

where c_2 and c_3 arbitrary constants or parameters. In short, the solution space (Observations 1.4.5 and 1.4.6) of this system of algebraic equations is given by

$$S = \left\{ c \in R^3 : c = c_2 \begin{bmatrix} -3 & 1 & 0 \end{bmatrix}^T + c_3 \begin{bmatrix} 0 & 0 & 1 \end{bmatrix}^T \right\}.$$

We note that the solution space S of this system of algebraic equations is generated by two linearly independent vectors, $c_1(1) = \begin{bmatrix} -3 & 1 & 0 \end{bmatrix}^T$ and $c_1(2) = \begin{bmatrix} 0 & 0 & 1 \end{bmatrix}^T$ (Definition 1.4.5). This solves part (b), and establishes the condition of (c). Moreover, the eigenvalue $\lambda_1 = 1$ is of multiplicity $3 = p = n$. Furthermore, the vectors $c_1(1)$ and $c_1(2)$ are ordinary eigenvectors corresponding to the eigenvalue $\lambda_1 = 1$. Here $q = 2$. Thus, the number $p - q = 1$ is the deficiency in the number of linearly independent eigenvectors corresponding to $\lambda_1 = 1$ of multiplicity $3 = p$. This completes the part (c)(ii). To find the rank r [part (c)(iii)] of the generalized eigenvector corresponding to each eigenvalue $\lambda_1 = 1$, we note that for $c \neq 0$, $(A - 1I)\, c \neq 0$. We consider $(A - 1I)^2 c$, and compute

$$(A-1I)^2 c = \begin{bmatrix} -3 & -9 & 0 \\ 1 & 3 & 0 \\ 1 & 3 & 0 \end{bmatrix} \begin{bmatrix} -3 & -9 & 0 \\ 1 & 3 & 0 \\ 1 & 3 & 0 \end{bmatrix} \begin{bmatrix} c_1 \\ c_2 \\ c_3 \end{bmatrix} = \begin{bmatrix} 0 & 0 & 0 \\ 0 & 0 & 0 \\ 0 & 0 & 0 \end{bmatrix} \begin{bmatrix} c_1 \\ c_2 \\ c_3 \end{bmatrix} = \begin{bmatrix} 0 \\ 0 \\ 0 \end{bmatrix}.$$

From the definition of the rank r of the generalized eigenvector (Definition 4.4.1), we conclude that any nonzero solution to the above system of linear homogeneous algebraic equations is a generalized eigenvector of rank 2 (Definition 4.4.1) corresponding to the eigenvalue $\lambda_1 = 1$. This completes the discussion about the solution process of (c)(iii).

Now, we pick an ordinary eigenvector corresponding to the eigenvalue $\lambda_1 = 1$, and compute the length k [(c)(iv)] chain of generalized eigenvectors with respect to each ordinary eigenvector in (c)(i). In this case. Using the above algorithm in (4.61), $c_1(1) = \begin{bmatrix} -3 & 1 & 0 \end{bmatrix}^T$ and we have

$$(A - 1I)\, c = c_1(l) = \begin{bmatrix} -3 & -9 & 0 \\ 1 & 3 & 0 \\ 1 & 3 & 0 \end{bmatrix} \begin{bmatrix} c_1(l,1) \\ c_2(l,1) \\ c_3(l,1) \end{bmatrix} = \begin{bmatrix} -3 \\ 1 \\ 0 \end{bmatrix}.$$

This system of linear nonhomogeneous algebraic is not consistent, i.e. $c_1 + 3c_2 = 0$ and $c_1 + 3c_2 = 1$. Therefore, it has no solution. This shows that for the ordinary eigenvector $\begin{bmatrix} -3 & 1 & 0 \end{bmatrix}^T$, there is no generalized eigenvector of A corresponding to the eigenvalue $\lambda_1 = 1$. We note that the above ordinary eigenvectors were determined by the choice of $c_2 = 1$, $c_3 = 0$ and $c_2 = 0$, $c_3 = 1$, respectively. Instead of this, we pick the ordinary eigenvectors by the choice of $c_2 = 1$, $c_3 = 1$ and $c_2 = 0$, $c_3 = 1$, respectively. With this choice, we have $c_1(1) = \begin{bmatrix} -3 & 1 & 1 \end{bmatrix}^T$ and $c_1(2) = \begin{bmatrix} 0 & 0 & 1 \end{bmatrix}^T$. Again, let us try to solve the above linear system of nonhomogeneous algebraic equations with $c_1(1) = \begin{bmatrix} -3 & 1 & 1 \end{bmatrix}^T$. In this case,

$$(A - 1I)\, c = c_1(1) = \begin{bmatrix} -3 & -9 & 0 \\ 1 & 3 & 0 \\ 1 & 3 & 0 \end{bmatrix} \begin{bmatrix} c_1(1,1) \\ c_2(1,1) \\ c_3(1,1) \end{bmatrix} = \begin{bmatrix} -3 \\ 1 \\ 1 \end{bmatrix},$$

i.e.

$$-3c_1(1,1) - 9c_2(1,1) = -3,$$
$$c_1(1,1) + 3c_2(1,1) = 1,$$
$$c_1(1,1) + 3c_2(1,1) = 1.$$

This system has infinitely many nonzero solutions. We choose a solution that is linearly independent of the ordinary eigenvectors $c_1(1) = [-3\ \ 1\ \ 1]^T$ and $c_1(2) = [0\ \ 0\ \ 1]^T$. One choice is $c(1,1) = [1\ \ 0\ \ 0]^T$. The length 2 chain of generalized eigenvector of matrix A with respect to the ordinary eigenvector $c_1(1) = [-3\ \ 1\ \ 1]^T$ is a set $\{[-3\ \ 1\ \ 1]^T, [1\ \ 0\ \ 0]^T\}$. This completes the solution process of (c)(iv). As noted, regarding $c_1(2) = [0\ \ 0\ \ 1]^T$, there is no generalized eigenvector of matrix A.

From Illustration 4.4.4, the complete set of solutions is $x_1(t) = e^t[-3\ \ 1\ \ 1]^T$, $x_2(t) = e^t[0\ \ 0\ \ 1]^T$, and $x_3(t) = e^t(t[-3\ \ 1\ \ 1]^T + [1\ \ 0\ \ 0]^T)$. A general fundamental matrix solution is $\Phi(t) = [x_1(t), x_2(t), x_3(t)]$. This completes the solution process of the problem (c)(v).

Observation 4.4.1. From Theorem 1.5.8(e), we note that $\Phi(t) = \exp[At]$ is a matrix solution to (4.35). In this case, for any c in R^n, $x(t) = \Phi(t)c = \exp[At]\,c$ is a solution to (4.35). In fact, $x(t) = \exp[At]\,c$ can be calculated explicitly in terms of the matrix A, its any eigenvalue λ, the multiplicity p of λ, and the generalized eigenvectors (Definition 4.4.1) associated with the eigenvalue λ [in particular, rank $r = 1 + (p - q)$, where $1 \leq r \leq n$]:

$$\begin{aligned}
x(t) &= \exp[At]\,c \\
&= \exp[\lambda It + (A - \lambda I)t]\,c \quad \text{(by algebraic manipulation)} \\
&= \exp[\lambda It]\exp[(A - \lambda I)t]\,c \quad \text{[by Theorem 1.5.8(d)]} \\
&= \exp[\lambda t]\, I \left[I + (A - \lambda I)t + \cdots + (A - \lambda I)^j \frac{t^j}{j!} + \cdots + (A - \lambda I)^r \frac{t^r}{r!} + \cdots\right] c \\
&\quad \text{[from (1.79)]} \\
&= \exp[\lambda t]\left[c + t(A - \lambda I)c + \cdots + \frac{t^j}{j!}(A - \lambda I)^j c + \cdots + \frac{t^r}{r!}(A - \lambda I)^r c\right] \\
&= \exp[\lambda t]\left[c_1 + tc_2 + \cdots + \frac{t^j}{j!}\, c_j + \cdots + \frac{t^{r-1}}{(r-1)!}\, c_r\right]
\end{aligned}$$

in view of Illustration 1.5.2, Theorems 1.3.1 and 1.5.1, and by setting

$$c = c_1,$$
$$(A - \lambda I)c_1 = c_2,$$
$$(A - \lambda I)^2 c = (A - \lambda I)c_2 = c_3,$$
$$\cdots\cdots\cdots\cdots$$

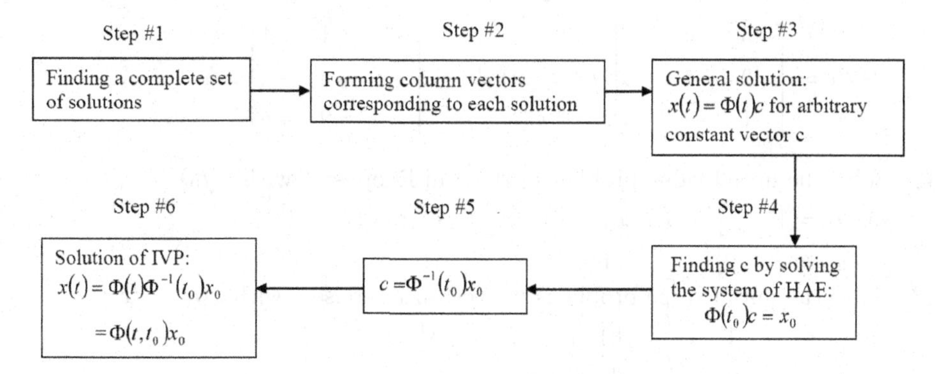

Fig. 4.14 Flowchart for finding fundamental solution.

$$(A-\lambda I)^{j-1}c = (A-\lambda I)c_{j-1} = c_j,$$
$$\cdots\cdots\cdots\cdots\cdots$$
$$(A-\lambda I)^{r-1}c = (A-\lambda I)c_{r-1} = c_r,$$
$$(A-\lambda I)^{r}c = (A-\lambda I)c_{r} = 0.$$

The above calculation is valid for every eigenvalue of A with or without its multiplicity. Hence, it determines all n linearly independent solutions to (4.35). This shows that $\Phi(t) = \exp[At]$ is indeed a fundamental solution to (4.35). We further note that the above calculation provides another algebraic (linear algebra) method for finding a fundamental matrix solution to (4.35). In addition, it provides a basis for seeking a solution form as described in (4.62). Figure 4.14 describes the procedure for finding a fundamental matrix solution and the solution of the IVP.

4.4 Exercises

(1) Find the general solution to each system of differential equations:

(a) $dx = \begin{bmatrix} -1 & 1 \\ 4 & -1 \end{bmatrix} xdt$ (b) $dx = \begin{bmatrix} -1 & 1 \\ 1 & -1 \end{bmatrix} xdt$ (c) $dx = \begin{bmatrix} 1 & -1 \\ -1 & 1 \end{bmatrix} xdt$

(d) $dx = \begin{bmatrix} 1 & -2 \\ -2 & 1 \end{bmatrix} xdt$ (e) $dx = \begin{bmatrix} 1 & -1 \\ 1 & -1 \end{bmatrix} xdt$ (f) $dx = \begin{bmatrix} -1 & -1 \\ 1 & -1 \end{bmatrix} xdt$

(g) $dx = \begin{bmatrix} -1 & 0 \\ 1 & -1 \end{bmatrix} xdt$ (h) $dx = \begin{bmatrix} 1 & -1 \\ 1 & 1 \end{bmatrix} xdt$ (i) $dx = \begin{bmatrix} -1 & -1 \\ 1 & -1 \end{bmatrix} xdt$

(j) $dx = \begin{bmatrix} -3 & 1 & 1 \\ 1 & -4 & 2 \\ 2 & 2 & -5 \end{bmatrix} xdt$ (k) $dx = \begin{bmatrix} -3 & -1 & 1 \\ 2 & -4 & 2 \\ 1 & -2 & -5 \end{bmatrix} xdt$

(l) $dx = \begin{bmatrix} -3 & 0 & 1 \\ 2 & -4 & 2 \\ 1 & 0 & -5 \end{bmatrix} x dt$ (m) $dx = \begin{bmatrix} -3 & 1 & 1 \\ -2 & -4 & -2 \\ 1 & 2 & -5 \end{bmatrix} x dt$

(2) Solve the initial value problems (a)–(i) in Exercise 1 with $x(t_0) = x_0$:
(a) $x_0 = \begin{bmatrix} 1 & 1 \end{bmatrix}^T$ (b) $x_0 = \begin{bmatrix} 2 & 1 \end{bmatrix}^T$ (c) $x_0 = \begin{bmatrix} 1 & 2 \end{bmatrix}^T$
(d) $x_0 = \begin{bmatrix} 2 & 2 \end{bmatrix}^T$ (e) any x_0

(3) Solve the initial value problems (j)–(m) in Exercise 1 with $x(t_0) = x_0$:
(a) $x_0 = \begin{bmatrix} 1 & 1 & 1 \end{bmatrix}^T$ (b) $x_0 = \begin{bmatrix} 2 & 2 & 1 \end{bmatrix}^T$
(c) $x_0 = \begin{bmatrix} 1 & 2 & 1 \end{bmatrix}^T$ (d) any x_0

(4) Use the solution procedure of Example 4.4.1 to find the expression for F, M and C:

(a) $F_0 = M_0 = 500$, $C_0 = 100$, $\beta = 0.5$, $\delta = 0.4$, and $\rho = 0.2$
(b) $F_0 = 500$, $M_0 = 100$, $C_0 = 100$, $\beta = 0.5$, $\delta = 0.4$, and $\rho = 0.2$
(c) $F_0 = 100$, $M_0 = 500$, $C_0 = 100$, $\beta = 0.5$, $\delta = 0.4$, and $\rho = 0.2$.

(5) ***Kendall's mathematical marriage model.*** Find the general solution to the linear homogeneous systems of differential equations of the marriage dynamic model

$$dF = \left[-\left(\delta_1 + \frac{1}{2}\rho\right) F - \frac{1}{2}\rho M + (\beta_1 + \delta_2)\, C\right] dt$$

$$dM = \left[-\frac{1}{2}\rho F - \left(\delta_2 + \frac{1}{2}\rho\right) M + (\beta_2 + \delta_1)\, C\right] dt$$

$$dC = \left[\frac{1}{2}\rho F + \frac{1}{2}\rho M - (\delta_1 + \delta_2)\, C\right] dt$$

in Exercise 5, Section 4.2.

(6) Using the Riemann–Stieltjes-type system of ordinary differential equations described in Exercise 8, Section 4.3 — $dx = Ax\, dF$, $x(t_0) = x_0$, where A is a 2×2 constant matrix whose real entries a_{ij} are any given real numbers; $\operatorname{tr}(A)$ and $\det(A)$ are as defined in Theorem 4.3.2 and Definition 1.4.2, respectively — find: (a) a general solution, and (b) the solution to the given initial value problem, whenever: (i) $d(A) = (\operatorname{tr}(A))^2 - 4\det(A) > 0$, (ii) $d(A) = (\operatorname{tr}(A))^2 - 4\det(A) < 0$, (iii) $d(A) = (\operatorname{tr}(A))^2 - 4\det(A) = 0$.

4.5 General Linear Homogeneous Systems

In this section, we utilize the eigenvalue-type method developed in Section 4.3 and the techniques of finding the fundamental matrix solution in Section 4.4 to compute a solution to a general first-order linear homogeneous system of differential equations with constant coefficients. This computational technique is feasible for a certain class of first-order systems of differential equations. Otherwise, it is not computationally feasible. However, it is possible to find solution information, conceptually. These types of systems of differential equations provide a mathematical model for dynamic processes with time varying deterministic and stochastic structural perturbations. In particular, the material presented in this section provides a motivation for finding solutions to linear stochastic differential equations of the Itô–Doob type [54].

4.5.1 *General problem*

Let us consider the following general first-order linear homogeneous system of differential equations with constant coefficients:

$$dx = Ax\,dt + Bx\,dt, \tag{4.67}$$

where A is as defined in (4.35), and $B = (b_{ij})_{n\times n}$ is an $n \times n$ constant matrix whose entries b_{ij} are in the set of real numbers R. The systems (4.35) and (4.67) are referred to as unperturbed and perturbed systems, respectively. Moreover, the coefficient matrix B is called a parametric perturbation (deterministic/stochastic) to the coefficient matrix A. In short, the rate coefficient matrix B is deterministic (uncertainty) error adjustments to the coefficient rate matrix A. This indeed is essential in studying the robustness (sensitivity analysis) of dynamic processes.

The goal is to present a procedure for finding a solution process of (4.67) (if possible). In addition, we are interested in solving an IVP associated with this problem. We note that the concept of the solution process of (4.67) can be defined similar to Definition 4.3.1.

4.5.2 *Procedure for finding a general solution*

We use the eigenvalue-type method developed in Sections 2.4, 4.3, and 4.4 to compute a solution to a general system of first-order linear homogeneous differential equations with constant coefficients.

Step 1 (decomposition process). We decompose (4.67) into its unperturbed and perturbed parts:

$$dx = Ax\,dt \quad \text{(unperturbed system)}, \tag{4.68}$$

$$dx = Bx\,dt \quad \text{(perturbed system)}, \tag{4.69}$$

respectively. By following the notations and definitions of Section 4.3, our goal of finding a general solution to (4.67) is decomposed into the following subgoals:

(a) To find fundamental matrix solution processes $\Phi_{\mathrm{U}}(t)$ and $\Phi_{\mathrm{P}}(t)$ of (4.68) and (4.69), respectively;

(b) To create a candidate for the fundamental matrix solution process of (4.67),

$$\Phi(t) = \Phi_{\mathrm{U}}(t)\Phi_{\mathrm{P}}(t); \tag{4.70}$$

(c) To test the validity of the fundamental matrix solution.

Here is a procedure to satisfy the stated goals in (a)–(c).

Step 2 (reduction process). We mimic the eigenvalue type method developed in Sections 4.3 and 4.4 to compute the fundamental matrix solution processes (Observations 4.3.3–4.3.5) of (4.68) and (4.69), respectively. Following the notations and definitions of Section 4.3, the fundamental matrix solution processes of (4.68) and (4.69) are denoted by $\Phi_U(t)$ and $\Phi_P(t)$, respectively. This fulfills the subgoal (a) in step 1.

Step 3. From Observation 4.3.4, we recall that $\Phi_U(t)$ and $\Phi_P(t)$ satisfy (4.68) and (4.69), i.e.

$$d\Phi_U(t) = A\Phi_U(t)dt, \tag{4.71}$$

$$d\Phi_P(t) = B\Phi_P(t)dt, \tag{4.72}$$

respectively. Now, we compute

$$\begin{aligned} d\Phi(t) &= d\Phi_U(t)\Phi_P(t) + \Phi_U(t)d\Phi_P(t) \quad \text{(Observation 1.5.2)} \\ &= A\Phi_U(t)dt\Phi_P(t) + \Phi_U(t)B\Phi_P(t)\,dt \quad \text{(4.71) and (4.72)} \\ &= A\Phi_U(t)\Phi_s(t)\,dt + \Phi_U(t)B\Phi_P(t)\,dt \quad \text{(simplifications).} \end{aligned} \tag{4.73}$$

If we assume that $\Phi_U(t)$ is the normalized fundamental matrix solution to (4.68), and the matrices A and B satisfy the commutative law for the multiplication, i.e.

$$AB = BA, \tag{4.74}$$

then from (4.70), (4.73) reduces to

$$\begin{aligned} d\Phi(t) &= A\Phi_U(t)\Phi_P(t)\,dt + B\Phi_U(t)\Phi_P(t)\,dt \quad \text{[Theorem 1.5.8(d)]} \\ &= A\Phi(t)\,dt + B\Phi(t)\,dt \quad \text{[from (4.70)].} \end{aligned} \tag{4.75}$$

This shows that $\Phi(t)$ defined in (4.70) is indeed a matrix solution to (4.67). Moreover, from Theorems 1.4.2 and 1.4.4, $\Phi(t)$ is a nonsingular matrix because $\Phi_U(t)$ and $\Phi_P(t)$ are nonsingular matrices. Hence, $\Phi(t)$ is the fundamental matrix solution to the perturbed system (4.67). This establishes the validity of both of the subgoals (b) and (c), whenever (4.74) is satisfied and $\Phi_U(t)$ is a normalized fundamental matrix solution to (4.68).

Step 4. From Observation 4.3.3, in particular (4.44), for arbitrary n-dimensional column vector a, we conclude that $x(t) = \Phi(t)a$ is the general solution to (4.67). This completes the procedure for finding a general solution to (4.67), provided that the matrices A and B satisfy (4.74) (the commutative property of the product of two matrices). The procedure for finding the general solution is summarized in the Figure 4.15.

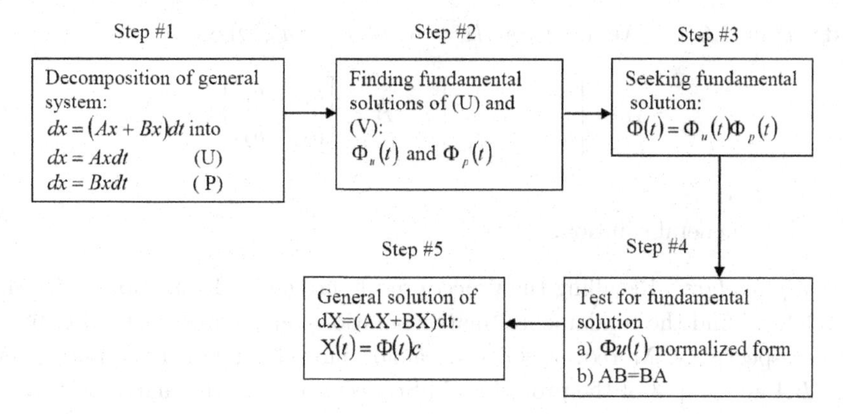

Fig. 4.15 Flowchart for finding general solution.

Observation 4.5.1

(i) We note that in step 3, it was assumed that the coefficient matrices A and B satisfy the commutative law for matrix multiplication and $\Phi_{\mathrm{U}}(t)$ is the normalized fundamental matrix solution to (4.68). This will be justified in this section. Moreover, we address the following question: What can we say about $\Phi(t)$ if

$$AB \neq BA? \tag{4.76}$$

(ii) In the above procedure, if $\Phi_{\mathrm{P}}(t)$ is the normalized fundamental matrix of (4.69) and $AB = BA$, then one can also replace (4.70) by

$$\Phi(t) = \Phi_{\mathrm{P}}(t)\Phi_{\mathrm{U}}(t), \tag{4.77}$$

and show that $\Phi(t)$ is the fundamental matrix solution to (4.67).

(iii) From Observation 4.4.1 and Theorem 1.5.8(d), we note that $\Phi_{\mathrm{U}}(t) = \exp[At]$ and $\Phi_{\mathrm{P}}(t) = \exp[Bt]$ are normalized fundamental matrix solutions to (4.68) and (4.69), respectively. In fact, $\Phi_{\mathrm{U}}(0) = \exp[A0] = I = \exp[B0] = \Phi_{\mathrm{P}}(0)$. Moreover, under the assumption of (4.74), the general solution and the fundamental matrix solutions are represented by

$$\begin{aligned} x(t) = \Phi(t)\,a &= \exp[At]\exp[Bt]\,a \text{ is the general solution to (4.67)} \\ &= \exp[(A+B)\,t]\,a \quad \text{[by Theorem 1.5.11(d)]}, \end{aligned} \tag{4.78}$$

$$\Phi(t) = \exp[(A+B)\,t], \tag{4.79}$$

respectively. These explicit or closed-form solutions to (4.67) are in terms of its coefficient matrices A and B. In general, this is not feasible.

Illustration 4.5.1. We are given $dx = Ax\,dt + Bx\,dt$, where

$$A = \begin{bmatrix} a_{11} & a_{12} \\ a_{21} & a_{22} \end{bmatrix}, \quad B = \begin{bmatrix} b_{11} & b_{12} \\ b_{21} & b_{22} \end{bmatrix},$$

and $AB = BA$.

Find the general solution.

Solution procedure. Recalling the procedures described in Illustrations 4.4.1, 4.4.3, and 4.4.4, we find the fundamental matrix solutions $\Phi_U(t)$ and $\Phi_P(t)$ to (4.68) and (4.69), respectively. Moreover, $\Phi_U(t)$ is a normalized fundamental matrix. From $AB = BA$ and step 3 of the procedure in Subsection 4.5.2, the matrix $\Phi(t)$ defined by $\Phi(t) = \Phi_U(t)\Phi_P$ is the fundamental matrix solution to the given perturbed system of differential equations in the illustration.

Observation 4.5.2. We observe that if the coefficient matrices A and B satisfy the condition (4.74), i.e. $AB = BA$. $AB = BA$ if and only if either

$$\text{(a)} \quad \frac{a_{12}}{b_{12}} = \frac{a_{21}}{b_{21}} = \frac{a_{11} - a_{22}}{b_{11} - b_{22}} = r$$

or

$$\text{(b)} \quad \frac{b_{12}}{a_{12}} = \frac{b_{21}}{a_{22}} = \frac{b_{11} - b_{22}}{a_{11} - a_{22}} = r,$$

for some real number r. Furthermore, we note that (a) $a_{12} = rb_{12}$, $a_{21} = rb_{21}$, and $a_{11} - a_{22} = r\,(b_{11} - b_{22})$ or (b) $b_{12} = ra_{12}$, $b_{21} = ra_{21}$, and $b_{11} - b_{22} = r\,(a_{11} - a_{22})$. Hence,

$$\begin{aligned} d(A) &= (\operatorname{tr}(A))^2 - 4\det(A) = (a_{11} - a_{22})^2 + 4a_{12}a_{21} \\ &= r^2 (b_{11} - b_{22})^2 + r^2 4b_{12}b_{21} = r^2 \left[(b_{11} - b_{22})^2 + 4b_{12}b_{21}\right] \\ &= r^2 \left[(\operatorname{tr}(B))^2 - 4\det(B)\right] = r^2 d(B) \end{aligned}$$

[or $d(A) = r^2 d(B)$]. This shows that $d(B)$ and $d(A)$ have the same sign, i.e. if $d(A) > 0$ ($d(A) < 0$), then $d(B) > 0$ ($d(B) < 0$). Moreover,

$$\begin{aligned} \lambda_{1A} - a_{11} &= \frac{\operatorname{tr}(A) + \sqrt{d(A)} - 2a_{11}}{2} = \frac{a_{22} - a_{11} + \sqrt{d(A)}}{2} \\ &= \frac{r\,(b_{22} - b_{11}) + |r|\sqrt{d(B)}}{2} = r\frac{(b_{22} - b_{11}) + \sqrt{d(B)}}{2} = r\,(\lambda_{1B} - b_{11}), \end{aligned}$$

for $r > 0$. A similar comment can be made for $r < 0$. The detailed justification about the validity of the statement is left to the reader. This observation provides a method of constructing numeral examples. Further extension of the method for a dimension $n \geq 2$ is left as a project for serious students.

Example 4.5.1. Let us consider $dx = Ax\,dt + Bx\,dt$, where

$$A = \begin{bmatrix} 3 & 1 \\ 1 & 4 \end{bmatrix}, \quad B = \begin{bmatrix} 2 & 1 \\ 1 & 3 \end{bmatrix}.$$

Find: (a) the fundamental, and (b) the general solutions to the given system of differential equations.

Solution procedure. The goal of the example is to find the general solution to the given system of differential equations. By using step 1, the decomposition of the given system of differential equations is $dx = Ax\,dt$ and $dx = Bx\,dt$.

Now, by following the ideas in step 2 of the procedure in Subsection 4.5.2 (Sections 4.3 and 4.4), the eigenvalues and corresponding eigenvectors of the matrices A and B are

$$\lambda_{1A} = \frac{7+\sqrt{5}}{2}, \quad c_{1A} = \begin{bmatrix} 1 & \frac{1+\sqrt{5}}{2} \end{bmatrix}^T,$$

$$\lambda_{2A} = \frac{7-\sqrt{5}}{2}, \quad c_{2A} = \begin{bmatrix} -\frac{1+\sqrt{5}}{2} & 1 \end{bmatrix}^T,$$

$$\lambda_{1B} = \frac{5+\sqrt{5}}{2}, \quad c_{1B} = \begin{bmatrix} 1 & \frac{1+\sqrt{5}}{2} \end{bmatrix}^T, \quad \text{and}$$

$$\lambda_{2B} = \frac{5-\sqrt{5}}{2}, \quad c_{2B} = \begin{bmatrix} -\frac{1+\sqrt{5}}{2} & 1 \end{bmatrix}^T,$$

respectively. Now, we observe that

$$AB = \begin{bmatrix} 3 & 1 \\ 1 & 4 \end{bmatrix} \begin{bmatrix} 2 & 1 \\ 1 & 3 \end{bmatrix} = \begin{bmatrix} 7 & 6 \\ 6 & 13 \end{bmatrix} = \begin{bmatrix} 2 & 1 \\ 1 & 3 \end{bmatrix} \begin{bmatrix} 3 & 1 \\ 1 & 4 \end{bmatrix} = BA.$$

This shows that matrices A and B satisfy the condition (4.74). Moreover, the fundamental matrix solutions of the unperturbed component and perturbed component of the differential equations are

$$\Phi_U(t) = \begin{bmatrix} e^{t\lambda_{1A}} & -\frac{1+\sqrt{5}}{2} e^{t\lambda_{2A}} \\ \frac{1+\sqrt{5}}{2} e^{t\lambda_{1A}} & e^{t\lambda_{2A}} \end{bmatrix},$$

$$\Phi_P(t) = \begin{bmatrix} e^{t\lambda_{1B}} & -\frac{1+\sqrt{5}}{2} e^{t\lambda_{2B}} \\ \frac{1+\sqrt{5}}{2} e^{t\lambda_{1B}} & e^{t\lambda_{2B}} \end{bmatrix},$$

respectively. This completes step 2 of the procedure in Subsection 4.5.2.

Let us compute $B\Phi_U(t)$ and $\Phi_U(t)B$:

$$B\Phi_U(t) = \begin{bmatrix} 2 & 1 \\ 1 & 3 \end{bmatrix} \begin{bmatrix} e^{t\lambda_{1A}} & -\frac{1+\sqrt{5}}{2}e^{t\lambda_{2A}} \\ \frac{1+\sqrt{5}}{2}e^{t\lambda_{1A}} & e^{t\lambda_{2A}} \end{bmatrix}$$

$$= \begin{bmatrix} \frac{5+\sqrt{5}}{2}e^{t\lambda_{1A}} & -\sqrt{5}e^{t\lambda_{2A}} \\ \frac{5+3\sqrt{5}}{2}e^{t\lambda_{1A}} & \frac{5-\sqrt{5}}{2}e^{t\lambda_{2A}} \end{bmatrix}$$

$$\Phi_U(t)B = \begin{bmatrix} e^{t\lambda_{1A}} & -\frac{1+\sqrt{5}}{2}e^{t\lambda_{2A}} \\ \frac{1+\sqrt{5}}{2}e^{t\lambda_{1A}} & e^{t\lambda_{2A}} \end{bmatrix} \begin{bmatrix} 2 & 1 \\ 1 & 3 \end{bmatrix}$$

$$= \begin{bmatrix} 2e^{t\lambda_{1A}} - \frac{1+\sqrt{5}}{2}e^{t\lambda_{2A}} & e^{t\lambda_{1A}} - 3\frac{1+\sqrt{5}}{2}e^{t\lambda_{2A}} \\ 2\frac{1+\sqrt{5}}{2}e^{t\lambda_{1A}} + e^{t\lambda_{2A}} & \frac{1+\sqrt{5}}{2}e^{t\lambda_{1A}} + 3e^{t\lambda_{2A}} \end{bmatrix}.$$

It is obvious that $B\Phi_U(t)$ and $\Phi_U(t)B$ are not equal. This shows that in the absence of the normalized fundamental matrix $\Phi_U(t)$, the commutative property of the matrices A and B is not enough to reduce (4.73) to (4.75). Now, let us compute the normalized fundamental matrix $\Phi_U(t,0)$, which corresponds to the fundamental matrix $\Phi_U(t)$:

$$\Phi_U(t,0) = \frac{1}{\det(\Phi_U(0))} \begin{bmatrix} e^{t\lambda_{1A}} & -\frac{1+\sqrt{5}}{2}e^{t\lambda_{2A}} \\ \frac{1+\sqrt{5}}{2}e^{t\lambda_{1A}} & e^{t\lambda_{2A}} \end{bmatrix} \begin{bmatrix} 1 & \frac{1+\sqrt{5}}{2} \\ -\frac{1+\sqrt{5}}{2} & 1 \end{bmatrix}$$

$$= \frac{4}{4+(1+\sqrt{5})^2} \begin{bmatrix} e^{t\lambda_{1A}} + \frac{1+\sqrt{5}^2}{4}e^{t\lambda_{2A}} & \frac{1+\sqrt{5}}{2}\left(e^{t\lambda_{1A}} - e^{t\lambda_{2A}}\right) \\ \frac{1+\sqrt{5}}{2}\left(e^{t\lambda_{1A}} - e^{t\lambda_{2A}}\right) & \frac{1+\sqrt{5}^2}{4}e^{t\lambda_{1A}} + e^{t\lambda_{2A}} \end{bmatrix}$$

$$= \frac{1}{4+\left(1+\sqrt{5}\right)^2} \begin{bmatrix} 4e^{t\lambda_{1A}} + (1+\sqrt{5})^2 e^{t\lambda_{2A}} & 2(1+\sqrt{5})(e^{t\lambda_{1A}} - e^{t\lambda_{2A}}) \\ 2(1+\sqrt{5})(e^{t\lambda_{1A}} - e^{t\lambda_{2A}}) & (1+\sqrt{5})^2 e^{t\lambda_{1A}} + 4e^{t\lambda_{2A}} \end{bmatrix}.$$

Let us compute $B\Phi_U(t,0)$ and $\Phi_U(t,0)B$:

$$(4+(1+\sqrt{5})^2)B\Phi_U(t,0)$$
$$= \begin{bmatrix} 2 & 1 \\ 1 & 3 \end{bmatrix} \begin{bmatrix} 4e^{t\lambda_{1A}} + (1+\sqrt{5})^2 e^{t\lambda_{2A}} & 2(1+\sqrt{5})(e^{t\lambda_{1A}} - e^{t\lambda_{2A}}) \\ 2(1+\sqrt{5})(e^{t\lambda_{1A}} - e^{t\lambda_{2A}}) & (1+\sqrt{5})^2 e^{t\lambda_{1A}} + 4e^{t\lambda_{2A}} \end{bmatrix}$$

$$= \begin{bmatrix} (10+2\sqrt{5})e^{t\lambda_{1A}} + 2(5+\sqrt{5})e^{t\lambda_{2A}} & (10+6\sqrt{5})e^{t\lambda_{1A}} - 4\sqrt{5}e^{t\lambda_{2A}} \\ (10+6\sqrt{5})e^{t\lambda_{1A}} - 4\sqrt{5}e^{t\lambda_{2A}} & (20+8\sqrt{5})e^{t\lambda_{1A}} + (10-2\sqrt{5})e^{t\lambda_{2A}} \end{bmatrix}$$

$$= (4+(1+\sqrt{5})^2)\Phi_{\mathrm{U}}(t,0)B.$$

If the matrices A and B commute, then this shows that the normalized fundamental matrix $\Phi_{\mathrm{U}}(t,0)$ also commutes with B. From this, $\Phi(t)$ in (4.70), and step 3 of the procedure in Subsection 4.5.2, we conclude that $\Phi(t)$ is a fundamental matrix solution to the given system of differential equations. It can be further written as

$$\Phi(t) = \Phi_{\mathrm{U}}(t,0)\Phi_{\mathrm{P}}(t) = \begin{bmatrix} \phi_{11}(t) & \phi_{12}(t) \\ \phi_{21}(t) & \phi_{22}(t) \end{bmatrix},$$

where

$$\phi_{11}(t) = \frac{4e^{t\lambda_{1A}} + (1+\sqrt{5})^2 e^{t\lambda_{2A}} + (1+\sqrt{5})^2 (e^{t\lambda_{1A}} - e^{t\lambda_{2A}})}{4+(1+\sqrt{5})^2} e^{t\lambda_{1B}}$$

$$= \frac{\left[\left(4+(1+\sqrt{5})^2\right)\right] e^{t\lambda_{1A}}}{4+(1+\sqrt{5})^2} e^{t\lambda_{1B}} \quad \text{(by simplification)}$$

$$= e^{(\lambda_{1A}+\lambda_{1B})t} \quad \text{(by simplification)},$$

$$\phi_{12}(t) = \frac{-(1+\sqrt{5})\left[4e^{t\lambda_{1A}} + (1+\sqrt{5})^2 e^{t\lambda_{2A}}\right] + 4(1+\sqrt{5})(e^{t\lambda_{1A}} - e^{t\lambda_{2A}})}{4+(1+\sqrt{5})^2}$$

$$\times \frac{1}{2} e^{t\lambda_{2B}}$$

$$= \frac{-(1+\sqrt{5})\left[(1+\sqrt{5})^2+4\right]}{4+(1+\sqrt{5})^2} \frac{1}{2} e^{(\lambda_{2A}+\lambda_{2B})t} \quad \text{(by simplification)}$$

$$= -\frac{1}{2}(1+\sqrt{5})e^{(\lambda_{2A}+\lambda_{2B})t} \quad \text{(by simplification)},$$

$$\phi_{21}(t) = \frac{(1+\sqrt{5})\left[4(e^{t\lambda_{1A}} - e^{t\lambda_{2A}}) + (1+\sqrt{5})^2 e^{t\lambda_{1A}} + 4e^{t\lambda_{2A}}\right]}{4+(1+\sqrt{5})^2} \frac{1}{2} e^{t\lambda_{1B}}$$

$$= \frac{(1+\sqrt{5})\left[4+(1+\sqrt{5})^2\right]}{4+(1+\sqrt{5})^2} \frac{1}{2} e^{(\lambda_{1A}+\lambda_{1B})t} \quad \text{(by simplification)}$$

$$= \frac{1}{2}(1+\sqrt{5})e^{(\lambda_{1A}+\lambda_{1B})t} \quad \text{(by simplification)},$$

$$\phi_{22}(t) = \frac{-(1+\sqrt{5})^2\left(e^{t\lambda_{1A}} - e^{t\lambda_{2A}}\right) + \left(1+\sqrt{5}\right)^2 e^{t\lambda_{1A}} + 4e^{t\lambda_{2A}}}{4 + \left(1+\sqrt{5}\right)^2} e^{t\lambda_{2B}}$$

$$= \frac{4 + \left(1+\sqrt{5}\right)^2}{4 + \left(1+\sqrt{5}\right)^2} e^{(\lambda_{2A}+\lambda_{2B})t} \quad \text{(by simplification)}$$

$$= e^{(\lambda_{2A}+\lambda_{2B})t} \quad \text{(by simplification)}.$$

Now, by the application of step 4 of the procedure in Subsection 4.5.2, we have

$$x(t) = \Phi(t)\,a.$$

This is the desired general solution to the given system of differential equations.

Example 4.5.2. Let us consider $dx = Ax\,dt + Bx\,dt$, where

$$A = \begin{bmatrix} 2 & 1 \\ 1 & 3 \end{bmatrix}, \quad B = \begin{bmatrix} 3 & 1 \\ 1 & 4 \end{bmatrix}.$$

Find: (a) the fundamental and (b) the general solutions to this system of perturbed differential equations.

Solution procedure. To avoid repetition, we just follow the mechanics of used in the solution process of Example 4.5.1. In this case, the eigenvalues and corresponding eigenvectors of matrices A and B are in Example 4.5.1. In fact, there is a change of notations and exchange of rate matrices:

$$\lambda_{1A} = \frac{5+\sqrt{5}}{2}, \quad c_{1A} = \begin{bmatrix} 1 & \frac{1+\sqrt{5}}{2} \end{bmatrix}^T,$$

$$\lambda_{2A} = \frac{5-\sqrt{5}}{2}, \quad c_{2A} = \begin{bmatrix} -\frac{1+\sqrt{5}}{2} & 1 \end{bmatrix}^T,$$

$$\lambda_{1B} = \frac{7+\sqrt{5}}{2}, \quad c_{1B} = \begin{bmatrix} 1 & \frac{1+\sqrt{5}}{2} \end{bmatrix}^T \quad \text{and}$$

$$\lambda_{2B} = \frac{7-\sqrt{5}}{2}, \quad c_{2B} = \begin{bmatrix} -\frac{1+\sqrt{5}}{2} & 1 \end{bmatrix}^T,$$

respectively. From this, the remaining argument is the direct imitation of the argument used in Example 4.5.1. Thus, we have

$$\Phi(t) = \Phi_{\mathrm{P}}(t,0)\Phi_{\mathrm{U}}(t), \quad x(t) = \Phi(t)a,$$

where the description of $\Phi(t)$ is exactly similar to the description of $\Phi(t)$ in Example 4.5.1.

In the following, we present a theoretical algorithm that provides a systematic approach to finding a general solution process of (4.67). This procedure does not require condition (4.74).

Theorem 4.5.1 (method of variation of constants/parameters). *Let us assume that: (H_1) $\Phi_{\mathrm{U}}(t)$ is a fundamental matrix solution to* (4.68); *and (H_2) let $x(t) = \Phi_{\mathrm{U}}(t)c(t)$ be a solution to* (4.67), *where $c(t)$ is an n-dimensional unknown vector function. Then $c(t)$ is a solution process of the following linear system of differential equations with the time-varying coefficient matrix process $\Phi_{\mathrm{U}}^{-1}(t)B\Phi_{\mathrm{U}}(t)$:*

$$dc = \Phi_{\mathrm{U}}^{-1}(t)B\Phi_{\mathrm{U}}(t)c\,dt. \tag{4.80}$$

Proof. From Observation 4.3.5, we assert that $\Phi_{\mathrm{U}}(t)$ is invertible, and hence $\Phi_{\mathrm{U}}^{-1}(t)$ exists on J. Moreover, from Theorem 4.3.3, we have

$$d\Phi_{\mathrm{U}}^{-1}(t) = -\Phi_{\mathrm{U}}^{-1}(t)A\,dt. \tag{4.81}$$

$x(t) = \Phi_{\mathrm{U}}(t)c(t)$ can be rewritten as

$$c(t) = \Phi_{\mathrm{U}}^{-1}(t)x(t). \tag{4.82}$$

By using the product rule (Theorem 1.5.11), we compute the following differential:

$$\begin{aligned}
dc(t) &= d\Phi_{\mathrm{U}}^{-1}(t)x(t) + \Phi_{\mathrm{U}}^{-1}(t)\,dx(t) \quad \text{(Theorem 1.5.11)}\\
&= \left[-\Phi_{\mathrm{U}}^{-1}(t)A\,dt\right]x(t) + \Phi_{\mathrm{U}}^{-1}(t)\left[Ax(t)\,dt + Bx(t)\,dt\right]\\
&\qquad \text{[Definition 4.3.1 and (4.81))}\\
&= \Phi_{\mathrm{U}}^{-1}(t)Bx(t)\,dt\\
&= \Phi_{\mathrm{U}}^{-1}(t)B\Phi_{\mathrm{U}}(t)c(t)\,dt \quad [\text{substitution: } x(t) = \Phi_{\mathrm{U}}(t)c(t)].
\end{aligned}$$

This shows that the unknown vector process $c(t)$ satisfies the linear system of differential equations (4.80) with the time-varying coefficient matrix $\Phi_{\mathrm{U}}^{-1}(t)B\Phi_{\mathrm{U}}(t)$. This completes the proof of the theorem. □

The following result is an alternate form of Theorem 4.5.1. It is obtained by interchanging the role of the pair ($\Phi_{\mathrm{U}}(t)$, B) with that of ($\Phi_{\mathrm{P}}(t)$, A).

Corollary 4.5.1 (method of variation of constants parameters). *Let us assume that: (H_1) $\Phi_{\mathrm{P}}(t)$ is a fundamental matrix solution to* (4.69); *and (H_2) let $x(t) = \Phi_{\mathrm{P}}(t)c(t)$ be a solution to* (4.67), *where $c(t)$ is an n-dimensional unknown vector function. Then $c(t)$ is a solution process of the following system of linear differential equations with the time-varying coefficient matrix process $\Phi_{\mathrm{P}}^{-1}(t)A\Phi_{\mathrm{P}}(t)$:*

$$dc = \Phi_{\mathrm{P}}^{-1}(t)A\Phi_{\mathrm{P}}(t)c\,dt. \tag{4.83}$$

Now, we present a theoretical algorithm that determines a general solution to (4.67) without assumption (4.74).

Theorem 4.5.2. *Let $\Phi_{\mathrm{U}}(t)$ be a fundamental matrix solution to (4.68) and let $\Phi_{\mathrm{D}}(t)$ be a fundamental matrix solution to $dc = D(t)c\,dt$, where $D(t) = \Phi_{\mathrm{U}}^{-1}(t)B\Phi_{\mathrm{U}}(t)$. Then, the fundamental matrix solution and the general solution to (4.67) are represented by* (a) $\Phi(t) = \Phi_{\mathrm{U}}(t)\Phi_{\mathrm{D}}(t)$ *and* (b) $x(t) = \Phi_{\mathrm{U}}(t)\Phi_{\mathrm{D}}(t)a$, *where a is an arbitrary constant vector.*

Proof. $\Phi_{\mathrm{U}}(t)$ and $\Phi_{\mathrm{D}}(t)$ are as described before. We define $\Phi(t) = \Phi_{\mathrm{U}}(t)\Phi_{\mathrm{D}}(t)$. We know that $\Phi(t)$ is a nonsingular and differentiable matrix function. Now, we compute

$$\begin{aligned} d\Phi(t) &= d\Phi_{\mathrm{U}}(t)\Phi_{\mathrm{D}}(t) + \Phi_{\mathrm{U}}(t)d\Phi_{\mathrm{D}}(t) \quad \text{(Observation 1.5.2)} \\ &= A\Phi_{\mathrm{U}}(t)\Phi_{\mathrm{D}}(t)\,dt + \Phi_{\mathrm{U}}(t)D(t)\Phi_{\mathrm{D}}(t)\,dt \quad \text{(assumption)} \\ &= A\Phi_{\mathrm{U}}(t)\Phi_{\mathrm{D}}(t)\,dt + \Phi_{\mathrm{U}}(t)\Phi_{\mathrm{U}}^{-1}(t)B\Phi_{\mathrm{U}}(t)\Phi_{\mathrm{D}}(t)\,dt \\ &= A\Phi(t)\,dt + B\Phi(t)\,dt. \end{aligned}$$

This proves part (a). Part (b) follows from the definition of a fundamental solution. Furthermore, the solution process of the system of perturbed differential equations $dx = Ax\,dt + Bx\,dt$ is the composition of the solution process of the unperturbed component of the differential equation and the solution process of the time-varying system with the coefficient matrix function $D(t)$. In addition, we do not need any assumptions about the matrices A, B, $\Phi_{\mathrm{U}}(t)$, and $\Phi_{\mathrm{P}}(t)$. □

The following corollary is parallel to Theorem 4.5.2. It is formulated by interchanging the role of $(\Phi_{\mathrm{U}}(t), B)$ with that of $(\Phi_{\mathrm{P}}(t), A)$.

Corollary 4.5.2. *Let $\Phi_{\mathrm{P}}(t)$ be a fundamental matrix solution to (4.69) and let $\Phi_{\mathrm{D}}(t)$ be a fundamental matrix solution to $dc = D(t)c\,dt$, where $D(t) = \Phi_{\mathrm{P}}^{-1}(t)A\Phi_{\mathrm{P}}(t)$. Then, the fundamental matrix solution and the general solution to (4.67) are represented by* (a) $\Phi(t) = \Phi_{\mathrm{P}}(t)\Phi_{\mathrm{D}}(t)$ *and* (b) $x(t) = \Phi_{\mathrm{P}}(t)\Phi_{\mathrm{D}}(t)\,a$, *where a is an arbitrary constant vector.*

Example 4.5.3. Find the (a) fundamental and (b) general solutions to the given linear system of differential equations $dx = (Ax + Bx)\,dt$, where

$$A = \begin{bmatrix} -\frac{1}{2} & 0 \\ 1 & -2 \end{bmatrix}, \quad B = \begin{bmatrix} 2 & 5 \\ 1 & 3 \end{bmatrix}.$$

Solution procedure. We note that the coefficient matrices in the example do not commute. However, imitating the argument used in Theorem 4.5.1 and Illustration 4.4.1, we have

$$\Phi_{\mathrm{U}}(t) = \begin{bmatrix} e^{-\frac{1}{2}t} & 0 \\ e^{-\frac{1}{2}t} & e^{-2t} \end{bmatrix}, \quad \det(\Phi_{\mathrm{U}}(t)) = e^{-\frac{5}{2}t}, \quad \Phi_{\mathrm{U}}^{-1}(t) = \begin{bmatrix} e^{\frac{1}{2}t} & 0 \\ -e^{2t} & e^{2t} \end{bmatrix}.$$

Let us define

$$D(t) = \Phi_{\mathrm{U}}^{-1}(t)B\Phi_{\mathrm{U}}(t) = \begin{bmatrix} e^{\frac{1}{2}t} & 0 \\ -e^{2t} & e^{2t} \end{bmatrix} \begin{bmatrix} 2 & 5 \\ 1 & 3 \end{bmatrix} \begin{bmatrix} e^{-\frac{1}{2}t} & 0 \\ e^{-\frac{1}{2}t} & e^{-2t} \end{bmatrix}$$

$$= \begin{bmatrix} e^{\frac{1}{2}t} & 0 \\ -e^{2t} & e^{2t} \end{bmatrix} \begin{bmatrix} 7e^{-\frac{1}{2}t} & 5e^{-2t} \\ 4e^{-\frac{1}{2}t} & 3e^{-2t} \end{bmatrix} = \begin{bmatrix} 7 & 5e^{-\frac{3}{2}t} \\ -3e^{\frac{3}{2}t} & -2 \end{bmatrix}.$$

Let $x(t) = \Phi_{\mathrm{U}}(t)c(t)$ be a solution to the system of differential equations, where $c(t)$ is a two-dimensional unknown continuous vector function. We compute

$$\begin{aligned} dx(t) &= d\Phi_{\mathrm{U}}(t)c(t) + \Phi_{\mathrm{U}}(t)\,dc(t) \\ &= A\Phi_{\mathrm{U}}(t)c(t)\,dt + \Phi_{\mathrm{U}}(t)\,dc(t) \\ &= Ax(t)\,dt + Bx(t)\,dt, \end{aligned}$$

which implies that

$$\Phi_{\mathrm{U}}(t)\,dc(t) = Bx(t)\,dt = B\Phi_{\mathrm{U}}(t)c(t)\,dt.$$

Hence, it reduces to (4.80):

$$\begin{aligned} dc(t) &= \Phi_{\mathrm{U}}^{-1}(t)B\Phi_{\mathrm{U}}(t)c(t)\,dt \\ &= D(t)c(t)\,dt. \end{aligned}$$

This is a linear system of differential equations with the time-varying coefficient matrix $D(t)$. The solution to this system has the representation $c(t) = \Phi_{\mathrm{D}}(t)\,a$, where a is an arbitrary constant vector. By the application of Theorem 4.5.2, $\Phi(t) = \Phi_{\mathrm{U}}(t)\Phi_{\mathrm{D}}(t)$ and $x(t) = \Phi_{\mathrm{U}}(t)\Phi_{\mathrm{D}}(t)\,a$ are the fundamental and general solutions, respectively.

Example 4.5.4. Find: (a) the fundamental and the general solutions to the given differential equation, and (b) the general solution to $dx = Ax\,dt + Bx\,dt$, where

$$A = \begin{bmatrix} -1 & 0 \\ 1 & -2 \end{bmatrix}, \quad B = \begin{bmatrix} 2 & 5 \\ 1 & 3 \end{bmatrix}.$$

Solution procedure. By following the argument used in Example 4.5.3 and Corollary 4.5.1, we have

$$\Phi_{\mathrm{P}}(t) = \begin{bmatrix} 5e^{\frac{5+\sqrt{21}}{2}t} & -\dfrac{1+\sqrt{21}}{2}e^{\frac{5-\sqrt{21}}{2}t} \\ \dfrac{1+\sqrt{21}}{2}e^{\frac{5+\sqrt{21}}{2}t} & e^{\frac{5-\sqrt{21}}{2}t} \end{bmatrix},$$

$$\det(\Phi_{\mathrm{P}}(t)) = \frac{20 + \left(1+\sqrt{21}\right)^2}{4}e^{5t},$$

$$\Phi_{\mathrm{P}}^{-1}(t) = \frac{4}{20 + \left(1+\sqrt{21}\right)^2} \begin{bmatrix} e^{-\frac{5+\sqrt{21}}{2}t} & \frac{1+\sqrt{21}}{2} e^{-\frac{5+\sqrt{21}}{2}t} \\ -\frac{1+\sqrt{21}}{2} e^{-\frac{5-\sqrt{21}}{2}t} & 5e^{-\frac{5-\sqrt{21}}{2}t} \end{bmatrix}.$$

Let us define

$$\begin{aligned}
\frac{20 + \left(1+\sqrt{21}\right)^2}{4} D(t) &= \frac{20 + \left(1+\sqrt{21}\right)^2}{4} \Phi_{\mathrm{P}}^{-1}(t) A \Phi_{\mathrm{P}}(t) \\
&= \begin{bmatrix} e^{-\frac{5+\sqrt{21}}{2}t} & \frac{1+\sqrt{21}}{2} e^{-\frac{5+\sqrt{21}}{2}t} \\ -\frac{1+\sqrt{21}}{2} e^{-\frac{5-\sqrt{21}}{2}t} & 5e^{-\frac{5-\sqrt{21}}{2}t} \end{bmatrix} \begin{bmatrix} -5e^{\frac{5+\sqrt{21}}{2}t} & \frac{(1+\sqrt{21})}{2} e^{\frac{5-\sqrt{21}}{2}t} \\ \left(4-\sqrt{21}\right) e^{\frac{5+\sqrt{21}}{2}t} & -\frac{5+\sqrt{21}}{2} e^{\frac{5-\sqrt{21}}{2}t} \end{bmatrix} \\
&= \begin{bmatrix} -5 + \frac{1+\sqrt{21}}{2}\left(4-\sqrt{21}\right) & \frac{1+\sqrt{21}}{2}\left[1 - \frac{5+\sqrt{21}}{2}\right] e^{-\sqrt{21}\,t} \\ 5\left[\frac{1+\sqrt{21}}{2} + \left(4-\sqrt{21}\right)\right] e^{\sqrt{21}\,t} & -\frac{\left(1+\sqrt{21}\right)^2}{4} + \frac{5\left(5+\sqrt{21}\right)}{2} \end{bmatrix} \\
&= \begin{bmatrix} \frac{3(-9+\sqrt{21})}{2} & \frac{-(24+4\sqrt{21})}{4} e^{-\sqrt{21}\,t} \\ \frac{5(9-\sqrt{21})}{2} e^{\sqrt{21}\,t} & -\left(18+3\sqrt{21}\right) \end{bmatrix}.
\end{aligned}$$

Thus,

$$\begin{aligned}
D(t) &= \Phi_{\mathrm{P}}^{-1}(t) A \Phi_{\mathrm{P}}(t) \\
&= \frac{4}{20 + \left(1+\sqrt{21}\right)^2} \begin{bmatrix} \frac{3(-9+\sqrt{21})}{2} & \frac{-(24+4\sqrt{21})}{4} e^{-\sqrt{21}\,t} \\ \frac{5(9-\sqrt{21})}{2} e^{\sqrt{21}t} & -(18+3\sqrt{21}) \end{bmatrix}.
\end{aligned}$$

Let $x(t) = \Phi_{\mathrm{P}}(t)c(t)$ be a solution to a system of given differential equations, where $c(t)$ is a two-dimensional unknown continuous vector function. In this case, (4.83) reduces to

$$\begin{aligned}
dc(t) &= \Phi_{\mathrm{P}}^{-1}(t) A \Phi_{\mathrm{P}}(t) c(t)\, dt \\
&= D(t)c(t)\, dt.
\end{aligned}$$

Now, by the application of Corollary 4.5.2, $\Phi(t) = \Phi_{\mathrm{P}}(t)\Phi_{\mathrm{D}}(t)$ and $x(t) = \Phi_{\mathrm{P}}(t)\Phi_{\mathrm{D}}(t)a$ are the desired fundamental and general solutions, respectively.

Observation 4.5.3

(i) Under the assumption (4.74) on the coefficient matrices (4.67), we have $\Phi_{\mathrm{U}}^{-1}(t)B\Phi_{\mathrm{U}}(t) = B$ and $\Phi_{\mathrm{P}}^{-1}(t)A\Phi_{\mathrm{P}}(t) = A$, provided that $\Phi_{\mathrm{U}}(t)$ and $\Phi_{\mathrm{P}}(t)$ are normalized matrices. Hence, the system of linear differential equations with time-varying coefficients in (4.80) and (4.83) reduces to the system of linear differential equations with constant coefficients in (4.69) and (4.68), respectively. Hence, the general solutions to (4.80) and (4.83) are the general solutions to (4.69) and (4.68), respectively. Thus, for $AB = BA$, $\Phi_{\mathrm{P}}(t)\Phi_{\mathrm{U}}(t) = \Phi_{\mathrm{U}}(t)\Phi_{\mathrm{P}}(t)$. Hence, $x(t)$ defined in (H_2) of Theorem 4.5.1 or Theorem 4.5.2 reduces to

$$\begin{aligned} x(t) &= \Phi_{\mathrm{U}}(t)c(t) \\ &= \Phi_{\mathrm{U}}(t)\Phi_{\mathrm{P}}(t)\, a \quad [\text{by substitution } c(t) = \Phi_{\mathrm{P}}(t)a] \\ &= \Phi(t)\, a \quad [\text{from (4.70)}] \\ &= \exp\left[(A+B)\,t\right] a. \quad [\text{from (4.79)}]. \end{aligned} \tag{4.84}$$

This is indeed the general solution process that was determined by the methods of finding solutions described in Sections 2.4, 4.3, and 4.4 under the condition (4.74) ($AB = BA$). Moreover, $\exp[At]$ and $\exp[Bt]$ are normalized fundamental matrices.

(ii) From the above observation (i), the problem of finding a general solution process of (4.67) with $AB \neq BA$ reduces to the problem of finding a general fundamental matrix solution to a system of linear differential equations with time-varying coefficients of the type (4.80) or (4.83). Unfortunately, in general, there is no computational procedure that would enable one to find a closed form representation of the fundamental matrix solution for the systems of linear differential equations with time-varying coefficients. In the literature, this issue has been addressed in the framework of the theoretical concepts. This topic is outlined in Section 4.7.

(iii) From the above observations (i) and (ii) and Observation 4.3.5, we further note that finding the general solution process of (4.67) depends on finding the fundamental solution to (4.67). Finding the fundamental solution to (4.67) depends on finding the fundamental solutions to (4.68) and (4.80) or (4.69) and (4.83). The task of justifying the existence of the fundamental solution for the systems of linear differential equations with time-varying coefficients is the topic of discussion in Section 4.7.

(iv) We further note that one can apply the method of solving a system of differential equations to (4.67), directly, by considering the aggregate rate matrix $(A+B)$ of the system as follows:

$$dx = Ax\,dt + Bx\,dt = (A+B)x\,dt.$$

However, this will not be able to shed any light, explicitly on the perturbation matrix (B). As a result of this, we may not be able to make any inferences about the robustness (sensitivity analysis) or structural changes in the system due to the parametric changes (deterministic/stochastic) in the coefficient of the system matrix. We further note that preserving the "identity" of "B" is a very important problem in the study of dynamic modeling under structural perturbations (stochastic/deterministic).

4.5.3 *Initial value problem*

Let us formulate an IVP for a general system of first-order linear homogeneous differential equations with constant coefficients. We consider

$$dx = Ax\,dt + Bx\,dt, \quad x(t_0) = x_0. \tag{4.85}$$

Here x, A, and B are as defined in (4.67). We note that the concept of the solution to the IVP (4.85) can be defined analogously to Definition 4.3.2. Again, its solution is represented by $x(t) = x(t, t_0, x_0)$, for $t \geq t_0$ and t in J.

4.5.4 *Procedure for solving the IVP*

In the following, we apply the procedure for finding a solution process of the IVP (4.41) to IVP (4.85) with $AB = BA$. The procedure is as follows:

Step 1. From the procedure for finding a general solution to (4.67) and Observation 4.5.1, we have

$$x(t) = \Phi(t)c = \exp\left[(A + B)\,t\right] c, \tag{4.86}$$

where $c \in R$, $t \in J = [a, b] \subseteq R$, and $\Phi(t)$ is as defined (4.79). Here, c is an arbitrary constant.

Step 2. This step deals with finding an arbitrary constant c in (4.86). For this purpose, we utilize given initial given data (t_0, x_0), and solve the system of algebraic equations

$$x(t_0) = \Phi(t_0)c = x_0 = \exp\left[(A + B)\,t_0\right] c, \tag{4.87}$$

for any given t_0 in J. The system of linear algebraic equations (4.87) has a unique solution, and it is given by

$$c = \Phi^{-1}(t_0)x_0 = \exp\left[-\,(A + B)\,t_0\right] x_0. \tag{4.88}$$

Step 3. The solution to the IVP (4.85) is determined by substituting the expression (4.88) into (4.86). Hence, we have

$$\begin{aligned}
x(t) &= \Phi(t)c \quad \text{[from (4.78)]} \\
&= \Phi(t)\Phi^{-1}(t_0)x_0 \quad \text{[substitution for } c \text{ from (4.88)]} \\
&= \exp\left[(A+B)(t-t_0)\right]x_0 \quad \text{(laws of exponents)} \\
&= \Phi(t,t_0)x_0 \quad \text{(by notation)},
\end{aligned} \tag{4.89}$$

where

$$\Phi(t)\Phi^{-1}(t_0) = \Phi(t,t_0) = \exp\left[(A+B)(t-t_0)\right]. \tag{4.90}$$

Thus, the solution $x(t) = x(t,t_0,x_0)$ of the IVP (4.85) is represented by

$$\begin{aligned}
x(t) &= x(t,t_0,x_0) \\
&= \Phi(t,t_0)x_0 \quad \text{[from (4.90)]} \\
&= \exp\left[(A+B)(t-t_0)\right]x_0 \quad \text{[from (4.90)]}.
\end{aligned} \tag{4.91}$$

The solution process in (4.91) is referred to as a particular solution to (4.67). $\Phi(t,t_0)$ is called the normalized fundamental solution to (4.67), because $\Phi(t_0,t_0) = I_{n\times n}$. The procedure for finding the solution to the IVP is summarized in Figure 4.16.

Illustration 4.5.2. We are given $dx = (Ax + Bx)\,dt$, $x(t_0) = x_0$, $x_0 \in R^2$, where

$$A = \begin{bmatrix} a_{11} & a_{12} \\ a_{21} & a_{22} \end{bmatrix}, \quad B = \begin{bmatrix} b_{11} & b_{12} \\ b_{21} & b_{22} \end{bmatrix}.$$

Find the solution to the IVP.

Solution procedure. By repeating the procedure outlined in Illustration 4.5.1, we arrive at the general solution $x(t) = \Phi(t)a$. From this, we follow the argument used in Illustrations 4.4.1, 4.4.3, and 4.4.4 (the procedure in Subsection 4.5.4), and obtain the solution to the given IVP. The details are left as an exercise. This completes the procedure for finding the general solution.

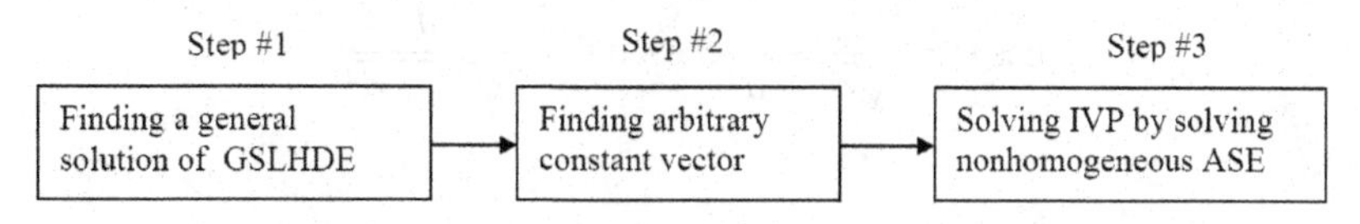

Fig. 4.16 Flowchart for finding particular solution.

Example 4.5.5. Solve the given IVP $dx = Ax\,dt + Bx\,dt$, $x(t_0) = x_0$, $x_0 \in R^2$, where

$$A = \begin{bmatrix} 5 & -\sqrt{5} \\ \sqrt{5} & 1 \end{bmatrix}, \quad B = \begin{bmatrix} 4\sigma & -\sqrt{5}\,\sigma \\ \sqrt{5}\sigma & 0 \end{bmatrix},$$

and $\sigma \neq 0$.

Solution procedure. From step 1 of the procedure in Subsection 4.5.4, we find the general solution to this IVP. For this purpose, we find the fundamental matrix solutions corresponding to the unperturbed and perturbed parts of the given system of differential equations. In this case, the eigenvalues of the matrices A and B are $3 \pm i$ and $2\sigma \pm \sigma i$, respectively. As per the procedures (DDC 1–DDC 3 of Section 4.4) for finding eigenvectors and solutions to differential equations with distinct complex eigenvalues, it is enough to find one eigenvector corresponding to one of the components of each pair of distinct complex eigenvalues. In this case, the eigenvectors and solutions corresponding to the eigenvalues of the matrices A and B are $\lambda_{1A} = 3 + i$ and $\lambda_{1B} = 2\sigma + \sigma i$,

$$c_{1A} = \begin{bmatrix} 2+i \\ \sqrt{5} \end{bmatrix}, \quad x_{1U}(t) = e^{3t}\begin{bmatrix} 2\cos t - \sin t \\ \sqrt{5}\cos t \end{bmatrix}, \quad x_{2U}(t) = e^{3t}\begin{bmatrix} \cos t + 2\sin t \\ \sqrt{5}\sin t \end{bmatrix},$$

$$c_{1P} = \begin{bmatrix} 2+i \\ \sqrt{5} \end{bmatrix}, \quad x_{1P}(t) = e^{2\sigma t}\begin{bmatrix} 2\cos\sigma t - \sin\sigma t \\ \sqrt{5}\cos\sigma t \end{bmatrix}, \quad x_{2P}(t) = e^{2\sigma t}\begin{bmatrix} \cos\sigma t + 2\sin\sigma t \\ \sqrt{5}\sin\sigma t \end{bmatrix},$$

respectively. Here, we have used the following trigonometric expressions: for nonzero real numbers a and b, from $a\sin\theta \pm b\cos\theta$, we recall that

$$a\sin\theta \pm b\cos\theta = \sqrt{a^2+b^2}\left(\frac{a}{\sqrt{a^2+b^2}}\sin\theta \pm \frac{b}{\sqrt{a^2+b^2}}\cos\theta\right).$$

By setting

$$\cos\psi = \frac{a}{\sqrt{a^2+b^2}}, \quad \sin\psi = \frac{b}{\sqrt{a^2+b^2}},$$

we have

$$a\sin\theta \pm b\cos\theta = \sqrt{a^2+b^2}\,(\sin\theta\cos\psi \pm \sin\psi\cos\theta) = \sqrt{a^2+b^2}\sin(\theta \pm \psi).$$

Similarly, by defining

$$\cos\psi = \frac{c}{\sqrt{c^2+d^2}}, \quad \sin\psi = \frac{d}{\sqrt{c^2+d^2}},$$

we have

$$c\cos\theta \pm d\sin\theta = \sqrt{c^2+d^2}\,(\cos\theta\cos\psi \pm \sin\theta\sin\psi) = \sqrt{c^2+d^2}\cos(\theta \mp \psi).$$

From these preliminary trigonometric identities and setting

$$\psi = \arctan\left(\frac{1}{2}\right),$$

we write the components of the above solution processes as follows:

$$2\cos t - \sin t = \sqrt{5}\cos(t+\psi), \quad 2\sin t + \cos t = \sqrt{5}\sin(t+\psi),$$

$$2\cos\sigma t - \sin\sigma t = \sqrt{5}\cos(\sigma t+\psi), \quad 2\sin\sigma t + \cos\sigma t = \sqrt{5}\sin(\sigma t+\psi).$$

With this mathematical compactification and from the above discussion, we write the fundamental matrices of the unperturbed and perturbed components of differential equations as

$$\Phi_{\mathrm{U}}(t) = e^{3t}\begin{bmatrix} \sqrt{5}\cos(t+\psi) & \sqrt{5}\sin(t+\psi) \\ \sqrt{5}\cos t & \sqrt{5}\sin t \end{bmatrix},$$

$$\Phi_{\mathrm{P}}(t) = e^{2\sigma t}\begin{bmatrix} \sqrt{5}\cos(\sigma t+\psi) & \sqrt{5}\sin(\sigma t+\psi) \\ \sqrt{5}\cos\sigma t & \sqrt{5}\sin\sigma t \end{bmatrix}.$$

Finding the fundamental matrix solution to the overall system of differential equations requires us to determine at least one of the fundamental matrices $\Phi_{\mathrm{U}}(t)$ and $\Phi_{\mathrm{P}}(t)$ to be of normalized form. Here, we find it with regard to $\Phi_{\mathrm{U}}(t)$. For this purpose, we use Observation 4.3.3(iv) and step 3 of the procedure in Subsection 4.3.4 with $t_0 = 0$. We compute $\Phi_{\mathrm{U}}(0)$, $\Phi_{\mathrm{U}}^{-1}(0)$ and the normalized fundamental matrix solution with respect to $\Phi_{\mathrm{U}}(t)$ as

$$\Phi_{\mathrm{U}}(0) = \begin{bmatrix} \sqrt{5}\cos\psi & \sqrt{5}\sin\psi \\ \sqrt{5} & 0 \end{bmatrix} = \begin{bmatrix} 2 & 1 \\ \sqrt{5} & 0 \end{bmatrix}, \quad \Phi_{\mathrm{U}}^{-1}(0) = \begin{bmatrix} 0 & \dfrac{1}{\sqrt{5}} \\ 1 & \dfrac{-2}{\sqrt{5}} \end{bmatrix},$$

$$\Phi_{\mathrm{U}}(t,0) = \Phi_{\mathrm{U}}(t)\Phi_{\mathrm{U}}^{-1}(0) = e^{3t}\begin{bmatrix} \sqrt{5}\cos(t+\psi) & \sqrt{5}\sin(t+\psi) \\ \sqrt{5}\cos t & \sqrt{5}\sin t \end{bmatrix}\begin{bmatrix} 0 & \dfrac{1}{\sqrt{5}} \\ 1 & \dfrac{-2}{\sqrt{5}} \end{bmatrix}$$

$$= e^{3t}\begin{bmatrix} \sqrt{5}\sin(t+\psi) & -\sqrt{5}\sin t \\ \sqrt{5}\sin t & -\sqrt{5}\sin(t-\psi) \end{bmatrix}.$$

From (4.70), the fundamental matrix solution to the overall system of differential equations (4.67) in the given example is

$$\Phi(t) = \Phi_{\mathrm{U}}(t,0)\Phi_{\mathrm{P}}(t) = e^{(3+2\sigma)t}\begin{bmatrix} \phi_{11}(t,\psi) & \phi_{12}(t,\psi) \\ \phi_{21}(t,\psi) & \phi_{22}(t,\psi) \end{bmatrix},$$

where

$$\phi_{11}(t,\psi) = 5\left[\sin(t+\psi)\cos(\sigma t+\psi) - \sin t\cos\sigma t\right],$$
$$\phi_{12}(t,\psi) = 5\left[\sin(t+\psi)\sin(\sigma t+\psi) - \sin t\sin\sigma t\right],$$
$$\phi_{21}(t,\psi) = 5\left[\sin t\cos(\sigma t+\psi) - \sin(t-\psi)\cos\sigma t\right],$$
$$\phi_{22}(t,\psi) = 5\left[\sin t\sin(\sigma t+\psi) - \sin(t-\psi)\sin\sigma t\right].$$

We further confirm our calculations by the following observations. First, we recall that

$$\sin\theta\sin\phi = \frac{1}{2}[\cos(\theta-\phi) - \cos(\theta+\phi)], \quad \cos 2\theta = 1 - 2\sin^2 t,$$

and compute

$$\begin{aligned}\sin(t+\psi)\sin(t-\psi) &= \frac{1}{2}\left[\cos 2\psi - \cos 2t\right]\\ &= \frac{1}{2}\left[1 - 2\sin^2\psi - (1-2\sin^2 t)\right]\\ &= \left[\sin^2 t - \sin^2\psi\right] \quad \text{(by simplification)}.\end{aligned}$$

Now, we compute

$$\begin{aligned}\det(\Phi_{\mathrm{U}}(t,0)) &= -5e^{6t}[\sin(t+\psi)\sin(t-\psi) - \sin^2 t]\\ &= -5e^{6t}[\sin^2 t - \sin^2\psi - \sin^2 t]\\ &= 5e^{6t}\sin^2\psi = e^{6t},\end{aligned}$$

and it fulfills the conclusion of Observation 4.3.3(iv). Similarly,

$$\begin{aligned}\det(\Phi_{\mathrm{P}}(t)) &= 5e^{4\sigma t}[\sin\sigma t\cos(\sigma t+\psi) - \sin(\sigma t+\psi)\cos\sigma t]\\ &= 5\sin(\sigma t - \psi - \sigma t)e^{4\sigma t}\\ &= 5e^{4\sigma t}\sin(-\psi) \quad \text{(simplification)}\\ &= -\sqrt{5}e^{4\sigma t} \quad [\sin(-B) = -\sin B],\end{aligned}$$

and hence $\det(\Phi_{\mathrm{P}}(t)) = -\sqrt{5}\,e^{4\sigma t}$, which satisfies conditions of Observation 4.3.3(iv). We further compute

$$\begin{aligned}&\det(\Phi(t))e^{-2(2\sigma+3)t}\\ &\quad = 25[(\sin(t+\psi)\cos(\sigma t+\psi) - \sin t\cos\sigma t)(\sin t\sin(\sigma t+\psi)\\ &\qquad - \sin(t-\psi)\sin\sigma t) - (\sin(t+\psi)\sin(\sigma t+\psi)\\ &\qquad - \sin t\sin\sigma t)(\sin t\cos(\sigma t+\psi) - \sin(t-\psi)\cos\sigma t)]\\ &\quad = 25[[\sin(t+\psi)\cos(\sigma t+\psi)\sin t\sin(\sigma t+\psi) + \sin t\cos\sigma t\\ &\qquad \times \sin(t-\psi)\sin\sigma t - \sin(t+\psi)\cos(\sigma t+\psi)\sin(t-\psi)\sin\sigma t\end{aligned}$$

$$
\begin{aligned}
&- \sin t \cos \sigma t \sin t \sin(\sigma t + \psi)] - [\sin(t + \psi) \sin(\sigma t + \psi) \sin t \\
&\times \cos(\sigma t + \psi) + \sin t \sin \sigma t \sin(t - \psi) \cos \sigma t - \sin(t + \psi) \\
&\times \sin(\sigma t + \psi) \sin(t - \psi) \cos \sigma t - \sin t \cos(\sigma t + \psi) \sin t \sin \sigma t]] \\
&= 25[[\sin(t + \psi) \sin(t - \psi)[\sin(\sigma t + \psi) \cos \sigma t - \cos(\sigma t + \psi) \sin \sigma t]] \\
&\quad + \sin^2 t[\cos(\sigma t + \psi) \sin \sigma t - \cos \sigma t \sin(\sigma t + \psi)]] \quad \text{(simplification)} \\
&= 25[\sin(t + \psi) \sin(t - \psi) \sin \psi - \sin^2 t \sin \psi] \\
&= 25 \sin \psi[\sin(t + \psi) \sin(t - \psi) - \sin^2 t] \quad \text{(further simplification)} \\
&= 25 \sin \psi[[\sin^2 t - \sin^2 \psi] - \sin^2 t] \\
&= 25 \sin \psi[- \sin^2 \psi] \\
&= -25 \sin^3 \psi = \frac{-25}{5\sqrt{5}} = -\sqrt{5} \quad \text{(definition of } \psi).
\end{aligned}
$$

Thus,

$$
\begin{aligned}
\det(\Phi(t)) &= -\sqrt{5}\, e^{2(2\sigma+3)t} = e^{6t}(-\sqrt{5}\, e^{4\sigma t}) \\
&= \det(\Phi_{\mathrm{U}}(t)) \det(\Phi_{\mathrm{P}}(t)) \quad \text{(Theorem 1.4.2)}.
\end{aligned}
$$

We further note that the coefficient matrices in this example commute, i.e. $AB = BA$. Therefore, under these considerations, the direct imitation of the argument used in Example 4.5.1, the fundamental matrix and the general solutions are defined by

$$
\Phi(t) = \Phi_{\mathrm{U}}(t, 0)\Phi_{\mathrm{P}}(t), \quad x(t) = \Phi(t)a,
$$

respectively. Thus, the solution to the IVP is given by $x(t) = \Phi(t)\Phi^{-1}(t_0)x_0$. This completes the solution process.

Observation 4.5.4. Note that the form of a general fundamental solution expression in (4.79) for the perturbed differential equation (4.67) is an extension of the form in Observation 4.4.1 of linear unperturbed systems of differential equations. In fact, for the choice of $B \equiv 0$, i.e. in the absence of perturbations/effects, it includes a fundamental solution expression in Observation 4.4.1 as a special case of the fundamental solution expression in (4.79).

4.5 Exercises

(1) Let us denote by a, b, and c arbitrary but given real numbers. For the following given pair of matrices (A, B), identify the pair of matrices that satisfy the commutative property of matrix multiplication:

(a) $A = \begin{bmatrix} 5 & \sqrt{5} \\ -\sqrt{5} & 1 \end{bmatrix}$, $B = \begin{bmatrix} 8 & -\sqrt{5} \\ -2\sqrt{5} & 0 \end{bmatrix}$ (b) $A = \begin{bmatrix} 2 & 0 \\ 0 & 1 \end{bmatrix}$, $B = \begin{bmatrix} a & 0 \\ 0 & a \end{bmatrix}$

(c) $A = \begin{bmatrix} 3 & 2 \\ -1 & 2 \end{bmatrix}$, $B = \begin{bmatrix} 1.5 & 1 \\ -0.5 & 1 \end{bmatrix}$ (d) $A = \begin{bmatrix} 1 & 0 \\ 0 & -1 \end{bmatrix}$, $B = \begin{bmatrix} \sqrt{3} & 0 \\ 0 & 2 \end{bmatrix}$

(e) $A = \begin{bmatrix} 4 & 1 \\ -4 & 8 \end{bmatrix}$, $B = \begin{bmatrix} 1 & 3 \\ 3 & 2 \end{bmatrix}$ (f) $A = \begin{bmatrix} 1 & 0 \\ 0 & 1 \end{bmatrix}$, $B = \begin{bmatrix} 3 & 0 \\ 0 & 2 \end{bmatrix}$

(g) $A = \begin{bmatrix} 3 & 2 \\ 1 & 4 \end{bmatrix}$, $B = \begin{bmatrix} 1 & 1 \\ 1 & 2 \end{bmatrix}$ (h) $A = \begin{bmatrix} a & 0 \\ 0 & b \end{bmatrix}$, $B = \begin{bmatrix} \sqrt{a} & 0 \\ 0 & \sqrt{b} \end{bmatrix}$

(i) $A = \begin{bmatrix} -1.1 & 1 \\ -1 & 0.1 \end{bmatrix}$, $B = \begin{bmatrix} 0.1 & -1 \\ 1 & 1.1 \end{bmatrix}$ (j) $A = \begin{bmatrix} a & 0 \\ 0 & b \end{bmatrix}$, $B = \begin{bmatrix} b\sqrt{a} & 0 \\ 0 & a\sqrt{b} \end{bmatrix}$

(k) $A = \begin{bmatrix} 3 & 2 \\ -1 & 4 \end{bmatrix}$, $B = \begin{bmatrix} 1 & 1 \\ 1 & 2 \end{bmatrix}$, (l) $A = \begin{bmatrix} 0 & 1 \\ 0 & 0 \end{bmatrix}$, $B = \begin{bmatrix} 0 & 0 \\ 1 & 0 \end{bmatrix}$

(2) For given

$$A = \begin{bmatrix} \alpha & a\delta \\ \delta & \beta \end{bmatrix}, \quad B = \begin{bmatrix} \gamma + b(\alpha - \beta) & ab\delta \\ b\delta & \gamma \end{bmatrix},$$

where a, b, α, β, δ, and γ are arbitrary real numbers such that a, b, and $\alpha - \beta$ are nonzero, verify that

$$AB = BA.$$

(3) Find the fundamental matrix solutions to the following system of differential equations: (U) $dx = Bx\,dt$ and (P) $dx = Ax\,dt$, from the matrices A and B found in Exercise 1.

(4) Find the fundamental matrix solutions to the following system of perturbed differential equations: $dx = Ax\,dt + Bx\,dt$, from the matrices A and B found in Exercise 1.

(5) Find the general solution to each system of differential equations:

(a) $dx = Ax\,dt + Bx\,dt$, where $A = \begin{bmatrix} -1 & 1 \\ 1 & -1 \end{bmatrix}$ and $B = \begin{bmatrix} 1 & -2 \\ -2 & 1 \end{bmatrix}$

(b) $dx = Ax\,dt + Bx\,dt$, where $A = \begin{bmatrix} -1 & 0 \\ 1 & -1 \end{bmatrix}$ and $B = \begin{bmatrix} 1 & -1 \\ 1 & 1 \end{bmatrix}$

(c) $dx = Ax\,dt + Bx\,dt$, where $A = \begin{bmatrix} -1 & -1 \\ 1 & -1 \end{bmatrix}$ and $B = \begin{bmatrix} 3 & 1 \\ 1 & 4 \end{bmatrix}$

(6) Solve the initial value problems (a)–(c) in Exercises 4 and 5, with $x(t_0) = x_0$: (a) $x_0 = \begin{bmatrix} 1 & 1 \end{bmatrix}^T$ (b) $x_0 = \begin{bmatrix} 2 & 1 \end{bmatrix}^T$ (c) $x_0 = \begin{bmatrix} 1 & 2 \end{bmatrix}^T$ (d) $x_0 = \begin{bmatrix} 2 & 2 \end{bmatrix}^T$ (e) any x_0

(7) Let us assume that stop-clock time is a random variable with a cumulative distribution function (cdf) F defined in Exercise 6 of Section 4.2, let us consider a differential equation of the Riemann–Stieltjes type,

$$dx = Ax\,dF + Bx\,dF, \quad x(t_0) = x_0,$$

where A and B are $n \times n$ constant matrices as described in (4.67):

(a) Develop a general solution procedure: $AB = BA$.
(b) Develop Theorems 4.5.1 and 4.5.2 and Corollaries 4.5.1 and 4.5.2.
(c) Moreover, solve the given IVP.

(8) Solve the IVP in Exercise 7 for the specified matrices A, B, and x_0 in Exercise 6.

4.6 Linear Nonhomogeneous Systems

In this section, we utilize the method of variation of constants/parameters to compute a particular solution to first order general linear nonhomogeneous systems of differential equations. We recall that this method is efficient and useful for solving first-order linear nonhomogeneous scalar ordinary differential equations. It is also applicable to both ordinary/stochastic time-invariant, time-varying linear and nonlinear systems of differential equations.

4.6.1 *General problem*

Let us consider the first-order linear nonhomogeneous systems of differential equations of the type

$$dx = [Ax + Bx + p(t)]dt, \tag{4.92}$$

where x, A, and B are as defined in (4.67) and satisfy the condition (4.74); and p is an n-dimensional continuous function defined on an interval $J \subseteq R$ into R^n.

The goal is to present a procedure for finding a solution process of (4.92). In addition, we are interested in solving an IVP associated with (4.92). The concept of the solution process defined in Definition 4.3.1 is directly applicable to (4.92).

4.6.2 *Procedure for finding a general solution*

To find a general solution to (4.92), we imitate the conceptual ideas of finding a general solution to linear nonhomogeneous system algebraic equations described in Theorem 1.4.7. In particular, we recast the procedure in Subsection 2.5.2 for finding a solution process of first-order linear homogeneous scalar differential equations corresponding to (2.41) in Section 2.5. The basic ideas are: (i) finding a general solution process x_c of a first-order linear homogeneous perturbed system of differential equations corresponding to (4.92), (ii) finding a particular solution process x_p of (4.92), and (iii) showing that $x = x_c + x_p$ is a general solution to (4.92). Here, we reintroduce the method of variation of constants/parameters for finding a particular solution process x_p of (4.92).

Step 1. This step deals with finding a fundamental matrix solution to a first-order linear homogeneous system of perturbed differential equations corresponding to (4.92), i.e.

$$dx = Ax\,dt + Bx\,dt. \tag{4.93}$$

We note that this perturbed system of differential equations is exactly the same as (4.67). Therefore, we imitate steps 1–4 described in the procedure in Subsection 4.5.2 for finding a general solution process of (4.67), and determine a fundamental matrix solution to (4.93) (Observation 4.5.1):

$$\begin{aligned}\Phi(t) &= \Phi_{\mathrm{U}}(t)\Phi_{\mathrm{P}}(t) \quad \text{[from (4.70)]}\\ &= \exp[(A+B)t] \quad \text{[from (4.79)]}.\end{aligned} \tag{4.94}$$

Step 2. (method of variation of constant parameters) From step 1, Observation 4.3.5, and Theorem 1.3.3(I1), we have

$$\Phi^{-1}(t) = \Phi_{\mathrm{P}}^{-1}(t)\Phi_{\mathrm{U}}^{-1}(t). \tag{4.95}$$

Moreover, from Theorem 4.3.3, we have

$$d\Phi_{\mathrm{U}}^{-1}(t) = -\Phi_{\mathrm{U}}^{-1}(t)A\,dt, \tag{4.96}$$

$$d\Phi_{\mathrm{P}}^{-1}(t) = -\Phi_{\mathrm{P}}^{-1}(t)B\,dt, \tag{4.97}$$

and applying the product formula (Theorem 1.5.11) to $\Phi^{-1}(t)$, we get

$$\begin{aligned}d\Phi^{-1}(t) &= d\Phi_{\mathrm{P}}^{-1}(t)\Phi_{\mathrm{U}}^{-1}(t) + \Phi_{\mathrm{P}}^{-1}(t)d\Phi_{\mathrm{U}}^{-1}(t) \quad \text{(Theorem 1.5.11)}\\ &= -\Phi_{\mathrm{P}}^{-1}(t)B\,dt\Phi_{\mathrm{U}}^{-1}(t) - \Phi_{\mathrm{P}}^{-1}(t)\Phi_{\mathrm{U}}^{-1}(t)A\,dt \quad \text{[(4.96) and (4.97)]}\\ &= -\Phi_{\mathrm{P}}^{-1}(t)\Phi_{\mathrm{U}}^{-1}(t)A\,dt - \Phi_{\mathrm{P}}^{-1}(t)\Phi_{\mathrm{U}}^{-1}(t)B\,dt \quad \text{[from (4.74)]}\\ &= -\Phi^{-1}(t)A\,dt - \Phi^{-1}(t)B\,dt \quad \text{[from (4.95)]}\\ &= -[\Phi^{-1}(t)A + \Phi^{-1}(t)B]dt \quad \text{(by simplification)}.\end{aligned} \tag{4.98}$$

With this discussion, we are ready to find a particular solution to (4.92). A particular solution process x_p of (4.92) is a function that satisfies the system of differential equations (4.92). Imitating the ideas of Section 2.5, we define a function:

$$x_p(t) = \Phi(t)c(t), \quad c(a) = c_a, \tag{4.99}$$

where $\Phi(t)$ is as determined in step 1, and $c(t)$ is an unknown function with $c(a) = c_a$. Our goal of finding a particular solution process $x_p(t)$ of (4.92) is equivalent to finding an unknown function $c(t)$ in (4.99) passing through c_a at $t = a$. For this purpose, we assume that $x_p(t)$ is a solution to (4.92). By using this assumption and applying the product formula (Theorem 1.5.11) to

$$c(t) = \Phi^{-1}(t)x_p(t), \tag{4.100}$$

we obtain

$$\begin{aligned}
dc(t) &= \Phi^{-1}(t)\,dx_p(t) + d\Phi^{-1}(t)x_p(t) \quad \text{(definition of } x_p \text{ and Theorem 1.5.11)} \\
&= \Phi^{-1}(t)[Ax_p(t) + Bx_p(t) + p(t)]dt \\
&\quad - [\Phi^{-1}(t)A + \Phi^{-1}(t)B]dtx_p(t) \quad \text{(by substitution)} \\
&= \Phi^{-1}(t)[Ax_p(t) + Bx_p(t)]dt \\
&\quad - [\Phi^{-1}(t)Ax_p(t) + \Phi^{-1}(t)Bx_p(t)]dt \\
&\quad + \Phi^{-1}(t)p(t)dt \quad \text{(by regrouping)} \\
&= \Phi^{-1}(t)p(t)]dt \quad \text{(by simplifications).}
\end{aligned} \tag{4.101}$$

Now, the problem of finding a particular solution $x_p(t)$ defined in (4.99) reduces to the problem of solving the following IVP:

$$dc(t) = \Phi^{-1}(t)p(t)\,dt, \quad c(a) = c_a. \tag{4.102}$$

This IVP is an integrable system of differential equations. This is an extension of the scalar IVP (2.18). Using the method of solving the IVP (2.18) and Definition 1.5.3, the solution to the IVP (4.102) is given by

$$c(t) = c_a + \int_a^t \Phi^{-1}(s)p(s)\,ds. \tag{4.103}$$

We substitute the expression for $c(t)$ in (4.103) into (4.99), and we have

$$\begin{aligned}
x_p(t) &= \Phi(t)c_a + \int_a^t \Phi^{-1}(s)p(s)\,ds \\
&= \Phi(t)c_a + \int_a^t \Phi(t)\Phi^{-1}(s)p(s)\,ds.
\end{aligned} \tag{4.104}$$

Step 3. Now, we follow step 3 of the procedure in Subsection 2.5.2, and justify the claim that a function defined by $x(t) = x_c(t) + x_p(t)$ is a general solution to (4.92). For this purpose, we consider

$$\begin{aligned}
x(t) &= x_c(t) + x_p(t) \\
&= \Phi(t)c_1 + \Phi(t)c_a + \int_a^t \Phi(t)\Phi^{-1}(s)p(s)\,ds \quad \text{(substitution)} \\
&= \Phi(t)(c_1 + c_a) + \int_a^t \Phi(t)\Phi^{-1}(s)p(s)\,ds \quad \text{[Theorem 1.3.1(M2)]}
\end{aligned}$$

$$= \Phi(t)c + \int_a^t \Phi(t)\Phi^{-1}(s)p(s)\,ds \quad [\text{rewriting } c \equiv c_1 + c_a]$$

$$= \Phi(t)[c + \int_a^t \Phi^{-1}(s)p(s)]ds \quad \text{(Theorem 1.3.1(M2))}, \tag{4.105}$$

where $c = c_1 + c_a$ is an arbitrary constant (generic constant), because of the fact that c_1 and c_a are arbitrary constants. Therefore, $\Phi(t)c$ is also the general solution to (4.92). We show that $x(t)$ is a general solution to (4.92):

$$\begin{aligned} dx(t) &= d\Phi(t)\left[c + \int_a^t \Phi^{-1}(s)p(s)\,ds\right] \\ &\quad + \Phi(t)d\left[c + \int_a^t \Phi^{-1}(s)p(s)\,ds\right] \quad \text{(Theorem 1.5.11)} \\ &= d\Phi(t)\left[c + \int_a^t \Phi^{-1}(s)p(s)\,ds\right] \\ &\quad + \Phi(t)d\left[\int_a^t \Phi^{-1}(s)p(s)\,ds\right] \quad \text{(Theorem 1.5.10)}. \end{aligned} \tag{4.106}$$

We know that

$$d\Phi(t) = A\Phi(t)\,dt + B\Phi(t)\,dt \quad [\text{from (4.75)}],$$

and also from the fundamental theorem of elementary integral calculus [85], we have

$$d\left[\int_a^t \Phi^{-1}(s)p(s)\,ds\right] = \Phi^{-1}(t)p(t)\,dt.$$

We substitute these expressions in (4.106), and simplify by using elementary calculus as

$$\begin{aligned} dx(t) &= [A\Phi(t)dt + B\Phi(t)\,dt]\left[c + \int_a^t \Phi^{-1}(s)p(s)\,ds\right] \\ &\quad + \Phi(t)\Phi^{-1}(t)p(t)\,dt \quad \text{(substitution)} \\ &= A\Phi(t)\left[c + \int_a^t \Phi^{-1}(s)p(s)\,ds\right]dt \\ &\quad + B(t)\Phi(t)\left[c + \int_a^t \Phi^{-1}(s)p(s)\,ds\right]dt \\ &\quad + \Phi(t)\Phi^{-1}(t)p(t)\,dt \quad [\text{Theorem 1.3.1(M2)}] \\ &= A\Phi(t)x(t)\,dt + B\Phi(t)x(t)\,dt + p(t)\,dt \quad [(4.105) \text{ and simplifications}] \\ &= [Ax(t) + Bx(t) + p(t)]dt \quad \text{(regrouping)}. \end{aligned} \tag{4.107}$$

Fig. 4.17 Flowchart for finding a general solution.

This shows that $x(t) = x_c(t) + x_p(t)$ is indeed a solution to (4.92). Moreover, from (4.79), the general solution to (4.92) is expressed as

$$\begin{aligned} x(t) &= \Phi(t)c + \left[\int_a^t \Phi^{-1}(s)p(s)\,ds\right] \\ &= \exp[(A+B)(t-a)]c + \int_a^t \exp[(A+B)(t-s)]p(s)\,ds. \end{aligned} \tag{4.108}$$

The process of finding the general solution is summarized in Figure 4.17.

Observation 4.6.1. An observation corresponding to Observation 2.5.1 can be reformulated. The details are left to the reader.

Illustration 4.6.1. Let us consider $dx = [Ax + Bx + p(t)]\,dt$, where

$$A = \begin{bmatrix} a_{11} & a_{12} \\ a_{21} & a_{22} \end{bmatrix}, \quad B = \begin{bmatrix} b_{11} & b_{12} \\ b_{21} & b_{22} \end{bmatrix},$$

and b_{ij}'s and a_{ij}'s are any given real numbers satisfying the following condition, (i) $d(A) = (\mathrm{tr}\,(A))^2 - 4\det(A) > 0$, (ii) $AB = BA$, and (iii) p is a two-dimensional continuous function defined on an interval $J \subseteq R$ into R^2. Find: (a) a general solution to the given differential equation, and also for (b) $B = 0$, (c) $B = 0$ and $p(t) \equiv 0$, (d) $A = 0$, and (e) $A = 0$ and $p(t) \equiv 0$.

Solution procedure. The main goal of the illustration is to find a general solution to the given system of differential equations. Prior to applying the procedure in Subsection 4.6.2, we need to (i) find a fundamental matrix solution corresponding to the overall linear system, (ii) find a particular solution to the given problem, and (iii) test the sum of the solutions in (i) and (ii). First, from $(\mathrm{tr}\,(A))^2 - 4\det(A) > 0$, and (ii) $AB = BA$ and imitating the solution procedure of Illustration 4.4.1, the expressions for the fundamental matrix solutions of the unperturbed and perturbed components of the above given system of differential equations can be obtained. Moreover, we need one of the fundamental matrix solutions to be normalized (step 2 of the procedure in Subsection 4.5.2 and Observation 4.5.1). Therefore, without loss of generality, we find the normalized fundamental matrix solution corresponding to the unperturbed part $\Phi_U(t)$. We use Observation 4.3.3(iv) and step 3 of the procedure in Subsection 4.3.4 with $t_0 = 0$ and compute $\Phi_U(0)$, $\Phi_U^{-1}(0)$, and the normalized fundamental matrix solution $\Phi_U(t, 0)$ with respect to $\Phi_U(t)$:

Case 1:

(i) $\Phi_U(0) = \begin{bmatrix} e^{a_{11}0} & 0 \\ 0 & e^{a_{22}0} \end{bmatrix} = \begin{bmatrix} 1 & 0 \\ 0 & 1 \end{bmatrix}$,

$$\Phi_U^{-1}(0) = \begin{bmatrix} 1 & 0 \\ 0 & 1 \end{bmatrix}, \quad \Phi_U(t,0) = \begin{bmatrix} e^{a_{11}t} & 0 \\ 0 & e^{a_{22}t} \end{bmatrix};$$

(ii) $\Phi_U(0) = \begin{bmatrix} (a_{11} - a_{22})e^{a_{11}0} & 0 \\ a_{21}e^{a_{11}0} & e^{a_{22}0} \end{bmatrix} = \begin{bmatrix} a_{11} - a_{22} & 0 \\ a_{21} & 1 \end{bmatrix}$,

$$\Phi_U^{-1}(0) = \begin{bmatrix} \dfrac{1}{a_{11} - a_{22}} & 0 \\ \dfrac{-a_{21}}{a_{11} - a_{22}} & 1 \end{bmatrix}, \quad \Phi_U(t,0) = \begin{bmatrix} e^{a_{11}t} & 0 \\ \dfrac{a_{21}(e^{a_{22}t} - e^{a_{11}t})}{a_{11} - a_{22}} & e^{a_{22}t} \end{bmatrix};$$

(iii) $\Phi_U(0) = \begin{bmatrix} e^{a_{11}0} & a_{12}e^{a_{22}0} \\ 0 & a_{22} - a_{11}e^{a_{22}0} \end{bmatrix} = \begin{bmatrix} 1 & a_{12} \\ 0 & a_{22} - a_{11} \end{bmatrix}$,

$$\Phi_U^{-1}(0) = \begin{bmatrix} 1 & \dfrac{-a_{12}}{a_{22} - a_{11}} \\ 0 & \dfrac{1}{a_{22} - a_{11}} \end{bmatrix}, \quad \Phi_U(t,0) = \begin{bmatrix} e^{a_{11}t} & \dfrac{a_{12}(e^{a_{11}t} - e^{a_{22}t})}{(a_{22} - a_{11})} \\ 0 & e^{a_{22}t} \end{bmatrix};$$

Case 2:

$$\Phi_U(0) = \begin{bmatrix} a_{12}e^{\lambda_1 0} & (\lambda_2 - a_{22})e^{\lambda_2 0} \\ (\lambda_1 - a_{11})e^{\lambda_1 0} & a_{21}e^{\lambda_2 0} \end{bmatrix} = \begin{bmatrix} a_{12} & \lambda_2 - a_{22} \\ \lambda_1 - a_{11} & a_{21} \end{bmatrix},$$

$$\Phi_U^{-1}(0) = \frac{2}{d(A) + (a_{22} - a_{11})\sqrt{d(A)}} \begin{bmatrix} a_{21} & a_{22} - \lambda_2 \\ a_{11} - \lambda_1 & a_{12} \end{bmatrix},$$

$$\Phi_U(t,0) = \begin{bmatrix} \phi_{11}(t) & \phi_{12}(t) \\ \phi_{21}(t) & \phi_{22}(t) \end{bmatrix},$$

where

$$\phi_{11}(t) = \frac{2[a_{21}a_{12}e^{\lambda_1 t} + (\lambda_2 - a_{22})(a_{11} - \lambda_1)e^{\lambda_2 t}]}{d(A) + (a_{22} - a_{11})\sqrt{d(A)}},$$

$$\phi_{12}(t) = \frac{2[a_{12}(a_{22} - \lambda_2)(e^{\lambda_1 t} - e^{\lambda_2 t})]}{d(A) + (a_{22} - a_{11})\sqrt{d(A)}},$$

$$\phi_{21}(t) = \frac{2[a_{21}(a_{11} - \lambda_1)(e^{\lambda_2 t} - e^{\lambda_1 t})]}{d(A) + (a_{22} - a_{11})\sqrt{d(A)}},$$

$$\phi_{22}(t) = \frac{2[(\lambda_1 - a_{11})(a_{22} - \lambda_2)e^{\lambda_1 t} + a_{21}a_{12}e^{\lambda_2 t}]}{d(A) + (a_{22} - a_{11})\sqrt{d(A)}}.$$

The computations of $\Phi_P(0)$, $\Phi_P^{-1}(0)$, and the normalized fundamental matrix solution $\Phi_P(t,0)$ with respect to $\Phi_P(t)$ can be reproduced, analogously. The details are left as an exercise.

Now, from step 2 of the procedure in Subsection 4.5.2, a suitable candidate for the fundamental matrix solution corresponding to the linear homogeneous system of differential equations is:

Case 1:

(i) $$\Phi(t,0) = \Phi_U(t,0)\Phi_P(t,0) = \begin{bmatrix} e^{a_{11}t} & 0 \\ 0 & e^{a_{22}t} \end{bmatrix}\begin{bmatrix} e^{b_{11}t} & 0 \\ 0 & e^{b_{22}t} \end{bmatrix} = \begin{bmatrix} e^{(a_{11}+b_{11})t} & 0 \\ 0 & e^{(a_{22}+b_{22})t} \end{bmatrix},$$

(ii) $$\Phi(t,0) = \Phi_U(t,0)\Phi_P(t,0)$$
$$= \begin{bmatrix} e^{a_{11}t} & 0 \\ \dfrac{a_{21}(e^{a_{22}t}-e^{a_{11}t})}{a_{11}-a_{22}} & e^{a_{22}t} \end{bmatrix}\begin{bmatrix} e^{b_{11}t} & 0 \\ \dfrac{b_{21}(e^{b_{22}t}-e^{b_{11}t})}{b_{11}-b_{22}} & e^{b_{22}t} \end{bmatrix}$$
$$= \begin{bmatrix} e^{(a_{11}+b_{11})t} & 0 \\ \dfrac{a_{21}e^{b_{11}t}(e^{a_{22}t}-e^{a_{11}t})}{a_{11}-a_{22}} + \dfrac{b_{21}e^{a_{22}t}(e^{b_{22}t}-e^{b_{11}t})}{b_{11}-b_{22}} & e^{(a_{22}+b_{22})t} \end{bmatrix},$$

(iii) $$\Phi(t,0) = \Phi_U(t,0)\Phi_P(t,0)$$
$$= \begin{bmatrix} e^{a_{11}t} & \dfrac{a_{12}(e^{a_{11}t}-e^{a_{22}t})}{a_{22}-a_{11}} \\ 0 & e^{a_{22}t} \end{bmatrix}\begin{bmatrix} e^{b_{11}t} & \dfrac{b_{12}(e^{b_{11}t}-e^{b_{22}t})}{b_{22}-b_{11}} \\ 0 & e^{b_{22}t} \end{bmatrix}$$
$$= \begin{bmatrix} e^{(a_{11}+b_{11})t} & \dfrac{a_{12}e^{b_{22}t}(e^{a_{11}t}-e^{a_{22}t})}{a_{22}-a_{11}} + \dfrac{b_{12}e^{a_{11}t}(e^{b_{11}t}-e^{b_{22}t})}{b_{22}-b_{11}} \\ 0 & e^{(a_{22}+b_{22})t} \end{bmatrix},$$

Case 2:

$$\Phi(t,0) = \Phi_U(t,0)\Phi_P(t,0) = \begin{bmatrix} \phi_{11}(t) & \phi_{12}(t) \\ \phi_{21}(t) & \phi_{22}(t) \end{bmatrix}\begin{bmatrix} \varpi_{11}(t) & \varpi_{12}(t) \\ \varpi_{21}(t) & \varpi_{22}(t) \end{bmatrix} = \begin{bmatrix} \varphi_{11}(t) & \varphi_{12}(t) \\ \varphi_{21}(t) & \varphi_{22}(t) \end{bmatrix},$$

where

$$\varpi_{11}(t) = \frac{2\left[b_{21}b_{12}e^{\lambda_1^P t} + \left(\lambda_2^P - b_{22}\right)\left(b_{11}-\lambda_1^P\right)e^{\lambda_2^P t}\right]}{d(B) + (b_{22}-b_{11})\sqrt{d(B)}},$$

$$\varpi_{12}(t) = \frac{2\left[b_{12}\left(b_{22}-\lambda_2^P\right)\left(e^{\lambda_1^P t} - e^{\lambda_2^P t}\right)\right]}{d(B) + (b_{22}-b_{11})\sqrt{d(B)}},$$

$$\varpi_{21}(t) = \frac{2\left[b_{21}\left(b_{11}-\lambda_1^{\mathrm{P}}\right)\left(e^{\lambda_2^{\mathrm{P}}t}-e^{\lambda_1^{\mathrm{P}}t}\right)\right]}{d(B)+(b_{22}-b_{11})\sqrt{d(B)}},$$

$$\varpi_{22}(t) = \frac{2\left[\left(\lambda_1^{\mathrm{P}}-b_{11}\right)\left(b_{22}-\lambda_2^{\mathrm{P}}\right)e^{\lambda_1^{\mathrm{P}}t}+b_{21}b_{12}e^{\lambda_2^{\mathrm{P}}t}\right]}{d(B)+(b_{22}-b_{11})\sqrt{d(B)}},$$

$$\varphi_{11}(t) = \phi_{11}(t)\varpi_{11}(t)+\phi_{12}(t)\varpi_{21}(t),$$

$$\varphi_{12}(t) = \phi_{11}(t)\varpi_{12}(t)+\phi_{12}(t)\varpi_{22}(t),$$

$$\varphi_{21}(t) = \phi_{21}(t)\varpi_{11}(w(t))+\phi_{22}(t)\varpi_{21}(t),$$

$$\varphi_{22}(t) = \phi_{21}(t)\varpi_{12}(t)+\phi_{22}(t)\varpi_{22}(t).$$

For $r>0$ (Observation 4.5.2),

$$\begin{aligned}
&\phi_{11}(t)\varpi_{11}(t)\\
&\quad= \frac{2[a_{21}a_{12}e^{\lambda_1 t}+(\lambda_2-a_{22})(a_{11}-\lambda_1)e^{\lambda_2 t}]}{d(A)+(a_{22}-a_{11})\sqrt{d(A)}}\\
&\qquad\times\frac{2[b_{21}b_{12}e^{\lambda_1^{\mathrm{P}}t}+(\lambda_2^{\mathrm{P}}-b_{22})(b_{11}-\lambda_1^{\mathrm{P}})e^{\lambda_2^{\mathrm{P}}t}]}{d(B)+b_{22}-b_{11}\sqrt{d(B)}}\\
&\quad= \frac{r^2 4[b_{21}b_{12}e^{\lambda_1 t}+(\lambda_2^{\mathrm{P}}-b_{22})(b_{11}-\lambda_1^{\mathrm{P}})e^{\lambda_2 t}]}{r^2(d(B)+(b_{22}-b_{11})\sqrt{d(B)})}\\
&\qquad\times\frac{[b_{21}b_{12}e^{\lambda_1^{\mathrm{P}}t}+(\lambda_2^{\mathrm{P}}-b_{22})(b_{11}-\lambda_1^{\mathrm{P}})e^{\lambda_2^{\mathrm{P}}t}]}{d(B)+(b_{22}-b_{11})\sqrt{d(B)}}\\
&\quad= \frac{4[b_{21}b_{12}e^{\lambda_1 t}+(\lambda_2^{\mathrm{P}}-b_{22})(b_{11}-\lambda_1^{\mathrm{P}})e^{\lambda_2 t}][b_{21}b_{12}e^{\lambda_1^{\mathrm{P}}t}+(\lambda_2^{\mathrm{P}}-b_{22})(b_{11}-\lambda_1^{\mathrm{P}})e^{\lambda_2^{\mathrm{P}}t}]}{(d(B)+(b_{22}-b_{11})\sqrt{d(B)})^2}\\
&\quad= \frac{4[b_{21}b_{12}(\lambda_2^{\mathrm{P}}-b_{22})(b_{11}-\lambda_1^{\mathrm{P}})e^{(\lambda_1+\lambda_2^{\mathrm{P}})t}+b_{21}b_{12}(\lambda_2^{\mathrm{P}}-b_{22})(b_{11}-\lambda_1^{\mathrm{P}})e^{(\lambda_2+\lambda_1^{\mathrm{P}})t}]}{(d(B)+(b_{22}-b_{11})\sqrt{d(B)})^2}\\
&\qquad+\frac{4[(b_{21}b_{12})^2e^{(\lambda_1+\lambda_1^{\mathrm{P}})t}+[(\lambda_2^{\mathrm{P}}-b_{22})(b_{11}-\lambda_1^{\mathrm{P}})]^2e^{(\lambda_2+\lambda_2^{\mathrm{P}})t}]}{(d(B)+(b_{22}-b_{11})\sqrt{d(B)})^2},
\end{aligned}$$

$$\begin{aligned}
&\phi_{12}(t)\varpi_{21}(t)\\
&\quad= \frac{2[a_{12}(a_{22}-\lambda_2)(e^{\lambda_1 t}-e^{\lambda_2 t})]}{d(A)+(a_{22}-a_{11})\sqrt{d(A)}}\,\frac{2[b_{21}(b_{11}-\lambda_1^{\mathrm{P}})(e^{\lambda_2^{\mathrm{P}}t}-e^{\lambda_1^{\mathrm{P}}t})]}{d(B)+(b_{22}-b_{11})\sqrt{d(B)}}\\
&\quad= \frac{4[b_{12}(b_{22}-\lambda_2^{\mathrm{P}})(e^{\lambda_1 t}-e^{\lambda_2 t})][b_{21}(b_{11}-\lambda_1^{\mathrm{P}})(e^{\lambda_2^{\mathrm{P}}t}-e^{\lambda_1^{\mathrm{P}}t})]}{(d(B)+(b_{22}-b_{11})\sqrt{d(B)})^2}\\
&\quad= \frac{\begin{array}{l}4[b_{21}b_{12}(b_{22}-\lambda_2^{\mathrm{P}})(b_{11}-\lambda_1^{\mathrm{P}})(e^{\lambda_1 t}-e^{\lambda_2 t})e^{\lambda_2^{\mathrm{P}}t}\\ -[b_{21}b_{12}(b_{22}-\lambda_2^{\mathrm{P}})(b_{11}-\lambda_1^{\mathrm{P}})(e^{\lambda_1 t}-e^{\lambda_2 t})e^{\lambda_1^{\mathrm{P}}t}]]\end{array}}{(d(B)+(b_{22}-b_{11})\sqrt{d(B)})^2}\\
&\quad= \frac{4[-b_{21}b_{12}(\lambda_2^{\mathrm{P}}-b_{22})(b_{11}-\lambda_1^{\mathrm{P}})e^{(\lambda_1+\lambda_2^{\mathrm{P}})t}-b_{21}b_{12}(\lambda_2^{\mathrm{P}}-b_{22})(b_{11}-\lambda_1^{\mathrm{P}})e^{(\lambda_2+\lambda_1^{\mathrm{P}})t}]}{(d(B)+(b_{22}-b_{11})\sqrt{d(B)})^2}\\
&\qquad+\frac{4[b_{21}b_{12}(\lambda_2^{\mathrm{P}}-b_{22})(b_{11}-\lambda_1^{\mathrm{P}})e^{(\lambda_2+\lambda_2^{\mathrm{P}})t}+b_{21}b_{12}(\lambda_2^{\mathrm{P}}-b_{22})(b_{11}-\lambda_1^{\mathrm{P}})e^{(\lambda_1+\lambda_1^{\mathrm{P}})t}]}{(d(B)+(b_{22}-b_{11})\sqrt{d(B)})^2}.
\end{aligned}$$

Hence,

$$
\begin{aligned}
&\varphi_{11}(t) \\
&\quad = \phi_{11}(t)\varpi_{11}(t) + \phi_{12}(t)\varpi_{21}(t) \\
&\quad = \frac{4[(b_{21}b_{12})^2 e^{(\lambda_1+\lambda_1^{\mathrm{P}}t)} + [(\lambda_2^{\mathrm{P}} - a_{22})(b_{11} - \lambda_1^{\mathrm{P}})]^2 e^{(\lambda_2+\lambda_2^{\mathrm{P}}t)}]}{(d(B) + (b_{22} - b_{11})\sqrt{d(B)})^2} \\
&\qquad + \frac{4[b_{21}b_{12}(\lambda_2^{\mathrm{P}} - b_{22})(b_{11} - \lambda_1^{\mathrm{P}})e^{(\lambda_2+\lambda_2^{\mathrm{P}})t} + b_{21}b_{12}(\lambda_2^{\mathrm{P}} - b_{22})(b_{11} - \lambda_1^{\mathrm{P}})e^{(\lambda_1+\lambda_1^{\mathrm{P}})t}]}{((d(B) + (b_{22} - b_{11})\sqrt{d(B)})^2} \\
&\quad = \frac{4[(b_{21}b_{12})^2 e^{(\lambda_1+\lambda_1^{\mathrm{P}})t} + b_{21}b_{12}(\lambda_2^{\mathrm{P}} - b_{22})(b_{11} - \lambda_1^{\mathrm{P}})e^{(\lambda_1+\lambda_1^{\mathrm{P}})t}]}{(d(B) + (b_{22} - b_{11})\sqrt{d(B)})^2} \\
&\qquad + \frac{4[b_{21}b_{12}(\lambda_2^{\mathrm{P}} - b_{22})(b_{11} - \lambda_1^{\mathrm{P}})e^{(\lambda_2+\lambda_2^{\mathrm{P}})t} + [(\lambda_2^{\mathrm{P}} - b_{22})(b_{11} - \lambda_1^{\mathrm{P}})]^2 e^{(\lambda_2+\lambda_2^{\mathrm{P}})t}]}{(d(B) + (b_{22} - b_{11})\sqrt{d(B)})^2} \\
&\quad = \frac{4[(b_{21}b_{12})^2 + b_{21}b_{12}(\lambda_2^{\mathrm{P}} - b_{22})(b_{11} - \lambda_1^{\mathrm{P}})]e^{(\lambda_1+\lambda_1^{\mathrm{P}})t}}{(d(B) + (b_{22} - b_{11})\sqrt{d(B)})^2} \\
&\qquad + \frac{4[b_{21}b_{12}(\lambda_2^{\mathrm{P}} - b_{22})(b_{11} - \lambda_1^{\mathrm{P}}) + [(\lambda_2^{\mathrm{P}} - b_{22})(b_{11} - \lambda_1^{\mathrm{P}})]^2]e^{(\lambda_2+\lambda_2^{\mathrm{P}})t}}{(d(B) + (b_{22} - b_{11})\sqrt{d(B)})^2},
\end{aligned}
$$

$$
\begin{aligned}
&\phi_{11}(t)\varpi_{12}(t) \\
&\quad = \frac{2[a_{21}a_{12}e^{\lambda_1 t} + (\lambda_2 - a_{22})(a_{11} - \lambda_1)e^{\lambda_2 t}]}{d(A) + (a_{22} - a_{11})\sqrt{d(A)}} \, \frac{2[b_{12}(b_{22} - \lambda_2^{\mathrm{P}})(e^{\lambda_1^{\mathrm{P}} t} - e^{\lambda_2^{\mathrm{P}} t})]}{d(B) + (b_{22} - b_{11})\sqrt{d(B)}} \\
&\quad = \frac{4[b_{21}b_{12}e^{\lambda_1 t} + (\lambda_2^{\mathrm{P}} - b_{22})(b_{11} - \lambda_1^{\mathrm{P}})e^{\lambda_2 t}]}{d(B) + (b_{22} - b_{11})\sqrt{d(B)}} \, \frac{[b_{12}(b_{22} - \lambda_2^{\mathrm{P}})(e^{\lambda_1^{\mathrm{P}} t} - e^{\lambda_2^{\mathrm{P}} t})]}{d(B) + (b_{22} - b_{11})\sqrt{d(B)}} \\
&\quad = \frac{4[b_{21}b_{12}^2(b_{22} - \lambda_2^{\mathrm{P}})e^{(\lambda_1+\lambda_1^{\mathrm{P}})t} - b_{12}(\lambda_2^{\mathrm{P}} - b_{22})^2(b_{11} - \lambda_1^{\mathrm{P}})e^{(\lambda_2+\lambda_2^{\mathrm{P}})t}]}{(d(B) + (b_{22} - b_{11})\sqrt{d(B)})^2} \\
&\qquad + \frac{4[b_{12}(\lambda_2^{\mathrm{P}} - b_{22})^2(b_{11} - \lambda_1^{\mathrm{P}})e^{(\lambda_2+\lambda_1^{\mathrm{P}})t} - b_{21}b_{12}^2(b_{22} - \lambda_2^{\mathrm{P}})e^{(\lambda_1+\lambda_2^{\mathrm{P}})t}]}{(d(B) + (b_{22} - b_{11})\sqrt{d(B)})^2},
\end{aligned}
$$

$$
\begin{aligned}
&\phi_{12}(t)\varpi_{22}(t) \\
&\quad = \frac{2[a_{12}(a_{22} - \lambda_2)(e^{\lambda_1 t} - e^{\lambda_2 t})]}{d(A) + (a_{22} - a_{11})\sqrt{d(A)}} \, \frac{2[(\lambda_1^{\mathrm{P}} - b_{11})(b_{22} - \lambda_2^{\mathrm{P}})e^{\lambda_1^{\mathrm{P}} t} + b_{21}b_{12}e^{\lambda_2^{\mathrm{P}} t}]}{d(B) + (b_{22} - b_{11})\sqrt{d(B)}} \\
&\quad = \frac{4[b_{12}(b_{22} - \lambda_2^{\mathrm{P}})(e^{\lambda_1 t} - e^{\lambda_2 t})]}{d(B) + (b_{22} - b_{11})\sqrt{d(B)}} \, \frac{[(\lambda_1^{\mathrm{P}} - b_{11})(b_{22} - \lambda_2^{\mathrm{P}})e^{\lambda_1^{\mathrm{P}} t} + b_{21}b_{12}e^{\lambda_2^{\mathrm{P}})t}]}{d(B) + (b_{22} - b_{11})\sqrt{d(B)}} \\
&\quad = \frac{4[b_{12}(\lambda_1^{\mathrm{P}} - b_{11})(b_{22} - \lambda_2^{\mathrm{P}})^2 e^{(\lambda_1+\lambda_1^{\mathrm{P}})t} - b_{21}b_{12}^2(b_{22} - \lambda_2^{\mathrm{P}})e^{(\lambda_2+\lambda_2^{\mathrm{P}})t}]}{(d(B) + (b_{22} - b_{11})\sqrt{d(B)})^2} \\
&\qquad + \frac{[b_{21}b_{12}^2(b_{22} - \lambda_2^{\mathrm{P}})e^{(\lambda_1+\lambda_2^{\mathrm{P}})t} - b_{12}(\lambda_1^{\mathrm{P}} - b_{11})(b_{22} - \lambda_2^{\mathrm{P}})^2 e^{(\lambda_2+\lambda_1^{\mathrm{P}})t}]}{(d(B) + (b_{22} - b_{11})\sqrt{d(B)})^2}.
\end{aligned}
$$

Hence,

$$\varphi_{12}(t) = \phi_{11}(t)\varpi_{12}(t) + \phi_{12}(t)\varpi_{22}(t)$$
$$= \frac{4[b_{21}b_{12}^2(b_{22}-\lambda_2^{\mathrm{P}})e^{(\lambda_1+\lambda_1^{\mathrm{P}})t} + b_{12}(\lambda_2^{\mathrm{P}}-b_{22})^2(b_{11}-\lambda_1^{\mathrm{P}})e^{(\lambda_2+\lambda_2^{\mathrm{P}})t}]}{\left(d(B)+(b_{22}-b_{11})\sqrt{d(B)}\right)^2}$$
$$+ \frac{4[b_{12}(\lambda_1^{\mathrm{P}}-b_{11})(b_{22}-\lambda_2^{\mathrm{P}})^2e^{(\lambda_1+\lambda_1^{\mathrm{P}})t} - b_{21}b_{12}^2(b_{22}-\lambda_2^{\mathrm{P}})e^{(\lambda_2+\lambda_2^{\mathrm{P}})t}]}{(d(B)+(b_{22}-b_{11})\sqrt{d(B)})^2}$$
$$= \frac{4[b_{21}b_{12}^2(b_{22}-\lambda_2^{\mathrm{P}})e^{(\lambda_1+\lambda_1^{\mathrm{P}})t} + b_{12}(\lambda_1^{\mathrm{P}}-b_{11})(b_{22}-\lambda_2^{\mathrm{P}})^2e^{(\lambda_1+\lambda_1^{\mathrm{P}})t}]}{(d(B)+(b_{22}-b_{11})\sqrt{d(B)})^2}$$
$$- \frac{4[b_{12}(b_{22}-\lambda_2^{\mathrm{P}})^2(\lambda_1^{\mathrm{P}}-b_{11})e^{(\lambda_2+\lambda_2^{\mathrm{P}})t} + b_{21}b_{12}^2(b_{22}-\lambda_2^{\mathrm{P}})e^{(\lambda_2+\lambda_2^{\mathrm{P}})t}]}{(d(B)+(b_{22}-b_{11})\sqrt{d(B)})^2}$$
$$= \frac{4b_{12}(b_{22}-\lambda_2^{\mathrm{P}})[b_{21}b_{12}+(\lambda_1^{\mathrm{P}}-b_{11})(b_{22}-\lambda_2^{\mathrm{P}})]e^{(\lambda_1+\lambda_1^{\mathrm{P}})t}}{(d(B)+(b_{22}-b_{11})\sqrt{d(B)})^2}$$
$$- \frac{4b_{12}(b_{22}-\lambda_2^{\mathrm{P}})[(\lambda_1^{\mathrm{P}}-b_{11})(b_{22}-\lambda_2^{\mathrm{P}})+b_{21}b_{12}]e^{(\lambda_2+\lambda_2^{\mathrm{P}})t}}{(d(B)+(b_{22}-b_{11})\sqrt{d(B)})^2}$$
$$= \frac{4b_{12}(b_{22}-\lambda_2^{\mathrm{P}})[b_{21}b_{12}+(\lambda_1^{\mathrm{P}}-b_{11})(b_{22}-\lambda_2^{\mathrm{P}})][e^{(\lambda_1+\lambda_1^{\mathrm{P}})t}-e^{(\lambda_2+\lambda_2^{\mathrm{P}})t}]}{(d(B)+(b_{22}-b_{11})\sqrt{d(B)})^2}.$$

We observe the fact that the coefficients of $e^{\lambda_1 t}$, $e^{\lambda_2 t}$, $e^{\lambda_1^{\mathrm{P}} t}$, and $e^{\lambda_2^{\mathrm{P}} t}$ in $\phi_{22}(t)\varpi_{22}(t)$, $\phi_{21}(t)\varpi_{12}(t)$ can be obtained by interchanging the corresponding coefficients of $e^{\lambda_1 t}$, $e^{\lambda_2 t}$, $e^{\lambda_1^{\mathrm{P}} t}$, and $e^{\lambda_2^{\mathrm{P}} t}$ in $\phi_{11}(t)\varpi_{11}(t)$ and $\phi_{12}(t)\varpi_{21}(t)$, respectively. Thus,

$$\varphi_{22}(t) = \phi_{21}(t)\varpi_{12}(t) + \phi_{22}(t)\varpi_{22}(t)$$
$$= \frac{4[(b_{21}b_{12})^2 + b_{21}b_{12}(\lambda_2^{\mathrm{P}}-b_{22})(b_{11}-\lambda_1^{\mathrm{P}})]e^{(\lambda_2+\lambda_2^{\mathrm{P}}t)}}{(d(B)+(b_{22}-b_{11})\sqrt{d(B)})^2}$$
$$+ \frac{4[b_{21}b_{12}(\lambda_2^{\mathrm{P}}-b_{22})(b_{11}-\lambda_1^{\mathrm{P}}) + [(\lambda_2^{\mathrm{P}}-b_{22})(b_{11}-\lambda_1^{\mathrm{P}})]^2]e^{(\lambda_1+\lambda_1^{\mathrm{P}})t}}{(d(B)+(b_{22}-b_{11})\sqrt{d(B)})^2}.$$

Similar comments are valid with regard to $\phi_{21}(t)\varpi_{11}(t)+\phi_{22}(t)\varpi_{21}(t) = \varphi_{21}(t)$ and $\varphi_{12}(t) = \phi_{11}(t)\varpi_{12}(t) + \phi_{12}(t)\varpi_{22}(t)$. Hence,

$$\varphi_{21}(t) = \phi_{21}(t)\varpi_{11}(t) + \phi_{22}(t)\varpi_{21}(t)$$
$$= \frac{4b_{21}(b_{11}-\lambda_1^{\mathrm{P}})[b_{21}b_{12}+(\lambda_1^{\mathrm{P}}-b_{11})(b_{22}-\lambda_2^{\mathrm{P}})][e^{(\lambda_2+\lambda_2^{\mathrm{P}})t}-e^{(\lambda_1+\lambda_1^{\mathrm{P}})t}]}{d(B)+(b_{22}-b_{11})\sqrt{d(B)^2}}.$$

This is the complete representation of the normalized fundamental matrix solution process of the corresponding first-order linear system of differential equations.

Now, by imitating steps 2 and 3 of the procedure in Subsection 4.6.2 as well as Example 2.5.1, one can easily find a particular solution. Then the general solution of the given illustration depends on a case. The form of the general solution is given by (4.108):

$$x(t) = \Phi(t)c + \int_a^t \Phi(t,s)p(s)ds,$$

where the fundamental matrix $\Phi(t)$ is as determined above. This completes the solution process of (a). For $B = 0$ (b) $\Phi_{\mathrm{P}}(t) = C$, where C is a constant nonsingular fundamental solution to (4.69). From (4.75), we conclude that $\Phi(t) = \Phi_{\mathrm{U}}(t)C$ is a fundamental solution to (4.93). Moreover, $\Phi_{\mathrm{U}}(t)$ does not need be a normalized fundamental solution to (4.68). With this observation, (4.108) reduces to

$$x(t) = \Phi_{\mathrm{U}}(t)c + \int_a^t \Phi_{\mathrm{U}}(t,s)p(s)\,ds,$$

where $c = Ca$, with a being an arbitrary constant vector. For the cases of (c), (d), and (e), we have (c) $x(t) = \Phi_{\mathrm{U}}(t)c$, (d) $x(t) = \Phi_{\mathrm{P}}(t)c + \int_a^t \Phi_{\mathrm{P}}(t,s)p(s)\,ds$, and $x(t) = \Phi_{\mathrm{P}}(t)c$, respectively. This completes the solution process.

Example 4.6.1. We are given $dx = [Ax + Bx + p(t)]\,dt$, with

$$A = \begin{bmatrix} 2\alpha & \alpha \\ \alpha & \alpha \end{bmatrix}, \quad B = \begin{bmatrix} 2\sigma & \sigma \\ \sigma & \sigma \end{bmatrix},$$

where σ and α are constants, and p is a 2-continuous function defined on an interval $J \subseteq R$ into R^2. Find: (a) a general solution to the given differential equation, and particularly for (b) $\sigma = 0$, (c) $\sigma = 0$ and $p(t) \equiv 0$, (d) $\alpha = 0$, and (e) $\alpha = 0$ and $p(t) \equiv 0$.

Solution procedure. We observe that $AB = BA$. Now, by following the procedure in Subsection 4.6.2, we find a general solution of the given example. We need to find (i) the fundamental matrix solution and (ii) a particular solution. First, we imitate the solution procedure of Illustrations 4.4.1 and 4.6.1. In this case, the eigenvalues and eigenvectors of the matrices of A and B are

$$\lambda_{1A} = \frac{3+\sqrt{5}}{2}\alpha, \quad c_{1A} = \begin{bmatrix} 1 & \dfrac{-1+\sqrt{5}}{2} \end{bmatrix}^T,$$

$$\lambda_{2A} = \frac{3-\sqrt{5}}{2}\alpha \quad \text{with } c_{2A} = \begin{bmatrix} \dfrac{1-\sqrt{5}}{2} & 1 \end{bmatrix}^T,$$

$$\lambda_{1B} = \frac{3+\sqrt{5}}{2}\sigma, \quad \text{with } c_{1B} = \begin{bmatrix} 1 & \dfrac{-1+\sqrt{5}}{2} \end{bmatrix}^T,$$

$$\lambda_{2B} = \frac{3-\sqrt{5}}{2}\sigma, \quad \text{with } c_{2B} = \begin{bmatrix} \dfrac{1-\sqrt{5}}{2} & 1 \end{bmatrix}^T,$$

respectively. The normalized fundamental matrix solution processes $\Phi_U(t,0)$ and $\Phi_P(t,0)$ of unperturbed and perturbed components are

$$\Phi_U(t,0)$$

$$= \frac{4}{\alpha^2[4+(1-\sqrt{5})^2]} \begin{bmatrix} e^{t\lambda_{1A}} & \dfrac{1-\sqrt{5}}{2} e^{t\lambda_{2A}} \\ \dfrac{-1+\sqrt{5}}{2} e^{t\lambda_{1A}} & e^{t\lambda_{2A}} \end{bmatrix} \begin{bmatrix} 1 & -\dfrac{1-\sqrt{5}}{2} \\ \dfrac{1-\sqrt{5}}{2} & 1 \end{bmatrix}$$

$$= \frac{1}{\alpha^2} \begin{bmatrix} \dfrac{4e^{t\lambda_{1A}} + (1-\sqrt{5})^2 e^{t\lambda_{2A}}}{4+(1-\sqrt{5})^2} & \dfrac{2(1-\sqrt{5})(e^{t\lambda_{2A}} - e^{t\lambda_{1A}})}{4+(1-\sqrt{5})^2} \\ \dfrac{-2(1-\sqrt{5})(e^{t\lambda_{1A}} - e^{t\lambda_{2A}})}{4+(1-\sqrt{5})^2} & \dfrac{(1-\sqrt{5})^2 e^{t\lambda_{1A}} + 4e^{t\lambda_{2A}}}{4+(1-\sqrt{5})^2} \end{bmatrix},$$

$$\Phi_P(t,0)$$

$$= \begin{bmatrix} \dfrac{4e^{t\lambda_{1B}} + (1-\sqrt{5})^2 e^{t\lambda_{2B}}}{4+(1-\sqrt{5})^2} & \dfrac{2(1-\sqrt{5})(e^{t\lambda_{2B}} - e^{t\lambda_{1B}})}{4+(1-\sqrt{5})^2} \\ \dfrac{-2(1-\sqrt{5})(e^{t\lambda_{1B}} - e^{t\lambda_{2B}})}{4+(1-\sqrt{5})^2} & \dfrac{(1-\sqrt{5})^2 e^{t\lambda_{1B}} + 4e^{t\lambda_{2B}}}{4+(1-\sqrt{5})^2} \end{bmatrix}.$$

From step 1 of the procedure in Subsection 4.6.2, the fundamental matrix solution to the corresponding homogeneous system $dx = Ax\,dt + Bx\,dt$ is given by

$$\Phi(t,0) = \Phi_U(t,0)\Phi_P(t,0) = \begin{bmatrix} \varphi_{11}(t) & \varphi_{12}(t) \\ \varphi_{21}(t) & \varphi_{22}(t) \end{bmatrix},$$

$$\varphi_{11}(t) = \frac{4e^{t\lambda_{1A}} + (1-\sqrt{5})^2 e^{t\lambda_{2A}}}{4+(1-\sqrt{5})^2} \frac{4e^{t\lambda_{1B}} + (1-\sqrt{5})^2 e^{t\lambda_{2B}}}{4+(1-\sqrt{5})^2}$$

$$+ \frac{2(1-\sqrt{5})(e^{t\lambda_{2A}} - e^{t\lambda_{1A}})}{4+(1-\sqrt{5})^2} \frac{-2(1-\sqrt{5})(e^{t\lambda_{1B}} - e^{t\lambda_{2B}})}{4+(1-\sqrt{5})^2}$$

$$= \frac{16e^{(\lambda_{1A}+\lambda_{1B})t} + (1-\sqrt{5})^4 e^{(\lambda_{2A}+\lambda_{2B})t}}{[4+(1-\sqrt{5})^2]^2}$$

$$+ \frac{4(1-\sqrt{5})^2[e^{(\lambda_{2A}+\lambda_{1B})t} + 4(1-\sqrt{5})^2 e^{(\lambda_{1A}+\lambda_{2B})t}]}{[4+(1-\sqrt{5})^2]^2}$$

$$+\frac{4(1-\sqrt{5})^2[e^{(\lambda_{2A}+\lambda_{2B})t}-e^{(\lambda_{2A}+\lambda_{1B})t}+e^{(\lambda_{1A}+\lambda_{1B})t}-e^{(\lambda_{1A}+\lambda_{2B})t}]}{[4+(1-\sqrt{5})^2]^2}$$

$$=\frac{16e^{(\lambda_{1A}+\lambda_{1B})t}+(1-\sqrt{5})^4e^{(\lambda_{2A}+\lambda_{2B})t}}{[4+(1-\sqrt{5})^2]^2}$$

$$+\frac{4(1-\sqrt{5})^2[e^{(\lambda_{2A}+\lambda_{2B})t}+e^{(\lambda_{1A}+\lambda_{1B})t}]}{[4+(1-\sqrt{5})^2]^2}$$

$$=\frac{[16+4(1-\sqrt{5})^2]e^{(\lambda_{1A}+\lambda_{1B})t}+[4(1-\sqrt{5})^2+(1-\sqrt{5})^4]e^{(\lambda_{2A}+\lambda_{2B})t}}{[4+(1-\sqrt{5})^2]^2}$$

$$=\frac{4e^{(\lambda_{1A}+\lambda_{1B})t}+(1-\sqrt{5})^4e^{(\lambda_{2A}+\lambda_{2B})t}}{4+(1-\sqrt{5})^2},$$

$$\varphi_{12}(t)=\frac{4e^{t\lambda_{1A}}+(1-\sqrt{5})^2e^{t\lambda_{2A}}}{4+(1-\sqrt{5})^2}\,\frac{2(1-\sqrt{5})(e^{t\lambda_{2B}}-e^{t\lambda_{1B}})}{4+(1-\sqrt{5})^2}$$

$$+\frac{2(1-\sqrt{5})(e^{t\lambda_{2A}}-e^{t\lambda_{1A}})}{4+(1-\sqrt{5})^2}\,\frac{(1-\sqrt{5})^2e^{t\lambda_{1B}}+4e^{t\lambda_{2B}}}{4+(1-\sqrt{5})^2}$$

$$=\frac{8(1-\sqrt{5})e^{(\lambda_{1A}+\lambda_{2B})t}+2(1-\sqrt{5})^3e^{(\lambda_{2A}+\lambda_{2B})t}}{[4+(1-\sqrt{5})^2]^2}$$

$$-\frac{8(1-\sqrt{5})e^{(\lambda_{1A}+\lambda_{1B})t}+2(1-\sqrt{5})^3e^{(\lambda_{2A}+\lambda_{1B})t}}{[4+(1-\sqrt{5})^2]^2}$$

$$+\frac{2(1-\sqrt{5})^3e^{(\lambda_{2A}+\lambda_{1B})t}+8(1-\sqrt{5})e^{(\lambda_{2A}+\lambda_{2B})t}}{[4+(1-\sqrt{5})^2]^2}$$

$$-\frac{2(1-\sqrt{5})^3e^{(\lambda_{1A}+\lambda_{1B})t}+8(1-\sqrt{5})e^{(\lambda_{1A}+\lambda_{2B})t}}{[4+(1-\sqrt{5})^2]^2}$$

$$=\frac{2(1-\sqrt{5})[(1-\sqrt{5})^2+4][e^{(\lambda_{2A}+\lambda_{2B})t}-e^{(\lambda_{1A}+\lambda_{1B})t}]}{[4+(1-\sqrt{5})^2]^2}$$

$$=\frac{2(1-\sqrt{5})[e^{(\lambda_{2A}+\lambda_{2B})t}-e^{(\lambda_{1A}+\lambda_{1B}\lambda)t}]}{[4+(1-\sqrt{5})^2]^2},$$

$$\varphi_{21}(t)=\frac{-2(1-\sqrt{5})(e^{t\lambda_{1A}}-e^{t\lambda_{2A}})}{4+(1-\sqrt{5})^2}\,\frac{4e^{t\lambda_{1B}}+(1-\sqrt{5})^2e^{t\lambda_{2B}}}{4+(1-\sqrt{5})^2}$$

$$+\frac{(1-\sqrt{5})^2e^{t\lambda_{1A}}+4e^{t\lambda_{2A}}}{4+(1-\sqrt{5})^2}\,\frac{-2(1-\sqrt{5})(e^{t\lambda_{1B}}-e^{t\lambda_{2B}})}{4+(1-\sqrt{5})^2}$$

$$= \frac{-8(1-\sqrt{5})e^{(\lambda_{1A}+\lambda_{1B})t} + 2(1-\sqrt{5})^3 e^{(\lambda_{2A}+\lambda_{2B})t}}{[4+(1-\sqrt{5})^2]^2}$$

$$+ \frac{8(1-\sqrt{5})e^{(\lambda_{2A}+\lambda_{1B})t} - 2(1-\sqrt{5})^3 e^{(\lambda_{1A}+\lambda_{2B})t}}{[4+(1-\sqrt{5})^2]^2}$$

$$+ \frac{2(1-\sqrt{5})^3 e^{(\lambda_{1A}+\lambda_{2B})t} + 8(1-\sqrt{5})e^{(\lambda_{2A}+\lambda_{2B})t}}{[4+(1-\sqrt{5})^2]^2}$$

$$- \frac{2(1-\sqrt{5})^3 e^{(\lambda_{1A}+\lambda_{1B})t} + 8(1-\sqrt{5})e^{(\lambda_{2A}+\lambda_{1B})t}}{[4+(1-\sqrt{5})^2]^2}$$

$$= \frac{2(1-\sqrt{5})[(1-\sqrt{5})^2+4][e^{(\lambda_{2A}+\lambda_{2B})t} - e^{(\lambda_{1A}+\lambda_{1B})t}]}{[4+(1-\sqrt{5})^2]^2}$$

$$= \frac{2(1-\sqrt{5})[e^{(\lambda_{2A}+\lambda_{2B})t} - e^{(\lambda_{1A}+\lambda_{1B})t}]}{4+(1-\sqrt{5})^2},$$

$$\varphi_{22}(t) = \frac{-2(1-\sqrt{5})(e^{t\lambda_{1A}} - e^{t\lambda_{2A}})}{4+(1-\sqrt{5})^2} \frac{2(1-\sqrt{5})(e^{t\lambda_{2B}} - e^{t\lambda_{1B}})}{4+(1-\sqrt{5})^2}$$

$$+ \frac{(1-\sqrt{5})^2 e^{t\lambda_{1A}} + 4e^{t\lambda_{2A}}}{4+(1-\sqrt{5})^2} \frac{(1-\sqrt{5})^2 e^{t\lambda_{1B}} + 4e^{t\lambda_{2B}}}{4+(1-\sqrt{5})^2}$$

$$= \frac{-4(1-\sqrt{5})^2 e^{(\lambda_{1A}+\lambda_{2B})t} + 4(1-\sqrt{5})^2 e^{(\lambda_{2A}+\lambda_{2B})t}}{[4+(1-\sqrt{5})^2]^2}$$

$$+ \frac{4(1-\sqrt{5})^2 e^{(\lambda_{1A}+\lambda_{1B})t} - 4(1-\sqrt{5})^2 e^{(\lambda_{2A}+\lambda_{1B})t}}{4+(1-\sqrt{5})^2}$$

$$+ \frac{(1-\sqrt{5})^4 e^{(\lambda_{1A}+\lambda_{1B})t} + 4(1-\sqrt{5})^2 e^{(\lambda_{1A}+\lambda_{2B})t}}{[4+(1-\sqrt{5})^2]^2}$$

$$+ \frac{4(1-\sqrt{5})^2 e^{(\lambda_{2A}+\lambda_{1B})t} + 16e^{(\lambda_{2A}+\lambda_{2B})t}}{[4+(1-\sqrt{5})^2]^2}$$

$$= \frac{[16+4(1-\sqrt{5})^2]e^{(\lambda_{2A}+\lambda_{2B})t} + [4(1-\sqrt{5})^2 + (1-\sqrt{5})^4]e^{(\lambda_{1A}+\lambda_{1B})t}}{[4+(1-\sqrt{5})^2]^2}$$

$$= \frac{4e^{(\lambda_{2A}+\lambda_{2B})t} + (1-\sqrt{5})^2 e^{(\lambda_{1A}+\lambda_{1B})t}}{4+(1-\sqrt{5})^2}.$$

This is the complete representation of the normalized fundamental matrix solution process of the corresponding first-order linear homogeneous system of differential equations. By following the argument used in Illustration 4.6.1, the general solution processes of (a), (b), (c), (d), and (e) reduce to

(a) $x(t) = \Phi(t,0)c + \int_a^t \Phi(t,s)p(s)\,ds$,
(b) $x(t) = \Phi_{\mathrm{U}}(t,0)c + \int_a^t \Phi_{\mathrm{U}}(t,s)p(s)\,ds$,
(c) $x(t) = \Phi_{\mathrm{U}}(t,0)c$,
(d) $x(t) = \Phi_{\mathrm{P}}(t,0)c + \int_a^t \Phi_{\mathrm{P}}(t,s)p(s)\,ds$,
(e) $x(t) = \Phi_{\mathrm{P}}(t,0)c$,

respectively, where $\Phi_{\mathrm{U}}(t,0)$ and $\Phi_{\mathrm{P}}(t,0)$ are the normalized solution processes of the corresponding unperturbed ($dx = Ax\,dt$) and perturbed parts ($dx = Bx\,dt$) of the given systems of differential equations. This completes the solution process.

Illustration 4.6.2. Let us consider $dx = [Ax + Bx + p(t)]\,dt$. Find: (a) a general solution to the following system differential equations particularly for (b) $B = 0$, (c) $B = 0$ and $p(t) \equiv 0$, (d) $A = 0$, and (e) $A = 0$ and $p(t) \equiv 0$, where

$$A = \begin{bmatrix} a_{11} & a_{12} \\ a_{21} & a_{22} \end{bmatrix}, \quad B = \begin{bmatrix} b_{11} & b_{12} \\ b_{21} & b_{22} \end{bmatrix},$$

a_{ij}'s and b_{ij}'s are any given real numbers satisfying the following conditions (i) $d(A) = (\mathrm{tr}\,(A))^2 - 4\det(A) < 0$, (ii) $AB = BA$, and (iii) p is a two-dimensional continuous function defined on an interval $J \subseteq R$ into R^n.

Solution procedure. We further recall that $d(A) = r^2 d(B)$. In this case, we have complex conjugate eigenvalues. When we repeat the procedures described in Illustrations 4.4.3, 4.5.1, and 4.6.1, we arrive at

$$\Phi_{\mathrm{U}}(t) = e^{\alpha_A t}\begin{bmatrix} a_{12}\cos\beta_A t & a_{12}\sin\beta_A t \\ [(\alpha_A - a_{11})\cos\beta_A t - \beta_A \sin\beta_A t] & [(\alpha_A - a_{11})\sin\beta_A t + \beta_A\cos\beta_A t] \end{bmatrix},$$

$$\Phi_{\mathrm{U}}(0) = \begin{bmatrix} a_{12} & 0 \\ \alpha_A - a_{11} & \beta_A \end{bmatrix}, \quad \Phi_{\mathrm{U}}^{-1}(0) = \begin{bmatrix} \dfrac{1}{a_{12}} & 0 \\ -\dfrac{\alpha_A - a_{11}}{a_{12}\beta_A} & \dfrac{1}{\beta_A} \end{bmatrix},$$

and hence

$$\Phi_{\mathrm{U}}(t,0) = e^{\alpha_A t}\begin{bmatrix} \dfrac{\beta_A\cos\beta_A t - (\alpha_A - a_{11})\sin\beta_A t}{\beta_A} & \dfrac{a_{12}\sin\beta_A t}{\beta_A} \\ \dfrac{[(\alpha_A - a_{11})] + \sin\beta_A t}{a_{12}\beta_A} & \dfrac{(\alpha_A - a_{11})\sin\beta_A t + \beta_A\cos\beta_A t}{\beta_A} \end{bmatrix},$$

where $\lambda_A = \alpha_A + i\beta_A$. Similarly, $\Phi_{\mathrm{P}}(t)$, $\Phi_{\mathrm{P}}(0)$, $\Phi_{\mathrm{P}}^{-1}(0)$, and $\Phi_{\mathrm{P}}(t,0)$ can be computed by replacing U with P. The details are left as an exercise. From step 1 of the procedure in Subsection 4.6.2, the fundamental matrix solution to the corresponding homogeneous system $dx = Ax\,dt + Bx\,dt$ is given by

$$\Phi(t,0) = \Phi_{\mathrm{U}}(t,0)\Phi_{\mathrm{P}}(t,0),$$

where $\Phi_{\mathrm{U}}(t,0)$ and $\Phi_{\mathrm{P}}(t,0)$ are as described before. We note that $\Phi(t,0)$ is a normalized fundamental matrix solution to the corresponding homogeneous system of perturbed differential equations of the given system of linear nonhomogeneous differential equations. To complete the remaining parts of the problem, the procedure of Illustration 4.6.1 can be imitated in the context of the given problem.

Example 4.6.2. We are given $dx = [Ax + Bx + p(t)]\,dt$. Find: (a) a general solution to the given differential equation, particularly for (b) $\sigma = 0$, (c) $\sigma = 0$ and $p(t) \equiv 0$, (d) $\alpha = 0$, and (e) $\alpha = 0$ and $p(t) \equiv 0$, where

$$A = \begin{bmatrix} 3\alpha & -\sqrt{5}\alpha \\ \sqrt{5}\alpha & -\alpha \end{bmatrix}, \quad B = \begin{bmatrix} 4\sigma & -2\sqrt{5}\sigma \\ 2\sqrt{5}\sigma & -4\sigma \end{bmatrix},$$

with σ and α being constants and p a two-dimensional continuous function defined on an interval $J \subseteq R$ into R^2.

Solution procedure. By following the solution procedures used in Illustrations 4.6.1 and 4.6.2 and Example 4.6.1, we conclude that for $n = 2$ the eigenvalues of the coefficient matrices A and B are complex conjugate numbers $(1 \pm i)\alpha$ and $\pm 2\sigma i$, respectively. Moreover, $\lambda_{1A} = (1+i)\alpha$, $c_{1A} = \begin{bmatrix} -\sqrt{5}\alpha & -(2-i)\alpha \end{bmatrix}^T$, $\lambda_{2A} = (1-i)\alpha$, and $c_{2A} = \begin{bmatrix} (2+i)\alpha & \sqrt{5}\alpha \end{bmatrix}^T$; similarly, $\lambda_{1B} = 2\sigma i$, $\begin{bmatrix} -\sqrt{5}\sigma & -(2-i)\sigma \end{bmatrix}^T = c_{1B}$, and $\lambda_{2B} = -2\sigma i$, $c_{2B} = \begin{bmatrix} (2+i)\sigma & \sqrt{5}\sigma \end{bmatrix}^T$. Here, we have followed the notations of Illustrations 4.4.3 and 4.6.2. The normalized fundamental matrices $\Phi_{\mathrm{U}}(t,0)$ and $\Phi_{\mathrm{P}}(t,0)$ of the unperturbed and perturbation components of the given problem are

$$\begin{aligned}
\Phi_{\mathrm{U}}(t,0) &= -\frac{e^{\alpha t}}{\alpha^2\sqrt{5}} \begin{bmatrix} -\alpha\sqrt{5}\cos\alpha t & -\alpha\sqrt{5}\sin\alpha t \\ -\alpha\,(2\cos\alpha t + \sin\alpha t) & \alpha\,(\cos\alpha t - 2\sin\alpha t) \end{bmatrix} \begin{bmatrix} \alpha & 0 \\ 2\alpha & -\alpha\sqrt{5} \end{bmatrix} \\
&= \frac{e^{\alpha t}}{\sqrt{5}} \begin{bmatrix} \sqrt{5}\cos\alpha t + 2\sqrt{5}\sin\alpha t & -5\sin\alpha t \\ 2\cos\alpha t + \sin\alpha t - 2\cos\alpha t + 4\sin\alpha t & \sqrt{5}\cos\alpha t - 2\sqrt{5}\sin\alpha t \end{bmatrix} \\
&= e^{\alpha t} \begin{bmatrix} \cos\alpha t + 2\sin\alpha t & -\sqrt{5}\sin\alpha t \\ \sqrt{5}\sin\alpha t & \cos\alpha t - 2\sin\alpha t \end{bmatrix} \\
&= e^{\alpha t} \begin{bmatrix} \sqrt{5}\cos(\alpha t - \theta) & -\sqrt{5}\sin\alpha t \\ \sqrt{5}\sin\alpha t & \sqrt{5}\cos(\alpha t + \theta) \end{bmatrix},
\end{aligned}$$

for $\theta = \arctan 2$, and by repeating the above computations we get

$$\Phi_{\mathrm{P}}(t,0) = \begin{bmatrix} \cos 2\sigma t + 2\sin 2\sigma t & -\sqrt{5}\sin 2\sigma t \\ \sqrt{5}\sin 2\sigma t & \cos 2\sigma t - 2\sin 2\sigma t \end{bmatrix}$$

$$= \begin{bmatrix} \sqrt{5}\cos(2\sigma t - \theta) & -\sqrt{5}\sin 2\sigma t \\ \sqrt{5}\sin 2\sigma t & \sqrt{5}\cos(2\sigma t + \theta) \end{bmatrix},$$

for $\theta = \arctan 2$.

The fundamental matrix solution to the corresponding homogeneous system is

$$\begin{aligned}\Phi(t,0) &= \Phi_{\mathrm{U}}(t,0)\Phi_{\mathrm{P}}(t,0) \\ &= e^{\alpha t}\begin{bmatrix} \cos(2\sigma+\alpha)t + 2\sin(2\sigma+\alpha)t & -\sqrt{5}\sin(2\sigma+\alpha)t \\ \sqrt{5}\sin(2\sigma+\alpha)t & \cos(2\sigma+\alpha)t - 2\sin(2\sigma+\alpha)t \end{bmatrix} \\ &= e^{\alpha t}\begin{bmatrix} \sqrt{5}\cos(2\sigma+\alpha t - \theta) & -\sqrt{5}\sin(2\sigma+\alpha)t \\ \sqrt{5}\sin(2\sigma+\alpha)t & \sqrt{5}\cos(2\sigma+\alpha t+\theta) \end{bmatrix}.\end{aligned}$$

We recall that $\Phi_{\mathrm{U}}(t,0)$, $\Phi_{\mathrm{P}}(t,0)$, and $\Phi(t,0)$ are the normalized fundamental matrices corresponding to the unperturbed ($dx = Ax\,dt$), the perturbation ($dx = Bx\,dt$), and the overall homogeneous perturbed system ($dx = Ax\,dt + Bx\,dt$) of the given nonhomogeneous system, respectively. By using these normalized fundamental matrices and following the argument used in Illustrations 4.6.1 and 4.6.2, the solution process of the example can be completed, analogously.

Illustration 4.6.3. Let us consider $dx = [Ax + Bx + p(t)]\,dt$. Find: (a) a general solution to the following system differential equations, particularly for (b) $B = 0$, (c) $B = 0$ and $p(t) \equiv 0$, (d) $A = 0$, and (e) $A = 0$ and $p(t) \equiv 0$, where

$$A = \begin{bmatrix} a_{11} & a_{12} \\ a_{21} & a_{22} \end{bmatrix}, \quad B = \begin{bmatrix} b_{11} & b_{12} \\ b_{21} & b_{22} \end{bmatrix},$$

with a_{ij}'s and b_{ij}'s being any given real numbers satisfying the following conditions: (i) $d(A) = (\mathrm{tr}\,(A))^2 - 4\det(A) = 0$, (ii) $AB = BA$, and (iii) p is a two-dimensional continuous function defined on an interval $J \subseteq R$ into R^n.

Solution procedure. Again, we note that $n = 2$. We imitate the arguments used in Illustrations 4.4.4, 4.5.1, and 4.6.1 and the procedure in Subsection 4.6.2 to find the general solution to the given system. Under the conditions (i) and (ii), we further note that $d(B) = (\mathrm{tr}\,(B))^2 - 4\det(B) = 0$, and

$$\begin{aligned}\lambda_1 - a_{11} &= \frac{\mathrm{tr}(A) - 2a_{11}}{2} = \frac{a_{22} - a_{11}}{2}\\ &= r\frac{b_{22} - b_{11}}{2} = r\frac{\mathrm{tr}(B) - 2b_{11}}{2}\\ &= r(\mu_1 - b_{11}) = -a_{11} \pm \sqrt{\det(A)} = -b_{11} \pm |r|\sqrt{\det(B)}.\end{aligned}$$

Thus, the quadratic polynomial equation has a root of multiplicity 2. Now, we repeat the procedures outlined in Illustrations 4.4.4 and 4.6.1, and obtain the general solution process with regard to the given system of differential equations as well as its particular cases (b)–(e). The details are left to the reader as an exercise.

4.6.3 *Initial value problem*

Let us formulate an IVP for the first-order linear nonhomogeneous perturbed systems of differential equations (4.92). We consider

$$dx = [Ax + Bx + p(t)]dt, \quad x(t_0) = x_0. \tag{4.109}$$

4.6.4 *Procedure for solving the IVP*

In the following, we apply the procedure for finding the general solution process of the first-order linear nonhomogeneous perturbed system of differential equations (4.92) to the problem of finding the solution to the IVP (4.109). This procedure is a direct extension of those described in Subsections 2.5.4, 4.3.4, and 4.5.4. Now, we briefly summarize the procedure for finding a solution to the IVP (4.109):

Step 1. Following the procedure for finding a general solution to (4.92), we have

$$x(t) = \Phi(t)c + \int_a^t \Phi^{-1}(s)p(s)\,ds = \Phi(t)c + \int_a^t \Phi(t,s)p(s)\,ds, \tag{4.110}$$

where $a \in R$, $t \in J = [a, b] \subseteq R$, $\Phi(t)$ is as defined in (4.79), and $\Phi(t, s) = \Phi(t)\Phi^{-1}(s)$. Here, c is an arbitrary n-dimensional constant vector.

Step 2. To find an arbitrary constant c in (4.110), we utilize given initial data (t_0, x_0), and solve the following system of nonhomogeneous algebraic equations:

$$x(t_0) = \Phi(t_0)c + \int_a^{t_0} \Phi^{-1}(s)p(s)\,ds = x_0, \tag{4.111}$$

for any given t_0 in J. Since $\det(\Phi(t_0)) = \det(\exp[(A+B)(t_0 - a)]) \neq 0$, the system of algebraic equations (4.111) has a unique solution (Observation 1.4.3). This solution is given by

$$c = \Phi^{-1}(t_0)x_0 - \int_a^{t_0} \Phi^{-1}(s)p(s)\,ds. \tag{4.112}$$

Step 3. The solution to the IVP (4.109) is determined by substituting the expression (4.112) into (4.110). Hence, we have

$$\begin{aligned}
x(t) &= \Phi(t)c + \int_a^t \Phi(t,s)ds \\
&= \Phi(t)\left[\Phi^{-1}(t_0)x_0 - \int_a^{t_0} \Phi^{-1}(s)p(s)\,ds\right] \\
&\quad + \int_a^t \Phi(t,s)p(s)\,ds \quad \text{[from (4.112)]} \\
&= \Phi(t,t_0)x_0 - \int_a^{t_0} \Phi(t,s)p(s)\,ds \\
&\quad + \int_a^t \Phi(t,s)p(s)\,ds \quad \text{(by simplifications)} \\
&= \Phi(t,t_0)x_0 + \int_{t_0}^t \Phi(t,s)p(s)\,ds \quad \text{(by grouping)},
\end{aligned} \tag{4.113}$$

where $\Phi(t, t_0)$ is as defined in (4.90).

From (4.79), (4.90), and (4.91), the solution process $x(t) = x(t, t_0, x_0)$ of the IVP (4.109) in (4.113) is also represented by

$$\begin{aligned}
x(t) &= x(t,t_0,x_0) \\
&= \exp[(A+B)(t-t_0)]x_0 + \int_{t_0}^t \exp[(A+B)(t-s)]p(s)\,ds.
\end{aligned} \tag{4.114}$$

This solution is referred to as a particular solution to (4.92). The procedure for finding the solution to the IVP is summarized in Figure 4.18.

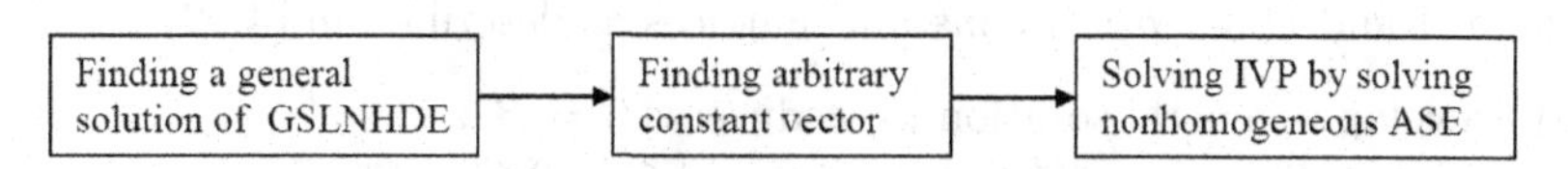

Fig. 4.18 Flowchart for finding particular solution.

4.6 Exercises

(1) Find the general solution to each system of differential equations:

(a) $dx = (Ax + Bx + p(t))\,dt$, where

$$A = \begin{bmatrix} -1 & 1 \\ 1 & -1 \end{bmatrix}, \quad B = \begin{bmatrix} 1 & -2 \\ -2 & 1 \end{bmatrix}, \quad \text{and} \quad p(t) = \begin{bmatrix} \sin t \\ \cos t \end{bmatrix}$$

(b) $dx = (Ax + Bx + p(t))\,dt$, where

$$A = \begin{bmatrix} -1 & 0 \\ 1 & -1 \end{bmatrix}, \quad B = \begin{bmatrix} 1 & -1 \\ 1 & 1 \end{bmatrix} \quad \text{and} \quad p(t) = \begin{bmatrix} t \\ t^2 \end{bmatrix}$$

(c) $dx = (Ax\,dt + Bx + p(t))\,dt$, where

$$A = \begin{bmatrix} -1 & -1 \\ 1 & -1 \end{bmatrix}, \quad B = \begin{bmatrix} 3 & 1 \\ 1 & 4 \end{bmatrix}, \quad \text{and} \quad p(t) = \begin{bmatrix} e^t \\ \ln t \end{bmatrix}.$$

(2) Solve the initial value problems (a)–(c) in Exercise 1, with $x(t_0) = x_0$: (a) $x_0 = \begin{bmatrix} 1 & 1 \end{bmatrix}^T$ (b) $x_0 = \begin{bmatrix} 2 & 1 \end{bmatrix}^T$ (c) $x_0 = \begin{bmatrix} 1 & 2 \end{bmatrix}^T$ (d) any x_0

(3) We are given

$$A = \begin{bmatrix} \alpha & a\delta \\ \delta & \beta \end{bmatrix}, \quad B = \begin{bmatrix} \gamma + b(\alpha - \beta) & ab\delta \\ b\delta & \gamma \end{bmatrix},$$

where a, b, α, β, δ, and γ are arbitrary real numbers such that a, b, and $\alpha - \beta$ are nonzero. Verify that

$$AB = BA.$$

(4) Find the general solution to $dx = (Ax + Bx + p(t))\,dt$, provided that: (a) $d(A) > 0$, (b) $d(A) < 0$, and (c) $d(A) = 0$, where $d(A) = (\mathrm{tr}(A))^2 - 4\det(A)$.

(5) Solve the initial value problems in Exercise 4 with $x(t_0) = x_0$.

(6) Let's assume that the time for a stop-clock is a random variable with a cumulative distribution function (cdf) F defined in Exercise 6 of Section 4.2. Let us consider a differential equation of the Riemann–Stieltjes type:

$$dx = (Ax + Bx + p(t))dF, \quad x(t_0) = x_0,$$

where A and B are $n \times n$ constant matrices as described in (4.92).

(a) Develop a general solution procedure: $AB = BA$.

(b) Develop Theorems 4.5.1 and 4.5.2 and Corollaries 4.5.1 and 4.5.2.

(c) Moreover, solve the given IVP.

(7) Solve the IVP in Exercise 6 for the specified matrices A, B, and x_0 in Exercise 2.

4.7 Fundamental Conceptual Algorithm Analysis

In Sections 4.3–4.6, we presented the computational procedures or methods for finding solution processes of linear systems of unperturbed and perturbed differential equations. The presented techniques provide the basis for addressing/raising several important questions in the modeling of dynamic processes in the biological, chemical, engineering, medical, physical, and social sciences. To avoid the repetition, we will not reformulate these questions. However, we will outline the three most important problems that were generated by the most fundamental questions 2, 3, and 10 posed in Section 2.6. The problems are (i) the existence problem, (ii) the uniqueness problem, and (iii) fundamental properties of solutions processes. In addition, a few other questions are partially answered in this chapter. These are part of the theory of systems of differential equations.

The presented procedures for finding solutions to IVPs in Sections 4.3–4.6 justify the existence of solutions to IVPs for the first-order linear system of differential equations. Without the existence of any solutions to differential equations, mathematical models of dynamic processes represented by differential equations are empty. We also note that these computational procedures were applicable to time-invariant coefficient matrices. In fact, the problem of finding solutions to certain time-invariant systems is reduced to the problem of solving linear time-varying systems (Theorems 4.5.1 and 4.5.2; Corollaries 4.5.1 and 4.5.2). The computational procedures are very limited in exhibiting the existence of the solution process. As a result of this, we need to find a theoretical procedure that assures the existence of the solution process for the systems of differential equations. This can be summarized by the following result. The existence result and the results presented thereafter play a very important role, and have a scope for both theoretical and practical points of view, particularly in the study of: (a) time-varying and time-invariant (nonstationary and stationary) nonlinear systems, (b) linear time-varying systems, and (c) nonlinear systems.

Theorem 4.7.1 (existence and uniqueness theorem). *Let us consider the following IVP with a linear time-varying nonhomogeneous perturbed system of differential equations:*

$$dy = [A(t)y + B(t)y + p(t)]dt, \quad y(t_0) = y_0, \tag{4.115}$$

where dy stands for the elementary differential [85]; *A and B are $n \times n$ continuous matrix functions defined on an interval $J = [a, b]$ $(J \subseteq R)$; p is an n-dimensional continuous vector function defined on an interval $J = [a, b]$, and initial data/conditions $(t_0, y_0) \in J \times R^n$.*

Then the IVP (4.115) *has a unique solution process, $y(t, t_0, y_0) = y(t)$, through any given initial data/conditions (t_0, y_0) for $t \in [t_0, b] = I$, $t_0 \in J$. Moreover,*

$$y(t) = y_0 + \int_{t_0}^{t} [A(s)y(s) + B(s)y(s) + p(s)]ds. \tag{4.116}$$

Proof. For $x, y \in R^n$, the distance between points associated with x and y is defined by $\|x - y\| = \sqrt{(x_1 - y_1)^2 + (x_2 - y_2)^2 + \cdots + (x_i - y_i)^2 + \cdots + (x_n - y_n)^2}$. From the continuity of rate functions A, B, and p on $J \times R^n$, the rate function defined by $C(t, y) = A(t)y + B(t)y + p(t)$ is continuous, and it satisfies the condition:

$$\|C(t, y)\| \leq K\|y\| + N \quad \text{(linear growth condition [7, 15, 42, 83, 84])}, \tag{4.117}$$

$$\|C(t, x) - C(t, y)\| \leq L\|x - y\| \quad \text{(Lipchitz condition [7, 15, 42, 83, 84])}, \tag{4.118}$$

for all $(t, x), (t, y) \in I \times R^n$ ($I = [t_0, b] \subseteq J \subseteq R$). Now, we define the approximation procedure

$$y_k(t) = \begin{cases} y_0, & \text{for } t_0 \leq t \leq b, \\ y_0 + \displaystyle\int_{t_0}^{t} [A(s)y_{k-1}(s) + B(s)y_{k-1}(s) + p(s)]\, ds, & \text{for } t_0 \leq t \leq b, \end{cases} \tag{4.119}$$

for $k \geq 1$. We recall that the integral term on the right-hand side of (4.119) is of the form of the Cauchy–Riemann integral [85]. By using the relations (4.117) and (4.118), the continuity of the rate coefficients on $[t_0, b] \times R^n$, and advanced real analysis techniques [120] (the concept of uniform convergence, Weierstrass's convergence test, etc.), one can show that the sequence of functions $\{y_k\}_{k=0}^{\infty}$ defined by (4.119) converges uniformly to a function y defined on I. Moreover, this function y satisfies the IVP. This establishes the basic scheme of the existence of the solution process of the IVP (4.115). Any further details require technical understanding of the advanced real analysis [120]. The validity of (4.116) follows from (4.119) and the uniform convergence of $\{y_k\}_{k=0}^{\infty}$ on I.

In view of the Lipchitz condition (4.118), the uniqueness of the solution process of the IVP (4.115) can be established by following the standard technical argument [15, 42, 83, 84]. The details are left to the reader in his or her desire to undertake the challenges in a future study of differential equations. □

Corollary 4.7.1. *Let us consider the following first-order linear time-varying homogeneous system of differential equations corresponding to* (4.115):

$$dx = A(t)x\, dt + B(t)x\, dt, \quad y(t_0) = y_0 = x_0, \tag{4.120}$$

where A and B are matrix functions; A, B, and initial data (t_0, x_0) satisfy the conditions of Theorem 4.7.1. *Then the IVP* (4.120) *has a unique solution process, $x(t, t_0, x_0) = x(t)$, through any given initial data/condition (t_0, x_0) for $t \in I$, $t_0 \in J$. Moreover,*

$$x(t) = x_0 + \int_{t_0}^{t} [A(s)x(s) + B(t)x(s)]\, ds, \quad y(t_0) = y_0 = x_0. \tag{4.121}$$

Observation 4.7.1

(i) We note that the IVP (4.109) is a special case of the IVP (4.115). As a result of this, the IVP (4.109) has two implicit representations/expressions, namely: (a) computationally explicit (4.113)/(4.114) and (b) conceptually implicit (4.116).
(ii) The uniqueness part of the theorem (Theorem 4.7.1) conceptualizes the solution process $y(t) = y(t, t_0, y_0)$ of the IVP (4.115) as a function of three variables (t, t_0, y_0). From the definition of function, for given (t, t_0, y_0), the output $y(t, t_0, y_0)$ [the value of the solution process at (t, t_0, y_0)] is uniquely determined under the uniqueness condition (4.118). This conceptualization plays a very important role in the advanced study of differential equations [7, 15, 19, 23, 42, 83, 84]. Moreover, the notation $y(t, t_0, y_0)$ is just a natural symbol for the solution process of (4.115).
(iii) We also note that the IVPs (4.41), (4.85), and (4.109) are special cases of the IVP (4.115). Furthermore, parts (i) and (ii) of the remark are also valid with regard to the IVP (4.120).

Based on our understanding and the knowledge of the "Basis and Scope for Conceptual Algorithms" (Chapter 2), we need the basic information on the fundamental solution process of the corresponding linear homogeneous differential equation. Therefore, prior to the study of the fundamental properties of (4.115), we need to establish first the basic properties of the solution process of linear homogeneous time-varying systems of differential equations (4.120).

Theorem 4.7.2 (fundamental property). *Let the hypotheses of Corollary* 4.7.1 *be satisfied. Let* $x_1(t), x_2(t), \ldots, x_k(t), \ldots, x_m(t)$ *be* m *solutions to* (4.120) *corresponding to* m *initial vectors* $x_{01}, x_{02}, \ldots, x_{0k}, \ldots, x_{0m}$ *in* R^n *defined on* I. *Then,*

$$\begin{aligned} x(t) &= a_1 x_1(t) + a_2 x_2(t) + \cdots + a_k x_k(t) + \cdots + a_m x_m(t) \\ &= \sum_{k=1}^{m} a_k x_k(t) \end{aligned} \tag{4.122}$$

is also a solution to (4.120) *on* I, *where* $a_1, a_2, \ldots, a_k, \ldots, a_m$ *are arbitrary scalar constants.*

Proof. The proof of the theorem mirrors the proof of Theorem 4.3.1. □

Remark 4.7.1.

(i) From Corollary 4.7.1, we note that if $x(t_0) = 0$, then $x(t, t_0, 0) = 0$.
(ii) We remark that an observation similar to Observation 4.3.2 can be reformulated with regard to the time-varying system (4.120). Furthermore, one of the immediate goals is to determine the condition(s) on initial vectors $\{x_{01}, x_{02}, \ldots, x_{0k}, \ldots, x_{0n}\}$ which ensures that the set of solutions

$\{x_1(t), x_2(t), \ldots, x_k(t), \ldots, x_m(t)\}$ of (4.120) is a complementary set of solutions.

(iii) For $m = n$, by following the arguments used in Observations 4.3.3, we can rewrite the expression of $x(t)$ in (4.122) as

$$x(t) = \Phi(t)a, \tag{4.123}$$

where Φ is an $n \times n$ matrix function defined on I and by $\Phi(t) = (x_{ik}(t))_{n\times n} \equiv (x_{ik}(t))_{n\times n}$; its columns are solutions $x_1(t), x_2(t), \ldots, x_k(t), \ldots, x_n(t)$ of the IVP (4.120) corresponding to initial vectors $x_{01}, x_{02}, \ldots, x_{0k}, \ldots, x_{0n}$ in R^n; these solutions are guaranteed by Corollary 4.7.1; an n-dimensional unknown constant column vector (or $n \times 1$ matrix) $a = [a_1, a_2, \ldots, a_k, \ldots, a_n]^T$ is formed by arbitrary coefficients of n solutions to (4.120) in (4.122). We further note that $\Phi(t)$ as defined in (4.123) depends on n arbitrary given initial vectors $x_{01}, x_{02}, \ldots, x_{0k}, \ldots, x_{0n}$ in R^n. Hence, $\Phi(t)$ is not uniquely determined by n any given initial vectors at a given initial time $t = t_0$. This suggests that an observation similar to Observation (4.3.3) can be rewritten by replacing a set of eigenvectors $\{c_1, \ldots, c_k, \ldots, c_n\}$ with the set of initial vectors $\{x_{01}, x_{02}, \ldots, x_{0k}, \ldots, x_{0n}\}$ corresponding to n solutions to (4.120). Again, our immediate task is to determine the condition(s) on initial vectors $\{x_{01}, x_{02}, \ldots, x_{0k}, \ldots, x_{0n}\}$ which ensures that $\Phi(t)$ as defined in (4.123) is nonsingular. Thus, the solution represented in (4.123) is called a general solution process of (4.120).

(iv) Employing the argument used in Observation 4.3.4, we remark that

$$d\Phi(t) = A(t)\Phi(t)\,dt + B(t)\Phi(t)\,dt, \tag{4.124}$$

where $\Phi(t)$ is as defined in (4.123). If $\Phi(t)$ defined in (4.123) is a nonsingular matrix, a normalized fundamental matrix solution to (4.120) at $t = t_0$ can be introduced in a natural way. In fact, one can use initial vectors in (ii) as $x_{0k} = e_k = (\delta_{ik})_{n\times 1}$, where $\delta_{ik} = 0$ for $i \neq k$, and $\delta_{ik} = 1$ for $i = k$, and $k = 1, 2, \ldots, n$. Then the solution to the IVP (4.120), $x(t, t_0, x_0)$, can be represented by

$$x(t, t_0, x_0) = x(t) = \Phi(t)\Phi^{-1}(t_0)x_0 = \Phi(t, t_0)x_0. \tag{4.125}$$

The detailed imitation of Observation 4.3.4 is left to the reader as an exercise.

In the following, we present a result similar to Theorem 4.3.2. This result provides an analytic test for nonsingularity of the matrix function Φ defined in (4.123) on I. This test connects a set of n initial vectors $\{x_{01}, x_{02}, \ldots, x_{0k}, \ldots, x_{0n}\}$ the set of n solution column vectors $\{x_1(t), x_2(t), \ldots, x_k(t), \ldots, x_n(t)\}$ of $\Phi(t)$.

Theorem 4.7.3 (Abel–Jacobi–Liouville formula [15, 42]). *Let the hypotheses of Corollary 4.7.1 be satisfied. Let $x_1(t), x_2(t), \ldots, x_k(t), \ldots, x_n(t)$ be any n solutions* *process of* (4.120) *on* I *corresponding to* n *initial vectors* $x_{01}, x_{02}, \ldots, x_{0k}, \ldots, x_{0n}$. *Let*

$$\Phi(t) = [x_1(t), x_2(t), \ldots, x_i(t), \ldots, x_n(t)] = \left[\Phi_1^T(t), \ldots, \Phi_i^T(t), \ldots, \Phi_n^T(t)\right]^T$$

be the $n \times n$ *matrix function defined on* I, *where* $\Phi_i(t) = [x_{i1}, x_{i2}, \ldots, x_{ik}, \ldots, x_{in}]$ *is the* i*th row of the matrix function* $\Phi(t)$. *Let*

$$\mathrm{tr}(A(t)) = \sum_{i=1}^{n} a_{ii}(t), \quad \mathrm{tr}(B(t)) = \sum_{i=1}^{n} b_{ii}(t).$$

Then,

$$d\det(\Phi(t)) = [\mathrm{tr}(A(t)) + \mathrm{tr}(B(t))]\det(\Phi(t))dt, \quad \textit{for } t \in I, \tag{4.126}$$

$$\exp\left[-\int_{t_0}^{t} [\mathrm{tr}(A(s)) + \mathrm{tr}(B(s))]\,ds\right]\det(\Phi(t)) = c \quad (\textit{constant}). \tag{4.127}$$

Moreover, $\det(\Phi(t)) \neq 0$ [*a general fundamental matrix solution process of* (4.120)] *if and only if* $c \neq 0$.

Proof. From (4.124), we conclude that $\Phi(t)$ is a matrix solution process of (4.120), i.e.

$$\begin{aligned} d\Phi(t) &= (dx_{ik}(t))_{n\times n} = A(t)\Phi(t)\,dt + B(t)\Phi(t)\,dt \\ &= \left(\sum_{j=1}^{n} a_{ij}(t)x_{jk}(t)\,dt + \sum_{j=1}^{n} b_{ij}(t)x_{jk}(t)\,dt\right)_{n\times n}. \end{aligned} \tag{4.128}$$

Applying Theorem 1.5.3 to $\det(\Phi(t))$ and following the proofs of Theorem 4.3.2, we arrive at

$$\begin{aligned} &W(\Phi_1(t), \ldots, d\Phi_i(t), \ldots, \Phi_n(t)) \\ &= \begin{vmatrix} x_{11} & \cdots & \cdots & x_{1n} \\ \cdots & \cdots & \cdots & \cdots \\ a_{ii}(t)x_{i1}(t)\,dt + b_{ii}(t)x_{i1}\,dt & \cdots\ \ \cdots & \cdots & a_{ii}(t)x_{i1}(t)\,dt + b_{ii}(t)x_{in}\,dt \\ \cdots & \cdots & \cdots & \cdots \\ x_{n1} & \cdots & \cdots & x_{nn} \end{vmatrix} \end{aligned}$$

This together with Theorem 1.4.1(D4) and (D6) gives us

$$W(\Phi_1(t), \ldots, d\Phi_i(t), \ldots, \Phi_n(t)) = a_{ii}(t)\det(\Phi(t))dt + b_{ii}(t)\det(\Phi(t))dt.$$

The rest of the proof can be constructed by employing elementary calculus [85] and the argument used in Theorem 4.3.2. The details are left as an exercise to the reader. □

Remark 4.7.2.

(i) We note that an observation similar to Observation 4.3.5 remains true with regard to the time-varying system (4.120).
(ii) From (4.127), we note that

$$c = \det(\Phi(t_0)) = \det\left([x_{01}, x_{02}, \ldots, x_{0k}, \ldots, x_{0n}]\right).$$

The set of n initial vectors $\{x_{01}, x_{02}, \ldots, x_{0k}, \ldots, x_{0n}\}$ is a set of bases of R^n if and only if (iff) $c = \det(\Phi(t_0)) \neq 0$ iff the set of column vectors of $\Phi(t)$, $\{x_1(t), x_2(t), \ldots, x_k(t), \ldots, x_m(t)\}$, is linearly independent (Theorem 4.7.3). Thus, the set forms a complete set of solutions to (4.120). Therefore, $\Phi(t)$ defined in (4.123) is indeed the fundamental solution process of (4.120) iff $c = \det(\Phi(t_0)) \neq 0$.
(iii) From (ii), we conclude that the solution defined in (4.123) is indeed the general solution to (4.120). Moreover, for given arbitrary initial data (t_0, x_0) in (4.120), an arbitrary vector a in (4.123) is uniquely determined iff $\det\left([x_{01}, x_{02}, \ldots, x_{0k}, \ldots, x_{0n}]\right) \neq 0$. Moreover, the solution to the IVP (4.120) is uniquely represented by (4.125), even though we neither know its explicit representation nor need the commutativity law for matrix multiplication ($AB = BA$). This signifies the power of conceptual understanding with regard to implicit versus explicit representation of the solution concept.

In the following, we present a result that is similar to Theorem 4.3.3 with regard to the time-varying system (4.120).

Theorem 4.7.4. *Let the hypotheses of Corollary* 4.7.1 *be satisfied. Let* Φ *be a fundamental matrix solution to* (4.120)*. Then,* Φ *is invertible and its inverse* Φ^{-1} *satisfies the following matrix differential equations*:

$$d\Phi^{-1}(t) = -\Phi^{-1}(t)A(t)\,dt - \Phi^{-1}(t)B(t)\,dt. \tag{4.129}$$

Moreover,

$$d(\Phi^{-1})^T = -A^T(t)\Phi^{-1}(t)^T\,dt - B^T(t)\Phi^{-1}(t)^T\,dt. \tag{4.130}$$

Proof. The proof of the theorem can be constructed by following the proof of Theorem 4.3.3. The details are left to the reader. □

Remark 4.7.3.

(i) We note that an observation similar to Observation 4.3.6 remains true with regard to the time-varying system (4.120). The verification of this remark is left as an exercise to the reader. In fact, $(\Phi^{-1})^T$ is a fundamental matrix solution to

$$dz = -A^T(t)z\,dt - B^T(t)z\,dt. \tag{4.131}$$

The system (4.131) is called the adjoint to the system (4.120). Moreover, this system is equivalent to the system

$$dz = -zA(t)\,dt - zB(t)\,dt, \tag{4.132}$$

where z is a row vector, while z in (4.131) is a column vector. In the light of this notional understanding, $\Phi^{-1}(t)$ is a fundamental matrix solution to (4.132). This can be justified from (4.129).

(ii) We further remark that (4.132) provides another simple calculus approach to finding the inverse of a general fundamental solution to (4.120).

Our experience in finding the closed-form solutions to the IVPs (4.41)/(4.85) (computational) and (4.120) (conceptual) is based on the construction of general fundamental matrix solutions [Observation 4.3.3 and Remark 4.7.1(iii)]. Moreover, the construction of a general fundamental matrix solution process depends on the choice of eigenvectors (Sections 4.3 and 4.5) and initial states of solutions (Corollary 4.7.1), respectively. In general, the fundamental matrix solution function is not uniquely determined. In fact, the following result provides the relationship between two fundamental matrix solution processes.

Lemma 4.7.1. *Let Φ_1 be a fundamental matrix solution process of* (4.120). *Φ_2 is any other general fundamental matrix solution process of* (4.120) *if and only if $\Phi_2(t) = \Phi_1(t)C$ on I for some nonsingular constant matrix C.*

Proof. First, we assume that $\Phi_1(t)$ is a given general fundamental matrix solution function of (4.120), and $\Phi_2(t)$ is any general fundamental matrix solution process of (4.120). First, we prove that $\Phi_2 = \Phi_1 C$ on I for some nonsingular constant matrix C. For this purpose, we define $\Psi = \Phi_1^{-1}\Phi_2$ on I. This is defined in view of Theorems 4.7.3 and 4.7.4 (Observation 4.3.5). Our goal reduces to showing that Ψ is a nonsingular constant matrix. For this purpose, we compute the elementary differential of Ψ as follows:

$$\begin{aligned}
d\Psi &= d\left(\Phi_1^{-1}\Phi_2\right) \quad \text{(by definition of } \Psi\text{)}\\
&= d\left(\Phi_1^{-1}\right)\Phi_2 + \Phi_1^{-1}d\Phi_2 \quad \text{(by Theorem 1.5.11)}\\
&= \left[-\Phi_1^{-1}A(t)\,dt - \Phi_1^{-1}B(t)\,dt\right]\Phi_2\\
&\quad + \Phi_1^{-1}\left[A(t)\Phi_2 dt + B(t)\Phi_2 dt\right] \quad \text{[from (4.124) and (4.129)]}
\end{aligned}$$

$$
\begin{aligned}
&= -\Phi_1^{-1}A(t)\Phi_2 dt - \Phi_1^{-1}B(t)\Phi_2 dt \\
&\quad + \Phi_1^{-1}A(t)\Phi_2 dt + \Phi^{-1}B(t)\Phi_2 dt \quad \text{(regrouping and simplification)} \\
&= 0 \quad \text{(final simplification).} \qquad (4.133)
\end{aligned}
$$

From (4.133), we conclude that $\Psi(t) = C$ is constant on I. This is due to the fact that if the differential of an arbitrary function of one variable is zero on an interval I for $t \geq t_0$, then the function is constant. Thus, by substitution and simplification, we conclude that $\Phi_2 = \Phi_1 C$ on I. From Theorems 1.4.2, 1.4.4 and 4.7.3, $\det(\Phi_2) = \det(\Phi_1)\det(C)$ and $\det(\Phi_2) \neq 0 \neq \det(\Phi_1)$, and hence $\det(C) \neq 0$. Therefore, C is a nonsingular constant matrix. This completes the proof of the "if" part.

To prove the "only if" part, we start with $\Phi_2 = \Phi_1 C$, where C is nonsingular constant matrix, and show that Φ_2 is a general fundamental solution process of (4.120). We find the differential of both sides of the expressions, and obtain

$$
\begin{aligned}
d\Phi_2 &= d\Phi_1 C \quad \text{(Observation 1.5.2)} \\
&= [A(t)\Phi_1 dt + B(t)\Phi_1 dt]\, C \quad \text{[substitution from (4.124)]} \\
&= A(t)\Phi_1 C dt + B(t)\Phi_1 C dt \quad \text{[Theorem 1.3.1(M2) and (M3)]} \\
&= A(t)\Phi_2 dt + B(t)\Phi_2 dt \quad \text{(by substitution for } \Phi_1 C\text{).}
\end{aligned}
$$

Hence, $\Phi_1 C$ is a solution to (4.120). Since $\det(\Phi_2) = \det(\Phi_1)\det(C) \neq 0$ for $C \neq 0$ (by Theorems 1.4.2, 1.4.4 and 4.7.3), we conclude that $\Phi_1 C$ is the general fundamental solution process of (4.120). This completes the proof of the theorem. □

Observation 4.7.2

(i) Obviously, two different linear homogeneous systems of differential equations cannot have the same general fundamental solution process. In fact, from (4.124), we note that $A(t)\,dt + B(t)\,dt = d\Phi(t)\Phi^{-1}(t)$. This is true in view of Remark 4.7.1(iii) and (iv).

(ii) Let us assume that Φ_1 and Φ_2 are two given general fundamental solution processes of (4.120). In addition, let $t_0 \in J$ be given. From Lemma 4.7.1, we have $\Phi_2(t) = \Phi_1(t)C$ on I. For $t_0 \in J$, we have $\Phi_2(t_0) = \Phi_1(t_0)C$. This implies that $C = \Phi_1^{-1}(t_0)\Phi_2(t_0)$. From this, we obtain $\Phi_2(t) = \Phi_1(t)\Phi_1^{-1}(t_0)\Phi_2(t_0)$, which implies that

$$\Phi_2(t)\Phi_2^{-1}(t_0) = \Phi_1(t)\Phi_1^{-1}(t_0),$$

in view of Theorem 4.7.4. From this and using the notation of the normalized fundamental solution process of (4.120), we have

$$\Phi(t, t_0) \equiv \Phi_2(t, t_0) = \Phi_1(t, t_0). \qquad (4.134)$$

This shows that the normalized fundamental solution process of (4.120) is uniquely determined by A, B, and t_0. The normalized fundamental solution process

$\Phi(t,t_0)$ of (4.120) defined in (4.134) is referred to as the normalized fundamental solution process of (4.120) at $t = t_0$.

In the following, we present a few algebraic properties of the normalized fundamental solution process of (4.120). The proof of the result is based on the observed properties of a general fundamental solution process, Observation 4.3.3, Theorem 4.3.3, and differential and integral calculus.

Lemma 4.7.2. *Let $\Phi(t,t_0)$ and $\Psi(t,t_0)$ be normalized fundamental matrix solution processes of* (4.120) *and* (4.132) *at $t = t_0$, respectively. In addition, let $\Phi_1(t,t_1)$ be a normalized fundamental matrix solution to* (4.120) *at $t = t_1$. Then, for all t_0, t_1, s, and t in I:*

(a)
$$\Psi(t,t_0)\Phi(t,t_0) = I_{n\times n} = \Phi(t_0,t)\,\Phi(t,t_0), \quad \textit{for } t_0 \le t; \tag{4.135}$$

where $I_{n\times n}$ is an $n \times n$ identity matrix, and

$$\Psi(t,t_0) = \Phi^{-1}(t,t_0) = \Phi(t_0,t), \quad \textit{for } t_0 \le t; \tag{4.136}$$

(b)
$$\Phi_1(t,t_1) = \Phi(t,t_0)\Phi_1(t_0,t_1), \quad \textit{for } t_0 \le t; \tag{4.137}$$

(c)
$$\Phi(t,t_0) = \Phi(t,\mathrm{s})\Phi(\mathrm{s},t_0) \quad \textit{for } t_0 \le t; \tag{4.138}$$

(d)
$$\Phi(t,\mathrm{s}) = \Phi(t,t_0)\Psi(\mathrm{s},t_0) = \Phi(t,t_0)\Phi(t_0,\mathrm{s}), \quad \textit{for } t_0 \le t; \tag{4.139}$$

(e)
$$\partial_s\Phi(t,s) = -\Phi(t,s)A(s)\,ds - \Phi(t,s)B(s)\,ds, \tag{4.140}$$

where $\partial_s\Phi(t,s)$ is the partial differential of $\Phi(t,s)$ with respect to s for fixed t.

Proof. To prove (a), by using the differential (Observation 1.5.2) we compute

$$\begin{aligned}
d(\Psi(t,t_0)\Phi(t,t_0)) &= d\Psi(t,t_0)\Phi(t,t_0) + \Psi(t,t_0)d\Phi(t,t_0)\\
&= [-\Psi(t,t_0)A(t)\,dt - \Psi(t,t_0)B(t)\,dt]\,\Phi(t,t_0)\\
&\quad + \Psi(t,t_0)\,[A(t)\Phi(t,t_0)dt\\
&\quad + B(t)\Phi(t,t_0)dt] \quad [(4.129) \text{ and } (4.124)]\\
&= -\Psi(t,t_0)A(t)\Phi(t,t_0)dt - \Psi(t,t_0)B(t)\Phi(t,t_0)dt\\
&\quad + \Psi(t,t_0)A(t)\Phi(t,t_0)dt + \Psi(t,t_0)B(t)\Phi(t,t_0)dt\\
&\qquad [\text{Theorem 1.3.1(M2) and (M3)}]\\
&= 0 \quad (\text{final simplification}).
\end{aligned}$$

This establishes the fact that $\Psi(t,t_0)\Phi(t,t_0) = C$, where C is an arbitrary constant matrix. This results from the fact that if the differential of an arbitrary function of one variable is zero on an interval I for $t \ge t_0$, then the function is constant. Hence, the matrix process $\Psi(t,t_0)\Phi(t,t_0)$ is a constant matrix on I. Moreover, $\Psi(t,t_0)$ and $\Phi(t,t_0)$ are the normalized solution processes (4.120) and (4.132) at $t = t_0$, respectively. Therefore,

$$\Psi(t,t_0)\Phi(t,t_0) = C = \Psi(t_0,t_0)\,\Phi(t_0,t_0) = I_{n\times n}. \tag{4.141}$$

This shows that $\Psi(t,t_0)$ is the algebraic inverse of $\Phi(t,t_0)$, and it is denoted by $\Phi(t_0,t)$. This statement is equivalent to other notations in (4.136). In view of these notations and (4.141), we have

$$\Psi(t,t_0)\Phi(t,t_0) = I_{n\times n} = \Phi(t_0,t)\,\Phi(t,t_0), \quad \text{for } t_0 \le t.$$

This completes the proof of (4.135).

To prove (b), from Lemma 4.7.1 and Observation 4.7.2(ii) we have $\Phi_1(t,t_1) = \Phi(t,t_0)C$. By using the assumption of the lemma and $t = t_1$, we get $I = \Phi_1(t_1,t_0)\,C$. From this and (a), we have $C = \Phi_1^{-1}(t_1,t_0) = \Phi_1(t_0,t_1)$. After substitution, we obtain

$$\Phi_1(t,t_1) = \Phi(t,t_0)\Phi_1(t_0,t_1), \quad \text{for } t_0 \le t.$$

This establishes the relation (4.137).

To prove (c), for $t_0 \le t$ and any s in I we utilize elementary algebraic properties of matrices and the definition of $\Phi(t,t_0)$ described in (4.125), and obtain

$$\begin{aligned}
\Phi(t,t_0) &= \Phi(t)\Phi^{-1}(t_0) \quad \text{[by notation (4.125)]}\\
&= \Phi(t)\Phi^{-1}(s)\Phi(s)\Phi^{-1}(t_0) \quad \text{[existence of } \Phi^{-1}(s)]\\
&= \Phi(t,s)\Phi(s,t_0) \quad \text{[by definitions (4.90) and (4.135)]}.
\end{aligned}$$

This completes the proof of (4.138).

For the proof of (d), from (4.138) and solving for $\Phi(t,s)$ we have

$$\begin{aligned}
\Phi(t,s) &= \Phi(t,t_0)\Phi^{-1}(s,t_0) \quad \text{[by (4.138)]}\\
&= \Phi(t,t_0)\Psi(\mathrm{s},t_0) \quad \text{[by notation (4.136)]}\\
&= \Phi(t,t_0)\Phi(t_0,s), \quad \text{for } t_0 \le t \quad \text{[notation (4.136)]}.
\end{aligned}$$

This proves (d).

For the proof of (e), we apply the differential (Theorem 1.5.11) on both sides with respect to s for fixed (t_0,t) to the expression (4.139), and obtain

$$\begin{aligned}
\partial_s\Phi(t,s) &= \Phi(t,t_0)\left[-\Psi(s,t_0)\,A(s)\,ds - \Psi(s,t_0)\,B(s)\,ds\right] \quad [(4.129)]\\
&= -\Phi(t,t_0)\Psi(s,t_0)\,A(s)\,ds - \Phi(t,t_0)\Psi(s,t_0)\,B(s)\,ds\\
&\qquad \text{[Theorem 1.3.1(M2)]}\\
&= -\Phi(t,s)A(s)\,ds - \Phi(t,s)B(s)\,ds \quad \text{[by (4.139)]}.
\end{aligned}$$

This establishes the result in (e), and completes the proof of the lemma. □

Observation 4.7.3

(i) We further emphasize that the proof of Lemma 4.7.2 depends heavily on Theorems 4.7.3 and 4.7.4. It is motivated by Theorems 4.3.2 and 4.3.3, and the closed-form representations of $\Phi(t, t_0)$ in Section 4.3.
(ii) An alternative proof of these expressions in Lemma 4.7.2 can be given by utilizing the conceptual approach. In fact, the proofs of parts (a), (c), and (e) will be illustrated at a later stage of this section.

In the following, we present a few more results which will shed light on the algebraic properties of the solution processes of the IVPs (4.120) and (4.132). We recall that the IVPs (4.120) and (4.132) include the IVPs (4.41), (4.54), (4.85), and (4.109) as special cases (Observation 4.7.1). These properties are utilized for studying fundamental properties of the solution process of the IVP (4.115). Moreover, these results provide an auxiliary conceptual tool and the insight to undertake the study of nonlinear and nonstationary scalar as well as systems of differential equations [7, 15, 23, 42, 83, 84].

Lemma 4.7.3 (principle of superposition). *Let $x_1(t, t_0, x_0) = x_1(t)$ and $x_2(t, t_0, x_0) = x_2(t)$ be solutions to the IVP* (4.120) *through* (t_0, x_0) *for* $t \geq t_0$, *and* $t \in I$, $t_0 \in J$. *In addition, let a_1 and a_2 be any two scalar real/complex numbers. Then,*

$$a_1 x_1(t, t_0, x_0) + a_2 x_2(t, t_0, x_0) \tag{4.142}$$

is a solution to (4.120) *through the initial data* $(t_0,\ a_1 x_0 + a_2 x_0)$ *for* $t \geq t_0$, *and* t, t_0 *in* I.

Proof. The proof of the lemma can be recasted by imitating the argument used in the proof of Lemma 2.6.3. To minimize repetition, we leave the details to the reader as an exercise. □

Observation 4.7.4

(i) From Illustration 1.4.4, Observation 1.4.5, and the conclusion of Lemma 4.7.3, we infer that an arbitrary linear combination of solutions to (4.120) is also a solution to the IVP (4.120).
(ii) In fact, by Observation 1.4.5, Theorem 1.4.5, and Illustration 1.4.5, we infer that the span $S(\{x_1, x_2, \ldots, x_k, \ldots, x_n\})$ of solution processes $x_1, x_2, \ldots, x_k, \ldots, x_n$ of (4.120) is a vector subspace of $C[[t_0, b], R[\Omega, R^n]]$.
(iii) From Definition 1.4.6, the dimension of subspace $S(\{x_1, x_2, \ldots, x_k, \ldots, x_n\})$ in (ii) is at most n. Thus, the general solution of (4.120) depends on n arbitrary constants. In fact, it is a linear combination of n functions.

The following result shows that the normalized fundamental matrix solution $\Phi(t, t_0)$ to (4.120) possesses certain basic algebraic properties as a function for each fixed (t, t_0) for $t \geq t_0$ and $t \in I$, $t_0 \in J$.

Lemma 4.7.4. *Let $x(t) = x(t, t_0, x_0)$ be a solution process of the IVP (4.120) through (t_0, x_0) for $t \geq t_0$, and $t \in I$, $t_0 \in J$. For fixed (t, t_0), the transformation/mapping $\Phi(t, t_0)$ [in (4.125)] is defined on R^n into R^n by*

$$\Phi(t, t_0)x_0 = x(t, t_0, x_0),$$

for x_0 in R^n. Then, for every x_0, x_1, x_2 and a in R,

$$\Phi(t, t_0)\,(x_1 + x_2) = \Phi(t, t_0)x_1 + \Phi(t, t_0)x_2, \tag{4.143}$$

$$\Phi(t, t_0)\,(ax_0) = a\,\Phi(t, t_0)x_0. \tag{4.144}$$

Proof. Again, the proof of the lemma can be reconstructed by following the argument used in the proof of Lemma 2.6.4. □

Observation 4.7.5

(i) We observe that mapping/transformation that satisfies the properties described in (4.143) and (4.144) is called a linear transformation/mapping. For fixed (t, t_0), the normalized fundamental solution $\Phi(t, t_0)$ to (4.120) at $t = t_0$ is a linear transformation defined on R^n into itself. In fact, it is a first-degree polynomial process in n variables. In the engineering literature, it is called a state transition matrix.

(ii) An observation, similar to the second part of Observation 2.6.6, can be easily reformulated.

In the following, we present a relationship between the solution processes of the linear homogeneous system of differential equations (4.120) and its adjoint system of differential equations (4.132).

Lemma 4.7.5. *Let $x(t) = \Phi(t, t_0)x_0$ and $z(t) = x_0^T\Psi(t, t_0)$ be solution processes of the IVP (4.120) and the corresponding IVP associated with its adjoint system of differential equations (4.132), respectively, through (t_0, x_0), where $\Phi(t, t_0)$ and $\Psi(t, t_0)$ are normalized fundamental solution processes of (4.120) and (4.132) for $t \geq t_0$, and $t \in I$, $t_0 \in J$. Then,*

$$z(t)x(t) = c, \tag{4.145}$$

where c is a constant number. Moreover, $c = x_0^Tx_0$ and

$$\Psi(t, t_0)\Phi(t, t_0) = I_{n\times n}, \tag{4.146}$$

for all $t \geq t_0$, and $t \in I$, $t_0 \in J$.

Proof. By imitating the proof of Lemma 2.6.5(a), we compute

$$\begin{aligned}
d\,(z(t)x(t)) &= dz(t)x(t) + z(t)\,dx(t) \quad \text{(Theorem 1.5.11)}\\
&= [-z(t)A(t)\,dt - z(t)B(t)\,dt]\,x(t)\\
&\quad + z(t)\,[A(t)x(t)\,dt\\
&\quad + B(t)x(t)\,dt] \quad \text{[from (4.120) and (4.132)]}\\
&= -z(t)A(t)x(t)\,dt - z(t)B(t)x(t)\,dt\\
&\quad + z(t)A(t)x(t)\,dt\\
&\quad + z(t)B(t)x(t)\,dt \quad \text{[by Theorem 1.3.1(M2)]}\\
&= 0 \quad \text{(by final simplification).}
\end{aligned}$$

This establishes $z(t)x(t) = c$, where c is a constant real number. This completes the proof of (4.145). The proof of $c = x_0^T x_0$ follows from the fact that

$$z(t_0)x(t_0) = c = x_0^T x_0, \quad \Phi(t_0, t_0) = I_{n\times n} = \Psi(t_0, t_0),$$

$$z(t)x(t) = x_0 \Psi(t, t_0)\Phi(t, t_0)x_0 = x_0^T x_0,$$

for any x_0 in R^n. The details are left to the reader. Finally, from

$$x_0 \Psi(t, t_0)\Phi(t, t_0)x_0 = x_0^T x_0,$$

we compute

$$\frac{\partial^2}{\partial x_0^2}\left(x_0 \Psi(t, t_0)\Phi(t, t_0)x_0\right) = \frac{\partial^2}{\partial x_0^2}\left(x_0^T x_0\right),$$

which implies that

$$2\Psi(t, t_0)\Phi(t, t_0) = 2I_{n\times n}.$$

Thus, $\Psi(t, t_0)\Phi(t, t_0) = I_{n\times n}$.

This completes the proof of (4.146). Thus, the proof of the lemma is complete. □

Observation 4.7.6

(i) The conclusions of Lemma 4.7.5 provide a relationship between the normalized fundamental matrix solution processes of (4.120) and (4.132). Moreover, Lemma 4.7.5 provides an alternative conceptual proof of Lemma 4.7.5(a).

(ii) In addition, Lemma 4.7.5, conceptually, confirms that the solution process provided by Theorem 4.7.4 is indeed the algebraic inverse of a fundamental matrix solution to (4.120). Of course, Theorem 4.7.4 provides an analytic method for finding an inverse of any fundamental matrix solution process of (4.120).

In the following, we provide an analytic (conceptual) alternative proof of Lemma 4.7.2(c) that was based on an elementary algebraic approach. This result offers a very general universal argument for establishing the validly of the relation (4.138). Moreover, the presented argument provides a conceptual reasoning to undertake the study of more complex properties of solutions to more general differential equations.

Lemma 4.7.6. *Let $\Phi(t, t_0)$ be a normalized fundamental matrix solution to* (4.120) *at $t = t_0$, and let $x(t, t_0, x_0)$ be the solution process of the IVP* (4.120) *through (t_0, x_0). Then, for all $t \geq t_0$, and $t \in I$, $t_0 \in J$,*

$$\text{(a)} \quad x\left(t, s, x\left(s, t_0, x_0\right)\right) = x(t, t_0, x_0), \tag{4.147}$$

$$\text{(b)} \quad \Phi(t, s)\Phi\left(s, t_0\right) = \Phi(t, t_0). \tag{4.148}$$

Proof. The proof of the lemma can be reproduced by imitating the proof of Lemma 2.6.6 by replacing R with R^n in the context of (4.120). The details are omitted. □

Prior to the presentation of several additional basic results concerning (4.115), we need to present a final result with respect to the system (4.120). This result deals with the differentiability of the solution process of (4.120) with respect to all three variables (t, t_0, x_0) via elementary calculus [85]. The byproduct of the result provides an analytic (conceptual) alternative proof of Lemma 4.7.2(e).

Lemma 4.7.7. *Let $x(t, t_0, x_0)$ be the solution process of the IVP* (4.120) *through (t_0, x_0). Then, for all $t \geq t_0$, and $t \in I$, $t_0 \in J$,*

$$\text{(a)} \quad \frac{\partial}{\partial x_0} x(t) = \frac{\partial}{\partial x_0} x(t, t_0, x_0)$$

exists and it satisfies the matrix differential equation

$$dX = A(t)X\,dt + B(t)X\,dt, \quad \frac{\partial}{\partial x_0} x(t_0) = I_{n\times n}, \tag{4.149}$$

where

$$\frac{\partial}{\partial x_0} x(t, t_0, x_0)$$

stands for a partial derivative of the solution process $x(t, t_0, x_0)$ of (4.120) with respect to x_0 at (t, t_0, x_0) for fixed (t, t_0);

$$\text{(b)} \quad \frac{\partial}{\partial t_0} x(t) = \frac{\partial}{\partial t_0} x(t, t_0, x_0)$$

exists and it satisfies the differential equation

$$dx = A(t)x\,dt + B(t)x\,dt, \quad \frac{\partial}{\partial t_0} x\left(t_0, t_0, x_0\right) = z_0, \tag{4.150}$$

where

$$\partial_{t_0} x(t, t_0, x_0) = \frac{\partial}{\partial t_0} x(t, t_0, x_0) dt_0$$

stands for the partial differential of the solution process $x(t, t_0, x_0)$ of (4.120) with respect to t_0 at (t, t_0, x_0) for fixed (t, x_0), and satisfies the system of differential equations

$$\partial_{t_0} x\,(t_0, t_0, x_0) \equiv \partial x(t_0) = -A(t_0)x_0 dt_0 - B(t_0)x_0\, dt_0 \tag{4.151}$$

or

$$\frac{\partial}{\partial t_0}\, x\,(t_0, t_0, x_0) = z_0 = -A(t_0)x_0 - B(t_0)x_0.$$

Proof. From Lemma 4.7.4, the solution $x(t, t_0, x_0) = \Phi(t, t_0)x_0$ to the IVP (4.120) through (t_0, x_0) is a linear function of x_0 defined R^n into R^n. By following the argument used in the proof of Lemma 2.6.8, one can establish the validity of the statement (a) in the theorem. In fact, one can conclude that

$$\frac{\partial}{\partial x_0}\, x(t) = \frac{\partial}{\partial x_0}\, x(t, t_0, x_0) = \Phi(t, t_0). \tag{4.152}$$

Moreover,

$$\frac{\partial}{\partial x_0}\, x(t, t_0, x_0)$$

is a continuous function in x_0. This is because of the fact that

$$\frac{\partial}{\partial x_0}\, x(t, t_0, x_0) = \Phi(t, t_0)$$

is independent of x_0 (i.e. constant) for each fixed (t, t_0) for $t \geq t_0$, and $t \in I$, $t_0 \in J$. From (4.152), it is clear that

$$\frac{\partial}{\partial x_0}\, x(t) = \frac{\partial}{\partial x_0}\, x(t, t_0, x_0)$$

is the normalized fundamental solution process of (4.120). Hence, it satisfies (4.149).

To prove the statement (b), for fixed (t, t_0), again from the definition of the solution $x(t, t_0, x_0) = \Phi(t, t_0)x_0$ in (4.125), we apply Lemma 4.7.2(e) to $x(t, t_0, x_0) = \Phi(t, t_0)x_0$ in (4.125), and we get

$$\begin{aligned}
\partial_{t_0} x(t, t_0, x_0) &= \partial_{t_0} \Phi(t, t_0)x_0 \quad \text{[from (4.125)]} \\
&= \Phi(t, t_0)[-A(t_0)x_0\, dt_0 \\
&\quad - B(t_0)x_0\, dt_0] \quad \text{[from (4.140) and Theorem 1.3.1(M2)]} \\
&= \Phi(t, t_0)\partial x(t_0) \quad \text{[from (4.151)]},
\end{aligned} \tag{4.153}$$

where $\partial_{t_0} x(t, t_0, x_0)$ denotes the partial differential of the solution process $x(t, t_0, x_0)$ of (4.120) with respect to t_0 at (t, t_0, x_0); $\partial x(t_0)$ is as defined in (4.151) in the lemma.

From (4.153), it is clear that $\partial_{t_0} x(t, t_0, x_0)$ satisfies the IVP (4.150) with $\partial x(t_0) = \partial_{t_0} x\,(t_0, t_0, x_0)$, which satisfies the system of differential equations (4.151).

Moreover, from the notation of $\partial_{t_0} x(t, t_0, x_0)$, we have

$$\frac{\partial}{\partial t_0} x(t, t_0, x_0) = -\Phi(t, t_0)A(t_0)x_0 + B(t_0)x_0.$$

This completes the proof of the theorem. For further details, see Lemma 2.6.8. □

Observation 4.7.7

(i) From the proof of Lemma 4.7.7, we observe that

$$\frac{\partial}{\partial x_0} x(t, t_0, x_0)$$

is independent of x_0, i.e.

$$\frac{\partial}{\partial x_0} x(t, t_0, x_0) = \Phi(t, t_0)$$

[Jacobian matrix of $x(t, t_0, x_0)$ for fixed (t, t_0)]. It is obvious that

$$\frac{\partial}{\partial x_0} x(t, t_0, x_0)$$

is continuous in the (t, t_0, x_0) function. Moreover, its second partial derivative of $x(t, t_0, x_0)$ is

$$\frac{\partial^2}{\partial x_0^2} x(t, t_0, x_0)$$

[Hessian matrix of $x(t, t_0, x_0)$ for fixed (t, t_0)], and

$$\frac{\partial^2}{\partial x_0^2} x(t, t_0, x_0) = 0.$$

(ii) Using (i) and Lemma 4.7.7, we introduce the partial differentials of the solution process $x(t, t_0, x_0)$ of (4.120) with respect to x_0 and t_0 at (t, t_0, x_0), as follows:

$$\partial_{x_0} x(t, t_0, x_0) = \frac{\partial}{\partial x_0} x(t, t_0, x_0)dx_0 = \Phi(t, t_0)dx_0, \tag{4.154}$$

$$\partial_{t_0} x(t, t_0, x_0) = \frac{\partial}{\partial t_0} x(t, t_0, x_0)\partial x(t_0) = \Phi(t, t_0)\partial x(t_0), \tag{4.155}$$

where dx_0 is the partial differential of $x(t, t_0, x_0)$ at t_0 with respect to x_0 for fixed t_0, i.e.

$$\frac{\partial}{\partial x_0} x(t_0)dx_0 = dx_0, \tag{4.156}$$

and

$$\frac{\partial}{\partial t_0} x\left(t_0, t_0, x_0\right) dt_0 = \frac{\partial}{\partial t_0} x(t_0)$$

is the partial differential of $x(t, t_0, x_0)$ at t_0 with respect to t_0 for fixed x_0. Note the difference between $\partial_{x_0}(x_0)$ and $\partial_{t_0} x(t_0)$.

(iii) From the notation, (4.154) and (4.155), the mixed partial differential of the solution process $x(t, t_0, x_0)$ of (4.120) is given by

$$\begin{aligned}\partial^2_{t_0 x_0} x(t, t_0, x_0) &= \partial_{t_0} (\partial_{x_0} x(t, t_0, x_0)) \quad \text{(partial differential)}\\ &= \partial_{t_0} \left(\frac{\partial}{\partial x_0} x(t, t_0, x_0) dx_0\right) \quad \text{[from (4.154)]}\\ &= \Phi(t, t_0)[-A(t_0)dt_0 - B(t_0)dt_0]dx_0 \quad \text{[from (4.140)]}. \end{aligned} \tag{4.157}$$

On the other hand, from (4.155) we have

$$\begin{aligned}\partial^2_{x_0 t_0} x(t, t_0, x_0) &= \partial_{x_0} (\partial_{t_0} x(t, t_0, x_0)) \quad \text{(partial differential)}\\ &= \partial_{x_0} (\Phi(t, t_0) \partial x(t_0)) \quad \text{[from (4.155)]}\\ &= \Phi(t, t_0) \partial_{x_0} (\partial x(t_0)) \quad \text{[Theorem 1.5.1(P2)]}\\ &= \Phi(t, t_0) [-A(t_0)dt_0\\ &\quad -B(t_0)dt_0]\, dx_0 \quad \text{[from (4.151)]}\\ &= -\Phi(t, t_0)[A(t_0)dt_0 + B(t_0)dt_0]dx_0. \end{aligned} \tag{4.158}$$

Hence, we conclude that its mixed second derivatives exist, and are equal,

$$\frac{\partial^2}{\partial x_0 \partial t_0} x(t, t_0, x_0) = \frac{\partial^2}{\partial t_0 \partial x_0} x(t, t_0, x_0),$$

and

$$\left(\frac{\partial^2}{\partial t_0 \partial x_0} x(t, t_0, x_0)\right)$$

is the normalized fundamental solution process of the adjoint differential equation (4.132).

(iv) We further observe that Lemma 4.7.7(b) provides a conceptual proof for Lemma 4.7.2(e). In fact, from (4.151) and (4.153) we have

$$\begin{aligned}\frac{\partial}{\partial x_0} (\partial_{t_0} x(t, t_0, x_0)) &= \frac{\partial}{\partial x_0} (\Phi(t, t_0) \partial x(t_0)) \quad \text{[(4.153)]}\\ &= \frac{\partial}{\partial x_0} (\Phi(t, t_0) [-A(t_0)dt_0 - B(t_0)dt_0]\, x_0) \quad \text{[(4.151)]}\\ &= \Phi(t, t_0) [-A(t_0)dt_0 - B(t_0)dt_0] \quad \left(\frac{\partial}{\partial x_0} x_0 = I_{n\times n}\right)\end{aligned}$$

$$= -\Phi(t,t_0)A(t_0)dt_0$$
$$-\Phi(t,t_0)B(t_0)dt_0 \quad \text{[Theorem 1.5.1(P2)]}. \qquad (4.159)$$

On the other hand,

$$\partial_{t_0}\left(\frac{\partial}{\partial x_0}x(t,t_0,x_0)\right) = \partial_{t_0}\left(\Phi(t,t_0)\right) \quad \text{[from (4.152)]}$$
$$= \partial_{t_0}\Phi(t,t_0) \quad \text{(by notation)}. \qquad (4.160)$$

From (iii), (4.159) and (4.160), the validity of (4.140) follows immediately.

In the following, we present a result that provides a relationship between the solution processes of (4.115) and (4.120). It gives rise to an alternative representation of (4.115), in the form of an integral equation. This result also provides an analytic tool for investigating the behavior solution process of (4.115) knowing the behavior of (4.120). This idea will be highlighted in Chapter 6.

Theorem 4.7.5 (method of variation of parameters). *Let the assumptions of Theorem* 4.7.1 *be satisfied. In addition, let* $y(t) = y(t,t_0,y_0)$ *and* $x(t) = x(t,t_0,y_0)$ *be the solution processes of* (4.115) *and* (4.120), *respectively, through the same initial data* (t_0, y_0), *for all* $t \geq t_0$, *and* $t \in I$, $t_0 \in J$. *Then*

$$y(t) = x(t) + \int_{t_0}^{t} \Phi(t,s)p(s)\,ds, \qquad (4.161)$$

for $t \in I$, $t_0 \in J$.

Proof. Let $y(t) = y(t,t_0,y_0)$ and $x(t) = x(t,t_0,y_0)$ be given solution processes of (4.115) and (4.120), respectively, through the same initial data (t_0, y_0), for all $t \geq t_0$, and $t \in I$, $t_0 \in J$. From (4.125) and Theorem 4.7.3, we recall that $x(t) = \Phi(t,t_0)y_0$, where $\Phi(t,t_0)$ is the normalized fundamental matrix solution process of (4.120).

By knowing the normalized fundamental matrix solution $\Phi(t,t_0)$ to (4.120), we imitate the procedure in Subsection 4.6.4 for finding the solution process of the IVP (4.115), and we arrive at (4.113), i.e.

$$y(t) = \Phi(t,t_0)y_0 + \int_{t_0}^{t} \Phi(t,s)p(s)\,ds. \qquad (4.162)$$

This together with (4.125) implies that

$$y(t) = x(t) + \int_{t_0}^{t} \Phi(t,s)p(s)\,ds,$$

for $t \in I$, $t_0 \in J$. This completes the proof of the theorem. □

A very important byproduct of the uniqueness theorem is that the knowledge of the solution process at a future time depends only on the knowledge of the initial data, and it is independent of its knowledge between the initial time and the future time. This is demonstrated by Lemma 4.7.6 in the context of the solution process of the IVP (4.120). The following result extends this idea to the IVP (4.115).

Lemma 4.7.8. *Let the hypotheses of Theorem* 4.7.1 *be satisfied. For* $t_0 \leq s \leq t$, *let* $y(t) = y(t, t_0, y_0)$ *and* $y(t, s, y(s)) = y(t, s, y(s, t_0, y_0))$ *be the solution processes of* (4.115) *through* (t_0, y_0) *and* $(s, y(s))$, *respectively. Then*

$$y(t) = y(t, t_0, y_0) = y(t, s, y(s, t_0, y_0)), \quad \textit{for } t_0 \leq s \leq t. \tag{4.163}$$

Proof. For $t_0 \leq s \leq t$, let $y(t, t_0, y_0)$ and $y(t, s, y(s))$ be the solutions to (4.115) through (t_0, y_0) and $(s, y(s)) = (s, y(s, t_0, y_0))$, respectively. From the representation of $y(t) = y(t, t_0, y_0)$ in (4.161), Lemma (4.7.2)(c) or Lemma 4.7.6, and imitating the proof of Lemma 2.6.7, one can easily establish the conclusion of the lemma. The details are left to the reader as an exercise. □

Observation 4.7.8. By utilizing Lemmas 2.6.6 and 4.7.8, the uniqueness part of Theorem 4.7.1, the implicit (4.116) and the explicit (4.161) representations of the solution process (4.115), one can reformulate an observation similar to Observation 2.6.8. The details are left to the reader as an exercise.

The following result establishes the continuity of the solution process $y(t, t_0, y_0)$ with respect to (t, t_0, y_0). We just state the result without its proof, and is left as an exercise to the reader.

Theorem 4.7.6 (continuous dependence of the solution on initial conditions). *Let the hypotheses of Theorem* 4.7.1 *be satisfied. Then the solution process* $y(t) = y(t, t_0, y_0)$ *is continuous with respect to* (t, t_0, y_0) *for* $t \geq t_0$, *and* $t \in I$, $t_0 \in J$. *In particular, it is continuous with respect to the initial conditions/data.*

In fact, in the following we present a result that exhibits the differentiability of the solution process of (4.115) with respect to all three variables (t, t_0, y_0).

Theorem 4.7.7 (differentiability of solutions with respect to initial conditions). *Let the hypotheses of Theorem* 4.7.1 *be satisfied. In addition, let* $y(t, t_0, y_0)$ *be the solution process of the IVP* (4.115). *Then, for all* $t \geq t_0$, *and* $t \in I$, $t_0 \in J$,

$$\text{(a)} \quad \frac{\partial}{\partial y_0} y(t) = \frac{\partial}{\partial y_0} y(t, t_0, y_0)$$

exists, and it satisfies the IVP (4.149);

$$\text{(b)} \quad \frac{\partial}{\partial t_0} y(t) = \frac{\partial}{\partial t_0} y(t, t_0, y_0)$$

exists, and it satisfies the IVP (4.150) *with* $(t_0, \partial y(t_0))$, *where*

$$\partial_{t_0} y(t, t_0, x_0) = \frac{\partial}{\partial t_0} y(t, t_0, y_0)\, dt_0$$

stands for the partial differential of the solution process $y(t,t_0,x_0)$ *to* (4.115) *with respect to* t_0 *at* (t,t_0,y_0) *for fixed* (t,y_0), *and satisfies the system of differential equations*

$$\frac{\partial}{\partial t_0} y\left(t_0,t_0,y_0\right) dt_0 \equiv \partial y(t_0) = \left[-A(t_0)y_0 - B(t_0)y_0 + p(t_0)\right] dt_0 \tag{4.164}$$

or

$$\frac{\partial}{\partial t_0} y\left(t_0,t_0,x_0\right) = z_0 = -A(t_0)y_0 - B(t_0)y_0 - p(t_0).$$

Proof. The proof of the theorem can be constructed by imitating the proof of Theorem 2.6.4. The details are left as an exercise to the reader. □

Observation 4.7.9. We note that the proof of Theorem 4.7.7 can also be reconstructed by utilizing the conceptual aspect of Lemma 4.7.7.

In the following, we present a very important result that connects the solution process of (4.115) with the corresponding homogeneous system of differential equations (4.120). The Alekseev-type approach [2] for the method of variation of constants/parameters is used to prove the following result. Furthermore, the result provides a conceptual tool for undertaking the study of both nonstationary and nonlinear systems of differential equations. In addition, it provides an alternative method for investigating the qualitative properties of more complex differential equations.

Theorem 4.7.8 (variation of constant formula — Alekseev-type). *Let the hypotheses of Theorem* 4.7.1 *be satisfied. In addition, let* $y(t) = y\left(t,t_0,y_0\right)$ *and* $x(t) = x\left(t,t_0,y_0\right)$ *be the solution processes of the IVPs* (4.115) *and* (4.120) *through the same initial data* (t_0,y_0) *on* I, *respectively. Then*

$$y\left(t,t_0,y_0\right) = x\left(t,t_0,y_0\right) + \int_{t_0}^{t} \Phi(t,s)p(s)\,ds, \tag{4.165}$$

for $t \in I$, $t_0 \in J$.

Proof. From the assumption of the theorem, for $t_0 \le s \le t$, we define a solution to the IVP (4.120) through s, $y(s)$, as follows:

$$x\left(t,s,y\left(s,t_0,y_0\right)\right) = x\left(t,s,y(s)\right), \tag{4.166}$$

where $y(s) = y\left(s,t_0,y_0\right)$ is a solution to the IVP (4.115) through (t_0,y_0). From (4.166) and the definition of the IVP, we observe that for $s = t_0$

$$x\left(t,t_0,y\left(t_0,t_0,y_0\right)\right) = x\left(t,t_0,y_0\right). \tag{4.167}$$

From the application of Lemma 4.7.7, the differential formula (Theorem 1.5.11), the chain rule, and Observation 4.7.7 with respect to s for $t_0 \le s \le t$, we have

$$\begin{aligned}
d_s x(t,s,y(s)) &= \partial_{x_0} x\,(t,s,y(s)) + \partial_{t_0} x\,(t,s,y(s)) \\
&= \frac{\partial}{\partial x_0} x\,(t,s,y(s))\,dy(s) + \frac{\partial}{\partial t_0} x\,(t,s,y(s))\,\partial_s x\,(s,s,y(s)) \\
&\qquad [(4.154),\ (4.155),\ (4.156) \text{ and } (4.157)] \\
&= \Phi(t,s)\,[A(s)y(s) + B(s)y(s) + B(s)y(s) + p(s)\,ds] \\
&\quad + \Phi(t,s)\,[-A(s)y(s)ds - B(s)y(s)\,ds] \\
&\qquad\qquad (\text{Lemma 4.7.7 } [(4.151) \text{ and } (4.156)]) \\
&= \Phi(t,s)A(s)y(s) + \Phi(t,s)B(s)y(s) + \Phi(t,s)p(s)\,ds \\
&\quad - \Phi(t,s)A(s)y(s)\,ds - \Phi(t,s)B(s)y(s)\,ds \quad [\text{from Theorem 1.3.1(M2)}] \\
&= \Phi(t,s)p(s)\,ds \quad (\text{by simplification}).
\end{aligned} \tag{4.168}$$

By integrating (4.168) on both sides with respect to s from t_0 to t, using Lemma 4.7.6, and the notations and the uniqueness part of the solution process of (4.120) (Corollary 4.7.1), we get

$$\begin{aligned}
&x\,(t,t,y\,(t,t_0,y_0)) - x\,(t,t_0,y(t_0)) \\
&\qquad = y\,(t,t_0,y_0) - x\,(t,t_0,y_0) = \int_{t_0}^{t} \Phi(t,s)p(s)\,ds,
\end{aligned}$$

which can be written as

$$y\,(t,t_0,y_0) = x\,(t,t_0,y_0) + \int_{t_0}^{t} \Phi(t,s)p(s)\,ds.$$

This completes the proof of the theorem. Moreover, from (4.125), we have

$$y\,(t,t_0,y_0) = \Phi(t,t_0)y_0 + \int_{t_0}^{t} \Phi(t,s)p(s)\,ds,$$

which is identical to (4.162). □

Observation 4.7.10

(i) From the conclusion of Theorem 4.7.8, it is obvious that Theorem 4.7.8 provides an alternative conceptual approach to the closed-form representation of the solution to (4.115). In fact, one can compare the solution expressions on the right-hand sides of (4.161) and (4.165). These expressions are identical.

(ii) We further observe that the solution representation of (4.115) in either (4.161) or (4.165) was derived based on the conceptual knowledge of the corresponding homogeneous perturbed system of differential equations (4.120). In particular, the smoothness (conceptual) properties of the solution process with respect to its initial data/conditions outlined in Observation 4.7.7 were utilized.

(iii) For the validity of Theorem 4.7.8, we merely needed the existence of the solution process of (4.115).

4.8 Notes and Comments

The development of the deterministic modeling procedure in Section 4.2 is modified from the classical modeling processes in sciences and engineering. It is a parallel extension of the single-state system dynamic processes of Section 2.2 to multivariate processes. In this chapter, a dozen illustrations and examples are given to exhibit multivariate dynamic modeling processes in the biological, social, and political sciences. For instance, Illustrations 4.2.1 (economic processes [3, 28, 96, 98]), 4.2.2 (multispecies competitive–cooperative processes [46, 55, 90, 116, 126]), 4.2.3 (multicomponent diffusion processes [46, 55, 90, 116, 119]), 4.2.4 (electric circuit network [35, 40, 55, 126]), 4.2.5 (cellular dynamics [38, 55, 58, 90, 116, 119]), 4.2.6 (multispecies ecosystems [55, 90, 94, 104, 114, 116, 118, 126, 129, 137]), 4.2.7 (multicompartment Systems [1, 46, 55, 58, 111, 112, 119]), 4.2.8 (group dynamics [44, 122, 126, 127]), and 4.2.9 (US government system [55, 92]), and Examples 4.2.1 (predator–prey model [55, 90, 94, 104, 116, 119, 126, 129]) 4.2.2 (pharmacokinetics [46, 55, 111, 112, 119, 134]), and 4.2.3 (Kendall's mathematical marriage model [36, 48, 106]) are formulated. A few notable exercises in Section 4.2 — Exercises 1 (reversible bacterial mutation [119]), 2 (pharmacokinetics — prompt injection model [134]), 3 (Kendall's two-sex model [48, 106]), 4 (Goodman's two-sex model [36, 106]), 5 (Kendall's marriage dynamic models [48, 106]), and the newly developed Exercises 6 and 7 — are also presented. An eigenvalue and eigenvector-type approach developed in Section 2.4 is systematically connected and extended to incorporate the existing approach to solving the system of linear homogeneous differential equations in Section 4.3. The concepts of fundamental, general, and particular solutions are introduced and determined in a coherent manner. To illustrate the concepts and to master the procedure, several numerical examples are worked out. Depending on the types of eigenvalues, the methods of finding the fundamental matrix solutions are outlined in Section 4.4. The computational algorithms in Illustrations 4.4.1 (real–distinct eigenvalue), 4.4.3 (distinct–complex eigenvalue), and 4.4.4 (multiple eigenvalue) for finding fundamental matrix solutions are outlined. These computational algorithms are applied to solve several numerical and applied examples and illustrations, namely Illustration 4.4.2 (dynamics of food passage in the ruminant model [8]) and Examples 4.4.1 [Kendall's mathematical marriage model (Example 4.2.3)], 4.4.2 (mixing problem [27, 55, 119]), 4.4.5 [predator–prey (Example 4.2.1)], 4.4.3 and 4.4.6 [electric network (Illustration 4.2.4)]. Again, to encourage the concepts, some applied exercises are added in the Section 4.4 Exercises. In Section 4.5, the method for solving a system of homogeneous differential equations is extended to a method for solving a general system of homogeneous differential equations composed of nominal and uncertain (deterministic) rate coefficient matrices. This kind of decomposition approach is motivated by the lack of knowledge of the coefficient rate matrix. The uncertainties characterize the internal changes or the lack of knowledge of certain parameters occurring in the coefficient rate matrix. Furthermore, this type of structural decomposition approach of dynamic processes is to investigate the deterministic/stochastic parametric uncertainties. The limitation of this approach is analyzed. The study of this type of dynamic model structure plays a very important role in the sensitivity analysis of dynamic processes. The limitation of the computational approach leads to a few conceptual algorithms. Several examples are presented to justify the claims and counterclaims. We note that the developments of both computational and conceptual results are new. We also note that these limitations open a new avenue to solving more complex problems in a systematic way (Vol. II). The method of variation of parameters is used to solve general first-order linear non-homogeneous differential equations in Section 4.6. This is achieved without compromising the study of solving the overall system of nonhomogeneous differential equations. Several numerical and applied examples are presented to illustrate the method and concepts. Section 4.7 outlines the role and scope of conceptual algorithms. This material is a natural extension of Section 2.6.

Chapter 5

Higher-Order Linear Differential Equations

5.1 Introduction

The objective of this chapter follows the theme and spirit of Chapters 2–4. Mathematical models of several dynamic processes influenced by not only their states but also their rates of change of states leading to higher-order differential equations are presented in Section 5.2. The computational techniques for higher-order differential equations outlined in Section 5.3 are in the same framework as in Chapter 2 or 4. Furthermore, by introducing a change of dependent variable, higher-order linear scalar differential equations reduce to systems of linear differential equations. Then the previously developed techniques for solving and analyzing linear systems of differential equations in Chapter 4 are applied to this reduced system of linear differential equations in Sections 5.4 and 5.5. Several numerical and applied examples are presented to illustrate the methods. Furthermore, using the concept of the Laplace transform, once again the problem of solving differential equations reduces to the problem of solving algebraic equations in Section 5.6. The byproduct of this leads to a solution to the IVPs in Section 5.7.

5.2 Mathematical Modeling

The material in this section is a special case of the material in Section 4.2. In particular, the development of a mathematical model is imbedded in the modeling process of Section 4.2. These statements will be easily justified by the development of this chapter's material, and we will present a few illustrations regarding the mathematical model building process.

Illustration 5.2.1 (spring problem [40]). Many mechanical systems arise when bodies are attached to the ends of springs. A simple prototype model for a force in a spring was introduced by Robert Hooke. A particle is attached to one end of the spring, and the other end is fixed. The motion of this particle is called simple harmonic motion (SHM).

SHM 1. The natural length of a spring coil is the length in the absence of an elongation or extension.

SHM 2. If we fix one end of the spring and pull the other end away from the fixed end to extend the natural length, then the spring exerts a force on the pull in the opposite direction to the pull.

SHM 3. If one pushes the end toward the fixed end of the spring to shorten the natural length, then the spring exerts a force on the push in the opposite direction to the push.

SHM 4. From SHM 2 and SHM 3, we observe that whether it is the extension or the compression of the spring, the force exerted by the spring acts to restore the spring to its natural length. This force exerted by the spring due to the pull or push is called a restoring force.

SHM 5. For a small extension and compression, the magnitude of the restoring force was discovered by Hooke. Hooke's law states: The magnitude of the restoring force in the spring is directly proportional to the length by which the spring is extended or compressed.

Step 1 (choice of coordinate system). Let y ft be the extension of the spring. To assign the correct sign to the exerted force due to the spring, we need to consider two cases: (1) the spring extension ($y > 0$) and (2) the spring compression ($y < 0$).

Let y be the position of the particle at a time t, and let l be the natural length of the spring. The length of $l+y$ depends on either the pull or the push. The natural length in the absence of the pull or the push is l for $y = 0$. Therefore, the origin of the y-coordinate system is at a distance l from the fixed end. If y is increasing, the particle is moving away from the fixed end. Therefore, the direction of motion is the positive direction for the y coordinate.

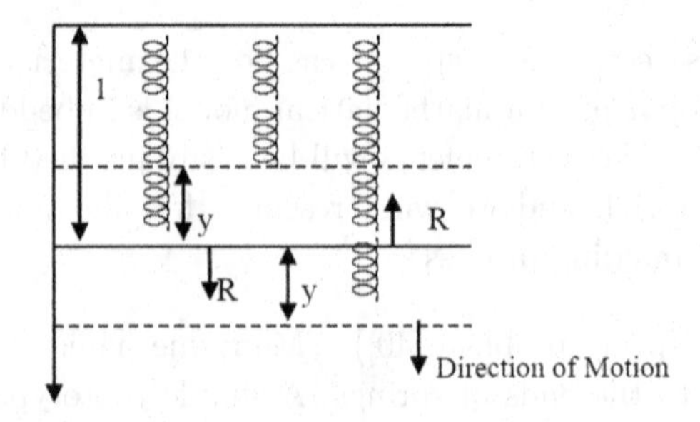

Fig. 5.1 Spring mass dynamic.

***Step* 2** (expression for restoring force). The representation of the restoring force is the goal of this step. Let R be a restoring force acting on the particle, and F be a resultant force acting on the particle. These forces are measured in the positive y direction. There are two cases:

***Case* 1** [the spring extension ($y > 0$)]. In this case, the spring is elongated or extended beyond the natural length by y ft. As a result of this, the spring pulls the particle backward toward the fixed end. This implies that restoring force acts in the negative y direction. Thus, $R < 0$. By Hooke's law,

$$\begin{aligned} &\text{(magnitude of the restoring force)} \\ &= \text{(constant of proportionality) (spring extension)}, \\ |R| &= ky, \end{aligned}$$

where $|R|$ stands for the magnitude of the restoring force; and k stands for the constant of proportionality, and is called the stiffness of the spring. In this case, from SHM 2, we conclude that

$$R = -ky. \tag{5.1}$$

***Case* 2** [the spring compression ($y < 0$)]. In this case, the spring is contracted or compressed, which shortens the natural length by $-y$ ft. As a result of this, the spring pushes the particle forward toward the origin. This implies that restoring force acts in the positive y direction. Thus, $R > 0$. By Hooke's law,

$$\begin{aligned} &\text{(magnitude of the restoring force)} \\ &= \text{(constant of proportionality) (spring extension)}, \\ |R| &= -ky, \end{aligned}$$

where $|R|$ and k are as defined before. Thus, we have

$$R = -ky. \tag{5.2}$$

In summary, from (5.1) and (5.2), the expression for the restoring force is the same, i.e.

$$R = -ky. \tag{5.3}$$

Step* *3 (resultant force). Assuming that there is no other force acting on the particle, the resultant force is given by

$$\begin{aligned} &\text{(the resultant force acting on the particle)} \\ &= \text{(the sum of all forces acting on the particle)}, \qquad (5.4) \\ F &= R. \end{aligned}$$

Step 4 (Newton's second law). From Newton's second law of motion, we have

$$\text{(the rate change of momentum)} = \text{(the resultant force acting on the particle)},$$

$$\begin{aligned}
\frac{d}{dt}[mv] &= F \quad (\text{momentum} = \text{mass} \times \text{velocity}),\\
\frac{d}{dt}\left[m\frac{d}{dt}y\right] &= R \quad \left[v = \frac{d}{dt}y = y^{(1)} \ (5.4)\right],\\
m\frac{d}{dt}\left[\frac{d}{dt}y\right] &= R \quad [\text{mass } (m) \text{ is constant}],\\
m\frac{d^2}{dt^2}y &= -ky \quad [(5.3) \text{ second derivative}].
\end{aligned}$$

Thus,

$$dy^{(1)} + \frac{k}{m}y\,dt = 0. \tag{5.5}$$

Illustration 5.2.2 (RLC circuit [40]). We recall the basic components and ideas of the circuit network (Illustration 4.2.4 and Exercise 6, Section 2.2). In this illustration, we consider an electric circuit consisting of a resistor (R), inductor (L), and capacitor (C).

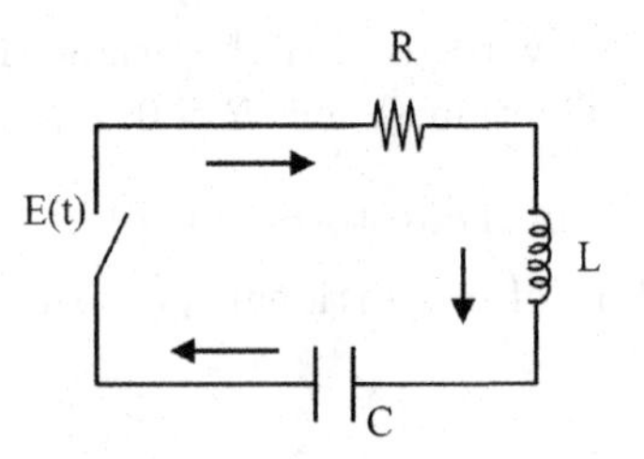

Fig. 5.2 RLC circuit.

As before, we denote by $I(t)$ and $Q(t)$ the electric current in the circuit and the charge on the capacitor, respectively. Let $E(t)$ be the external voltage applied to the circuit. Now, we apply ECN 5 (Kirchhoff's voltage law), and we have

$$\begin{aligned}
E(t) &= L\frac{d}{dt}I + \frac{Q}{C} + RI\\
&= L\frac{d^2}{dt^2}Q + \frac{Q}{C} + R\frac{d}{dt}Q \quad (\text{ECN 2: Coulomb's law})\\
&= L\frac{d^2}{dt^2}Q + R\frac{d}{dt}Q + \frac{Q}{C} \quad (\text{rearrangement}).
\end{aligned} \tag{5.6}$$

Thus, the given circuit is described by the differential equation

$$dQ^{(1)} + \left[\frac{R}{L}Q^{(1)} + \frac{1}{LC}Q - \frac{1}{L}E(t)\right]dt = 0. \tag{5.7}$$

Illustration 5.2.3 (planetary motion [40]). We consider a prototype model of two interacting bodies. The interaction can be generated by the magnetic, electric, or gravitational field. For the sake of illustration, we consider the motion of the planet moving around the sun. The dynamic of the motion is described by Newton's gravitational law. It depends on both the mass and the distance.

NLM 1 (Newton's law of universal gravitation). Each pair of bodies in the universe exerts a force of mutual attraction of magnitude

$$\frac{Gm_1m_2}{r^2},$$

where m_1 and m_2 are the masses of the bodies, r is the distance between them, and G is the constant independent of the bodies.

NLM 2 (Newton's second law of motion). The force produces an acceleration in the direction in which it acts. The magnitude of the force is jointly proportional to the acceleration and the mass of the body. In short, the rate of change of momentum (momentum = mass × velocity) is equal to the resultant force acting on the body.

NLM 3. The motion of the planet is assumed to be planar motion. The plane is described by polar coordinate (r, θ).

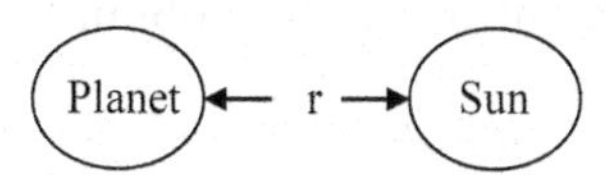

Fig. 5.3 Interactions between planet and sun.

Let M_p be the mass of a planet and M_s be the mass of the sun. We assume that $\mathbf{r} = \mathbf{r}(t)$ and $\mathbf{u}_r$ are the position vector of the planet and the unit vector in the direction of $\mathbf{r}$, respectively. Under this notation and NLM 1 and NLM 2, we have

$$M_p\frac{d^2\mathbf{r}}{dt^2} = -\frac{\gamma M_pM_s}{r^2}\mathbf{u}_r, \tag{5.8}$$

and hence

$$\frac{d^2\mathbf{r}}{dt^2} = -\frac{k}{r^2}\mathbf{u}_r, \tag{5.9}$$

where $|\mathbf{r}| = r$, $\mathbf{u}_r = \cos\theta i + \sin\theta j$, $\mathbf{r} = r\mathbf{u}_r = r(\cos\theta i + \sin\theta j)$, r is the magnitude of the position vector $\mathbf{r}$, γ is the universal gravitational constant, and $k = \gamma M_s$.

Now, we have

$$\begin{aligned}\frac{d\mathbf{r}}{dt} &= \frac{dr}{dt}\,\mathbf{u}_r + r\,\frac{d\theta}{dt}\,\mathbf{u}_\theta,\\ \frac{d^2\mathbf{r}}{dt^2} &= \left(\frac{d^2r}{dt^2} - r\left(\frac{d\theta}{dt}\right)^2\right)\mathbf{u}_r + \left(2\frac{dr}{dt}\,\frac{d\theta}{dt} + r\frac{d^2\theta}{dt^2}\right)\mathbf{u}_\theta,\end{aligned} \tag{5.10}$$

where $\mathbf{u}_\theta = -\sin\theta i + \cos\theta j$ and

$$\frac{d\mathbf{u}_\theta}{dt} = -(\cos\theta i + \sin\theta j).$$

We substitute the expressions in (5.10) into (5.9), and we have

$$\left(\frac{d^2r}{dt^2} - r\left(\frac{d\theta}{dt}\right)^2\right)\mathbf{u}_r + \left(2\frac{dr}{dt}\frac{d\theta}{dt} + r\frac{d^2\theta}{dt^2}\right)\mathbf{u}_\theta = -\frac{k}{r^2}\,\mathbf{u}_r. \tag{5.11}$$

This implies (using the method of undetermined coefficients) that

$$\frac{d^2r}{dt^2} - r\left(\frac{d\theta}{dt}\right)^2 = -\frac{k}{r^2}, \tag{5.12}$$

$$2\,\frac{dr}{dt}\,\frac{d\theta}{dt} + r\,\frac{d^2\theta}{dt^2} = 0. \tag{5.13}$$

Multiplying both sides of (5.13) by r, we obtain

$$\begin{aligned}0 &= 2r\,\frac{dr}{dt}\,\frac{d\theta}{dt} + r^2\frac{d^2\theta}{dt^2}\\ &= \frac{d}{dt}\left[r^2\frac{d\theta}{dt}\right] \quad \text{(product rule).}\end{aligned} \tag{5.14}$$

Thus, upon the integration in (5.14), we have

$$r^2\frac{d\theta}{dt} = c \quad (c \text{ is a constant of integration}). \tag{5.15}$$

We substitute for $\frac{d\theta}{dt}$ into (5.12), and get

$$\frac{d^2r}{dt^2} - \frac{rc^2}{r^4} = -\frac{k}{r^2}.$$

Hence,

$$-\frac{r^2}{c^2}\,\frac{d^2r}{dt^2} + \frac{1}{r} = \frac{k}{c^2}. \tag{5.16}$$

This is a second-order nonlinear differential equation. However, by using the substitution $p = 1/r$ and solving for p as a function of θ instead of t,

$$\begin{aligned}\frac{dp}{dt} &= \frac{d}{dt}(1/r) = -r^{-2}\frac{dr}{dt}\\ &= \frac{dp}{d\theta}\frac{d\theta}{dt} \quad \text{(chain rule)}\\ &= \frac{c}{r^2}\frac{dp}{d\theta} \quad \text{[substitution from (5.15)]}.\end{aligned} \tag{5.17}$$

From (5.17), we have

$$\frac{dp}{d\theta} = -\frac{1}{c}\frac{dr}{dt}.$$

Hence,

$$\frac{d^2p}{d\theta^2} = -\frac{1}{c}\frac{d^2r}{dt^2}\frac{dt}{d\theta} = -\frac{r^2}{c^2}\frac{d^2r}{dt^2} \quad \text{(chain rule)}. \tag{5.18}$$

From (5.18) and $p = 1/r$, (5.16) reduces to

$$\begin{aligned}\frac{d^2p}{d\theta^2} + \frac{1}{r} &= \frac{k}{c^2} \quad \text{[from (5.16) and (5.18)]}\\ \frac{d^2p}{d\theta^2} + p &= \frac{k}{c^2} \quad \text{[by substitution: } p = 1/r\text{]}.\end{aligned} \tag{5.19}$$

This is the second order linear nonhomogeneous differential equation with the independent variable θ.

Illustration 5.2.4 (intravenously injected creatinine clearance [123]). This illustration is based on the study conducted by Sapirstein, Vidit, Mandel, and Hanusek (1955). A single dose of creatinine (2.0 g in 50 cc of saline) was rapidly injected, intravenously. The plasma disappearance curve was determined by taking a blood sample at 5 min intervals for 60 min. The plasma creatinine was determined by the method of Bones and Taussky. The data were plotted on semilog paper. Observing the sketch, it was concluded that the curve is described by two straight lines. This shows that it is described by $y = c_1 \exp[-\alpha t] + c_2 \exp[-\beta t]$. To justify the above discussion, we follow the derivation of Ref. 123.

ICC 1. It is assumed that plasma and body tissue can be considered to be two compartments. The diffusion of creatinine through the boundary (membrane) follows Fick's law (Illustration 2.5.3).

ICC 2. It is further assumed that the plasma compartment is considered to be a "leaky" compartment. This leak is represented through the kidneys. It is assumed that the creatinine loss through this pathway is directly proportional to its concentration in the blood compartment.

ICC 3. Furthermore, it is assumed that the creatinine injection in the blood compartment is "instantaneous," and it is instantaneously and homogeneously mixed in the blood.

ICC 4. The creatinine is homogeneously distributed in the body tissue compartment.

ICC 5. It is further assumed the exogenous creatinine is not metabolized. It is removed from the body by one and only one way of the kidney.

ICC 6. Finally, it is assumed that the volumes of the compartments are fixed constants.

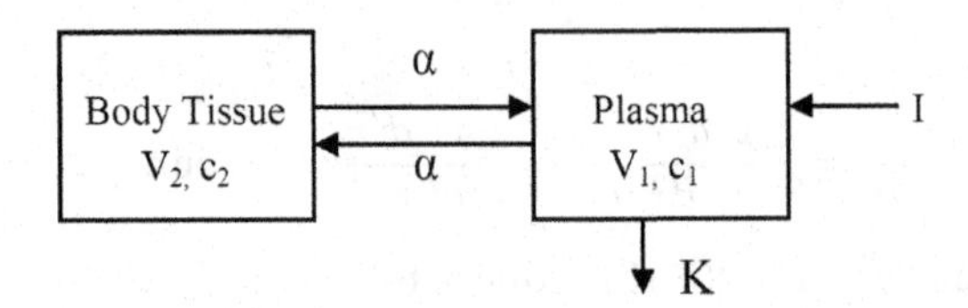

Fig. 5.4 Creatine clearance dynamic process.

Let V_1 and V_2 be the volumes of the plasma and body tissue compartments, respectively. Let c_1 and c_2 be the concentrations of creatinine in the respective compartments at a time t. It is assumed that I is the amount of creatinine injected into the blood, intravenously. Let $m = c_1V_1$ be the amount in the plasma at a time t, and $e(t)$ be the amount of creatinine excreted at time t. From the assumptions, we have

$$dm = [-\kappa c_1 - \alpha(c_1 - c_2)]\,dt \quad \text{(from ICC 1 and ICC 2)}, \tag{5.20}$$

$$de = \kappa c_1 dt \quad \text{(from ICC 2)}, \tag{5.21}$$

$$I = c_1V_1 + c_2V_2 + e(t) \quad \text{(from ICC 1, ICC 2, ICC 3 and ICC 5)}. \tag{5.22}$$

From the definitions of c_1, c_2, and $e(t)$ and assumption ICC 6, (5.20)–(5.22) can be reduced to

$$V_1dc_1 = [-\kappa c_1 - \alpha(c_1 - c_2)]\,dt \quad \text{(from definitions of } c\text{'s and ICC 6)}, \tag{5.23}$$

$$e(t) = \int_0^t \kappa c_1(s)ds \quad \text{(by integration)}, \tag{5.24}$$

$$c_2 = \frac{I - \int_0^t \kappa c_1(s)ds - c_1V_1}{V_2} \quad \text{[from (5.22) and (5.24)]}. \tag{5.25}$$

Now, we substitute the expression for c_2 in (5.25) into (5.23), and we get

$$\begin{aligned} V_1 dc_1 &= [-\kappa c_1 - \alpha (c_1 - c_2)]\, dt \\ &= \left[-\kappa c_1 - \alpha \left(c_1 - \frac{I - \int_0^t \kappa c_1(s)ds - c_1 V_1}{V_2}\right)\right] dt, \end{aligned}$$

which upon differentiation reduces to

$$\begin{aligned} V_1 dc_1^{(1)} &= \left[-\kappa dc_1 - \alpha dc_1 + \alpha d\frac{I - \int_0^t \kappa c_1(s)ds - c_1 V_1}{V_2}\right] dt \\ &= -\left[(\kappa + \alpha)\, c_1^{(1)} + \frac{\alpha\kappa}{V_2} c_1 + \frac{\alpha V_1}{V_2} c_1^{(1)}\right] dt \quad \text{(notation)}. \end{aligned}$$

Thus,

$$dc^{(1)} + \left[\left(\frac{\kappa + \alpha}{V_1} + \frac{\alpha}{V_2}\right) c_1^{(1)} + \frac{\alpha\kappa}{V_1 V_2} c_1\right] dt = 0. \tag{5.26}$$

This second-order linear differential equation with constant coefficients describes the dynamic of creatinine clearances.

5.2 Exercises

(1) ***Angular motion of the cupula*** [135]. It is known that the deflection of the cupula is related to the sensory perception of rotation. If a blindfolded person sitting on a revolving chair is abruptly stopped, then he or she will experience the sensation of rotation with an angular velocity opposite to the original direction of rotation. This sensation gradually decays. The physical structure that plays this after-sensation is the horizontal semicircular canal. It consists of the wide portion, the utriculus, and a continuous hollow circular tube (hollow ring). The ring is filled with a viscous fluid, namely endolymph. A kind of gelatinous partition attached to a bony rim is known as the cupula. It protrudes into the hollow tube and fills the whole cross-section, and its function is analogous to the swinging door with elastic properties. Upon the impulse, the cupula deviates from its equilibrium position, and then slowly returns to its rest position. From this observation, one concludes that the liquid ring together with the cupula represents an overdamped harmonic oscillator. Let us consider that a person is constantly rotated on a chair (Barany chair). The rotation is used for deflection of the cupula from its rest position. It is stopped at $t = 0$. The motion is stopped and the cupula is jerked "forward" by the flow of endolymph due to the inertia of the endolymph. At $t = 0$, x_0 is the maximum deflection of the cupula, and for $t \geq 0$ it is assumed that there is no external torque acting on the cupula. It is also assumed that $x(t)$ is the deflection of the cupula at $t \geq 0$. The cupula acts like a spring with one end fixed. The torque acts like a restoring force ($R = -\alpha x$) in the spring. Moreover, the motion of the cupula is in fluid,

and this motion generates a force due to the fluid. This force is called the drag force. The drag force $\left(F_d = -\beta \frac{dx}{dt}\right)$ has the same function as the frictional force acting on the body attached to the spring. Since we are dealing with the angular moment, let m be the moment of inertia instead of mass. Employing Newton's second law of motion, show that the dynamic of the cupula is described by

$$m dx^{(1)} + [\beta x^{(1)} + \alpha x] dt = 0.$$

(2) Using Illustration 5.2.2, show that an alternate mathematical model of an RLC circuit is described by

$$dI^{(1)} + \left[\frac{R}{L} I^{(1)} + \frac{1}{LC} I - \frac{1}{L}\frac{d}{dt}E(t)\right] dt = 0.$$

In particular, if E is a constant function, then

$$dI^{(1)} + \left[\frac{R}{L} I^{(1)} + \frac{1}{LC} I\right] dt = 0.$$

(3) ***Circular motion of the body*** [27]. The Cartesian coordinates and the corresponding position vector of the particle are given by $(x, y) = (a\cos\theta, a\sin\theta)$ and $\mathbf{X} = a\cos\theta\mathbf{i} + a\sin\theta\mathbf{j}$, respectively, where a is a positive constant, and $\mathbf{i}$ and $\mathbf{j}$ are unit vectors in the directions of the x and y axes, respectively. Show that the velocity $\left(\frac{d}{dt}\mathbf{X}\right)$ and the acceleration $\left(\frac{d^2}{dt^2}\mathbf{X}\right)$ vectors, represented by

$$\left(\frac{d}{dt}\mathbf{X}\right) = a\frac{d}{dt}\theta\tau, \quad \frac{d^2}{dt^2}\mathbf{X} = -a\left(\frac{d}{dt}\theta\right)^2 \nu + a\frac{d^2}{dt^2}\theta\tau,$$

respectively, where $\nu = \cos\theta\mathbf{i} + \sin\theta\mathbf{j}$ and $\tau = \sin\theta\mathbf{i} + \cos\theta\mathbf{j}$ are unit normal (pointing outward) and tangent (pointing anticlockwise) vectors to a circle with the radius a, respectively; $\frac{d}{dt}\theta$ is an angular velocity;

$$-a\left(\frac{d}{dt}\theta\right)^2, \quad a\frac{d^2}{dt^2}\theta,$$

are normal and tangential components of the accelerations, respectively. Moreover, $\frac{d^2}{dt^2}\theta$ is referred to as the angular acceleration of the particle. The motion is called uniform circular motion, if $\frac{d}{dt}\theta = \omega$ is constant.

(4) The Earth moves around the sun in an orbit which is almost circular, and the radius of the Earth is 1.49×10^8 km. Find: (i) the angular velocity of the Earth around the sun, (ii) the speed of the Earth, and (iii) the magnitude of the acceleration toward the sun.

(5) Let us consider a light rod of length l pivoted smoothly at one end and with a particle of mass m attached to the other end. The particle is called the bob of the pendulum. The motion of the bob is in the vertical plane. Let θ be the angle made by the rod with the vertical line at a time t. The bob is free to move

in the vertical plane under the gravitational force. Show that the motion of the bob in the vertical plane is described by

$$l\frac{d^2}{dt^2}\theta = -mg\sin\theta.$$

5.3 Linear Homogeneous Equations

In this section, we extend the eigenvalue-type method described in Sections 2.4, 4.3, and 4.4 to find a solution to linear higher-order homogeneous differential equations with constant coefficients.

5.3.1 *General problem*

Let us consider a higher-order linear homogeneous differential equation of the type

$$L(d)y \equiv dy^{(n-1)} + [a_{n-1}y^{(n-1)} + a_{n-2}y^{(n-2)} + \cdots + a_1 y^{(1)} + a_0 y]dt = 0, \tag{5.27}$$

where $a_{n-1}, a_{n-2}, \ldots, a_1$, and a_0 are given real numbers; $y^{(k)}$ stands for the kth derivative of y; and L is a linear differential operator defined by

$$L(d) = d\frac{d^{(n-1)}}{dt^{n-1}} + a_{n-1}d\frac{d^{(n-2)}}{dt^{n-2}} + \cdots + \cdots + a_1 d + a_0.$$

We note that

$$d\frac{d^{(k-1)}}{dt^{k-1}}y = \frac{d^k}{dt^k}\,y\,dt = y^{(k)}dt,$$

for $k = 1, 2, \ldots, k, \ldots, n$ with $d^{(0)}y = ydt(d^{(0)} = I)$.

In this subsection, our goal is to discuss a procedure for finding a general solution to (5.27). We are also interested in solving an IVP associated with this equation. This provides a basis for solving real-world problems. The highest derivative of the state variable occurring in the differential equation is referred to as the order of the differential equation. We recall that if the exponents of the state variable (dependent variable) and derivatives occurring in the differential equation are "unity," then the differential equation is called a linear differential equation.

Definition 5.3.1. Let $J = [a, b]$, for $a, b \in R$; and hence let $J \subseteq R$ be an interval. A solution to an nth-order linear differential equation of the type (5.27) is an n times continuously differentiable function y defined on J into R such that it satisfies (5.27) on J in the sense of deterministic calculus [85]. In short, the differential $dy^{(n-1)}$ of the $(n-1)$th derivative of y, i.e.

$$dy^{(n-1)} = d\frac{d^{n-1}}{dt^{n-1}}y,$$

together with other lower-order derivatives, namely $y^{(n-2)}, \ldots, y^{(1)}$, and y, satisfy the given differential equation (5.27).

Example 5.3.1. Let $L(d)y \equiv dy^{(1)} + y\,dt = 0$. Show that $y(t) = \sin t$ is the solution to the given the differential equation.

Solution procedure. First, we note that this is a second-order differential equation. We apply Definition 5.3.1 to show that $y(t) = \sin t$ is the solution to the given differential equation. For this purpose, we differentiate and obtain

$$y(t) = \sin t, \quad y^{(1)}(t) = -\cos t, \quad dy^{(1)}(t) = -\sin t\,dt.$$

By substituting these expressions into the equation, we have

$$L\left(y(t)\right)dt \equiv dy^{(1)}(t) + y(t)\,dt = (-\sin t + \sin t)\,dt = 0.$$

This exhibits the fact that $y(t) = \sin t$ satisfies the given differential equation. Therefore, $y(t) = \sin t$ is the solution to the given differential equation.

5.3.2 *Procedure for finding a general solution*

We present a three-step procedure for determining a solution to (5.27). Again, the idea and rationale are the same as what is outlined in Sections 2.4 and 4.3, i.e. "to reduce the problem of solving linear differential equations to the problem of solving linear algebraic equations." Here are the steps:

Step 1. Let us seek a solution to (5.27) of the following form:

$$y(t) = \exp\left[\lambda t\right]c, \tag{5.28}$$

where c is an unknown real number and λ is an unknown scalar quantity (a real or complex number). We note that if $c = 0$, then $y(t)$ in (5.27) is a trivial solution [zero solution, i.e. zero function $x(t) \equiv 0$ on J] of (5.27).

We seek a nontrivial solution (nonzero solution) to (5.27). For this purpose, we need to compute λ and c. This is achieved as follows. We differentiate y in (5.28) with respect to t, as

$$\begin{aligned} y^{(1)} &= \lambda \exp\left[\lambda t\right]c, y^{(2)} = \lambda^2 \exp\left[\lambda t\right]c, \ldots, \\ &= y^{(n-1)} = \lambda^{(n-1)} \exp\left[\lambda t\right]c, y^{(n)} = \lambda^n \exp\left[\lambda t\right]c, \end{aligned}$$

and substituting in (5.27) we obtain

$$\begin{aligned} &\lambda^n \exp\left[\lambda t\right]c\,dt + [a_{n-1}\lambda^{(n-1)} \exp\left[\lambda t\right]c + \cdots \\ &\qquad + a_1\lambda \exp\left[\lambda t\right]c + a_0 \exp\left[\lambda t\right]c]dt = 0. \end{aligned}$$

Hence,

$$\exp\left[\lambda t\right][\lambda^n + a_{n-1}\lambda^{(n-1)} + \cdots + a_1\lambda + a_0]c = 0.$$

From this and the fact that $\exp\left[\lambda t\right] \neq 0$ on J, we have

$$[\lambda^n + a_{n-1}\lambda^{(n-1)} + \cdots + a_1\lambda + a_0]c = 0. \tag{5.29}$$

Step 2. Now, the problem of finding a solution to the nth-order linear differential equation (5.27) reduces to the problem of solving the nth-degree polynomial equation in λ (real or complex number) with real coefficients for an arbitrary nonzero constant c in (5.29). Thus,

$$\Lambda(\lambda) = \lambda^n + a_{n-1}\lambda^{n-1} + \cdots + a_1\lambda + a_0 = 0. \tag{5.30}$$

Equation (5.30) is referred to as the auxiliary equation of (5.27). Moreover, it is an nth-degree polynomial equation in λ. Its leading coefficient is always 1. Hence, from the fundamental theorem of algebra, the auxiliary equation (5.30) always has n roots $\lambda_1, \ldots, \lambda_k, \ldots, \lambda_n$ (real or complex number), which are not necessarily all distinct. We further remark that these roots are uniquely determined by the coefficients in (5.27).

Step 3. For n roots $\lambda_1, \lambda_k, \ldots, \lambda_n$ of (5.30) in step 2, from (5.28) we can find n solutions,

$$x_1(t) = \exp[\lambda_1 t]\, c_1, \ldots, x_k(t) = \exp[\lambda_k t]\, c_k, \ldots, x_n(t) = \exp[\lambda_n t]\, c_n,$$

to the differential equation (5.27), where, $c_1, \ldots, c_k, \ldots, c_n$ are arbitrary constants. This completes the procedure for finding solutions to (5.27).

Example 5.3.2. Let us consider $L(d)y \equiv [dy^{(1)} - 5y^{(1)} + 6y]dt = 0$. Find solutions to the given differential equation.

Solution procedure. This is the second-order linear homogeneous differential equation. To find solutions to this differential equation, we imitate the procedure in Subsection 5.3.2. We seek a solution of the form described in step 1, i.e. $y(t) = \exp[\lambda t]\, c$. The primary goal is to find a scalar number λ for an arbitrary nonzero real number c. For this purpose, we use (5.29), and we have

$$\left[\lambda^2 - 5\lambda + 6\right] c = 0.$$

The auxiliary equation (5.30) corresponding to the given example is

$$0 = \lambda^2 - 5\lambda + 6 = (\lambda - 3)(\lambda - 2) \quad \text{(by factoring)}.$$

The roots are $\lambda_1 = 3$ and $\lambda_2 = 2$. This completes step 2 of the procedure in Subsection 5.3.2. By using the argument of step 3 in the procedure in Subsection 5.3.2, we found two solutions,

$$y_1(t) = e^{3t}, \quad y_2(t) = e^{2t},$$

corresponding to two roots, $\lambda_1 = 3$ and $\lambda_2 = 2$, of the auxiliary equation. This completes the solution procedure.

The following result is similar to Theorem 4.3.1, and the proof is left as an exercise for the reader.

Theorem 5.3.1 (Principle of superposition — fundamental property). *Let $y_1(t), y_2(t), \ldots, y_k(t), \ldots, y_m(t)$ be m solutions to (5.27) on J. Then*

$$y(t) = c_1 y_1(t) + c_2 y_2(t) + \cdots + c_k y_k(t) + \cdots + c_m y_m(t) = \sum_{k=1}^{m} c_k y_k(t) \tag{5.31}$$

is also a solution to (5.27) on J, where $c_1, \ldots, c_k, \ldots, c_m$ are arbitrary constants (real numbers).

Observation 5.3.1. A set of solutions $\{y_1(t), y_2(t), \ldots, y_k(t), \ldots, y_m(t)\}$ is said to be a complete set of solutions if: (i) $m = n$ and (ii) any solution to (5.27) can be found by a linear combination of the members of this set. This is also called the maximal linearly independent set (Definition 1.4.6) of solutions to (5.27).

5.3.3 *Initial value problem*

Let us formulate an IVP relative to (5.27). This will motivate us to find a complete set of solutions to (5.27). The complete set of solutions is used to solve any given IVP. We consider

$$L(d)y \equiv dy^{(n-1)} + [a_{n-1}y^{(n-1)} + a_{n-2}y^{(n-2)} + \cdots + a_1 y^{(1)} + a_0 y]dt = 0, \tag{5.32}$$

with

$$y^{(n-1)}(t_0) = y_0^{(n-1)}, \ldots, y^{(k)}(t_0) = y_0^{(k)}, \ldots, y^{(1)}(t_0) = y_0^{(1)}, \quad y(t_0) = y_0. \tag{5.33}$$

Here, for $i = 0, 1, \ldots, n-1$, a_i's and y are as defined in (5.27); t_0 is in J, and $y_0^{(n-1)}, \ldots, y_0^{(k)}, \ldots, y_0^{(1)}$, and y_0 belong to R. $(t_0, y_0^{(n-1)}, \ldots, y_0^{(k)}, \ldots, y_0^{(1)}, y_0)$ is called initial data or initial conditions. The problem of finding a solution to (5.32) and (5.33) is referred to as the initial value problem (IVP). Its solution is represented by

$$y(t) = y(t, t_0, y_0^{(n-1)}, \ldots, y_0^{(k)}, \ldots, y_0^{(1)}, y_0),$$

for $t \geq t_0$ and t in J.

Definition 5.3.2. A solution to the (IVP) (5.32)–(5.33) is the n times continuously differentiable function y defined on J into R such that it satisfies: (i) the given nth-order differential equation (5.32) on J in the sense of deterministic calculus, and (ii) any given initial condition $(t_0, y_0^{(n-1)}, \ldots, y_0^{(k)}, \ldots, y_0^{(1)}, y_0)$ in (5.33). In short, (i) $y(t)$ is a solution to (5.27) (Definition 5.3.1) defined on J, and (ii) $y^{(n-1)}(t_0) = y_0^{(n-1)}, \ldots, y^{(k)}(t_0) = y_0^{(k)}, \ldots, y^{(1)}(t_0) = y_0^{(1)}$, and $y(t_0) = y_0$.

Example 5.3.3. We are given $dy^{(1)} - 5y^{(1)} + 6y = 0$, $y^{(1)}(t_0) = y_0^{(1)}$, and $y(t_0) = y_0$. Show that for $t \in J$ and $t \geq t_0$,

$$y(t) = (y_0^{(1)} - 2y_0)\exp[3(t - t_0)] + (3y_0 - y_0^{(1)})\exp[2(t - t_0)]$$

is the solution to the given IVP.

Solution procedure. By using the argument in Example 5.3.1, we need to show that

(i) $y(t) = (y_0^{(1)} - 2y_0)\exp[3(t - t_0)] + (3y_0 - y_0^{(1)})\exp[2(t - t_0)]$

satisfies the given differential equation for $t \in J$ and $t \geq t_0$, and

(ii) $y(t_0) = y_0$ and $y^{(1)}(t_0, t_0) = (y_0^{(1)})$.

We compute $dy(t, t_0)$ and $dy^{(1)}(t, t_0)$ as follows:

$$\begin{aligned} dy(t) &= d[(y_0^{(1)} - 2y_0)\exp[3(t - t_0)] + (3y_0 - y_0^{(1)})\exp[2(t - t_0)]] \\ &= [3(y_0^{(1)} - 2y_0)\exp[3(t - t_0)] \\ &\quad + 2(3y_0 - y_0^{(1)})\exp[2(t - t_0)]dt \quad \text{(Observation 1.5.2)}, \\ dy^{(1)}(t) &= d[3(y_0^{(1)} - 2y_0)\exp[3(t - t_0)] + 2(3y_0 - y_0^{(1)})\exp[2(t - t_0)]] \\ &= [9(y_0^{(1)} - 2y_0)\exp[3(t - t_0)] + 4(3y_0 - y_0^{(1)})\exp[2(t - t_0)]]dt. \end{aligned}$$

On the other hand, we compute $L(d)y(t)$ as

$$\begin{aligned} L(d)y(t) &= (d^2 - 5d^{(1)} + 6)y(t, t_0)dt \quad \text{[definition of } L(d)] \\ &= dy^{(1)}(t, t_0) - 5y^{(1)}(t, t_0) + 6y(t, t_0) \quad \text{(Observation 1.5.2)} \\ &= [9(y_0^{(1)} - 2y_0)\exp[3(t - t_0)] + 4(3y_0 - y_0^{(1)})\exp[2(t - t_0)]]dt \\ &\quad - 5[3(y_0^{(1)} - 2y_0)\exp[3(t - t_0)] + 2(3y_0 - y_0^{(1)})\exp[2(t - t_0)]]dt \\ &\quad + 6[(y_0^{(1)} - 2y_0)\exp[3(t - t_0)] + (3y_0 - y_0^{(1)})\exp[2(t - t_0)]]dt \\ &= 0 \quad \text{(simplification)}. \end{aligned}$$

From the above discussion, we conclude that $y(t)$ satisfies the given differential equation. Finally, it remains for us to prove that $y(t_0) = y_0$ and $y^{(1)}(t_0) = y_0^{(1)}$. We observe that

$$\begin{aligned} y(t_0) &= (y_0^{(1)} - 2y_0)\exp[3(t_0 - t_0)] + (3y_0 - y_0^{(1)})\exp[2(t_0 - t_0)] \\ &= (y_0^{(1)} - 2y_0) + (3y_0 - y_0^{(1)}) = y_0, \\ y^{(1)}(t_0) &= 3(y_0^{(1)} - 2y_0)\exp[3(t_0 - t_0)] + 2(3y_0 - y_0^{(1)})\exp[2(t_0 - t_0)] \\ &= 3(y_0^{(1)} - 2y_0) + 2(3y_0 - y_0^{(1)}) = y_0^{(1)}. \end{aligned}$$

This completes the solution process of the example.

5.3.4 *Procedure for solving the IVP*

Now, we discuss a procedure for finding a solution to the IVPs (5.32) and (5.33). The procedure is almost parallel to the procedure in Subsection 4.3.4. For the sake of completeness, we summarize it by the following:

Step 1. We imitate the procedure in Subsection 4.3.4, and assume that m solutions

$$y_1(t), y_2(t), \ldots, y_k(t), \ldots, y_m(t),$$

defined in Theorem 5.3.1, were determined by the three-step procedure in Subsection 5.3.2 for finding a solution to (5.27). Therefore, from Theorem 5.3.1, we conclude that a linear combination of m solutions $y_1(t), y_2(t), \ldots, y_k(t), \ldots, y_m(t)$ to (5.27),

$$y(t) = c_1 y_1(t) + c_2 y_2(t) + \cdots + c_k y_k(t) + \cdots + c_m y_m(t), \tag{5.34}$$

is also a solution to (5.27).

Step 2. By using the information in step 1, we try to solve the IVP (5.32)–(5.33). For this purpose, we simply need to verify that a solution to (5.27) defined in (5.34) satisfies the initial conditions (5.33), i.e.

$$y\,(t_0) = c_1 y_1\,(t_0) + \cdots + c_k y_k\,(t_0) + \cdots + c_m y_m\,(t_0) = y_0. \tag{5.35}$$

We note that we have used only one initial condition. We need to use the remaining conditions to obtain a system of algebraic equations so as to find the arbitrary constants $c_1, \ldots, c_k, \ldots, c_m$. For this purpose, we need to generate at least $m-1$ more algebraic equations similar to (5.35). This is achieved by differentiating both sides of (5.34) and evaluating at $t = t_0$:

$$\begin{aligned}
y^{(1)}(t) &= c_1 y_1^{(1)}(t) + \cdots + c_k y_k^{(1)}(t) + \cdots + c_m y_m^{(1)}(t) \\
&\cdots\cdots\cdots\cdots\cdots\cdots\cdots\cdots\cdots\cdots\cdots\cdots \\
y^{(i)}(t) &= c_1 y_1^{(i)}(t) + \cdots + c_k y_k^{(i)}(t) + \cdots + c_m y_m^{(i)}(t) \\
&\cdots\cdots\cdots\cdots\cdots\cdots\cdots\cdots\cdots\cdots\cdots\cdots \\
y^{(n-1)}(t) &= c_1 y_1^{(n-1)}(t) + \cdots + c_k y_k^{(n-1)}(t) + \cdots + c_m y_m^{(n-1)}(t).
\end{aligned} \tag{5.36}$$

Hence,

$$\begin{aligned}
y^{(1)}\,(t_0) &= c_1 y_1^{(1)}\,(t_0) + \cdots + c_k y_k^{(1)}\,(t_0) + \cdots + c_m y_m^{(1)}\,(t_0) = y_0^{(1)} \\
&\cdots\cdots\cdots\cdots\cdots\cdots\cdots\cdots\cdots\cdots\cdots\cdots \\
y^{(i)}\,(t_0) &= c_1 y_1^{(i)}\,(t_0) + \cdots + c_k y_k^{(i)}\,(t_0) + \cdots + c_m y_m^{(i)}\,(t_0) = y_0^{(i)} \\
&\cdots\cdots\cdots\cdots\cdots\cdots\cdots\cdots\cdots\cdots\cdots\cdots \\
y^{(n-1)}\,(t_0) &= c_1 y_1^{(n-1)}\,(t_0) + \cdots + c_k y_k^{(n-1)}\,(t_0) + \cdots + c_m y_m^{(n-1)}\,(t_0) = y_0^{(n-1)}.
\end{aligned} \tag{5.37}$$

Now, we have n nonhomogeneous algebraic equations in (5.35) and (5.37) and m unknown coefficients. Repeating the rest of the argument in step 2 of the procedure in Subsection 4.3.4, we need to assume that (i) $m = n$ and (ii) the determinant of the coefficient matrix of the system in (5.35) and (5.37) is different from zero, i.e. $\det(y_k^{(i)}(t_0)) \neq 0$, where $y_k^{(i)}(t_0)$ is the ith derivative of the kth solution, for $i = 0, 1, \ldots, (n-1)$ and $k = 1, 2, \ldots, n$.

Step 3. Under the assumption $\det(y_k^{(i)}(t_0)) \neq 0$, the coefficient matrix corresponding to the system of algebraic equations in (5.35) and (5.37), $\Phi(t_0) = (y_k^{(i)}(t_0))_{n\times n}$, is invertible (Theorem 1.4.4). Again, repeating the argument used in step 3 of the procedure in Subsection 4.3.4, we have (Theorem 1.4.4)

$$c = \Phi^{-1}(t_0)\, x_0, \tag{5.38}$$

where

$$[y_0, y_0^{(1)}, \ldots, y_0^{(k)}, \ldots, y_0^{(n-1)}]^T = [x_{10}, x_{20}, \ldots, x_{k0}, \ldots, x_{n0}] = x_0,$$
$$c = [c_1, \ldots, c_k, \ldots, c_n]^T.$$

Step 4. The solution to the IVPs (5.32) and (5.33) is obtained by substituting the values of $c_1, \ldots, c_k, \ldots, c_n$ in (5.38) into (5.34). This completes the solution procedure of the IVPs (5.32) and (5.33).

Observation 5.3.2

(i) Step 4 of the procedure in Subsection 5.3.4 can be further clarified by the following mathematical description in the framework of matrix representation. From Definition 1.3.10, (5.34) can be rewritten as

$$y(t) = x^T(t)c, \tag{5.39}$$

where

$$x(t) = [y(t), y^{(1)}(t), \ldots, y^{(k)}(t), \ldots, y^{(n-1)}(t)]^T,$$
$$c = [c_1, \ldots, c_k, \ldots, c_n]^T.$$

Thus, the solution to the IVPs (5.32) and (5.33) can be expressed as

$$y(t) = x^T(t)\Phi^{-1}(t_0)\, x_0, \tag{5.40}$$

where $\Phi(t_0) = (y_k^{(i)}(t_0))_{n\times n}$ and x_0 is as defined in (5.38).

(ii) We further note that the solution to the IVPs (5.32) and (5.33) is the first component of the following n-dimensional function:

$$\begin{aligned} x(t, t_0, x_0) &= \Phi(t)\Phi^{-1}(t_0)\, x_0 \\ &= \Phi(t, t_0)\, x_0, \end{aligned} \tag{5.41}$$

where $\Phi(t) = (y_k^{(i)}(t))_{n\times n}$, and $\Phi^{-1}(t_0)$ and x_0 are as defined in (5.40) and (5.38), respectively.

(iii) We further observe that $\det(y_k^{(i)}(t_0)) \neq 0$ if and only if the solution set $\{y(t), y^{(1)}(t), \ldots, y^{(k)}(t), \ldots, y^{(n-1)}(t)\}$ is a complete set of linearly independent solutions to (5.27). This statement will be further justified at a later stage.

(iv) In view of (i)–(iii), one can introduce the following concepts. The solution representation in (5.39) is referred to as the general solution to (5.27). The solution representation of the IVP (5.32)–(5.33) in (5.40) is called the particular solution to (5.27). Det $(y_k^{(i)}(t)) = W(t)$ is known as the Wronskian of the solution set in (iii).

(v) The procedure for finding a complete set of linearly independent solutions to (5.27) depends on roots of the auxiliary equation (5.30). This problem is exactly the same as the problem of finding eigenvalues of the characteristic equation (4.38). Therefore, the goal outlined in Section 4.4 is applicable. Thus, we have to deal with the following three cases. These cases depend on the nature of the roots of (5.30), namely: (1) distinct real numbers, (2) distinct complex numbers, and (3) repeated, real or complex numbers. The ideas and algorithms for finding n linearly independent solutions are almost similar to those presented in Section 4.4. Instead of repeating the presentation, we discuss the ideas through examples.

(vi) This observation suggests that there is indeed a very close connection between higher-order linear differential equations and a linear system of differential equations. In fact, a higher-order differential equation is special case of a system of differential equations. This statement will be proved later in the chapter.

Illustration 5.3.1 (spring–dashpot system [10, 55]). Let l be the natural length of a light spring with the coefficient stiffness k, and one end is attached to the fixed point. A dashpot is a cylinder filled with a fluid, and one end is sealed and attached to the fixed point. The other end of the spring and dashpot with a loose plunger is attached to a body of mass m measured in kilograms lying on a smooth fixed surface. Let y denote the extension of the spring measured in meters. Describe the motion of the body under the following conditions: (a) y_0, $y_0^{(1)} = v_0$, and $t = 0$; (b) y_0, $y_0^{(1)} = 0$; and $t = 0$; and (c) $y_0 = 0$, $y_0^{(1)} = v_0$, and $t = 0$.

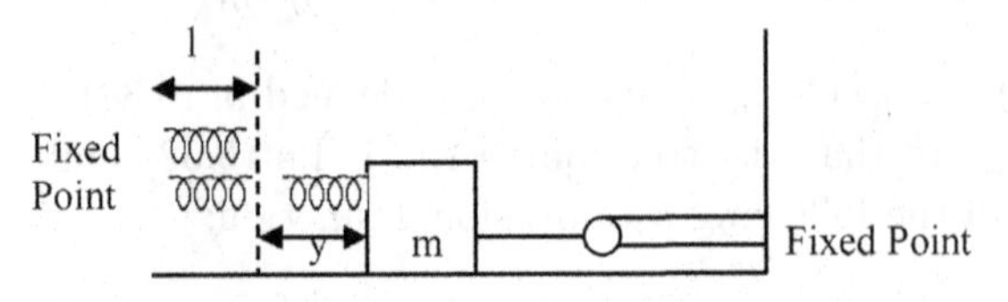

Fig. 5.5 Spring–dashpot system.

Solution process. We use the basic elements of the prototype model described in Illustration 5.2.1.

Step 1. First, we find the resultant force acting on the body of mass m. The body is moving in a straight line in the plane. The component of the resultant force in the horizontal direction is composed of: (i) the restoring force due to the spring R and the force generated by the motion of the body due to the dashpot F_d. Thus, the resultant force, $R + F_d$, is in the horizontal direction. The vertical component of the force acting on the body is zero, because the motion of the body is on the fixed surface.

Step 2. From Hooke's law, we have $R = -ky$. The magnitude and the direction of the force that is exerted by the dashpot on the body depend on the magnitude and the direction of the velocity of the body. In fact, the direction of F_d is always opposite to the direction of the velocity of the body. The magnitude of F_d is generally assumed to be proportional to the magnitude of the velocity of the body. Thus, $|F_d| = \gamma|v|$, where $v = y^{(1)}$, $\gamma > 0$, and γ is called the damping constant. In this case,

$$R + F_d = -ky - \gamma y^{(1)}.$$

On the other hand, by Newton's second law of motion,

$$my^{(2)} = \frac{d}{dt}[mv] = R + F_d = -ky - \gamma y^{(1)}.$$

Hence,

$$dy^{(1)} + \left[\frac{\gamma}{m}y^{(1)} + \frac{k}{m}y\right]dt = 0, \quad y_0, y_0^{(1)} = v_0, \quad t = 0.$$

Step 3. To describe the motion of the body, we need to solve the second-order linear differential equation with constant coefficients. Therefore, we apply the procedure in Subsection 5.3.4. For this purpose, we use the procedure in Subsection 5.3.2 to find the general solution to the above IVP. In this case, the auxiliary equation is

$$\Lambda(\lambda) = \lambda^2 + \frac{\gamma}{m}\lambda + \frac{k}{m} = 0.$$

The roots corresponding to this quadratic equation are

$$\lambda_1 = -\frac{\gamma}{2m} + \frac{1}{2}\sqrt{\left(\frac{\gamma}{m}\right)^2 - \frac{4k}{m}} = -\frac{\gamma}{2m} + \left(\frac{k}{m}\right)^{\frac{1}{2}}\sqrt{\frac{\gamma^2}{4mk} - 1},$$

$$\lambda_2 = -\frac{\gamma}{2m} - \frac{1}{2}\sqrt{\left(\frac{\gamma}{m}\right)^2 - \frac{4k}{m}} = -\frac{\gamma}{2m} - \left(\frac{k}{m}\right)^{\frac{1}{2}}\sqrt{\frac{\gamma^2}{4mk} - 1}.$$

From this and Observation 5.3.2(v), we have the following three descriptions of the motion of the body.

Underdamped motion $[\gamma < 2(mk)^{\frac{1}{2}}]$. In this case, $\frac{\gamma^2}{4mk} - 1 < 0$, the roots are complex. We set

$$\omega^2 = 1 - \frac{\gamma^2}{4m} \quad \tau = \frac{2m}{\gamma}.$$

From Theorem 5.3.1 and the procedure regarding the complete set of solutions in Section 4.4, the general solution is given by

$$\begin{aligned} y(t) &= \exp\left[\lambda_1 t\right] c_1 + \exp\left[\lambda_2 t\right] c_2 \\ &= \exp\left[-\frac{t}{\tau}\right] \cos \omega t c_1 + \exp\left[-\frac{t}{\tau}\right] \sin \omega t c_2. \end{aligned}$$

The solution to the given IVP determined by finding the values of c_1 and c_2 is

$$y_0 = \cos \omega 0 c_1 = c_1 \quad \text{and} \quad y_0^{(1)} = v_0 = -\frac{1}{\tau} c_1 + \omega c_2.$$

Thus, $y_0 = c_1$, $\frac{y_0 + \tau v_0}{\tau \omega} = c_2$ and the solution to the given IVP is

$$\begin{aligned} y(t) &= y_0 \exp\left[-\frac{t}{\tau}\right] \cos \omega t + \frac{y_0 + \tau v_0}{\tau \omega} \exp\left[-\frac{t}{\tau}\right] \sin \omega t \\ &= \exp\left[-\frac{t}{\tau}\right] \left[y_0 \cos \omega t + \frac{y_0 + \tau v_0}{\tau \omega} \sin \omega t\right] \\ &= \exp\left[-\frac{t}{\tau}\right] \frac{\sqrt{(y_0 \tau \omega)^2 + (y_0 + \tau v_0)^2}}{\tau \omega} \left[\sin \phi \cos \omega t + \cos \phi \sin \omega t\right] \\ &= \exp\left[-\frac{t}{\tau}\right] \frac{\sqrt{(y_0 \tau \omega)^2 + (y_0 + \tau v_0)^2}}{\tau \omega} \sin\left(\omega t + \phi\right), \end{aligned}$$

where

$$\begin{aligned} &\left[\frac{\tau \omega y_0}{\sqrt{(y_0 \tau \omega)^2 + (y_0 + \tau v_0)^2}} \cos \omega t + \frac{y_0 + \tau v_0}{\sqrt{(y_0 \tau \omega)^2 + (y_0 + \tau v_0)^2}} \sin \omega t\right] \\ &= \sin \phi \cos \omega \tau + \cos \phi \sin \omega \tau \quad \left(\phi = \tan^{-1} \frac{\tau \omega y_0}{y_0 + \tau v_0}\right). \end{aligned}$$

The motion is simple harmonic motion. Its period, frequency, and phase shift are $\frac{2\pi}{\omega}$, ω, and ϕ, respectively. The amplitude is

$$\exp\left[-\frac{t}{\tau}\right] \frac{\sqrt{(y_0 \tau \omega)^2 + (y_0 + \tau v_0)^2}}{\tau \omega},$$

and it decreases steadily with t until the oscillations are finally damped out. The factor $\frac{1}{\tau} = \frac{\gamma}{2m}$ is called the damping factor. We further note that

$$\begin{aligned}
y\left(t + \frac{2\pi}{\omega}\right) &= \exp\left[-\left(\frac{t}{\tau} + \frac{2\pi}{\tau\omega}\right)\right] \frac{\sqrt{(y_0\tau\omega)^2 + (y_0 + \tau v_0)^2}}{\tau\omega} \sin\left(\omega\left(t + \frac{2\pi}{\omega}\right) + \phi\right) \\
&= \exp\left[-\left(\frac{t}{\tau} + \frac{2\pi}{\tau\omega}\right)\right] \frac{\sqrt{(y_0\tau\omega)^2 + (y_0 + \tau v_0)^2}}{\tau\omega} \sin(\omega t + 2\pi + \phi) \\
&= y(t)\exp\left[-\frac{2\pi}{\tau\omega}\right].
\end{aligned}$$

Thus,

$$\ln y\left(t + \frac{2\pi}{\omega}\right) - \ln y(t) = -\frac{2\pi}{\tau\omega}.$$

Because of the expression, $\frac{2\pi}{\tau\omega}$ is called the logarithmic decrement of the damped oscillations.

Overdamped motion $[\gamma > 2(mk)^{\frac{1}{2}}]$. In this case, $\frac{\gamma^2}{4mk} - 1 > 0$, the roots are real and distinct. The solution to the IVP is given as before:

$$\begin{aligned}
y(t) &= \exp[\lambda_1 t]\, c_1 + \exp[\lambda_2 t]\, c_2, \\
y^{(1)}(t) &= \lambda_1 \exp[\lambda_1 t]\, c_1 + \lambda_2 \exp[\lambda_2 t]\, c_2.
\end{aligned}$$

From this, we obtain

$$y_0 = c_1 + c_2 \quad \text{and} \quad v_0 = \lambda_1 c_1 + \lambda_2 c_2.$$

Thus,

$$\frac{\lambda_2 y_0 - v_0}{\lambda_2 - \lambda_1} = c_1, \quad \frac{\lambda_1 y_0 - v_0}{\lambda_1 - \lambda_2} = c_2,$$

and the solution to the given IVP is

$$y(t) = \frac{\lambda_2 y_0 - v_0}{\lambda_2 - \lambda_1}\exp[\lambda_1 t] + \frac{\lambda_1 y_0 - v_0}{\lambda_1 - \lambda_2}\exp[\lambda_2 t].$$

In this case, the solution process is never equal to zero, however, as time approaches infinity, the solution approaches zero. Thus, $t = 0$ is the horizontal asymptote.

Critically damped motion $[\gamma = 2(mk)^{\frac{1}{2}}]$ **:** In this case, $\frac{\gamma^2}{4mk} - 1 = 0$, the roots are real and are repeated with $\lambda_1 = \lambda_2 = -\frac{\gamma}{2m}$. The general solution is

$$\begin{aligned}
y(t) &= \exp[\lambda_1 t]\, c_1 + t\exp[\lambda_1 t]\, c_2, \\
y^{(1)}(t) &= \lambda_1 \exp[\lambda_1 t]\, c_1 + \exp[\lambda_1 t]\, c_2 + t\lambda_1 \exp[\lambda_1 t]\, c_2.
\end{aligned}$$

From this, we obtain

$$y_0 = c_1 \quad \text{and} \quad v_0 = \lambda_1 c_1 + c_2.$$

Thus,

$$y_0 = c_1, \quad v_0 + \frac{\gamma y_0}{2m} = c_2,$$

and the solution to the given IVP is

$$\begin{aligned} y(t) &= y_0 \exp\left[-\frac{\gamma}{2m}t\right] + \left(v_0 + \frac{\gamma y_0}{2m}\right) t \exp\left[-\frac{\gamma}{2m}t\right] \\ &= \left(y_0 + \left(v_0 + \frac{\gamma y_0}{2m}\right) t\right) \exp\left[-\frac{\gamma}{2m}t\right]. \end{aligned}$$

Again, in this case, the solution process is never equal to zero; however, as time approaches infinity, the solution approaches zero. Thus, $t = 0$ is the horizontal asymptote.

This completes the solution process of (a). The remaining parts of the illustration are left as an exercise for the reader.

Example 5.3.4. Using the cupula dynamic equation,

$$dx^{(1)} + [10x^{(1)} + 0.99x]dt = 0,$$

$x(0) = x_0$, $x^{(1)}(0) = 0$ (Exercise 1, Section 5.2). Find: (a) the displacement of the cupula at t, (b) the displacement of the cupula by ignoring the dominant decaying term, (c) t_e, where t_e is the time it takes to cease the feeling of rotation, and $x_{\min}$ is the corresponding position of the cupula and is referred to as the "threshold," and (d) the graph of t_e knowing that x_0 is proportional to the angular velocity ω of the chair.

Solution process. From Exercise 1, Section 5.2, we note that x_0 and $x_0^{(1)}$ are the deflection (angular displacement) and velocity (angular velocity) at $t = 0$, respectively. By following the procedure in Subsection 5.3.4, we have

$$\Lambda(\lambda) = \lambda^2 + 10\lambda + 0.99 = 0.$$

The roots of this quadratic equation are $\lambda_1 = -9.9$ and $\lambda_2 = -0.1$, and hence

$$\begin{aligned} x(t) &= \exp[\lambda_1 t]\, c_1 + \exp[\lambda_2 t]\, c_2, \\ x^{(1)}(t) &= \lambda_1 \exp[\lambda_1 t]\, c_1 + \lambda_2 \exp[\lambda_2 t]\, c_2. \end{aligned}$$

From this, we obtain

$$x_0 = c_1 + c_2 \quad \text{and} \quad 0 = \lambda_1 c_1 + \lambda_2 c_2.$$

Thus,

$$-\frac{0.1x_0}{9.8} = c_1, \quad \frac{9.9x_0}{9.8} = c_2,$$

and the solution to the given IVP is

$$x(t) = \frac{x_0}{9.8}\left[9.9\exp\left[-0.1t\right] - 0.1\exp\left[-9.9t\right]\right].$$

This is the displacement of the cupula at a time t. Hence, this completes the solution process of (a). To discuss the solution process of part (b), we note that the second term is the dominant decaying term, because $\exp[-9.9t]$ decays faster than $\exp[-0.1t]$. Therefore, we neglect the second term and replace $\frac{9.9}{9.8}$ by 1, and we have the approximation of cupula displacement

$$x(t) \approx x_0 \exp[-0.1t].$$

Now, to find t_e that corresponds to the $x_{\min}$ as described in (c), we have

$$x_{\min} = x_0 \exp\left[-0.1t_e\right],$$

which implies that

$$t_e = 10\ln\left(\frac{x_0}{x_{\min}}\right).$$

This completes the solution process of (c). To complete the problem, we note that at the maximum displacement of the cupula x_0, we are given $x_0 = k\omega$. We substitute this in the above expression of t_e, and obtain

$$\begin{aligned} t_e &= 10\ln\left(\frac{x_0}{x_{\min}}\right) = 10\ln\left(\frac{k\omega}{x_{\min}}\right) \\ &= 10\ln\omega + 10\ln\left(\frac{k}{x_{\min}}\right) \\ &= 10\ln\omega + b, \end{aligned}$$

where

$$b = 10\ln\left(\frac{k}{x_{\min}}\right)$$

is a constant. The graph of t_e against $\ln\omega$ on the semilognormal paper is a straight line with slope 10 and t_e intercept b. This completes the solution process of the example.

5.3 Exercises

(1) Find the general solution to the following differential equations:
(a) $dy^{(1)} - ydt = 0$ (b) $dy^{(3)} - ydt = 0$ (c) $dy^{(1)} + 4ydt = 0$
(d) $2dy^{(1)} - [9y^{(1)} + 5y]dt = 0$ (e) $dy^{(3)} - [13y^{(2)} - 36y]dt = 0$

(f) $dy^{(2)} + [3y^{(2)} + 3y^{(1)} + y]dt = 0$ (g) $4dy^{(1)} - [12y^{(1)} - 9y]dt = 0$
(h) $dy^{(3)} - y^{(2)}dt = 0$ (i) $dy^{(3)} - [3y^{(2)} - y]dt = 0$

(2) Solve the following initial value problems:

(a) $dy^{(1)} + [2y^{(1)} + y]dt = 0$, $y(0) = 1$, $y^{(1)}(0) = 2$
(b) $dy^{(2)} - [3y^{(2)} - 3y^{(1)} + y]dt = 0$, $y(0) = 0$, $y^{(1)}(0) = 0$, $y^{(2)}(0) = 2$
(c) $dy^{(1)} - [12y^{(1)} - 9y]dt = 0$, $y(0) = 0$, $y^{(1)}(0) = 2$

(3) Find the general solutions to the differential equations in Exercises 5.2.
(4) Using Illustration 5.2.2, solve the differential equations for $L = 1\,\text{H}$, $R = 100\,\Omega$, $C = 10^{-4}\,\text{F}$: (i) $E_e = 1000\,\text{V}$, and (ii) $E_e(t) = 500 \sin 50t$ V with $t = 0$, $I(0) = 0$, $Q(0) = 0$. Moreover, find $Q(t)$.

5.4 Companion System

This section is devoted to the representation of the nth-order linear nonhomogeneous differential equation as a system of linear nonhomogeneous differential equations. As a result of this representation, the procedures and conceptual analysis developed in Chapter 4 are directly applicable to the higher-order linear differential equations in a systematic and unified way. Moreover, an attempt is made to utilize the particular structure of the coefficient matrix of the linear system of differential equations corresponding to the higher-order linear homogeneous differential equations, so as to systematically simplify the procedures for finding the general solution to higher-order differential equations. In short, this section deals with an alternative way of finding solution processes of higher-order linear nonhomogeneous differential equations with constant coefficients.

Let us consider a general nth-order linear nonhomogeneous differential equation and its corresponding higher-order linear homogeneous differential equation as follows:

$$dy^{(n-1)} + [a_{n-1}y^{(n-1)} + a_{n-2}y^{(n-2)} + \cdots + a_1y^{(1)} + a_0y - g(t)]dt = 0, \quad (5.42)$$

$$dy^{(n-1)} + [a_{n-1}y^{(n-1)} + a_{n-2}y^{(n-2)} + \cdots + a_1y^{(1)} + a_0y]dt = 0, \quad (5.43)$$

where $a_{n-1}, a_{n-2}, \ldots, a_1$, and a_0 are given real numbers, and $g(t)$ is a given continuous function defined on an interval $J = [a, b]$, $a, b \in R$.

Theorem 5.4.1. *Every nth-order linear nonhomogeneous differential equation of the type* (5.42) *is equivalent to the following first-order linear system of differential equations:*

$$dx = (Ax + p(t))\,dt, \quad (5.44)$$

where A is an $n \times n$ constant matrix, x and $p(t)$ are $n \times 1$ matrix functions;

$$A = \begin{bmatrix} 0 & 1 & 0 & \cdots & \cdots & 0 \\ 0 & 0 & 1 & \cdots & \cdots & 0 \\ \cdots & \cdots & \cdots & \cdots & \cdots & \cdots \\ 0 & \cdots & \cdots & 0 & 1 & 0 \\ 0 & \cdots & \cdots & \cdots & 0 & 1 \\ -a_0 & -a_1 & \cdots & \cdots & -a_{n-2} & -a_{n-1} \end{bmatrix}, \tag{5.45}$$

$x^T = \left[x_1\, x_2\, \dots\, x_{n-1}\, x_n\right]$, $[p(t)]^T = [0, 0, \dots, g(t)]^T$, *respectively, and T stands for a transpose of the matrix.*

Proof. Let the nth-order differential equation of the type (5.42) be given. Let us denote by

$$y = x_1, y^{(1)} = x_2, \dots, y^{(i-1)} = x_i, \dots, y^{(n-2)} = x_{n-1}, y^{(n-1)} = x_n. \tag{5.46}$$

With this notation and the concept of the differential, we rewrite (5.42) as

$$\begin{aligned} dy &= dx_1 = x_2\, dt, \\ dy^{(1)} &= dx_2 = x_3\, dt, \\ dy^{(2)} &= dx_3 = x_4\, dt, \\ &\dots\dots\dots\dots \\ dy^{(i-1)} &= dx_i = x_{i+1}\, dt, \\ &\dots\dots\dots\dots \\ dy^{(n-2)} &= dx_{n-1} = x_n\, dt, \\ dy^{(n-1)} &= dx_n = -\left(a_0 x_1 + a_1 x_2 + \cdots + a_{n-2} x_{n-1} + a_{n-1} x_n - g(t)\right) dt. \end{aligned} \tag{5.47}$$

By using the definition, the notations, a derivative of a matrix function, and the matrix algebra, the first-order system of differential equations (5.47) can be rewritten in form (5.44). Conversely, by reversing the above argument, any system of first-order differential equations of the type (5.44) can be rewritten as (5.42). The details are left to the reader. The proof of the theorem is complete. □

Definition 5.4.1. A matrix of the form defined in (5.45) is called a companion matrix. It is denoted by A_c.

Definition 5.4.2. The system (5.44) is called the companion system for the nth-order differential equation of the type (5.42).

Example 5.4.1. Let us consider $dy^{(2)} - [4y^{(2)} - 4y^{(1)} + t^2]dt = 0$. Write: (i) the companion matrix and (ii) the companion system.

Solution process. To write the companion matrix and the companion system corresponding to the differential equation, we follow the procedure described in the proof of Theorem 5.4.1:

$$dy = dx_1 = x_2\,dt, \quad dy^{(1)} = dx_2 = x_3\,dt, \quad dy^{(2)} = dx_3 = [4x_3 - 4x_2 + t^2]dt.$$

Hence, the companion system is:

$$dx = [A_c x + p(t)]\,dt,$$

where the companion matrix A_c is defined by

$$A_c = \begin{bmatrix} 0 & 1 & 0 \\ 0 & 0 & 1 \\ 0 & -4 & 4 \end{bmatrix}, \quad p(t) = \begin{bmatrix} 0 \\ 0 \\ t^2 \end{bmatrix}.$$

This completes the solution process of the example.

The following result illustrates the importance of the companion system in a natural way.

Theorem 5.4.2. *Let $y(t)$ and $x(t)$ be any solutions to* (5.42) *and* (5.44), *respectively. Then*

(a) $[x_y(t)]^T = [y(t), y^{(1)}(t), \ldots, y^{(i-1)}(t), \ldots, y^{(n-1)}(t)]$ *is the solution process of* (5.44);
(b) $dx_1^{(n-1)} = -[a_0 x_1 + a_1 x_1^{(1)} + \cdots + a_{n-2} x_1^{(n-2)} + a_{n-1} x_1^{(n-1)} - g(t)]dt$, *where* $x_1(t)$ *is the first component of the solution process* $x(t)$ *of* (5.44).

Proof. To prove (a), let us assume that $y(t)$ is any solution process of (5.42). By following the argument used in the proof of Theorem 5.4.1, we arrive at (5.47). From (5.46) and (5.47), using the matrix algebra and calculus, we conclude that the n-dimensional vector function defined by

$$\begin{aligned}[x_y(t)]^T &= [y(t), y^{(1)}(t), \ldots, y^{(i-1)}(t), \ldots, y^{(n-1)}(t)]^T \\ &= [x_1(t), x_2(t), \ldots, x_i(t), \ldots, x_n(t)]^T\end{aligned}$$

is also the solution process of (5.44). This completes the proof of (a).

To prove (b), we assume that $x(t)$ is any solution to (5.44). This implies that $x(t)$ and $dx(t)$ satisfy the companion system. Again from (5.47), we have

$$dx_n = -[a_0 x_1 + a_1 x_2 + \cdots + a_{n-2} x_{n-1} + a_{n-1} x_n - g(t)]dt,$$

and thus, by (5.46),

$$dx_1^{(n-1)} = -[a_0 x_1 + a_1 x_1^{(1)} + \cdots + a_{n-2} x_1^{(n-2)} + a_{n-1} x_1^{(n-1)} - g(t)]dt.$$

This establishes the proof of (b). □

Theorem 5.4.2 provides a natural justification for defining the general solution to (5.42).

Definition 5.4.3. The general solution to the nth-order differential equation (5.42) is defined as the first component of the general solution to the corresponding companion system of the first order differential equations (5.44).

Theorem 5.4.2 is also useful for establishing the existence and uniqueness result. Moreover, the "principle of superposition" for the higher order differential equation can be constructed.

Theorem 5.4.3 (existence and uniqueness). *Let us consider the following initial value problem:*

$$dy^{(n-1)} + a_{n-1}y^{(n-1)} + a_{n-2}y^{(n-2)} + \cdots + a_1 y^{(1)} + a_0 y - g(t))\,dt = 0, \quad (5.48)$$

$$y(t_0) = y_1,\ y^{(1)}(t_0) = y_2,\ y^{(2)}(t_0) = y_3,\ y^{(i-1)}(t_0) = y_i,\ y^{(n-1)}(t_0) = y_n. \quad (5.49)$$

The IVP (5.48)–(5.49) *have unique solutions.*

Theorem 5.4.4. *Let* $\Lambda(\lambda) = \lambda^n + a_{n-1}\lambda^{n-1} + \cdots + a_1\lambda + a_0$. *Then for any number* λ,

$$\begin{bmatrix} 0 & 1 & \cdots & \cdots & 0 \\ 0 & 0 & 1 & \cdots & 0 \\ \cdots & \cdots & \cdots & \cdots & \cdots \\ 0 & \cdots & \cdots & 0 & 1 \\ -a_0 & -a_1 & \cdots & \cdots & -a_{n-1} \end{bmatrix} \begin{bmatrix} 1 \\ \lambda \\ \cdot \\ \lambda^{n-2} \\ \lambda^{n-1} \end{bmatrix} = \lambda \begin{bmatrix} 1 \\ \lambda \\ \cdot \\ \lambda^{n-2} \\ \lambda^{n-1} \end{bmatrix} - \Lambda(\lambda) \begin{bmatrix} 0 \\ 0 \\ \cdot \\ 0 \\ 1 \end{bmatrix}. \quad (5.50)$$

Proof. The proof follows by direct multiplication of matrices on the left-hand side expression. However, an alternative proof of this result is as follows. We consider

$$(A_c - \lambda I)\,c = \begin{bmatrix} -\lambda & 1 & \cdots & \cdots & 0 \\ 0 & -\lambda & 1 & \cdots & 0 \\ \cdots & \cdots & \cdots & \cdots & \cdots \\ 0 & \cdots & \cdots & -\lambda & 1 \\ -a_0 & -a_1 & \cdots & \cdots & -a_{n-1} - \lambda \end{bmatrix} \begin{bmatrix} c_1 \\ c_2 \\ \vdots \\ c_{n-1} \\ c_n \end{bmatrix}$$

$$= \begin{bmatrix} -\lambda c_1 + c_2 \\ -\lambda c_2 + c_3 \\ \vdots \\ -\lambda c_{n-1} + c_n \\ -[a_0 c_1 + \cdots + \lambda c_n] \end{bmatrix} = -\lambda \begin{bmatrix} c_1 \\ c_2 \\ \vdots \\ c_{n-1} \\ c_n \end{bmatrix} + \begin{bmatrix} c_2 \\ c_3 \\ \vdots \\ c_n \\ -[a_0 c_1 + \cdots + a_{n-1}c_{n-1} + c_n] \end{bmatrix}, \quad (5.51)$$

where $c = [c_1, \ldots, c_k, \ldots, c_n]^T$ is an arbitrary n-dimensional vector and λ is any number. In particular, we choose c so that

$$-\lambda c_1 + c_2 = 0, \quad -\lambda c_2 + c_3 = 0, \ldots, -\lambda c_{n-1} + c_n.$$

This implies that

$$c_1 = 1c_1, c_2 = \lambda c_1, c_3 = \lambda c_2 = \lambda^2 c_1, \ldots, c_n = \lambda c_{n-1} = \lambda^{n-1} c_1.$$

Hence,

$$\begin{aligned} &-[a_0 c_1 + \cdots + a_{n-1} c_n + \lambda c_n] \\ &\quad = -c_1[\lambda^n + a_{n-1}\lambda^{n-1} + \cdots + a_1 \lambda + a_0] = -c_1 \Lambda(\lambda). \end{aligned} \tag{5.52}$$

From (5.52), (5.51) reduces to the desired result in (5.50). This completes the proof of the theorem. □

Observation 5.4.1

(i) By using Principle of Mathematical Induction 1.2.3, one can show that the characteristic equation (4.38) associated with the companion matrix A_c in (5.44) reduces to

$$\det(A_c - \lambda I) = (-1)^n \Lambda(\lambda) = 0, \tag{5.53}$$

where $\Lambda(\lambda) = 0$ is the auxiliary equation in (5.30) of (5.27). This partially justifies Observation 5.3.2(vi).

(ii) From (5.30) and Theorem 5.4.4, we observe that any solution λ to (5.30) is an eigenvalue of the companion matrix A_c with the eigenvector $c(\lambda) = [1, \lambda, \ldots, \lambda^i, \ldots, \lambda^{n-1}]^T$. This shows that knowing the eigenvalue of A_c, the corresponding eigenvector is easily determined.

(iii) From Theorem 5.4.2, the problem of finding the fundamental set of solutions and the general solution to (5.27) is equivalent to the problem of finding the fundamental matrix and the general solutions to the companion system associated with (5.27)/(5.43):

$$dx = A_c x \, dt. \tag{5.54}$$

(iv) From (i)–(iii), we note that we can imitate the procedures developed in Sections 4.3 and 4.4 to solve the problems with regard to the linear homogeneous companion system (5.54). In short, this observation completely justifies Observation 5.3.2(vi). Moreover, this provides an alternative way of solving the linear higher-order differential equations with constant coefficients.

Illustration 5.4.1. We are given $dy^{(2)} + [ay^{(1)} + by]dt = 0$, $y(0) = y_0$, $y^{(1)}(0) = y_0^{(1)}$, where a and b are any given real numbers. By using the companion system representation, solve the IVP.

Solution process. To solve this type of initial problems, we follow the steps outlined below.

Step 1. For $n = 2$, we follow the proof of Theorem 5.4.1, and we arrive at the companion representation of the given second-order differential equation:

$$dy = dx_1 = y^{(1)}dt = x_2 dt,$$
$$dy^{(1)} = dx_2 = -\left[ay^{(1)} + by\right] dt = -\left[ax_2 + bx_1\right] dt.$$

We rewrite this in its matrix representation,

$$dx = A_c x\, dt, \quad x(0) = x_0,$$

where $[x_1\, x_2]^T$, $x_0 = [y_0\, y_0^{(1)}]^T$, and the companion matrix

$$A_c = \begin{bmatrix} 0 & 1 \\ -b & -a \end{bmatrix}.$$

Step 2. From Observation 5.4.1(i)–(iv) and the procedure in Subsection 4.3.2, we obtain

$$0 = \det(A_c - \lambda I) = (-1)^2 \Lambda(\lambda) = \Lambda(\lambda)$$
$$= -\lambda(-a - \lambda) + b = \lambda^2 + a\lambda + b.$$

Thus, the eigenvalues of the companion matrix are

$$\lambda_1 = \frac{-a + \sqrt{a^2 - 4b}}{2}, \quad \lambda_2 = \frac{-a - \sqrt{a^2 - 4b}}{2}.$$

Depending on the sign of the discriminant $a^2 - 4b$, we have three cases (Section 4.4):

(a) **Real and distinct roots** ($a^2 - 4b > 0$). From Observation 5.4.1(ii), the eigenvectors corresponding to the eigenvalues λ_1 and λ_2 are $c_1 = [1 \lambda_1]^T$ and $c_2 = [1 \lambda_2]^T$, respectively. Thus, the general solution is

$$x(t) = \Phi(t)a,$$

where $\Phi(t)$ and a are the fundamental matrix solution to the companion system and unknown arbitrary constant vector, respectively. Moreover,

$$\Phi(t) = \begin{bmatrix} \exp[\lambda_1 t] & \exp[\lambda_2 t] \\ \lambda_1 \exp[\lambda_1 t] & \lambda_2 \exp[\lambda_2 t] \end{bmatrix}.$$

We recall (Procedures 4.3 and 4.4) that to solve the IVP we need to find the normalized fundamental matrix at $t = 0$, and it is given by

$$\Phi(t,0) = \Phi(t)\Phi^{-1}(0) = \frac{1}{\lambda_2 - \lambda_1}\begin{bmatrix} \exp[\lambda_1 t] & \exp[\lambda_2 t] \\ \lambda_1 \exp[\lambda_1 t] & \lambda_2 \exp[\lambda_2 t]\end{bmatrix}\begin{bmatrix} \lambda_2 & -1 \\ -\lambda_1 & 1\end{bmatrix}$$

$$= \frac{1}{\lambda_2 - \lambda_1}\begin{bmatrix} \lambda_2 \exp[\lambda_1 t] - \lambda_1 \exp[\lambda_2 t] & \exp[\lambda_2 t] - \exp[\lambda_1 t] \\ \lambda_1\lambda_2 [\exp[\lambda_1 t] - \exp[\lambda_2 t]] & \lambda_2 \exp[\lambda_2 t] - \lambda_1 \exp[\lambda_1 t]\end{bmatrix}.$$

Thus, the solution to the IVP associated with the companion system is

$$x(t) = \Phi(t,0)\, x_0.$$

From this and Definition 5.4.3, we have

$$y(t) = \frac{y_0}{\lambda_2 - \lambda_1}[\lambda_2 \exp[\lambda_1 t] - \lambda_1 \exp[\lambda_2 t]] + \frac{y_0^{(1)}}{\lambda_2 - \lambda_1}[\exp[\lambda_2 t] - \exp[\lambda_1 t]].$$

In this case, this is the solution to the given IVP.

(b) **Complex and distinct roots** ($a^2 - 4b < 0$). Thus, the eigenvalues of the companion matrix are

$$\lambda = \frac{-a \pm \sqrt{4b - a^2}i}{2} = \alpha \pm \beta i.$$

By imitating Procedure 4.4 with regard to complex roots, the eigenvector and solution corresponding to the eigenvalue λ_1 are $c_1 = [1\lambda_1]^T$ and

$$\begin{aligned}&\exp[\lambda_1 t][1\lambda_1]^T \\ &\quad = \exp[\alpha t](\cos\beta t + i\sin\beta t)[1\alpha + \beta i]^T \\ &\quad = \exp[\alpha t][(\cos\beta t + i\sin\beta t)(\alpha\cos\beta t - \beta\sin\beta t) + i(\beta\cos\beta t + \alpha\sin\beta t)]^T,\end{aligned}$$

where

$$\alpha + \beta i = \lambda_1, \quad \alpha = \frac{-a}{2}, \quad \beta = \frac{\sqrt{4b - a^2}}{2}.$$

Thus, the general solution is

$$x(t) = \Phi(t)\, a,$$

where $\Phi(t)$ and a are the fundamental matrix solution to the companion system and unknown arbitrary constant vector, respectively; moreover,

$$\Phi(t) = \begin{bmatrix} \exp[\alpha t]\cos\beta t & \exp[\alpha t]\sin\beta t \\ (\alpha\cos\beta t - \beta\sin\beta t)\exp[\alpha t] & (\beta\cos\beta t + \alpha\sin\beta t)\exp[\alpha t]\end{bmatrix}$$

where $r = \sqrt{\alpha^2 + \beta^2}$ and $\tan^{-1}\frac{\beta}{\alpha}$, $\alpha \neq 0$.

We recall (procedures of Sections 4.3 and 4.4) that to solve the IVP that we need to find the normalized fundamental matrix at $t = 0$, and it is given by

$$\begin{aligned}\Phi(t,0) &= \Phi(t)\Phi^{-1}(0)\\ &= \frac{1}{\beta}\begin{bmatrix}\exp[\alpha t]\cos\beta t & \exp[\alpha t]\sin\beta t\\ (\alpha\cos\beta t - \beta\sin\beta t)\exp[\alpha t] & (\beta\cos\beta t + \alpha\sin\beta t)\exp[\alpha t]\end{bmatrix}\\ &\quad\times\begin{bmatrix}\beta & 0\\ -\alpha & 1\end{bmatrix}\\ &= \frac{1}{\beta}\begin{bmatrix}\exp[\alpha t](\beta\cos\beta t - \alpha\sin\beta t) & \exp[\alpha t]\sin\beta t\\ -\left(\alpha^2+\beta^2\right)\sin\beta t\exp[\alpha t] & (\beta\cos\beta t + \alpha\sin\beta t)\exp[\alpha t]\end{bmatrix}.\end{aligned}$$

Thus, the solution to the IVP associated with the companion system is

$$x(t) = \Phi(t,0)\,x_0.$$

From this and Definition 5.4.3, we have

$$y(t) = \frac{y_0}{\beta}\exp[\alpha t](\beta\cos\beta t - \alpha\sin\beta t) + \frac{y_0^{(1)}}{\beta}\exp[\alpha t]\sin\beta t.$$

In this case, this is the solution to the given IVP.

(c) **Repeated roots** ($a^2 - 4b = 0$). In this case, $\lambda = \frac{-a}{2}$ repeated root. By following the argument used in Section 4.4, $\Phi(t)$:

$$\Phi(t) = \begin{bmatrix}\exp\left[-\frac{a}{2}t\right] & t\exp\left[-\frac{a}{2}t\right]\\ -\frac{a}{2}\exp\left[\frac{-a}{2}t\right] & \left(1-\frac{a}{2}t\right)\exp\left[-\frac{a}{2}t\right]\end{bmatrix},$$

and

$$\begin{aligned}\Phi(t,0) &= \Phi(t)\Phi^{-1}(0)\\ &= \begin{bmatrix}\exp\left[-\frac{a}{2}t\right] & t\exp\left[-\frac{a}{2}t\right]\\ -\frac{a}{2}\exp\left[\frac{-a}{2}t\right] & \left(1-\frac{a}{2}t\right)\exp\left[-\frac{a}{2}t\right]\end{bmatrix}\begin{bmatrix}1 & 0\\ \frac{a}{2} & 1\end{bmatrix}\\ &= \begin{bmatrix}\exp\left[-\frac{a}{2}t\right]\left(1+\frac{a}{2}t\right) & t\exp\left[-\frac{a}{2}t\right]\\ -\frac{1}{4}a^2 t\exp\left[\frac{-a}{2}t\right] & \left(1-\frac{a}{2}t\right)\exp\left[-\frac{a}{2}t\right]\end{bmatrix}.\end{aligned}$$

Thus, the solution to the IVP associated with the companion system is

$$x(t) = \Phi(t, 0)\, x_0.$$

From this and Definition 5.4.3, we have

$$y(t) = y_0 \exp\left[-\frac{a}{2}t\right]\left(1 + \frac{a}{2}t\right) + y_0^{(1)} t \exp[-\frac{a}{2}t].$$

This is the solution to the given IVP. This completes the solution process of the given problem.

5.4 Exercises

(1) Write the companion matrices and companion systems corresponding to the differential equations in Exercises 1–4 of Section 5.3.

(2) Given the system matrices for the companion system, write the higher-order differential equations (if possible):

(a) $\begin{bmatrix} 0 & 1 \\ 1 & -1 \end{bmatrix}$ (b) $\begin{bmatrix} 1 & -1 \\ 1 & 0 \end{bmatrix}$ (c) $\begin{bmatrix} 0 & 1 \\ 0 & -1 \end{bmatrix}$ (d) $\begin{bmatrix} 0 & 1 \\ 1 & 0 \end{bmatrix}$

(e) $\begin{bmatrix} 1 & 0 \\ 1 & -1 \end{bmatrix}$ (f) $\begin{bmatrix} 1 & 2 & 4 \\ 0 & 0 & 1 \\ 0 & 1 & 0 \end{bmatrix}$ (g) $\begin{bmatrix} 0 & 1 & 0 \\ 0 & 0 & 1 \\ 1 & 2 & 4 \end{bmatrix}$ (h) $\begin{bmatrix} 0 & 0 & 0 \\ 0 & 0 & 1 \\ 2 & 1 & 3 \end{bmatrix}$

(i) $\begin{bmatrix} 0 & 1 & 0 \\ 0 & 0 & 1 \\ 0 & 1 & 2 \end{bmatrix}$

(3) By finding the fundamental matrix solution to the homogeneous companion system, find the general solutions to higher-order differential equations of Exercise 1, Section 5.3.

(4) By finding the fundamental matrix solution to the homogeneous companion system, solve the initial problems in Exercise 2, Section 5.3.

(5) By writing the companion systems, solve the following initial value problems:

(a) $dy^{(1)} - 4y\,dt = 0$, $y(0) = 1$, $y^{(1)}(0) = -1$

(b) $dy^{(1)} + 4y\,dt = 0$, $y(0) = 1$, $y^{(1)}(0) = 0$

(c) $dy^{(2)} + 8y\,dt = 0$, $y(0) = 1$, $y^{(1)}(0) = 0$, $y^{(2)}(0) = 1$

(d) $dy^{(1)} + 4\left[y^{(1)} + y\right] dt = 0$, $y(0) = 0$, $y^{(1)}(0) = 1$

(e) $dy^{(3)} - y\,dt = 0$, $y(0) = -1$, $y^{(1)}(0) = 1$, $y^{(2)}(0) = y^{(3)}(0) = 0$

5.5 Higher-Order Linear Nonhomogeneous Equations

The basic ideas of solving nonhomogeneous differential equations, described in Sections 2.5 and 4.6, are exactly the same for solving higher-order linear nonhomogeneous differential equations. Using the companion system of linear nonhomogeneous differential equations (5.44) and the method of variation of parameters, we solve the higher-order linear nonhomogeneous differential equations (5.42). This method is superior to any other method of finding a particular solution. As a result of this, we merely present several examples to illustrate this method.

Illustration 5.5.1. Let us consider $dy^{(2)} + [ay^{(1)} + by]dt = g(t)$, $y(t_0) = y_0$, $y^{(1)}(t_0) = y_0^{(1)}$, where a and b are any given real numbers, and g is a Riemann–Cauchy integrable function. Applying the companion system representation, solve the IVP.

Solution process. Following the procedure described in Illustration 5.4.1, the companion form of the system is:

$$dx = [A_c x + p(t)]\,dt, \quad x(t_0) = x_0,$$

where x and A_c are as defined in Illustration 5.4.1, and $p(t) = [0, g(t)]^T$. From Illustration 5.4.1, we have three possible fundamental matrix solutions to the homogeneous companion system. We consider any one of the three fundamental matrix solutions to the companion system in Illustration 5.4.1 as

$$\Phi(t) = \begin{bmatrix} \exp[\lambda_1 t] & \exp[\lambda_2 t] \\ \lambda_1 \exp[\lambda_1 t] & \lambda_2 \exp[\lambda_2 t] \end{bmatrix}.$$

Thus, the general solution to the homogeneous companion system (the complementary solution to the corresponding nonhomogeneous system) is

$$x_c(t) = \Phi(t)a,$$

where a is an arbitrary constant vector.

We assume that $x_p(t) = \Phi(t)c$ is a particular solution to the given nonhomogeneous companion system, where c is an unknown vector function. We follow the procedure in Subsection 4.6.4, and obtain

$$\Phi(t)\,dc = p(t)\,dt,$$

which implies that

$$c(t) = c(a) + \int_a^t \Phi^{-1}(s)p(s)\,ds,$$

where $c(a)$ is an arbitrary constant. The general solution to the nonhomogeneous companion system is given by

$$\begin{aligned} x(t) &= x_c(t) + x_p(t) \\ &= \Phi(t)a + \Phi(t)c = \Phi(t)a + \Phi(t)\left[c(a) + \int_a^t \Phi^{-1}(s)p(s)\,ds\right] \end{aligned}$$

$$= \Phi(t)\left[(a + c(a)) + \int_a^t \Phi^{-1}(s)p(s)\,ds\right]$$

$$= \Phi(t)\left[a + \int_{t_0}^t \Phi^{-1}(s)p(s)\,ds\right].$$

The solution to the IVP corresponding to the companion system is

$$x(t) = \Phi(t, t_0)\, x + \int_{t_0}^t \Phi(t, s)p(s)ds.$$

The solution to the given second-order system is given by the first component of this solution (Definition 5.4.3).

In particular, in the case of (a) real and distinct roots ($a^2 - 4b > 0$) and using Illustration 5.4.1, the solution to the given IVP (Definition 5.4.3) is

$$\begin{aligned} y(t) &= \frac{y_0}{\lambda_2 - \lambda_1}\left[\lambda_2 \exp\left[\lambda_1 (t - t_0)\right] - \lambda_1 \exp\left[\lambda_2 (t - t_0)\right]\right] \\ &\quad + \frac{y_0^{(1)}}{\lambda_2 - \lambda_1}\left[\exp\left[\lambda_2 (t - t_0)\right] - \exp\left[\lambda_1 (t - t_0)\right]\right] \\ &\quad - \frac{1}{\lambda_2 - \lambda_1}\exp\left[\lambda_1 t\right]\int_{t_0}^t \exp\left[-\lambda_1 s\right] g(s)\,ds \\ &\quad + \frac{1}{\lambda_2 - \lambda_1}\exp\left[\lambda_2 t\right]\int_{t_0}^t \exp\left[-\lambda_2 s\right] g(s)\,ds. \end{aligned}$$

We recall (Procedures 4.3 and 4.4) that to solve the IVP we need to find the normalized fundamental matrix $\Phi(t, t_0)$ at $t = t_0$, and it is given by

$$\Phi(t, t_0) = \frac{1}{\lambda_2 - \lambda_1}\begin{bmatrix} \exp[\lambda_1 t] & \exp[\lambda_2 t] \\ \lambda_1 \exp[\lambda_1 t] & \lambda_2 \exp[\lambda_2 t] \end{bmatrix}\begin{bmatrix} \lambda_2 \exp[-\lambda_1 t_0] & -\exp[-\lambda_1 t_0] \\ -\lambda_1 \exp[-\lambda_2 t_0] & \exp[-\lambda_2 t_0] \end{bmatrix}$$

$$= \frac{1}{\lambda_2 - \lambda_1}\begin{bmatrix} \lambda_2 \exp[\lambda_1 (t - t_0)] - \lambda_1 \exp[\lambda_2 (t - t_0)] & \exp[\lambda_2 (t - t_0)] - \exp[\lambda_1 (t - t_0)] \\ \lambda_1\lambda_2 \left[\exp[\lambda_1 (t - t_0)] - \exp[\lambda_2 (t - t_0)]\right] & \lambda_2 \exp[\lambda_2 (t - t_0)] - \lambda_1 \exp[\lambda_1 (t - t_0)] \end{bmatrix}.$$

(b) Complex and distinct roots ($a^2 - 4b < 0$); from Illustration 5.4.1 and Definition 5.4.3, the solution to the given IVP is

$$\begin{aligned} y(t) &= \frac{1}{\beta}\exp[\alpha (t - t_0)][\gamma y_0 \cos(\beta (t - t_0) + \psi) + y_0^{(1)} \sin\beta (t - t_0)] \\ &\quad + \frac{1}{\beta}\exp[\alpha t]\left[-\cos\beta t\int_{t_0}^t \exp[-\alpha s]\sin\beta s g(s)\,ds\right. \\ &\quad \left. + \sin\beta t\int_{t_0}^t \exp\left[-\alpha s\right]\cos\beta s g(s)\,ds\right]. \end{aligned}$$

Again, we recall (Procedures 4.3 and 4.4) that the normalized fundamental matrix $\Phi(t, t_0)$ at $t = t_0$ is given by

$$\begin{aligned}
&\Phi(t)\Phi^{-1}(t_0)\\
&\quad = \frac{1}{\beta}\exp\left[\alpha(t-t_0)\right]\begin{bmatrix}\cos\beta t & \sin\beta t\\ -\gamma\sin(\beta t-\psi) & \gamma\cos(\beta t-\psi)\end{bmatrix}\\
&\quad\times\begin{bmatrix}\gamma\cos(\beta t_0-\psi) & -\sin\beta t_0\\ \gamma\sin(\beta t_0-\psi) & \cos\beta t_0\end{bmatrix}\\
&\quad = \frac{1}{\beta}\exp\left[\alpha(t-t_0)\right]\begin{bmatrix}\gamma\cos(\beta(t-t_0)+\psi) & \sin\beta(t-t_0)\\ -\gamma^2\sin\beta(t-t_0) & \cos(\beta(t-t_0)-\psi)\end{bmatrix},
\end{aligned}$$

where

$$\begin{aligned}
\cos(\beta(t-t_0)+\psi) &= \gamma\cos(\beta t_0-\psi)\cos\beta t+\gamma\sin\beta t\sin(\beta t_0-\psi),\\
\sin\beta(t-t_0) &= -\sin\beta t_0\cos\beta t+\sin\beta t\cos\beta t_0,\\
\sin\beta(t-t_0) &= \sin(\beta t-\psi)\cos(\beta t_0-\psi)-\cos(\beta t-\psi)\sin(\beta t_0-\psi),\\
\cos(\beta(t-t_0)-\psi) &= \cos(\beta t-\psi)\cos\beta t_0+\gamma\sin\beta t_0\sin(\beta t-\psi),
\end{aligned}$$

$$\gamma = \sqrt{\alpha^2+\beta^2} \quad\text{and}\quad \psi = \tan^{-1}\frac{\alpha}{\beta}, \quad \text{for } \beta \text{ not equal to zero.}$$

(c) Repeated roots ($a^2 - 4b = 0$): from Illustration 5.4.1 and Definition 5.4.1, the solution process of the illustration is

$$\begin{aligned}
y(t) &= \exp\left[-\frac{a}{2}(t-t_0)\right]\left[y_0\left(1+\frac{a}{2}(t-t_0)\right)+y_0^{(1)}(t-t_0)\right]\\
&\quad + \exp\left[-\frac{a}{2}t\right]\left[-\int_{t_0}^{t} s\exp\left[\frac{a}{2}s\right]g(s)\,ds + t\int_{t_0}^{t}\exp\left[\frac{a}{2}s\right]g(s)\,ds\right].
\end{aligned}$$

Here,

$$\begin{aligned}
\Phi(t,t_0) &= \Phi(t)\Phi^{-1}(t_0)\\
&= \exp\left[-\frac{a}{2}(t-t_0)\right]\begin{bmatrix}1 & t\\ -\frac{a}{2} & 1-\frac{a}{2}t\end{bmatrix}\begin{bmatrix}1-\frac{a}{2}t_0 & -t_0\\ \frac{a}{2} & 1\end{bmatrix}\\
&= \exp\left[-\frac{a}{2}(t-t_0)\right]\begin{bmatrix}1+\frac{a}{2}(t-t_0) & t-t_0\\ -\frac{1}{4}a^2(t-t_0) & 1-\frac{a}{2}(t-t_0)\end{bmatrix}.
\end{aligned}$$

This completes the solution process of the illustration.

5.5 Exercises

(1) Find the general solution to the following differential equations:

(a) $dy^{(1)} + 4[y^{(1)} + y]dt = \exp[-2t]dt$ (b) $dy^{(3)} + y\,dt = \ln(t = 1)\,dt$
(c) $dy^{(2)} + [3y^{(2)} + 3y^{(1)} + y]dt = \sin t\,dt$ (d) $dy^{(3)} - y\,dt = t^2\,dt$
(e) $dy^{(1)} - [6y^{(1)} - 5y]dt = \exp[5t]dt$ (f) $dy^{(5)} - y^{(4)}\,dt = 1\,dt$

(2) Solve the following initial value problems:

(a) $dy^{(1)} - 4y\,dt = \exp[2t]dt,\ y(0) = 1,\ y^{(1)}(0) = -1$
(b) $dy^{(1)} + 4y\,dt = \cos 2tdt,\ y(0) = 1,\ y^{(1)}(0) = 0$
(c) $dy^{(2)} + 8y\,dt = (1 + t)dt,\ y(0) = 1,\ y^{(1)}(0) = 0,\ y^{(2)}(0) = 1$
(d) $dy^{(1)} + 4[y^{(1)} + y]dt = \sin tdt,\ y(0) = 0,\ y^{(1)}(0) = 1$
(e) $dy^{(3)} - y\,dt = \cosh tdt,\ y(0) = -1,\ y^{(1)}(0) = 1,\ y^{(2)}(0) = y^{(3)}(0) = 0$

5.6 The Laplace Transform

In this section, we present the concept of the Laplace transform and its applications to find Laplace transforms of well-known elementary functions. Several basic results of finding a class of Laplace transforms are outlined. The presented results were applied to find the Laplace transform of linear transformed functions, derivatives, and integrals, as well as linear integral equations.

Definition 5.6.1. Let f be a real-valued function defined for all real numbers $t \geq 0$. The Laplace transform of f in the sense of the Cauchy–Riemann integral is defined by

$$F(s) = \mathfrak{L}(f)(s) = \int_0^\infty e^{-st} f(t)\,dt = \lim_{T\to\infty}\left[\int_0^T e^{-st} f(t)\,dt\right], \tag{5.55}$$

for all values of s for which this improper integral exists. It is denoted by $F(s) = \mathfrak{L}(f)(s)$.

Example 5.6.1. Find the Laplace transform of $f(t) = c$ in the sense of the Cauchy–Riemann integral for $c \neq 0$.

Solution process. Let us define the Cauchy–Riemann improper integral

$$\mathfrak{L}(f)(s) = \int_0^\infty e^{-st} c\,dt = c\lim_{T\to\infty}\left[\int_0^T e^{-st}\,dt\right] = -c\lim_{T\to\infty}\left[\frac{e^{-st}}{s}\,|_0^T\right]$$
$$= \frac{c}{s}\lim_{T\to\infty}[1 - e^{-sT}] = \frac{c}{s},\ \text{provided that } s > 0.$$

Thus, for $s > 0$ and $c \neq 0$,

$$\mathfrak{L}(c)(s) = \frac{c}{s} \quad \text{if and only if } \mathfrak{L}(1)(s) = \frac{1}{s}. \tag{5.56}$$

Example 5.6.2. Find the Laplace transform of $f(t) = t$ in the sense of the Cauchy–Riemann improper integral.

Solution process. Let us define the improper integral

$$\begin{aligned}
\mathfrak{L}(f)(s) &= \int_0^\infty e^{-st}t\,dt = \lim_{T\to\infty}\left[\int_0^T e^{-st}t\,dt\right] \\
&= -\lim_{T\to\infty}\left[\frac{e^{-st}}{s}t\,\Big|_0^T + \frac{1}{s}\int_0^T e^{-st}\,dt\right] \quad \text{(integration by parts)} \\
&= \frac{1}{s^2}\lim_{T\to\infty}\left[1 - e^{-sT}\right] \quad \text{(limit concept and integration)} \\
&= \frac{1}{s^2}, \text{ provided that } s > 0.
\end{aligned}$$

Thus, for $s > 0$,

$$\mathfrak{L}(t)(s) = \frac{1}{s^2}. \tag{5.57}$$

Definition 5.6.2. Let $L[f]$ be a class of smooth functions that satisfy the exponential growth condition

$$|f(t)| \leq Ke^{Mt}, \quad s > M, \tag{5.58}$$

for some $M > 0$ and $K \geq 0$.

From Definition 5.6.1 and the properties of the integral and limit, the Laplace transform obeys the following property:

Theorem 5.6.1. *Let $f_1, f_2 \in L[f]$, and let c_1 and c_2 be arbitrary given constants. Then*

$$\mathfrak{L}\left(c_1 f_1 + c_2 f_2\right)(s) = c_1 \mathfrak{L}\left(f_1\right)(s) + c_2 \mathfrak{L}\left(f_2\right)(s). \tag{5.59}$$

Observation 5.6.1. From Theorem 5.6.1 and the definition of a linear function/operator, we conclude that the Laplace transform is defined on the space of functions into itself. In fact, the integral, differential, and limit are linear operations. For example, the integral of sum of two functions is the sum of integrals of the functions, and the integral of a constant times a function is a constant times the integral of the function.

Example 5.6.3. Find the Laplace transform of $f(t) = e^{at}$ in the sense of the Cauchy–Riemann integral.

Solution process. Let us define the improper Cauchy–Riemann integral

$$\begin{aligned}\mathfrak{L}(f)(s) &= \int_0^\infty e^{-st}e^{at}dt = \lim_{T\to\infty}\left[\int_0^T e^{-st+at}\,dt\right] \quad \text{(law of exponents)}\\ &= \lim_{T\to\infty}\left[\left(-\frac{e^{-(s-a)t}}{s-a}\right)\Big|_0^T\right] \quad \text{(by integration)}\\ &= \lim_{T\to\infty}\left[-\frac{e^{-(s-a)T}}{s-a}+\frac{1}{s-a}\right] = \frac{1}{s-a}, \quad \text{for } s \geq a.\end{aligned}$$

Thus,

$$\mathfrak{L}(f)(s) = \mathfrak{L}(e^{at})(s) = \frac{1}{s-a}. \tag{5.60}$$

Example 5.6.4. Find the Laplace transform of $f(t) = \sin at$ in the sense of the Cauchy–Riemann sense integral for $a \neq 0$.

Solution process. Let us define the improper Cauchy–Riemann integral

$$\begin{aligned}&\mathfrak{L}(\sin at)(s)\\ &= \int_0^\infty e^{-st}\sin at\,dt = \lim_{T\to\infty}\left[\int_0^T e^{-st}\sin at\,dt\right] \quad \text{(definition)}\\ &= \lim_{T\to\infty}\left[\left(-\frac{e^{-st}}{a}\cos at\right)\Big|_0^T - s\int_0^T e^{-st}\cos at\,dt\right] \quad \text{(integration by parts)}\\ &= \lim_{T\to\infty}\left[\frac{1}{a}-\frac{s}{a^2}e^{-st}\sin at\,\Big|^T - \frac{s^2}{a^2}\int_0^T e^{-st}\sin at\,dt\right] \quad \text{(integration by parts)}\\ &= \lim_{T\to\infty}\left[\frac{1}{a}-\frac{s^2}{a^2}\int_0^T e^{-st}\sin at\,dt\right] \quad \text{(simplification)}\\ &= \frac{1}{a}-\frac{s^2}{a^2}\mathfrak{L}\left(\sin at\right)(s) \quad \text{(Definition 5.6.1).}\end{aligned}$$

Thus,

$$\mathfrak{L}(\sin at)(s) = \frac{a}{s^2+a^2}. \tag{5.61}$$

In the following, we present a Laplace transform of a derivative of a function, an indefinite integral, and the convolution integral of two functions. First, let us introduce the concept of the convolution integral of two functions.

Definition 5.6.3. Let f and g be piecewise continuous functions defined on $t \geq 0$. The Cauchy–Riemann convolution integrals of f and g we defined by

$$(f * g)(t) = \int_0^t g(t-u)f(u)\,du. \tag{5.62}$$

Observation 5.6.2. From (5.62), we observe that

$$\begin{aligned}(f * g)(t) &= \int_0^t g(t-u)f(u)\,du \\ &= \int_t^0 g(z)f(t-z)\,(-dz) \quad \text{(substitution } u = t-z) \\ &= -\int_t^0 g(z)f(t-z)\,dz \quad \text{(simplification)} \\ &= \int_0^t g(z)f(t-z)\,dz \quad \text{(indefinite integral property)} \\ &= (g * f)(t) \quad \text{[from (5.62)].}\end{aligned}$$

This shows that the convolution integral defined in (5.62) satisfies the commutative law.

Theorem 5.6.2 (Laplace transform of a derivative). *Let us suppose that f has $n-1$ continuous derivatives on $[0,\infty)$, and for each i, $0 \le i \le n-1$, let $f^{(i)} \in L[f]$. Further assume that $f^{(n)}$ is piecewise continuous in every subinterval $0 \le t < b$. Then $f^{(n)} \in L[f]$ and*

$$\mathfrak{L}\left(f^{(n)}\right)(s) = s^n \mathfrak{L}(f)(s) - s^{n-1}f(0) - s^{n-2}f^{(1)}(0) - \cdots - f^{(n-1)}(0). \qquad (5.63)$$

Proof. We prove this theorem by the method of induction. For $n = 1$, we first assume that f and f' are continuous and $f \in L[f]$. Let us integrate the following definite integral:

$$\begin{aligned}\int_0^T e^{-st} f^{(1)}(t)\,dt &= e^{-st} f(t) \,\Big|_0^T + s\int_0^T e^{-st} f(t)\,dt \quad \text{(integrating by parts)} \\ &= e^{-sT} f(T) - f(0) + s\int_0^T e^{-st} f(t)\,dt.\end{aligned}$$

Since

$$\lim_{T\to\infty} e^{-sT} f(T) = 0 \quad \text{and} \quad \lim_{T\to\infty} \left[\int_0^T e^{-st} f(t)\,dt\right] = \mathfrak{L}(f)(s),$$

we have

$$\lim_{T\to\infty} \left[\int_0^T e^{-st} f^{(1)}(t)\,dt\right] = s\mathfrak{L}(f)(s) - f(0).$$

In the case of continuity of $f^{(1)}$, this establishes the validity of (5.64). If $f^{(1)}$ is discontinuous but still piecewise continuous with discontinuities at $0 < t_1 < t_2 < \cdots < t_n \cdots$, the proof is broken into parts corresponding to the intervals of continuity of $f^{(1)}$. The details are left to the reader. By Principle of Mathematical Induction 1.2.3, one can prove the expression (5.63) for any positive integer n. □

Theorem 5.6.3 (Laplace transform of an indefinite integral). *Let us assume that $f \in L[f]$. In addition, let I be Cauchy–Riemann indefinite integrals of f. Then $I \in L[f]$,*

$$\mathfrak{L}(I)(s) = \mathfrak{L}\left(\int_0^t f(u)\,du\right)(s) = \frac{\mathfrak{L}(f)(s)}{s}. \tag{5.64}$$

Proof. From the smoothness (such as piecewise continuity) of the function f and the properties of definite integrals (Cauchy–Riemann indefinite integrals), $I \in L[f]$. We also prove (5.64). From Definition 5.6.1, we have

$$\begin{aligned}
\mathfrak{L}(I)(s) &= \int_0^\infty e^{-st}\left(\int_0^t f(u)\,du\right)dt \\
&= \lim_{T\to\infty}\left[\int_0^T e^{-st}\left(\int_0^t f(u)\,du\right)dt\right] \quad \text{(Definition 5.6.1)} \\
&= \lim_{T\to\infty}\left[-\frac{e^{-st}}{s}\left(\int_0^t f(u)\,du\right)\Big|_0^T \right. \\
&\quad \left. + \frac{1}{s}\int_0^T e^{-st} f(t)\,dt\right] \quad \text{(by integration by parts)} \\
&= \lim_{T\to\infty}\left[-\frac{e^{-sT}}{s}\left(\int_0^T f(u)\,du\right) + \frac{e^{-s0}}{s}\left(\int_0^0 f(u)\,du\right)\right] \\
&\quad + \frac{1}{s}\lim_{T\to\infty}\int_0^T e^{-st} f(t)\,dt \quad \text{(by notations)} \\
&= \frac{1}{s}\lim_{T\to\infty}\int_0^T e^{-st} f(t)\,dt \quad \text{(by notations)} \\
&= \frac{1}{s}\int_0^\infty e^{-st} f(t)\,dt = \frac{\mathfrak{L}(f)(s)}{s}.
\end{aligned}$$

This completes the proof of the theorem. □

Definition 5.6.4. A real-valued function $\kappa_{R_+} \equiv \kappa$ is called a characteristic function with respect to a set $[0, \infty)$ if

$$\kappa_{R_+}(t) \equiv \kappa(t) = \begin{cases} 0, & \text{if } t < 0, \\ 1, & \text{if } t \geq 0, \end{cases} \qquad \kappa(t-c) = \begin{cases} 0, & \text{if } t < c, \\ 1, & \text{if } t \geq c, \end{cases} \tag{5.65}$$

for any $c \in R$.

Theorem 5.6.4. *For $0 \leq c$, a real-valued function g is defined on $-c \leq t < \infty$, and its $\mathfrak{L}(g)(s)$ exists. Then*

$$\mathfrak{L}(g(t-c)\kappa(t-c))(s) = e^{-sc}\mathfrak{L}(g)(s). \tag{5.66}$$

Proof. From Definition 5.6.1, we have

$$\begin{aligned}
\mathfrak{L}(g(t-c)\kappa(t-c))(s) &= \int_0^\infty e^{-st} g(t-c)\kappa(t-c)\,dt \\
&= \int_0^c e^{-st} g(t-c)\kappa(t-c)\,dt \\
&\quad + \int_c^\infty e^{-st} g(t-c)\kappa(t-c)\,dt \\
&= \int_c^\infty e^{-st} g(t-c)\,dt \quad ((5.65)) \\
&= \int_0^\infty e^{-s(u+c)} g(u)\,du \quad (\text{substitution: } u = t-c) \\
&= e^{-sc}\int_0^\infty e^{-su} g(u)\,du \quad (\text{substitution: } u = t-c) \\
&= e^{-sc}\mathfrak{L}(g)(s) \quad (\text{Definition 5.6.1}).
\end{aligned}$$

This establishes the validity of (5.66). The proof of the theorem is complete. □

Example 5.6.5. Let us consider

$$g(t) = \begin{cases} 3, & \text{if } 0 < t < 1, \\ t, & \text{if } t \geq 1. \end{cases}$$

Find the Laplace transform of g.

Solution process. First, we rewrite $g(t)$ in terms of the unit step function, as:

$$g(t) = 3(1 - \kappa(t-1)) + t\kappa(t-1).$$

This is the expression of $g(t)$. Hence, we can apply Theorem 5.6.4, and we have

$$\mathfrak{L}g(t)(s) = \frac{3}{s} - \frac{3}{s}e^{-s} + \frac{e^{-s}}{s^2}.$$

Theorem 5.6.5 (Laplace transform of the convolution integral). *Let us assume that $f, g \in L[f]$. Then $f * g \in L[f]$,*

$$\mathfrak{L}(f * g)(s) = \mathfrak{L}\left(\int_0^t g(t-u)f(u)du\right)(s) = \mathfrak{L}(f)(s) \cdot \mathfrak{L}(g)(s), \tag{5.67}$$

Proof. From Definitions 5.6.2 and 5.6.3, we can conclude that $f * g \in L[f]$. Furthermore, from Definition 5.6.1, we have

$$\begin{aligned}
\mathfrak{L}(f*g)(s) &= \int_0^\infty e^{-st}\left[\int_0^t g(t-u)f(u)\,du\right]dt \quad \text{(Definition 5.6.3)}\\
&= \int_0^\infty e^{-st}\left[\int_0^t g(t-u)f(u)\right]du\,dt \quad \text{(Fubini's theorem)}\\
&= \int_0^\infty \int_u^\infty e^{-st} g(t-u)f(u)\,dt\,du \quad \text{(change of order of integration)}\\
&= \int_0^\infty \int_0^\infty e^{-st} g(t-u)\kappa(t-u)f(u)\,dt\,du \quad [\text{definition of } \kappa(t-u)]\\
&= \int_0^\infty e^{-su}\mathfrak{L}(g)(s)f(u)\,du \quad \text{(Theorem 5.6.4)}\\
&= \mathfrak{L}(g)(s)\int_0^\infty e^{-su}f(u)\,du \quad \text{(by simplification)}\\
&= \mathfrak{L}(g)(s)\mathfrak{L}(f)(s) \quad \text{(Definition 5.6.1).}
\end{aligned}$$

This completes the proof of the theorem. □

Example 5.6.6. Solve the given equation $g(t) = t + \int_0^t \sin(t-u)g(u)\,du$.

Solution process. Let $\mathfrak{L}(g)(s)$. Then

$$\mathfrak{L}(\sin t)(s) = \frac{1}{1+s^2}.$$

Now, applying Theorem 5.6.5, we have

$$\begin{aligned}
\mathfrak{L}(g)(s) = \mathfrak{L}(f*g))(s) &= \mathfrak{L}(t)(s) + \mathfrak{L}(g)(s)\mathfrak{L}(f)(s)\\
&= \frac{1}{s^2} + \frac{\mathfrak{L}(g)(s)}{1+s^2}.
\end{aligned}$$

By solving for $\mathfrak{L}(g)(s)$, we obtain

$$\mathfrak{L}(g)(s) = \frac{1+s^2}{s^4} = \frac{1}{s^2} + \frac{1}{s^4}.$$

The inverse Laplace transform of this is given by

$$\begin{aligned}
g(t) &= \mathfrak{L}^{-1}\left[\left(\frac{1}{s^2}+\frac{1}{s^4}\right)\right] = \mathfrak{L}^{-1}\left[\frac{1}{s^2}\right] + \mathfrak{L}^{-1}\left[\frac{1}{s^4}\right]\\
&= \mathfrak{L}^{-1}\left[\frac{1}{s^2}\right] + \frac{1}{6}\mathfrak{L}^{-1}\left[\frac{6}{s^4}\right] = t + \frac{1}{6}t^3.
\end{aligned}$$

Example 5.6.7. Find the Laplace transform of $f(t) = \cos at$.

Solution process. From Definition 5.6.1, we have

$$\begin{aligned}
&\int_0^{\infty} e^{-st}\cos at\,dt \\
&\quad = \lim_{T\to\infty}\left[-\frac{1}{s}e^{-st}\cos(at)\,|_0^T + \frac{a}{s}\int_0^T e^{-st}\sin at\,dt\right] \quad \text{(integrating by parts)} \\
&\quad = \lim_{T\to\infty}\left[-\frac{1}{s}e^{-st}\cos(at)\,|_0^T - \frac{a}{s^2}e^{-st}\sin at\,|_0^T - \frac{a^2}{s^2}\int_0^T e^{-st}\cos at\,dt\right].
\end{aligned}$$

We know that

$$\lim_{T\to\infty} e^{-sT}\cos aT - e^{-s0}\cos a0 = 1 \quad \text{and} \quad \lim_{T\to\infty} e^{-sT}\sin aT - e^{-s0}\sin a0 = 0.$$

Thus,

$$\begin{aligned}
\mathfrak{L}(\cos at)(s) &= \int_0^{\infty} e^{-st}\cos at\,dt = \frac{1}{s} - \frac{a^2}{s^2}\lim_{T\to\infty}\int_0^T e^{-st}\cos at\,dt \\
&= \frac{1}{s} - \frac{a^2}{s^2}\mathfrak{L}(\cos at)(s) \quad \text{(Definition 5.6.1)}.
\end{aligned}$$

Now, by solving for $\mathfrak{L}(\cos at)(s)$, we obtain

$$\mathfrak{L}(\cos at)(s) = \frac{s}{s^2+a^2}.$$

This completes the solution process of the example.

A short table of Laplace transforms:

$f(t)$	$\mathfrak{L}(f(t))(s)$	$f(t)$	$\mathfrak{L}(f(t))(s)$
c	$\frac{c}{s},\ s>0$	t^n	$\frac{n!}{s^{n+1}},\ s>0$
e^{at}	$\frac{1}{s-a},\ s>a$	$t^n\,e^{at}$	$\frac{n!}{(s-a)^{n+1}},\ s>0$
$\sin at$	$\frac{a}{s^2+a^2}$	$t\sin at$	$\frac{2as}{(s^2+a^2)^2}$
$\cos at$	$\frac{s}{s^2+a^2}$	$t\cos at$	$\frac{s^2-a^2}{(s^2+a^2)^2}$
$\sinh at$	$\frac{a}{s^2-a^2}$	$e^{-at}\sin bt$	$\frac{b}{(s+a)^2+b^2}$
$\cosh at$	$\frac{s}{s^2-a^2}$	$e^{-at}\cos bt$	$\frac{s+a}{(s+a)^2+b^2}$

5.6 Exercises

(1) Find the Laplace transform of the following functions:

(a) $f(t) = \cos^2 at - \sin^2 at$ (b) $f(t) = \sin at \cos at$
(c) $p(t) = a_m t^m + a_q t^{2p-1}$ (d) $f(t) = t^m \sin at$
(e) $f(t) = t^m \cos at$ (f) $f(t) = t^m \cosh at$

(2) Find $\mathfrak{L}(f(t))(s)$ if $f(t)$ is given by

(a) $f(t) = \begin{cases} 3, & \text{if } 0 < t < 2, \\ t, & \text{if } t \geq 2 \end{cases}$

(b) $g(t) = \begin{cases} 0, & \text{if } 0 < t < 1, \\ \sin t, & \text{if } 1 \leq t < 2, \\ t, & \text{if } 2 \leq t \end{cases}$

(c) $f(t) = \begin{cases} 3, & \text{if } 0 < t < 2, \\ 1, & \text{if } t \geq 2 \end{cases}$

(3) Find the inverse Laplace transform of the following:

(a) $F(s) = \frac{2}{s^2+4s+8}$ (b) $G(s) = \frac{s^3+1}{(s^2+1)s^5}$
(c) $H(s) = \frac{s^2+1}{(s^2+a^2)(s^3+b^3)}$ (d) $G(s) = \frac{s+2}{s^2+2s+8}$
(e) $M(s) = \frac{s}{s^2+4} - \frac{1}{s^2} - \left(\frac{s}{s^2+4} - \frac{1}{s^2}\right) e^{-s}$

(4) Solve the given integral equations:

(a) $g(t) = t^2 + \int_0^t \cos(t-u) g(u)\, du$
(b) $x(t) = 1 + \int_0^t \exp[-2(t-u)]\, x(u)\, du$
(c) $y(t) = \cos t + \int_0^t \exp[(t-u)]\, y(u)\, du$
(d) $h(t) = t^2 + \int_0^t \sin(t-u) h(u)\, du$

5.7 Applications of the Laplace Transform

In this section, the Laplace transform is applied to solve IVPs. The Laplace transform transforms linear differential equations with constant coefficients into linear algebraic equations. The techniques for solving the algebraic equations are easier than the methods for solving the IVPs.

Example 5.7.1. Use the Laplace transform to solve the given IVP:

$$dy^{(1)} + y\, dt = 0, \quad y(0) = 0, \quad y'(0) = 1.$$

Solution procedure. We note that the differential equation is equivalent to the integral equation

$$y^{(1)}(t) = y^{(1)}(0) - \int_0^t y(u)du.$$

Now, we apply the Laplace transform to both sides, and obtain

$$\begin{aligned}
\mathfrak{L}(y^{(1)}(t)) &= \mathfrak{L}\left(e^{-st}\left[y^{(1)}(0) - \int_0^t y(u)\,du\right]\right) \quad \text{(Definition 5.6.1)} \\
&= \mathfrak{L}\left(e^{-st}y^{(1)}(0)\right) - \mathfrak{L}\left(\int_0^t y(u)\,du\right) \quad \text{(Theorem 5.6.1)} \\
&= \frac{1}{s} - \frac{\mathfrak{L}(y(t))}{s} \quad \text{(Theorem 5.6.3 and initial data).}
\end{aligned}$$

Moreover, from Theorem 5.6.2 we have $\mathfrak{L}(y^{(1)}(t)) = s\mathfrak{L}(y(t)) - y(0)$. From this and using the initial conditions, the above expression reduces to

$$s\mathfrak{L}(y(t)) = \frac{1}{s} - \frac{\mathfrak{L}(y(t))}{s}.$$

Now, we solve for $\mathfrak{L}(y(t))$, and its solution is

$$\mathfrak{L}(y(t)) = \frac{s}{1+s^2}\left(\frac{1}{s}\right) = \frac{1}{1+s^2}.$$

By applying the inverse Laplace transform to both sides, we get

$$\begin{aligned}
y(t) &= \mathfrak{L}^{-1}\left(\frac{1}{1+s^2}\right) \\
&= \mathfrak{L}^{-1}\left(\frac{1}{1+s^2}\right) \quad (\mathfrak{L}^{-1} \text{ is a linear operator}) \\
&= \sin t.
\end{aligned}$$

This completes the solution process of the given problem.

Example 5.7.2. Use the Laplace transform to solve the IVP

$$dy^{(1)} + \beta y^{(1)}dt = 0, \quad y(0) = y_0, \quad y^{(1)}(0) = v_0, \quad \text{for } \beta > 0.$$

Solution process. We note that the differential equation is equivalent to the integral equation

$$y^{(1)}(t) = y^{(1)}(0) - \beta\int_0^t y^{(1)}(u)\,du.$$

Applying the Laplace transform to both sides, we obtain

$$\begin{aligned}
\mathfrak{L}(y^{(1)}(t)) &= \mathfrak{L}\left(e^{-st}y^{(1)}(0) - \beta\int_0^t y^{(1)}(u)\,du\right) \quad \text{(Definition 5.6.1)}\\
&= \mathfrak{L}\left(e^{-st}y^{(1)}(0)\right) - \beta\mathfrak{L}\left(\int_0^t y^{(1)}(u)\,du\right) \quad \text{(Theorem 5.6.1)}\\
&= \frac{v_0}{s} - \frac{\beta\mathfrak{L}\left(y^{(1)}(t)\right)}{s} \quad \text{(Theorem 5.6.3 and the initial data).}
\end{aligned}$$

Moreover, from Theorem 5.6.2 we have $\mathfrak{L}(y^{(1)}(t)) = s\mathfrak{L}(y(t)) - y(0)$. From this and using the initial conditions, the above expression reduces to

$$\begin{aligned}
s\mathfrak{L}(y(t)) - y_0 &= \frac{v_0}{s} - \frac{s\beta\mathfrak{L}(y(t)) - \beta y_0}{s}\\
&= \frac{v_0}{s} - \beta\mathfrak{L}(y(t)) + \frac{\beta y_0}{s}.
\end{aligned}$$

Now, we solve for $\mathfrak{L}(y(t))$, and the solution is

$$\begin{aligned}
\mathfrak{L}(y(t)) &= \frac{v_0}{\beta s + s^2} + \frac{y_0}{s}\\
&= \frac{v_0}{s(\beta + s)} + \frac{y_0}{s} \quad \text{(simplification)}\\
&= \frac{v_0}{\beta}\left[\frac{1}{s} - \frac{1}{\beta + s}\right] + \frac{y_0}{s} \quad \text{(method of partial fractions).}
\end{aligned}$$

By applying the inverse Laplace transform to both sides, we get

$$\begin{aligned}
y(t) &= \mathfrak{L}^{-1}\left(\frac{v_0}{\beta}\left[\frac{1}{s} - \frac{1}{\beta + s}\right] + \frac{y_0}{s}\right)\\
&= y_0 + \frac{v_0}{\beta}\left(1 - e^{-\beta t}\right) \quad \text{(table and Theorem 5.6.5).}
\end{aligned}$$

Thus, the solution to the IVP is given by

$$y(t) = y_0 + \frac{v_0}{\beta}\left(1 - e^{-\beta t}\right).$$

This completes the solution process of the given problem.

Example 5.7.3. Use the Laplace transform to solve the IVP

$$dy^{(1)} + (\beta y^{(1)} + \nu^2 y)dt = 0, \quad y(0) = y_0, \quad y^{(1)}(0) = v_0, \quad \beta > 0.$$

Solution process. We note that the differential equation is equivalent to the integral equation

$$y^{(1)}(t) = y^{(1)}(0) - \beta\int_0^t y^{(1)}(u)\,du - \nu^2\int_0^t y(u)\,du.$$

Now, we apply the Laplace transform to both sides, and obtain

$$
\begin{aligned}
&\mathfrak{L}\left(y^{(1)}(t)\right) \\
&= \mathfrak{L}\left(e^{-st}\left[y^{(1)}(0) - \beta\int_0^t y^{(1)}(u)\,du - \nu^2\int_0^t y(u)\,du\right]\right) \quad \text{(Definition 5.6.1)} \\
&= \mathfrak{L}(e^{-st}y^{(1)}(0)) - \beta\mathfrak{L}\left(\int_0^t y^{(1)}(u)\,du\right) - \nu^2\mathfrak{L}\left(\int_0^t y(u)\,du\right) \quad \text{(Theorem 5.6.3)} \\
&= \frac{v_0}{s} - \frac{\beta\mathfrak{L}\left(y^{(1)}(t)\right)}{s} - \frac{\nu^2\mathfrak{L}(y(t))}{s} \quad \text{(Theorem 5.6.3 and initial data).}
\end{aligned}
$$

Moreover, from Theorem 5.6.2 we have $\mathfrak{L}(y^{(1)}(t)) = s\mathfrak{L}(y(t)) - y(0)$, and imitating the argument used in Example 5.7.2, we obtain

$$
\begin{aligned}
s\mathfrak{L}(y(t)) - y_0 &= \frac{v_0 - \beta\left[s\mathfrak{L}(y(t)) - y(0)\right] - \nu^2\mathfrak{L}(y(t))}{s} \\
&= \frac{v_0 - \left(\beta s + \nu^2\right)\mathfrak{L}\left(y(t)\right) + \beta y(0)}{s} \\
&= \frac{v_0 + \beta y(0)}{s} - \frac{(\beta s + \nu^2)\mathfrak{L}(y(t))}{s}.
\end{aligned}
$$

After algebraic manipulations we have

$$
\begin{aligned}
\mathfrak{L}(y(t)) &= \frac{v_0 + \beta y(0))}{\nu^2 + \beta s + s^2} + \frac{sy(0)}{\nu^2 + \beta s + s^2} \\
&= \frac{v_0 + \frac{1}{2}\beta y(0)}{\left(s + \frac{1}{2}\beta\right)^2 + \frac{4\nu^2 - \beta^2}{4}} + \frac{\left(\frac{1}{2}\beta + s\right)y_0}{\left(s + \frac{1}{2}\beta\right)^2 + \frac{4\nu^2 - \beta^2}{4}}.
\end{aligned}
$$

By applying the inverse Laplace transform to both sides, we get

$$
y(t) = \mathfrak{L}^{-1}\left(\frac{v_0 + \frac{1}{2}\beta y_0}{\left(s + \frac{1}{2}\beta\right)^2 + \frac{4\nu^2 - \beta^2}{4}} + \frac{\left(\frac{1}{2}\beta + s\right)y_0}{\left(s + \frac{1}{2}\beta\right)^2 + \frac{4\nu^2 - \beta^2}{4}}\right).
$$

Depending on the magnitudes of ν^2 and β^2, and the sign of $4\nu^2 - \beta^2$, the representation of the solution to the given differential equation is

$$
y(t) = \begin{cases}
e^{-\frac{1}{2}\beta t}\left(\dfrac{v_0 + \frac{1}{2}\beta y_0}{b}\sin bt + y_0\cos bt\right), & \text{if } b^2 = \dfrac{1}{4}(4\nu^2 - \beta^2) > 0, \\
2v_0 t e^{-\frac{1}{2}\beta t} + y_0 e^{-\frac{1}{2}\beta t}, & \text{if } 4\nu^2 - \beta^2 = 0, \\
\dfrac{v_0\left[e^{-(\beta - b)t} - e^{-(\beta + b)t}\right]}{2b} + \dfrac{y_0\left[e^{-(\beta - b)t} + e^{-(\beta + b)t}\right]}{2}, & \text{if } \dfrac{1}{4}(4\nu^2 - \beta^2) < 0.
\end{cases}
$$

This completes the solution process of the given problem.

Example 5.7.4. A spring with a spring constant of 0.5 lb/ft lies on a smooth table. A 4 lb weight is attached to the spring and is at rest at the equilibrium position. A 2 lb force is applied to the support along the line of action of the spring for 2 s and is then removed. Discuss the motion of the system.

Solution procedure. By following the discussion in Illustration 5.2.1, we have

$$m dx^{(1)} + kx\,dt = f(t), \quad x(0) = x_0, \quad x^{(1)}(0) = v_0,$$

where

$$m = \frac{w}{g} = \frac{4}{32}, \quad k = \frac{1}{2}, \quad x_0 = 0, \quad v_0 = 0.$$

The external force is described by

$$f(t) = \begin{cases} 2, & \text{if } 0 \le t < 2, \\ 0, & \text{if } t \ge 2. \end{cases}$$

Using this information and Definition 5.6.4, the IVP reduces to

$$dx^{(1)} + 4x\,dt = 8f(t) = 16\left[1 - \kappa(t-2)\right],$$
$$x(0) = x_0, \quad x^{(1)}(0) = 0,$$

where

$$\kappa(t-2) = \begin{cases} 0, & \text{if } t < 2, \\ 1, & \text{if } t \ge 2. \end{cases}$$

Now, by using the Laplace transform and imitating the problem-solving process in Example 5.7.3, we have

$$s^2\mathfrak{L}(x(t)) + 4\mathfrak{L}(y(t)) = \frac{16}{s}\left(1 - e^{-2s}\right),$$

which implies that

$$\begin{aligned}\mathfrak{L}(x(t)) &= \frac{16}{s(s^2+4)}(1 - e^{-2s}) \\ &= 4\left[\frac{1}{s} - \frac{s}{s^2+4}\right](1 - e^{-2s}).\end{aligned}$$

Thus,

$$\begin{aligned} x(t) &= \mathfrak{L}^{-1}4\left[\frac{1}{s} - \frac{s}{s^2+4}\right] - \mathfrak{L}^{-1}\left(4\left[\frac{1}{s} - \frac{s}{s^2+4}\right]e^{-2s}\right) \\ &= 4\mathfrak{L}^{-1}\left(\frac{1}{s}\right) - 4\mathfrak{L}^{-1}\left(\frac{s}{s^2+4}\right) - 4\mathfrak{L}^{-1}\left(\frac{1}{s}e^{-2s} - \frac{s}{s^2+4}e^{-2s}\right) \\ &= 4 - 4\cos 2t - 4\left[1 - \cos 2(t-2)\right]\kappa(t-2). \end{aligned}$$

Hence,

$$x(t) = \begin{cases} 4\left[1 - \cos 2t\right], & \text{if } t < 2, \\ 4\left[1 - \cos 2t\right] - 4\left[1 - \cos 2(t-2)\right], & \text{if } t \geq 2 \end{cases}$$

$$= \begin{cases} 4\left(1 - \cos 2t\right), & \text{if } t < 2, \\ 4\left[\cos 2(t-2) - \cos 2t\right], & \text{if } t \geq 2. \end{cases}$$

Example 5.7.5. Use the Laplace transform to solve the IVP

$$dx = \left[Ax + f(t)\right] dt,$$

where

$$A = \begin{bmatrix} 4 & 2 \\ 3 & -1 \end{bmatrix}, \quad f(t) = \begin{bmatrix} 1 \ t \end{bmatrix}^T, \quad x(0) = \begin{bmatrix} 1 \ 1 \end{bmatrix}^T.$$

Solution procedure. From Corollary 4.7.1, we have

$$\begin{aligned}
\left[X_1(s)\ X_2(s)\right]^T &= \mathfrak{L}(x(t)) \\
&= \mathfrak{L}\left[x(0) + \int_0^t \left[Ax(s) + f(s)\right] ds)\right] \quad \text{[Theorem 1.5.2(A1)]} \\
&= \mathfrak{L}[x(0)] + \mathfrak{L}\left[\int_0^t Ax(s)\, ds\right] + \mathfrak{L}\left[\int_0^t f(s)\, ds\right] \quad \text{(Theorem 5.6.1)} \\
&= \mathfrak{L}[x(0)] + A\mathfrak{L}\left[\int_0^t x(s)\, ds\right] + \mathfrak{L}\left[\int_0^t f(s) ds\right] \quad \text{[Theorem 1.5.2(P2)]} \\
&= \left[\frac{1}{s}\ \frac{1}{s}\right]^T + A\mathfrak{L}\left[\int_0^t x(s)\, ds\right] + \frac{1}{s}\left[\frac{1}{s}\ \frac{1}{s^2}\right]^T \quad \text{(from table)} \\
&= \left[\frac{1}{s}\ \frac{1}{s}\right]^T + A\left[\frac{\mathfrak{L}(x_1(t))}{s}\ \frac{\mathfrak{L}(x_2(t))}{s}\right]^T + \frac{\left[\frac{1}{s}\ \frac{1}{s^2}\right]^T}{s} \quad \text{(Theorem 5.6.3)} \\
&= \frac{\left[1\ 1\right]^T + A\left[X_1(s)\ X_2(s)\right]^T + \left[\frac{1}{s}\ \frac{1}{s^2}\right]^T}{s} \quad \text{(simplification)}.
\end{aligned}$$

We solve for $\left[X_1(s)\ X_2(s)\right]^T$, and obtain

$$\begin{aligned}
\left[X_1(s)\ X_2(s)\right]^T &= (sI - A)^{-1}\left[1 + \frac{1}{s} \quad 1 + \frac{1}{s^2}\right]^T \\
&= \frac{1}{s^2 - 3s - 10}\begin{bmatrix} s+1 & 2 \\ 3 & s-4 \end{bmatrix}\left[1 + \frac{1}{s} \quad 1 + \frac{1}{s^2}\right]^T \\
&= \left[\frac{s^3 + 4s^2 + s + 2}{s^2\left(s^2 - 3s - 10\right)}\ \frac{s^3 - s^2 + 4s - 4}{s^2\left(s^2 - 3s - 10\right)}\right]^T.
\end{aligned}$$

Now, we apply the inverse Laplace transform to the above system, and we have

$$\begin{aligned}
\left[x_1(t)\ x_2(t)\right]^T &= \mathfrak{L}^{-1}\left(\begin{bmatrix}\frac{s^3+4s^2+s+2}{s^2(s^2-3s-10)} & \frac{s^3-s^2+4s-4}{s^2(s^2-3s-10)}\end{bmatrix}^T\right)\\
&= \left[\mathfrak{L}^{-1}\left(\frac{s^3+4s^2+s+2}{s^2(s^2-3s-10)}\right)\ \mathfrak{L}^{-1}\left(\frac{s^3-s^2+4s-4}{s^2(s^2-3s-10)}\right)\right]^T\\
&\quad \times (\text{properties of } \mathfrak{L}^{-1})\\
&= \frac{1}{25}\left[-1-5t-8e^{5t}+34e^{-2t}\ \ -13+10t+20e^{5t}+18e^{-2t}\right]^T\\
&\quad \times (\text{partial fractions}),
\end{aligned}$$

where

$$\begin{aligned}
\mathfrak{L}^{-1}\left(\frac{s^3+4s^2+s+2}{s^2(s^2-3s-10)}\right) &= \mathfrak{L}^{-1}\left(-\frac{\frac{1}{25}}{s}-\frac{\frac{1}{5}}{s^2}-\frac{\frac{8}{25}}{s-5}+\frac{\frac{34}{25}}{s+2}\right)\\
&= \frac{1}{25}[-1-5t-8e^{5t}+34e^{-2t}],\\
\mathfrak{L}^{-1}\left(\frac{s^3-s^2+4s-4}{s^2(s^2-3s-10)}\right) &= \mathfrak{L}^{-1}\left(-\frac{\frac{13}{25}}{s}+\frac{\frac{2}{5}}{s^2}+\frac{\frac{18}{25}}{s-5}+\frac{\frac{20}{25}}{s+2}\right)\\
&= \frac{1}{25}[-13+10t+20e^{5t}+18e^{-2t}].
\end{aligned}$$

5.7 Exercises

(1) Use the Laplace transform to solve the following initial value problems:

(a) $dx^{(1)} + \alpha^2 x\, dt = 2\sin bt dt$, $x(0) = x_0$, $x^{(1)}(0) = v_0$
(b) $dx^{(2)} + [3\alpha x^{(2)} + 3\alpha x^{(2)} + \alpha^3 x]dt = 0$, $x(0) = 1$, $x^{(1)}(0) = 0$, $x^{(2)}(0) = 1$
(c) $dx^{(1)} + [6x^{(1)} + 9x - te^{-3t}]dt = 0$, $x(0) = 1$, $x^{(1)}(0) = 0$
(d) $dx^{(1)} + [6x^{(1)} + 5x - e^{2t}]dt = 0$, $x(0) = 1$, $x^{(1)}(0) = 1$

(2) Use the Laplace transform to solve the following initial value problems:

(a) $dx = Ax\, dt$, where

$$A = \begin{bmatrix} 0 & -\beta \\ \beta & 0 \end{bmatrix}, \quad x(0) = [1\ 1]$$

(b) $dx = Ax\, dt$, where

$$A = \begin{bmatrix} -3 & 2 \\ 1 & -2 \end{bmatrix}, \quad x(0) = [1\ 1]$$

(c) $dx = Ax\,dt$, where

$$A = \begin{bmatrix} -1 & 0 \\ \beta & -2 \end{bmatrix}, \quad x(0) = \begin{bmatrix} 2 & 1 \end{bmatrix}$$

5.8 Notes and Comments

In this chapter, the development of a deterministic modeling procedure involving higher-order rates of the state of a system is demonstrated with several applied examples in Section 5.2. The examples, exercises, and illustrations are based on the second author's class notes over a forty-five plus year period. The development of the eigenvalue and eigenvector-type approach of Sections 2.4 and 4.3 is systematically connected and extended to include the existing approach to solving higher-order linear homogeneous differential equations in Section 5.3. The techniques for solving and analyzing linear systems of differential equations in Chapter 4 are directly applied to the transformed system of linear differential equations associated with higher-order differential equations in Sections 5.4 and 5.5. Section 5.6 deals with the classical ideas and properties of the Laplace transform. Finally, Section 5.7 exhibits the usage of the Laplace transform to solve linear differential and integral equations. The material presented in this chapter is based on the second author's notes [55], as well as existing textbooks [24, 35, 107].

Chapter 6

Topics in Differential Equations

6.1 Introduction

Brief outlines regarding, methods, analysis, and new trends in the modeling of dynamic processes are presented via concepts and examples. In the classical modeling approach, the evolution of states of dynamic systems can be considered to be a dynamic flow evolving on a given timescale. The goal is to find information (qualitative/quantitative) about the dynamic flows for planning and/or improving the quality of the performance of an underlying dynamic process. In general, a closed/exact form representation of a time evolution solution flow described by a nonlinear or linear, nonstationary interconnected system is not always feasible. Moreover, the closed-form solution of the dynamic process is not essential to finding information about the process. In applications, the qualitative and/or quantitative properties of dynamic processes are most significant for understanding the processes. During the past eighty-plus years, a few analytic nonlinear techniques have been developed that do not require the closed-form solution of dynamic processes. In Section 6.2, we briefly describe the fundamental conceptual results similar to the results outlined in Sections 2.6 and 4.7. Brief outlines of two of the best-known nonlinear analytic methods, namely (i) nonlinear variation of constants/parameters and (ii) the comparison method coupled with differential inequalities, are presented. Sections 6.3 and 6.4 deal with a method of variation of constants/parameters and its extensions for the nonlinear system of differential equations. A comparison method and its extensions in the context of differential inequalities are highlighted in Sections 6.5–6.7. Section 6.8 introduces linear hybrid systems through drug prescription and administration problems. Hereditary effects are explored via examples in Section 6.9. Next, certain qualitative properties of the solution processes of differential equations are summarized in Section 6.10. Section 6.11 introduces stochastic dynamic models described by Itô–Doob-type stochastic differential equations and the effects of random environmental presentations.

6.2 Fundamental Conceptual Algorithms and Analysis

Mathematical models of complex dynamic processes are nonlinear, nonstationary, and interconnected. The first and the most fundamental test for the validity of a mathematical model is the "problem of the existence of the solution process." Prior to using a mathematical model, one needs to validate the existence of the solution process of the proposed model. We present a brief outline of this fundamental issue to validate dynamic models.

Let us consider a mathematical description of a nonlinear dynamic process described by a system of nonlinear/nonstationary differential equations:

$$dx = f(t, x)\, dt, \quad x(t_0) = x_0, \tag{6.1}$$

where $f \in C[J \times D, R^n]$, $D \subseteq R^n$, is an open connected set, with $C[J \times D, R^n]$ standing for the class of continuous functions defined on $J \times D$ into R^n, and $J = [t_0, t_0 + a)$ for some real number $a > 0$.

Problem 6.2.1 (existence of a solution process [15, 23, 42, 75, 77, 83, 84]). The first and most basic test for validating a proposed mathematical dynamic model of the linear or nonlinear dynamic phenomena is the existence of the time-evolving flow that satisfies the IVP (6.1). This is one of the first basic mathematical problems in the area of dynamic modeling of nonlinear processes in the biological, chemical, engineering, medical, physical, and social sciences. In the modeling of the nonlinear dynamic phenomena, it is very important to note that the IVP (6.1) has a solution process. This can be as one of the first fundamental mathematical problems in modeling of the dynamic processes. It is well known [15, 42, 83, 84] that under the above-stated assumption on f, the IVP (6.1) has a solution. We further note that other than conceptual information, f and x_0 may not be known, completely.

Problem 6.2.2 (uniqueness of the solution process [15, 23, 42, 75, 77, 83, 84]). In many chemical, engineering, physical, and social systems, the uniqueness of a solution to (6.1) is a very important problem. Otherwise, for the users of this type of dynamic model, it creates ambiguities in the decision making process. In addition to the conditions specified in (6.1), f needs to satisfy certain conditions, such as, the Lipschitz condition:

$$||f(t, x) - f(t, y)|| \leq L||x - y||, \quad \text{for } (t, x), (t, y) \in J \times D, \tag{6.2}$$

where L is some positive constant. We note that the Lipschitz condition (6.2) is global. However, if (6.2) is satisfied in the neighborhood of a given point (t_0, x_0), then it is a local Lipschitz condition. The significance of the uniqueness of the solution process is the fact that the future knowledge of the process depends only on the initial knowledge (initial data) of the process.

A sufficient condition for the local Lipschitz condition on f is the boundedness of the derivative of f with respect to x in the neighborhood of a given point (t_0, x_0). This is proven in the following lemma.

Lemma 6.2.1. *Let $f \in C[J \times D, R^n]$ and its derivative $\frac{\partial}{\partial x} f$ exist and be continuous on a closed ball $\overline{\mathrm{B}}(x_0, \rho)$ with a center at x_0 and a radius ρ for each $t \in J$. Then, for $(t,x), (t,y) \in J \times \overline{\mathrm{B}}(x_0, \rho)$,*

$$f(t,x) - f(t,y) = \int_0^1 \left[\frac{\partial}{\partial x} f(t, \theta x + (1-\theta)y)\right](x-y)\, d\theta, \tag{6.3}$$

$$||f(t,x) - f(t,y)|| \leq L||x-y||,$$

where

$$L = \sup \left\|\frac{\partial}{\partial x} f(t,z)\right\|$$

$$(t,z) \in J \times \overline{\mathrm{B}}(x_0, \rho).$$

Proof. From the convexity of $\overline{\mathrm{B}}(x_0, \rho)$ [15, 42, 75, 83, 84], we have

$$F(\theta) = f(t, \theta x + (1-\theta)y), \quad 0 \leq \theta \leq 1. \tag{6.4}$$

It is obvious that F is continuously differentiable on $[0, 1]$, and hence

$$\frac{d}{d\theta} F(\theta) = \frac{\partial}{\partial x} f(t, \theta x + (1-\theta)y)(x-y). \tag{6.5}$$

By integrating both sides of (6.5) with respect to θ over an interval $[0,1]$, we obtain

$$F(1) - F(0) = \int_0^1 \left[\frac{\partial}{\partial x} f(t, \theta x + (1-\theta)y)\right](x-y)\, d\theta.$$

This together with the fact that $F(1) = f(t,x)$ and $F(0) = f(t,y)$ yields the desired relation (6.3). The proof of the local Lipschitz condition follows from the properties of the linear operator

$$\int_0^1 \left[\frac{\partial}{\partial x} f(t, \theta x + (1-\theta)y)\right] d\theta$$

and the continuity of Jacobian matrix f_x on the closed and bounded ball $\overline{\mathrm{B}}(x_0, \rho)$. This completes the proof of the lemma. □

In the absence of the uniqueness assumption, the IVP can have infinitely many solutions. This is illustrated by the following simple example.

Example 6.2.1. We consider the very simple IVP

$$dx = 2x^{\frac{1}{2}}\, dt, \quad x(0) = 0. \tag{6.6}$$

Solution procedure. We note $f(t,x) = 2x^{\frac{1}{2}}$. By employing the method of variable separables (Section 3.6), the given IVP can be solved. One of the solutions is $x(t) = t^2$; another is $x(t) \equiv 0$. In fact,

$$x(t) = \begin{cases} 0, & \text{if } 0 \leq t \leq c, \\ (t-c)^2, & \text{if } c \leq t \end{cases}$$

is also a solution for any $0 \leq c$. This shows that the IVP (6.6) has infinitely many solutions. We note that $f(t,x) = 2x^{\frac{1}{2}}$ is differentiable on $(0,\infty)$. However, its derivative is not bounded in the neighborhood of "0."

Problem 6.2.3 (dependence of the solution process on system parameters [15, 23, 42, 75, 77, 83, 84]). In the development of a mathematical model for dynamic processes, the rate functions as well as the initial data are subject to errors. These errors arise in the choice of system parameters including the initial data (t_0, x_0). And, so a solution flow depends on these system parameters. From this, the basic question becomes to what extent do these parametric changes influence the errors in the solution flows with respect to the nominal system?. Are they harmless in some sense? We try to discuss the basic properties of the solution flow as a function of parameters. To address this problem, we need to introduce the system parameters, explicitly, in the mathematical description (6.1). For the sake of preciseness, we modify the dynamic system description (6.1) as

$$dx = f(t,x,\lambda)dt, \quad x(t_0) = x_0, \tag{6.7}$$

and the corresponding nominal system

$$dx = f(t,x,\lambda_0)dt, \quad x(t_0) = x_0, \tag{6.8}$$

where $(t_0, x_0, \lambda_0) \in R \times R^n \times \Lambda$, and the system parameter set Λ is open in R^m.

In the following, we present a very simple result that exhibits the continuous dependence of the solution process of (6.7) with respect to $(t_0, x_0, \lambda_0) \in R \times R^n \times \Lambda$.

Lemma 6.2.2. *Assume that $f \in C[J \times D \times \Lambda, R^n]$. Further assume that f satisfies the conditions*

$$\lim_{\lambda \to \lambda_0} f(t,x,\lambda) = f(t,x,\lambda_0) \tag{6.9}$$

uniformly in (t,x); and for $(t,x_1,\lambda), (t,x_2,\lambda) \in J \times D \times \Lambda$,

$$\|f(t,x_1,\lambda) - f(t,x_2,\lambda)\| \leq L\|x_1 - x_2\|. \tag{6.10}$$

Let $x(t,t_1,y_0,\lambda)$ and $x(t,t_0,x_0,\lambda_0)$ be the solution processes of (6.7) and (6.8) through (t_1,y_0) and (t_0,x_0), respectively. Then, given $\epsilon > 0$, there exists a $\delta(\epsilon) > 0$

such that

$$\|x(t, t_1, y_0, \lambda) - x(t, t_0, x_0, \lambda_0)\| < \epsilon \tag{6.11}$$

whenever

$$|t_1 - t_0| + \|y_0 - x_0\| + \|\lambda - \lambda_0\| < \delta(\epsilon).$$

Proof. Without the loss of generality we assume that $t_1 > t_0$. Let us define

$$\begin{aligned} m(t) &= \|x(t, t_1, y_0, \lambda) - x(t, t_0, x_0, \lambda_0)\|, \\ m(t_1) &= \|y_0 - x(t_1, t_0, x_0, \lambda_0)\|. \end{aligned} \tag{6.12}$$

We note that

$$\begin{aligned} D^+m(t) &\leq \|f(t, x(t, t_1, y_0, \lambda), \lambda) - f(t, x(t, t_0, x_0, \lambda_0), \lambda_0)\| \\ &\leq \|f(t, x(t, t_1, y_0, \lambda), \lambda) - f(t, x(t, t_0, x_0, \lambda_0), \lambda)\| \\ &\quad + \|f(t, x(t, t_0, x_0, \lambda_0), \lambda) - f(t, x(t, t_0, x_0, \lambda_0), \lambda_0)\|. \end{aligned}$$

This along with (6.10) and (6.12) yields

$$D^+m(t) \leq Lm(t) + p(t), \quad m(t_1) = \|y_0 - x(t_1, t_0, x_0, \lambda_0)\|, \tag{6.13}$$

where

$$p(t) = \|f(t, x(t, t_0, x_0, \lambda_0), \lambda) - f(t, x(t, t_0, x_0, \lambda_0), \lambda_0)\|.$$

By solving (6.13), we obtain

$$m(t) \leq m(t_1)\exp[L(t - t_1)] + \int_{t_1}^{t} p(s)\exp[L(t - s)], \quad t \geq t_1 > t_0. \tag{6.14}$$

From the definitions, the solution to (6.7) and p in (6.13), (6.9), (6.14), and noting that

$$m(t_1) = \|y_0 - x(t_1, t_0, x_0, \lambda_0)\| \leq \|y_0 - x_0\| + \|x_0 - x(t_1, t_0, x_0, \lambda_0)\|,$$

for $t_1, t \in [t_0, t_0 + a]$, and given $\epsilon > 0$, there exists a $\delta(\epsilon) > 0$, so that

$$\|x(t, t_1, y_0, \lambda) - x(t, t_0, x_0, \lambda_0)\| < \epsilon$$

whenever

$$|t_1 - t_0| + \|y_0 - x_0\| + \|\lambda - \lambda_0\| < \delta(\epsilon).$$

This completes the proof of the lemma. □

The following result provides the differentiability of the solution process of (6.1) with respect to the initial data.

Theorem 6.2.1. *Assume that the hypotheses of Lemma 6.2.1 are satisfied. Let $x(t, t_0, x_0)$ be the solution process of (6.1), existing for $t \geq t_0$, and let*

$$H(t, t_0, x_0) = \frac{\partial}{\partial x} f(t, x(t, t_0, x_0)). \tag{6.15}$$

Then

$$\text{(a)} \quad \Phi(t, t_0, x_0) = \frac{\partial}{\partial x} x(t, t_0, x_0) \tag{6.16}$$

exists, and is the solution to

$$dy = H(t, t_0, x_0) y \, dt \tag{6.17}$$

such that $\Phi(t_0, t_0, x_0)$ is the $n \times n$ identify matrix;

$$\text{(b)} \quad \frac{\partial}{\partial t_0} x(t, t_0, x_0)$$

exists, and is the solution to (6.17) and satisfies the relation

$$\frac{\partial}{\partial t_0} x(t, t_0, x_0) = -\Phi(t, t_0, x_0) f(t_0, x_0), \quad t \geq t_0. \tag{6.18}$$

Proof. First, we shall show that (a) holds. From the hypotheses of the theorem, and Lemma 6.2.1, it is obvious that $x(t) = x(t, t_0, x_0)$ is unique solution of (6.1). Moreover, by Lemma 6.2.2, the solution is continuous with respect to the initial data (t_0, x_0). For small λ in R, $x(t, \lambda) = x(t, t_0, x_0 + \lambda e_k)$ and $x(t) = x(t, t_0, x_0)$ are solution processes of (6.1) through $(t_0, x_0 + \lambda e_k)$ and (t_0, x_0), respectively, where $e_k = (0, 0, \ldots, 1, \ldots, 0)^T$ and 1 is the kth component of the n-dimensional unit vector in R^n. From Lemma 6.2.2, it clear that

$$\lim_{\lambda \to 0} x(t, \lambda) = x(t) \quad \text{uniformly on } J. \tag{6.19}$$

Set

$$\Delta x_\lambda(t) = \frac{x(t, \lambda) - x(t)}{\lambda}, \quad \Delta x_\lambda(t_0) = e_k, \quad \lambda \neq 0, \tag{6.20}$$

$x = x(t)$ and $x(t, \lambda)$. This together with the application of Lemma 6.2.1 yields

$$\Delta x'_\lambda(t) = \int_0^1 \left[\frac{\partial}{\partial x} f(t, \theta x(t) + (1 - \theta) x(t, \lambda)) \right] \Delta x_\lambda(t) \, d\theta. \tag{6.21}$$

Define

$$\frac{\partial}{\partial x} f(t, x(t), \lambda) = \int_0^1 \left[\frac{\partial}{\partial x} f(t, \theta x(t) + (1 - \theta) x(t, \lambda)) \right]. \tag{6.22}$$

We note that the integral is the Cauchy–Riemann integral. From the hypotheses of the theorem, $f_x(t, x, \lambda)$ is continuous in (t, x, λ). Furthermore, from (6.19), we have

$$\lim_{\lambda \to 0} \frac{\partial}{\partial x} f(t, \theta x(t, \lambda) + (1 - \theta)x(t)) = \frac{\partial}{\partial x} f(t, x(t)) \text{ uniformly in } \theta \in [0, 1]. \tag{6.23}$$

This along with (6.22) establishes

$$\lim_{\lambda \to 0} \frac{\partial}{\partial x} f(t, x(t), \lambda) = \frac{\partial}{\partial x} f(t, x(t)). \tag{6.24}$$

From (6.20) and (6.22), the relation (6.21) reduces to

$$d\Delta x_\lambda(t) = \frac{\partial}{\partial x} f(t, x(t), \lambda)\Delta x_\lambda(t)\, dt, \quad \Delta x_\lambda(t_0) = e_k. \tag{6.25}$$

In view of (6.23), the system (6.17) can be considered as the nominal system corresponding to (6.25) with initial data $y(t_0) = e_k$, i.e.

$$dy = H(t, t_0, x_0)y\, dt, \quad y(t_0) = e_k. \tag{6.26}$$

It is obvious that the IVP (6.25) satisfies all the hypotheses of Lemma 6.2.2, and hence by the application of Lemma 6.2.2, we have

$$\lim_{\lambda \to 0} \Delta x_\lambda(t) = y_k(t) \text{ uniformly on } J, \tag{6.27}$$

where $y_k(t)$ is the solution process of (6.26). Because of (6.20), the limit of $\Delta x_\lambda(t)$ in (6.27) is equivalent to the derivative

$$\frac{\partial}{\partial x_k} x(t, t_0, x_0),$$

which is the solution process of (6.26). From the uniqueness,

$$y_k(t) = \frac{\partial}{\partial x_k} x(t, t_0, x_0),$$

for each $k \in I(1, n)$. Thus,

$$\frac{\partial}{\partial x} x(t, t_0, x_0)$$

is the fundamental matrix solution process of (6.26) and it satisfies the matrix differential equation (6.17). Moreover,

$$\frac{\partial}{\partial x} x(t, t_0, x_0) \text{ is denoted by } \Phi(t, t_0, x_0).$$

This completes the proof of (a).

To prove the first part of (b), we define

$$\triangle\overline{x}_\lambda(t) = \frac{x(t,\lambda) - x(t)}{\lambda}, \quad \triangle\overline{x}_\lambda(t) = \frac{x(t_0,\lambda) - x_0}{\lambda}, \tag{6.28}$$

where $x(t,\lambda) = x(t,t_0+\lambda,x_0)$ and $x(t) = x(t,t_0,x_0)$ are the solution processes of (6.1) through $(t_0+\lambda, x_0)$ and (t_0, x_0), respectively. Again, by following the proof of part (a) and by replacing $\triangle x_\lambda(t)$ with $\triangle\overline{x}_\lambda(t)$, one can conclude that the derivative

$$\frac{\partial}{\partial t_0} x(t,t_0,x_0)$$

exists and is the solution to (6.17) through $(t_0, -f(t_0,x_0))$, provided that $\lim_{\lambda\to 0}\triangle\overline{x}_\lambda(t_0)$ exists and is equal to $-f(t_0,x_0)$. We need to prove that this statement is true. By using the uniqueness of the solution process of (6.1), we have

$$\begin{aligned}\triangle\overline{x}_\lambda(t_0) &= \frac{x(t_0,t_0+\lambda,x_0) - x(t_0+\lambda,t_0+\lambda,x_0)}{\lambda}\\ &= -\frac{x(t_0+\lambda,t_0+\lambda,x_0) - x(t_0,t_0+\lambda,x_0)}{\lambda}\\ &= -\frac{1}{\lambda}\int_{t_0}^{t_0+\lambda} f(s,x(s,t_0+\lambda,x_0))ds.\end{aligned} \tag{6.29}$$

From this and the uniform convergence theorem [120], it is implied that $\lim_{\lambda\to 0}\triangle\overline{x}_\lambda(t_0)$ exists and is equal to $-f(t_0,x_0)$. Now the rest of the proof follows from the uniqueness of the solution process of (6.1), part (a), and the application of Lemma 6.2.1 to $x(t,t_0,x_0)$, i.e.,

$$\begin{aligned}\triangle\overline{x}_\lambda(t) &= x(t,t_0+\lambda,x_0) - x(t,t_0,x_0)\\ &= x(t,t_0,x(t_0,t_0+\lambda,x_0)) - x(t,t_0,x_0)\\ &= \int_0^1 \left[\frac{\partial}{\partial x} x(t,\theta y_0 + (1-\theta)x_0)\right] \triangle x_\lambda(t_0)d\theta,\end{aligned} \tag{6.30}$$

where $y_0 = x(t_0,t_0+\lambda,x_0)$ and $x(t_0) = x_0$. From (6.29) and (6.30), we have

$$\begin{aligned}&\triangle\overline{x}_\lambda(t)\\ &\quad = -\int_0^1 \left[\frac{\partial}{\partial x} x(t,\theta y_0 + (1-\theta)x_0)\right] d\theta \left(\frac{1}{\lambda}\int_{t_0}^{t_0+\lambda} f(s,x(s,t_0+\lambda,x_0))ds\right).\end{aligned}$$

□

In the following, we present a simple example.

Example 6.2.2. Consider a scalar nonlinear autonomous differential equation:

$$dx = -\frac{1}{2}x^3 dt, \quad x(t_0) = x_0. \tag{6.31}$$

(a) Is the solution to the IVP differentiable? Find

$$\text{(b)}\quad \frac{\partial}{\partial x_0}x(t,t_0,x_0), \quad \text{(c)}\quad \frac{\partial}{\partial t_0}x(t,t_0,x_0).$$

Solution process. We note that here

$$f(t,x) = -\frac{1}{2}x^3$$

is continuously differential with respect to x. The closed-form solution to (6.30) is

$$x(t,t_0,x_0) = |x_0|\left[1 + x_0^2(t-t_0)\right]^{-\frac{1}{2}}. \tag{6.32}$$

It is clear that the nontrivial solution to (6.31) is continuously differentiable with respect to (t, t_0, x_0). Moreover, in this case, the expressions (6.16) and (6.18) reduce to

$$\frac{\partial}{\partial x_0}x(t,t_0,x_0) = \Phi(t,t_0,x_0) = \text{signum}(x_0)\left[1 + x_0^2(t-t_0)\right]^{-\frac{3}{2}}, \tag{6.33}$$

$$\frac{\partial}{\partial t_0}x(t,t_0,x_0) = -\Phi(t,t_0,x_0)f(t_0,x_0) = \frac{1}{2}|x|_0^3\left[1 + x_0^2(t-t_0)\right]^{-\frac{3}{2}}, \tag{6.34}$$

respectively.

Illustration 6.2.1 (population dynamics [55, 90, 94, 104, 116, 126, 129]). We discuss a simple mathematical model in population dynamics. This mathematical model is one of competing species. It is based on a number of simplified assumptions:

PDM 1. The density of a species, i.e. the number of individuals per unit area, can be adequately represented by a single variable. This variable ignores differences in age, sex, and genotype.

PDM 2. Crowding affects all population members, equally. This is unlikely to be true if the members of the species occur in clumps, rather than being evenly distributed throughout the available space.

PDM 3. The effects of interactions within and between species are instantaneous. This means that there is no delayed action on the dynamics of the population.

PDM 4. Abiotic environmental factors are sufficiently constant.

PDM 5. Population growth is density-dependent even at the lowest densities. It may be more reasonable to suppose that there is some threshold density below which individuals do not interfere with one another.

PDM 6. The females in a sexually reproducing population always find mates, even though the density may be low.

The assumptions relative to the density-dependent and crowding effects reflect the fact that the growth of any species in a restricted environment must eventually be limited by a shortage of resources.

Under the assumptions of PDM 1–PDM 6, a mathematical model of an n-species community model is described by

$$dN_i = N_i F_i(t, N)\, dt, \quad N_i(t_0) = N_{i0}, \tag{6.35}$$

where N_i is the population density of the ith species in the model ecosystem for $i \in I(1, n)$; for each $i \in I(1, n)$, $F_i \in C\,[J \times D, R^n]$, and it describes the per capita growth rate of the ith species. Moreover, the sign of

$$\frac{\partial}{\partial N_j} F_i$$

describes the interactions between the ith and jth species in the community. For example, competition $(-, -)$, commensalism $(+, +)$, and predator $(+,\ -)$, i.e. for $i \neq j$,

$$\frac{\partial}{\partial N_j} F_i, \quad \frac{\partial}{\partial N_i} F_j$$

are negative,

$$\frac{\partial}{\partial N_j} F_i, \quad \frac{\partial}{\partial N_i} F_j$$

are positive, and

$$\frac{\partial}{\partial N_j} F_i > 0, \quad \frac{\partial}{\partial N_i} F_j < 0,$$

respectively. For $i = j$,

$$\frac{\partial}{\partial N_j} F_i$$

describes the interspecific effects within the species. In particular,

$$F_i(t, N) = a_i(t) - b_{ii}(t) N_i - \sum_{j \neq 1}^{n} b_{ij}(t) N_j, \tag{6.36}$$

where, for $i,\ j \in I(1, n)$, a_i and b_{ij} can be considered as system parameters. It is obvious that $F_i(t, N)$ in (6.36) are continuous with respect to a_i and b_{ij} uniformly on a closed and bounded ball. It is also obvious that in the absence of cross-interactions between the species in the community, that the mathematical model of isolated members of the community obeys the Verhulst–Pearl logistic equation [55, 94, 104]:

$$dN_i = N_i(a_i - b_{ii} N_i) dt, \quad N_i(t_0) = N_{i0}, \tag{6.37}$$

for each $i \in I(1,n)$. In the competition model, b_{ii} describes the inhibiting effects of each species. There are several examples in literature that show the close correspondence between the natural and the laboratory growth of the population.

6.2 Exercises

(1) Verify that the theoretical curve associated with the solution to the differential equation (6.37) is

$$N_i(t, t_0, N_{i0}) = \frac{a_i}{b_{ii}} \left[1 + \frac{\frac{a_i}{b_{ii}} - N_{i0}}{N_{i0}} \exp\left[-a_i(t - t_0)\right] \right]^{-1}.$$

(2) Find

$$\frac{\partial}{\partial N_0} N_i(t, t_0, N_{i0}), \quad \frac{\partial}{\partial t_0} N_i(t, t_0, N_{i0})$$

(if possible).

(3) Solve the nonlinear differential equations

$$dx = -\frac{1}{2n} x^{2n+1} dt, \quad x(t_0) = x_0,$$

for $n \in I(1, \infty) = \{1, 2, 3, \dots\}$. Moreover, determine

$$\frac{\partial}{\partial x} x(t, t_0, x_0), \quad \frac{\partial}{\partial t_0} x(t, t_0, x_0).$$

(4) If possible, solve the differential equation

$$dx = -\frac{a(t)}{2n} x^{2n+1} dt, \quad x(t_0) = x_0,$$

where $a(t)$ is a continuous function and $n \in I(1, \infty)$. Moreover, find

$$\frac{\partial}{\partial x} x(t, t_0, x_0), \quad \frac{\partial}{\partial t_0} x(t, t_0, x_0).$$

(5) Let us consider the linear time-varying system of differential equations

$$dx = A(t)x\, dt, \quad x(t_0) = x_0,$$

where $A(t)$ is an $n \times n$ continuous matrix function. Find

$$\frac{\partial}{\partial x} x(t, t_0, x_0), \quad \frac{\partial}{\partial t_0} x(t, t_0, x_0).$$

(6) (a) Solve the following IVPs:
(i) $dx = 4x^{\frac{3}{4}} dt$, $x(0) = 0$, and (ii) $dx = 4x^{\frac{3}{4}} dt$, $x(1) = 1$.
(b) Is $x(t) \equiv 0$ a solution to (i)? Justify your answer.

(c) Is

$$x(t) = \begin{cases} 0, & \text{if } 0 \le t \le c, \\ (t-c)^4, & \text{if } c \le t \end{cases}$$

a solution to (i)? Justify your answer.

(d) Are the solutions to the IVPs in (i) and (ii) uniquely determined? Justify your answer.

6.3 Method of Variation of Parameters

In this section, one of the commonly used methods for finding information on the dynamic processes in sciences and engineering is the method of nonlinear variation of constants parameters [2]. Having knowledge of the existence of the solution process and utilizing this method, one can gain information about the behavior of the dynamic flow.

We also recall (Chapters 2, 4, and 5) that the method of variation of parameters has played a very powerful role in finding a solution representation of a differential equation. The idea is very simple. One decomposes a complex system of differential equations into two parts such that a differential equation corresponding to the simpler part is either easily solvable in a closed form or analytically analyzable, but that the original complex system of differential equations is neither easily solvable in a closed form nor analytically analyzable. The method of variations of parameters provides a formula for a solution to the original complex system (perturbed system) — generally, not easily solvable in a closed form — in terms of the solution process of the differential equation (unperturbed system) corresponding to the simpler part of the rate function. The following theorem is an extension of Lagrange's formula [42].

Theorem 6.3.1. *Let us consider a decomposition of the system of differential equations* (6.1). *Further, assume that* $f = F + R$, *and*

$$dx = F(t,x)\,dt + R(t,x)\,dt, \quad x(t_0) = x_0, \tag{6.38}$$

$$dy = F(t,y)\,dt, \quad y(t_0) = x_0, \tag{6.39}$$

where

(i) $F, R \in C\,[J \times R^n, R^n]$, *where* $y, x, x_0 \in R^n$ *and* $J = [t_0, t_0 + a]$;
(ii) *The IVP* (6.39) *has a unique solution* $y(t, t_0, x_0)$ *existing for* $t \ge t_0$;
(iii) $\frac{\partial}{\partial x_0} y(t, t_0, x_0) = \Phi(t, t_0, x_0)$ *exists and it is continuous in* (t_0, x_0) *for fixed* t;
(iv) *The inverse of* $\Phi(t, t_0, x_0)$ *exists and it is continuous in* (t, x_0) *for fixed* t_0.

Then, for any solution $x(t, t_0, x_0)$ to the IVP (6.38) satisfies the relation

$$x(t, t_0, x_0) = y(t, t_0, x_0 + \int_{t_0}^{t} \Phi^{-1}(s, t_0, z(s)) R(s, x(s)) ds), \tag{6.40}$$

where $z(t)$ is a solution to the IVP

$$dz = \Phi^{-1}(t, t_0, z) R(t, y(t, t_0, z)) dt, \quad z(t_0) = x_0. \tag{6.41}$$

Proof. From the hypotheses of the theorem, the IVPs (6.38) and (6.39) have the solution processes $x(t) = x(t, t_0, x_0)$ and $y(t) = y(t, t_0, x_0)$, respectively. The elementary method of variation of parameters (Sections 2.4, 4.6, and 5.5) requires the determination of a function $z(t)$ so that

$$x(t, t_0, x_0) = y(t, t_0, z(t)), \quad z(t_0) = x_0, \tag{6.42}$$

is a solution process of (6.38). From Theorem 6.2.1 and 1.5.9, we have

$$F(t, x(t)) dt + R(t, x(t)) dt = dy(t, t_0, z(t)) + \frac{\partial}{\partial x_0} y(t, t_0, z(t)) dz(t).$$

This, together with (iii), (iv), and (6.42) gives

$$dz(t) = \Phi^{-1}(t, t_0, z) R(t, x(t, t_0, x_0)) dt, \quad z(t_0) = x_0,$$

which implies that

$$z(t) = x_0 + \int_{t_0}^{t} \Phi^{-1}(s, t_0, z(s)) R(s, x(s)) ds. \tag{6.43}$$

From (6.42) and (6.43), we have

$$x(t, t_0, x_0) = y(t, t_0, x_0 + \int_{t_0}^{t} \Phi^{-1}(s, t_0, z(s)) R(s, x(s)) ds),$$

which proves the theorem. □

Another version of the variation-of-constants formula for a solution process of (6.38) is presented in the following corollary.

Corollary 6.3.1. *Under the hypotheses of Theorem 6.3.1, the following relation is also valid:*

$$x(t, t_0, x_0) = y(t, t_0, x_0) + \int_{t_0}^{t} \Phi(t, t_0, z(s)) \Phi^{-1}(s, t_0, z(s)) R(s, x(s)) ds, \tag{6.44}$$

where $z(t)$ is any solution to (6.41).

Proof. For $t_0 \leq s \leq t$, from (6.41), we obtain

$$\begin{aligned} dy(t,t_0,z(s)) &= \frac{\partial}{\partial x_0} y(t,t_0,z(s))dz(s) \\ &= \Phi(t,t_0,z(s))\Phi^{-1}(s,t_0,z(s))R(s,y(s,t_0,z(s))ds. \end{aligned} \tag{6.45}$$

From (6.45), we have

$$y(t,t_0,z(t)) = y(t,t_0,x_0) + \int_{t_0}^{t} \Phi(t,t_0,z(s))\Phi^{-1}(s,t_0,z(s))R(s,y(s,t_0,z(s))ds.$$

Together with (6.42) this yields

$$x(t,t_0,x_0) = y(t,t_0,x_0) + \int_{t_0}^{t} \Phi(t,t_0,z(s))\Phi^{-1}(s,t_0,z(s))R(s,x(s))ds,$$

establishing the relation (6.44). □

In the following, we present a variation-of-constants formula which is analogous to the known result due to Alekseev [2]. Moreover, we show that this obtained nonlinear variation-of-constants formula is equivalent to the formula in (6.44).

Theorem 6.3.2. *Assume that F and R satisfy the hypotheses of Theorems* 6.2.1 *and* 6.3.1, *respectively. Then, any solution $x(t) = x(t,t_0,x_0)$ to* (6.38) *satisfies the formulas* (6.40), (6.44), *and the formula*

$$x(t,t_0,x_0) = y(t,t_0,x_0) + \int_{t_0}^{t} \Phi(t,s,x(s))R(s,x(s))ds, \tag{6.46}$$

where $y(t,t_0,x_0)$ is the solution to (6.39). *Moreover,* (6.44) *and* (6.46) *are equivalent if*

$$\Phi(t,s,x(s)) = \Phi(t,t_0,z(s))\Phi^{-1}(s,t_0,z(s)). \tag{6.47}$$

Proof. From Theorem 6.2.1, $y(t) = y(t,t_0,x_0)$ is a unique solution to (6.39),

$$\frac{\partial}{\partial x_0} y(t,t_0,x_0) = \Phi(t,t_0,x_0),$$

exists and $\Phi(t,t_0,x_0)$ is the fundamental matrix solution to (6.17). Moreover, $\Phi^{-1}(t,t_0,x_0)$ exists and is continuous in (t,t_0,x_0). Thus, F, R, $\Phi(t,t_0,x_0)$ and $\Phi^{-1}(t,t_0,x_0)$ satisfy hypotheses (ii)–(iv) of Theorem 6.3.1. Therefore, by the application of Theorem 6.3.1, any solution $x(t) = x(t,t_0,x_0)$ to the IVP (6.1) satisfies the formulas (6.40) and (6.44). To show $x(t,t_0,x_0)$ satisfies (6.46), for $t_0 \leq s \leq t$, we consider $x(s) = x(s, t_0, x_0)$ and $y(t, s, x(s))$ as the solution processes of (6.38) and (6.39) through (t_0, x_0) and $(s, x(s))$, respectively. Then

$$dy(t, s, x(s)) = \frac{\partial}{\partial x_0} y(t, s, x(s))dx(s) + \frac{\partial}{\partial t_0} y(t, s, x(s))ds.$$

Along with (6.17), (6.18), and (6.38), this yields

$$dy(t, s, x(s)) = \Phi(t, s, x(s))R(s, x(s))ds. \tag{6.48}$$

Upon integration from t_0 to t, (6.48) reduces to

$$y(t, t, x(t)) = y(t, t_0, x_0) + \int_{t_0}^{t} \Phi(t, s, x(s))R(s, x(s))ds. \tag{6.49}$$

By the uniqueness of the solution process of (6.39), we have

$$x(t, t_0, x_0) = y(t, t_0, x_0) + \int_{t_0}^{t} \Phi(t, s, x(s))R(s, x(s))ds.$$

This proves the relation (6.46).

We next show that the formulas (6.44) and (6.46) are equivalent if and only if

$$\Phi(t, s, x(s)) = \Phi(t, t_0, z(s))\Phi^{-1}(s, t_0, z(s)).$$

For $t_0 \leq s \leq t$, from (6.42) and the uniqueness of solution to (6.39), we arrive at

$$y(t, s, x(s)) = y(t, s, y(s, t_0, z(s))) = y(t, t_0, z(s)), \tag{6.50}$$

where $y(t, s, x(s))$ and $x(s) = y(s, t_0, z(s))$ are as defined above, and $z(s)$ is a solution process of (6.41) through (t_0, x_0). Under these considerations, by differentiating with respect to s, (6.50), after substitution of the expressions of $dx(s)$ and $dz(s)$, we get

$$\Phi(t, s, x(s))R(s, x(s)) = \Phi(t, t_0, z(s))\Phi^{-1}(s, t_0, z(s))R(s, x(s)),$$

which is equivalent to

$$\Phi(t, s, x(s)) = \Phi(t, t_0, z(s))\Phi^{-1}(s, t_0, z(s)).$$

From this, (6.45), and (6.48), (6.44) and (6.46) are equivalent if and only if

$$\Phi(t, s, x(s))R(s, x(s)) = \Phi(t, t_0, z(s))\Phi^{-1}(s, t_0, z(s)).$$

This completes the proof of the theorem. □

In the following, we provide a couple of examples to illustrate the presented results.

Example 6.3.1. We consider the scalar nonlinear autonomous differential equation (6.31) in Example 6.2.2 as an unperturbed differential equation,

$$dy = -\frac{1}{2}y^3, \quad y(t_0) = x_0, \tag{6.51}$$

and its corresponding perturbed differential equation is

$$dx = -\frac{1}{2}x^3 + R(t,x), \quad x(t_0) = x_0. \tag{6.52}$$

By employing

$$\frac{\partial}{\partial t_0}y(t,t_0,x_0) = -\Phi(t,t_0,x_0)F(t_0,x_0) = \frac{1}{2}|x|_0^3\left[1 + x_0^2(t-t_0)\right]^{-\frac{3}{2}},$$

$\Phi(t,t_0,x_0) = \text{sgn}(x_0)\left[1 + x_0^2(t-t_0)\right]^{-\frac{3}{2}}$, and imitating the proof of Theorem 6.3.2, we arrive at

$$\begin{aligned} x(t) &= y(t) + \int_{t_0}^{t} \Phi(t,s,x(s))R(s,x(s))ds \\ &= |x_0|\left[1 + x_0^2(t-t_0)\right]^{-\frac{1}{2}} + \int_{t_0}^{t}\left[1 + x^2(s)(t-s)\right]^{-\frac{3}{2}}\text{sign}(x(s))R(s,x(s))ds. \end{aligned} \tag{6.53}$$

Example 6.3.2. Again, we consider a simple scalar differential equation,

$$dy = \frac{y}{1+t}dt, \quad y(t_0) = x_0, \tag{6.54}$$

and its corresponding perturbed equation is

$$dx = \left[\frac{x}{1+t} + R(t,x)\right]dt, \quad x(t_0) = x_0. \tag{6.55}$$

By employing

$$\begin{aligned} \frac{\partial}{\partial t_0}y(t,t_0,x_0) &= -\Phi(t,t_0,x_0)F(t_0,x_0) = -\frac{1+t}{(1+t_0)^2}x_0, \\ \Phi(t,t_0,x_0) &= \frac{1+t}{1+t_0}, \end{aligned}$$

and imitating the proof of Theorem 6.3.2, we arrive at

$$\begin{aligned} x(t) &= y(t) + \int_{t_0}^{t} \Phi(t,s,x(s))R(s,x(s))ds \\ &= \frac{1+t}{1+t_0}x_0 + \int_{t_0}^{t}\frac{1+t}{(1+s)^2}x(s)R(s,x(s))ds. \end{aligned} \tag{6.56}$$

6.3 Exercises

(1) By treating the system (6.35) in the context of (6.36) as a perturbed system of (6.37), find the variation-of-constants formula (6.46).

(2) Let us consider the following nonlinear differential equations:

$$dx = -\frac{1}{2n}x^{2n+1}dt + R(t,x)dt, \quad x(t_0) = x_0,$$

$$dy = -\frac{1}{2n}y^{2n+1}dt, \quad y(t_0) = x_0,$$

where $R \in C[J \times R, R]$ and $n \in I(1,\infty)$. Find the variation-of-constants formula (6.46) for a perturbed scalar differential equation with a corresponding unperturbed scalar differential equation.

(3) We are given the following nonlinear differential equations:

$$dx = -\frac{a(t)}{2n}x^{2n+1}dt + R(t,x)dt, \quad x(t_0) = x_0,$$

$$dy = -\frac{a(t)}{2n}y^{2n+1}dt, \quad y(t_0) = x_0,$$

where $R \in C[J \times R, R]$, $a \in C[J, R]$ is a continuous function, and $n \in I(1,\infty)$. Find the variation-of-constants formula (6.46) for a perturbed scalar differential equation with a corresponding unperturbed scalar differential equation.

(4) Let us consider the linear time-varying system of differential equations

$$dx = A(t)x\,dt + R(t,x)dt, \quad x(t_0) = x_0,$$
$$dy = A(t)y\,dt, \quad y(t_0) = x_0,$$

where $A(t)$ is an $n \times n$ continuous matrix function and $R \in C[J \times R^n, R^n]$. Find the variation-of-constants formula (6.46) for the perturbed system of differential equations with the corresponding unperturbed system of differential equations.

6.4 Generalized Method of Variation of Parameters

By following the classical modeling approach, an evolution of the state of a dynamic system can be considered to be a dynamic flow described by (6.38). The goal is to find the information (qualitative/quantitative) on a dynamic flow. In general, a closed-form representation of a flow [a solution process (6.38)] of a nonlinear and nonstationary dynamic system is not possible. Historically, a well-known technique is to measure a dynamic flow by means of a suitable auxiliary measurement device, and then to use this measurement to determine the desired information about the original dynamic solution flow. This idea also overcomes the complexity caused by the high dimensionality of the multivariate processes. In the following, an energy/Lyapunov-like function is used as a measurement device. This provides a relationship between a solution process of the complex system in terms of the solution process of the corresponding simpler model of the dynamic system. This result is called the generalized variation-of-constants formula [56].

Theorem 6.4.1. *Assume that all the hypotheses of Theorem 6.3.2 are satisfied. Further assume that* $V \in C\left[R_+ \times R^n, R^N\right]$, *and its partial derivatives,* V_t *and* $\frac{\partial}{\partial x}V$, *exist and are continuous on* $R_+ \times R^n$. *Let* $x(t)$ *and* $y(t)$ *be the solution processes of* (6.38) *and* (6.39), *respectively. Then*

$$V(t, x(t)) = V(t, y(t)) + \int_{t_0}^{t} LV(s, x(s))\, ds, \tag{6.57}$$

where

$$\begin{aligned} LV(s, x(s)) &= V_t(s, y(t, s, x(s))) \\ &\quad + \frac{\partial}{\partial x} V(s, y(t, s, x(s)))\Phi(t, s, x(s))R(s, x(s)). \end{aligned} \tag{6.58}$$

Proof. Let $x(t)$ and $y(t)$ be the solution processes of (6.38) and (6.39) through (t_0, x_0), respectively. For $t_0 \le s \le t$, we consider $V(s, y(t, s, x(s)))$, differentiate with respect to s for fixed t, and obtain

$$\begin{aligned} dV(s, y(t, s, x(s))) &= V_t(s, y(t, s, x(s)))ds + \frac{\partial}{\partial x} V(s, y(t, s, x(s))) \\ &\quad \times \left[\frac{\partial}{\partial x_0} y(t, s, x(s))dx(s) + \frac{\partial}{\partial t_0} y(t, s, x(s))ds\right]. \end{aligned}$$

This together with (6.17) and (6.18) gives us

$$\begin{aligned} dV(s, y(t, s, x(s))) &= V_t(s, y(t, s, x(s)))ds \\ &\quad + \frac{\partial}{\partial x} V(s, y(t, s, x(s)))\Phi(t, s, x(s))R(s, x(s))ds. \end{aligned} \tag{6.59}$$

Now, using (6.58) and integrating both sides of (6.59) with respect to s from t_0 to t, we obtain

$$V(t, y(t, t, x(t))) - V(t, y(t)) = \int_{t_0}^{t} LV(s, x(s))ds. \tag{6.60}$$

From (6.60) and the uniqueness of the solution to (6.39), the validity of (6.57) follows, immediately. This completes the proof of the theorem. □

***Remark* 6.4.1.** We note that if $V(t, x) = x$, then the generalized variation-of-constants formula (6.57) reduces to the well-known result (6.46) [2].

Example 6.4.1. If

$$V(t, x) = \frac{1}{2}\|x\|^2,$$

then $V_t(t, x) \equiv 0$ and

$$\frac{\partial}{\partial x} V(t, x) = x^T.$$

In this case, (6.57) reduces to

$$\|x(t)\|^2 = \|y(t)\|^2 + \int_{t_0}^{t} y^T(t, s, x(s))\Phi(t, s, x(s))R(s, x(s)). \tag{6.61}$$

The following result provides a deviation of the solution process of the perturbed system (6.38) with respect to the solution to the unperturbed system (6.39).

Theorem 6.4.2. *Assume that all the hypotheses of Theorem* 6.4.1 *hold. Then*

$$V(t, x(t) - y(t)) - V(t, 0) = \int_{t_0}^{t} LV(s, x(s) - y(s))ds. \tag{6.62}$$

Proof. By following the proof of Theorem 6.4.1, we have the relation

$$\begin{aligned} dV(s, y(t, s, x(s)) - y(t, s, y(s))) &= V_t(s, y(t, s, x(s)) - y(t, s, y(s)))ds \\ &\quad + \frac{\partial}{\partial x} V(s, y(t, s, x(s)) - y(t, s, y(s))) \\ &\quad \times \Phi(t, s, x(s))R(s, x(s))ds. \end{aligned}$$

After integration, this relation yields (6.62).

In the following, we present a result that establishes a relationship between the solution processes of (6.39) with the fundamental solution process of the corresponding linear system of differential equations (referred to as a variational system [75, 83, 84]):

$$dz = H(t, t_0, x_0)z\,dt, \tag{6.63}$$

where

$$H(t, t_0, x_0) = \frac{\partial}{\partial x} F(t, y(t, t_0, x_0)), \tag{6.64}$$

with $y(t, t_0, x_0)$ being the solution process of (6.39) through (t_0, x_0). □

Lemma 6.4.1. *Assume that F in* (6.39) *satisfies the hypotheses of Theorem* 6.2.1. *Then the fundamental solution*

$$\Phi(t, t_0, x_0) = \frac{\partial}{\partial x_0} y(t, t_0, x_0)$$

to (6.63), *and the processes $y(t, t_0, x_0)$ and $y(t, t_0, y_0)$ of* (6.39) *through (t_0, x_0) and (t_0, y_0), respectively, satisfy the following relation:*

$$y(t, t_0, x_0) = y(t, t_0, y_0) + \int_0^1 \Phi(t, t_0, \theta x_0 + (1 - \theta)y_0)(x_0 - y_0)d\theta. \tag{6.65}$$

Proof. We set

$$\Psi(\theta) = y(t, t_0, \theta x_0 + (1 - \theta)y_0), \quad 0 \le \theta \le 1. \tag{6.66}$$

It is obvious that Ψ is continuously differentiable on $[0, 1]$, and hence

$$d\Psi(\theta) = \frac{\partial}{\partial x_0} y(t, t_0, \theta x_0 + (1 - \theta)y_0)(x_0 - y_0)d\theta. \tag{6.67}$$

By integrating both sides of (6.67) with respect to θ over an interval $[0, 1]$, we obtain

$$\Psi(1) - \Psi(0) = \int_0^1 \Phi(t, t_0, \theta x_0 + (1 - \theta)y_0)(x_0 - y_0)d\theta. \tag{6.68}$$

This together with the fact that $\Psi(1) = y(t, t_0, x_0)$ and $\Psi(0) = y(t, t_0, y_0)$ yields the desired relation (6.65). □

***Remark* 6.4.2.** Assume that $F(t, 0) \equiv 0$ on J. Then, under the assumption of Theorem 6.2.1, (6.39) has a unique trivial solution, $y(t) \equiv 0$, through $(t_0, 0)$. In view of this, the relation (6.65) reduces to

$$y(t, t_0, x_0) = \int_0^1 \Phi(t, t_0, \theta x_0) x_0 \, d\theta.$$

$$= \left(\int_0^1 \frac{\partial}{\partial x_{0j}} y_i(t, t_0, \theta x_0) d\theta \right)_{n \times n} x_0 \qquad (6.3.13)$$

Moreover,

$$\begin{aligned}
\Phi(t, t_0, x_0) &= \int_0^1 \frac{\partial}{\partial x_0} \Phi(t, t_0, \theta x_0) \cdot x_0 d\theta x_0 + \int_0^1 \Phi(t, t_0, \theta x_0) x_0 d\theta \\
&= \int_0^1 \frac{\partial}{\partial x_0} \Phi(t, t_0, x_0) \cdot x_0 d\theta x_0 + \int_0^1 \Phi(t, t_0, \theta x_0) x_0 d\theta \\
&= \left(\int_0^1 \frac{\partial^2}{\partial x_0 \partial x_{0j}} y_i(t, t_0, \theta x_0) x_0 d\theta \right)_{n \times n} x_0 + \int_0^1 \Phi(t, t_0, \theta x_0) x_0 d\theta,
\end{aligned}$$

provided that $\frac{\partial}{\partial x_0} \Phi(t, t_0, x_0)$ exists.

6.4 Exercises

(1) By treating the system (6.35) in the context of (6.36) as a perturbed system of (6.37), find the generalized variation-of-constants formula (6.57) in the context of a quadratic type of energy function of the type

$$V(t, N) = (N_1^2, N_2^2, \ldots, N_i^2, \ldots, N_n^2)^T.$$

(2) Given the nonlinear differential equations

$$dx = \left[-\frac{1}{2n} x^{2n+1} + R(t, x) \right] dt, \quad x(t_0) = x_0,$$

$$dy = -\frac{1}{2n} y^{2n+1} dt, \quad y(t_0) = x_0,$$

where $R \in C[J \times R, R]$ and $n \in I(1, \infty)$, determine the generalized variation-of-constants formula (6.57) with respect to the energy functions (a) $V(t, x) = x^{2n}$ and (b) $V(t, x) = x^{2m}$ for $m \in I(2, \infty)$.

(3) Let us consider the following nonlinear differential equations:

$$dx = \left[-\frac{a(t)}{2n} x^{2n+1} + R(t, x) \right] dt, \quad x(t_0) = x_0,$$

$$dy = -\frac{a(t)}{2n} y^{2n+1} dt, \quad y(t_0) = x_0,$$

where $R \in C[J \times R, R]$, $a \in C[J, R]$ is a continuous function, and $n \in I(1, \infty)$. Determine the generalized variation-of-constants formula (6.57) with respect to the energy functions (a) $V(t,x) = x^{2n}$ and (b) $V(t,x) = x^{2m}$ for $m \in I(2, \infty)$.

(4) Let us consider the following system of linear time-varying differential equations:

$$dx = [A(t)x + R(t,x)]\,dt, \quad x(t_0) = x_0,$$
$$dy = A(t)y\,dt, \quad y(t_0) = x_0,$$

where $A(t)$ is an $n \times n$ continuous matrix function and $R \in C[J \times R^n, R^n]$. Determine the generalized variation-of-constants formula (6.57) with respect to the energy functions (a) $V(t,x) = \|x\|^2$ and (b) $V(t,x) = \|x\|^{2m}$ for $m \in I(2, \infty)$.

6.5 Differential Inequalities and Comparison Theorems

In general, a closed/exact-form representation of a time evolution flow described by a nonlinear/linear, nonstationary interconnected system is not always feasible. However, having the existence of a dynamic flow and using features of the mathematical description (6.1), the time evolution or a functional of a flow, a lower-order system or scalar dynamic differential inequalities, is highly feasible. Solving these differential inequalities, an estimate on either the time evolution or a functional of a flow, is determined by the corresponding measured dynamic flow. In particular, it is well known [75, 83, 84, 126] that an arbitrary measured dynamic flow satisfying differential inequalities is estimated by the extremal solution to the corresponding comparison system of differential equations. This technique is called the comparison method [126]. In this section, the above-stated technique is presented in its simplified form. For this purpose, let m be an arbitrary measured dynamic flow.

Prior to the presentation of the basic differential inequality results and knowing the existence of more than one solution to a given differential equation (Example 6.2.1), we introduce the concepts of maximal and minimal solutions to the following IVP.

For simplicity, we consider the simpler differential equation corresponding to a functional of flow. It is described by a scalar differential equation with an initial condition:

$$du = g(t,u)\,dt, \quad u(t_0) = u_0, \tag{6.70}$$

where $g \in C[E, R]$; with E being an open (t,u) set in R^2. This type of differential equation is called a comparison differential equation.

Definition 6.5.1 ([75, 83, 126]). A solution $r(t)$ to the scalar differential equation (6.70) on $[t_0, t_0 + a)$ is said to be the maximal solution to (6.70) if every solution $u(t)$ to (6.70) existing on $[t_0, t_0 + a)$ satisfies the inequality

$$u(t) \le r(t), \quad \text{for } t \in [t_0, t_0 + a). \tag{6.71}$$

A solution $\rho(t)$ to the scalar differential equation (6.70) on $[t_0, t_0 + a)$ is said to be the minimal solution to (6.70) if $\rho(t)$ and any solution $u(t)$ to (6.70) satisfy the above inequality in reverse order, i.e.

$$\rho(t) \le u(t), \quad \text{for } t \in [t_0, t_0 + a). \tag{6.72}$$

Example 6.5.1. We consider the following very simple IVP:

$$du = 2u^{\frac{1}{2}}\,dt, \quad u(0) = 0.$$

Solution procedure. From Example 6.2.1, we remark that the solution flow of this IVP is

$$u(t) = \begin{cases} 0, & \text{if } 0 \le t \le c, \\ (t-c)^2, & \text{if } c \le t. \end{cases}$$

We further note that the maximal and minimal solutions processes are $r(t) = t^2$ and $\rho(t) = 0$, respectively, for $t \ge 0$.

The concepts of the maximal and minimal solutions to differential equations are the backbone for the study of nonlinear nonstationary dynamic systems. Therefore, the existence of extremal solutions is a very important mathematical problem [75, 83, 84]. However, we do not undertake this task here. In our further discussion, we will be considering very simple comparison differential equations whose extremum solutions are easily guaranteed.

Theorem 6.5.1 (comparison theorem [83, 84]). *Assume that*

($H_{6.5}$): (a) *$g \in C[E, R]$, where $E = [t_0, t_0 + a) \times D$, D open in $R \times R$;*

(b) *$r(t)$ is the maximal solution to (6.70) existing on $[t_0, t_0 + a)$;*

(c) *$m \in C[[t_0, t_0 + a), R]$, $(t, m(t)) \in E$, m is differentiable and*

$$dm(t) \le g(t, m(t))dt, \quad \textit{for } t \in [t_0, t_0 + a), \tag{6.73}$$

and

$$m(t_0) \le u_0, \tag{6.74}$$

where $\Delta t = dt(\Delta t > 0)$. Then

$$m(t) \le r(t), \quad \textit{for } t \in [t_0, t_0 + a). \tag{6.75}$$

Proof. The proof is left to the reader. □

Corollary 6.5.1 (Comparison Theorem [83, 84]). *Assume that all hypotheses of Theorem 6.5.1 are satisfied, except the maximal solution $r(t)$ is replaced by the minimal solution $\rho(t)$ and the inequalities (6.73)–(6.74) and are reversed. Then,*

$$\rho(t) \le m(t), \quad \textit{for } t \in [t_0, t_0 + a). \tag{6.76}$$

In the following, we present a few very useful integral inequalities that are reducible to differential inequalities.

Lemma 6.5.1 ([7, 23, 42, 83, 84]). *Let $\phi, \lambda \in C[[t_0, t_0 + a), R,]$, where λ is a nonnative function defined on R. Assume that for some nonnegative constant c, we have*

$$\phi(t) \le c + \int_{t_0}^{t} \lambda(s)\phi(s)\, ds, \quad t \in [t_0, t_0 + a). \tag{6.77}$$

Then

$$dm(t) \le \lambda(t)m(t)\, dt, \text{ for positive differential } dt, \tag{6.78}$$

$$m(t_0) = c, \tag{6.79}$$

$$du = \lambda(t)u\, dt, \quad u(t_0) = c = u_0, \tag{6.80}$$

$$\phi(t) \le c \exp\left[\int_{t_0}^{t} \lambda(s)\, ds\right]. \tag{6.81}$$

Proof. We define

$$m(t) = c + \int_{t_0}^{t} \lambda(s)\phi(s)\, ds, \quad \text{for } t \in [t_0, t_0 + a).$$

We note that

$$m(t_0) = c,\ dm(t) = \lambda(t)\phi(t)dt, \quad \text{and} \quad \phi(t) \le m(t). \tag{6.82}$$

From this, (6.77) and the nonnegativity of λ and the differential dt, we have

$$dm(t) = \lambda(t)\phi(t) \le \lambda(t)m(t)\, dt.$$

This establishes (6.78) and also (6.80). This comparison differential equation has a unique solution,

$$u(t) = \exp\left[\exp \int_{t_0}^{t} \lambda(s)\, ds\right] u_0,$$

through (t_0, u_0) (Section 2.4). Therefore, its maximal solution is

$$r(t) = \left[\exp \int_{t_0}^{t} \lambda(s)\, ds\right] u_0.$$

From this discussion, it is clear that all the hypotheses of Theorem 6.5.1 are satisfied. Thus, (6.81) is established, in view of the fact that $\phi(t) \le m(t)$ and $m(t) \le r(t)$ with $u_0 = c$. This completes the proof of the lemma. □

Lemma 6.5.2. *Let $\phi, \mu \in C[[t_0, t_0+a), R]$, and μ be a nonnegative function defined on R. Assume that for some nonnegative constant c, we have*

$$\phi(t) \ge c + \int_{t_0}^{t} \mu(s)\phi(s)\, ds, \quad t \in [t_0, t_0 + a). \tag{6.83}$$

Then, for positive differential dt,

$$dm(t) \geq \mu(t)m(t)\,dt, \tag{6.84}$$

$$m(t_0) = c, \tag{6.85}$$

$$du = \mu(t)u\,dt, \quad u(t_0) = c = u_0, \tag{6.86}$$

$$\phi(t) \geq c\exp\left[\int_{t_0}^{t} \mu(s)\,ds\right]. \tag{6.87}$$

Proof. The proof of this lemma can be established by imitating the proof of Lemma 6.5.1. The details are left to the reader as an exercise. □

***Remark* 6.5.1.** Under the hypotheses of Lemmas 6.5.1 and 6.5.2, we have

$$\left[\exp\int_{t_0}^{t} \mu(s)\,ds\right] u_0 \leq \phi(t) \leq \exp\left[\int_{t_0}^{t} \lambda(s)\,ds\right] u_0. \tag{6.88}$$

6.5 Exercises

(1) Let $E = (t_0, t_0 + a) \times (c, d) \subseteq R^2$ and $g \in C[E, R]$. Assume that α and β are continuously differentiable functions with $(t, \alpha(t))$, $(t, \beta(t)) \in E$. In addition, assume that for $t \in (t_0, t_0 + a)$, α and β satisfy the following differential inequalities for (positive differential of t, dt):

$$d\alpha(t) < g(t, \alpha(t))dt,$$
$$d\beta(t) \geq g(t, \beta(t))dt,$$
$$\alpha(t_0) < \beta(t_0).$$

Then,

$$\alpha(t) < \beta(t), \quad t \in (t_0, t_0 + a).$$

(2) Prove the conclusions of Lemma 6.5.2.

(3) Let $\phi, \psi, \lambda \in C[[t_0, t_0 + a), R]$, where λ a is nonnegative function defined on R. Further, assume that ϕ, ψ, and λ satisfy the following integral inequality:

$$\phi(t) \leq \psi(t) + \int_{t_0}^{t} \lambda(s)\phi(s)\,ds, \quad t \in [t_0, t_0 + a).$$

(a) Show that

$$\phi(t) \leq \psi(t) + \int_{t_0}^{t} \lambda(s)\psi(s)\exp\left[\int_{s}^{t} \lambda(\theta)\,d\theta\right] ds.$$

(b) If ψ is continuously differentiable on $[t_0, t_0 + a)$, then the relation in (a) reduces to

$$\phi(t) \leq \psi(a)\exp\left[\int_{t_0}^{t} \lambda(\theta)\,d\theta\right] + \int_{t_0}^{t} \psi'(s)\exp\left[\int_{s}^{t} \lambda(\theta)\,d\theta\right] ds.$$

(4) Let $\phi, \psi, \lambda \in C[[t_0, t_0 + a), R]$, where λ is a nonnegative function on R. Further, assume that ϕ, ψ, and λ satisfy the following integral inequality:

$$\phi(t) \geq \varphi(t) + \int_{t_0}^{t} \mu(s)\phi(s)\, ds, \quad t \in [t_0, t_0 + a).$$

(a) Show that

$$\phi(t) \geq \varphi(t) + \int_{t_0}^{t} \mu(s)\varphi(s) \exp\left[\int_{s}^{t} \mu(\theta)\, d\theta\right] ds.$$

(b) If φ is continuously differentiable on $[t_0, t_0 + a)$, then the relation (a) reduces to

$$\phi(t) \geq \varphi(a) \exp\left[\int_{t_0}^{t} \mu(\theta) d\theta\right] + \int_{t_0}^{t} \varphi'(s) \exp\left[\int_{s}^{t} \mu(\theta) d\theta\right] ds.$$

(5) Let $\phi \in C[[t_0, t_0 + a), R]$, and let

$$\phi(t) \leq c + \gamma \int_{t_0}^{t} \phi(s)\, ds,$$

$t \in [t_0, t_0 + a)$. Then:

(a) For $c = 1$, show that $\phi(t) \leq \gamma \exp[\gamma(t - t_0)]$, $t \in [t_0, t_0 + a)$;
(b) For $c = 0$, show that $\phi(t) \leq 0, t \in [t_0, t_0 + a)$;
(c) If $0 \leq \phi(t)$ on $[t_0, t_0 + a)$, what can you say about (a) and (b)?

6.6 Energy/Lyapunov Function and Comparison Theorems

It is well known that the energy/Lyapunov's second method [75, 83, 84, 126] has played a very significant role in the qualitative and quantitative analysis of nonlinear nonstationary systems of dynamic systems in the biological, engineering, physical, and social sciences. The concept of the Lyapunov/energy function is considered to be a suitable auxiliary measurement device for measuring a dynamic flow determined by (6.1).

Let us consider a Lyapunov-like/energy function (Section 3.3) that satisfies $(H_{6.6.1})$:

(a) Let $V : R_+ \times R^n \to R$ be a function that is continuous on $R_+ \times R^n$;
(b) $V(t, x)$ is locally Lipschitzian in x for each $t \in R_+$.
We define $d^+V(t, x)$ as follows:

$$d^+_{(6.1)}V(t, x) = \limsup[V(t + \Delta t, x + \Delta t f(t, x)) - V(t, x)]dt, \tag{6.89}$$

for $(t, x) \in R_+ \times R^n$. We denote $d^+V(t, x)$ by $d^+_{(6.1)}V(t, x)$.

We present a very basic comparison theorem in the framework of the Lyapunov-like function.

Theorem 6.6.1. *Let $V \in C[R_+ \times R^n, R_+]$ and satisfy hypothesis $(H_{6.6.1})$. Assume that $(H_{6.6.2})$:*

(a) *The function $d^+V(t,x)$ defined in (6.89) satisfies the following inequality:*

$$d^+V(t,x) \le g(t, V(t,x))dt, \quad (t,x) \in [R_+ \times R^n], \tag{6.90}$$

where $g \in C\left[J \times R_+, R\right]$, and $\triangle t = dt$ is a positive differential of t;

(b) *Let $r(t) = r(t, t_0, u_0)$ be the maximal solution of the scalar differential equation*

$$du = g(t,u)dt, \quad u(t_0) = u_0, \tag{6.91}$$

existing to the right of t_0;

(c) *$x(t) = x(t, t_0, x_0)$ is any solution to (6.1) existing for $t \ge t_0$ such that*

$$V(t_0, x_0) \le u_0. \tag{6.92}$$

Then

$$V(t, x(t)) \le r(t), \quad t \ge t_0. \tag{6.93}$$

Proof. We set

$$m(t) = V(t, x(t)), \quad m(t_0) = V(t_0, x_0). \tag{6.94}$$

From the continuity of V, for sufficiently small $\triangle t > 0$, we have

$$\begin{aligned} m(t+\triangle t) - m(t) &= V(t+\triangle t, x(t+\triangle t)) - V(t, x(t)) \\ &= V(t+\triangle t, x(t) + \triangle t f(t, x(t))) - V(t, x(t)) \\ &\quad + V(t+\triangle t, x(t+\triangle t)) - V(t+\triangle t), x(t) + \triangle t f(t, x(t)). \end{aligned}$$

Now, using the Lipschitzian property of V together with x being any solution to (6.1), we have

$$\begin{aligned} m(t+\triangle t) - m(t) &\le L\|x(t+\triangle t) - x(t) - \triangle t f(t, x(t))\| \\ &+ V(t+\triangle t), x(t) + \triangle t f(t, x(t)) - V(t, x(t)), \end{aligned}$$

where L is the local Lipschitz constant relative to $V(t, x(t))$. Together with (6.89), for $\triangle t > 0$, this yields the inequality

$$d^+m(t) \le d^+V(t, x(t)). \tag{6.95}$$

From (6.90), (6.92), (6.94), and (6.95), we arrive at

$$\begin{aligned} &d^+m(t) \le g(t, m(t))dt, \quad t \ge t_0, \\ &m(t_0) \le u_0. \end{aligned} \tag{6.96}$$

From (6.95), (6.96), and under the assumption of the theorem, all the assumptions of Theorem 6.5.1 are satisfied. Hence, by the application of Theorem 6.5.1, we have

$$m(t) \le r(t), \quad t \ge t_0.$$

Thus,

$$V(t, x(t)) \leq r(t), \quad t \geq t_0.$$

This completes the proof of the theorem. □

Several special cases of Theorem 6.6.1 in the literature are exhibited by the following corollary. The proof is left as an exercise.

Corollary 6.6.1.

(a) *For* $g(t,u) \equiv 0$. *The inequality* (6.90) *reduces to* $dV(t,x) \leq 0$. *Then, the function* $V(t,x(t))$ *is nonincreasing in* t, *and the relation* (6.93) *reduces to*

$$V(t, x(t)) \leq V(t_0, x_0), \quad t \geq t_0.$$

(b) *For* $g(t,u) \equiv \lambda(t)u$. *The inequality* (6.90) *reduces to*

$$d^+V(t,x) \leq \lambda(t)V(t,x)dt.$$

Then, the relation (6.93) *becomes*

$$V(t, x(t)) \leq V(t_0, x_0) \left[\exp \int_{t_0}^{t} \lambda(s)\, ds\right], \quad t \geq t_0.$$

The following variant of Theorem 6.6.1 is more useful in several applications [75, 83, 84].

Theorem 6.6.2. *Assume that the hypotheses of Theorem* 6.6.1 *hold except that the inequality* (6.90) *is replaced by*

$$[A(t)d^+V(t,x) + V(t,x)d^+A(t)] \leq g(t, A(t)V(t,x))dt, \tag{6.97}$$

$(t,x) \in [R_+ \times R^n]$, *where* $A \in C[J,R]$, $A(t) > 0$ *on* J, *and*

$$d^+A(t) = \limsup[A(t+\Delta t) - A(t)]dt.$$

Then

$$A(t)V(t, x(t)) \leq r(t), \quad t \geq t_0, \tag{6.98}$$

whenever

$$A(t_0)V(t_0, x_0) \leq u_0. \tag{6.99}$$

Proof. We define

$$W(t,x) = A(t)V(t,x).$$

For sufficiently small $\Delta t > 0$, we consider

$$\begin{aligned} &W(t+\Delta t, x+\Delta t f(t,x)) - W(t,x) \\ &= A(t+\Delta t)V(t+h, x+\Delta t f(t,x)) - A(t)V(t,x) \\ &= A(t+\Delta t)V(t+\Delta t, x+\Delta t f(t,x)) \\ &\quad - A(t)V(t+\Delta t, x+\Delta t f(t,x)) \\ &\quad + A(t)V(t+\Delta t, x+\Delta t f(t,x)) - A(t)V(t,x) \\ &= A(t)\left[V(t+\Delta t, x+\Delta t f(t,x)) - V(t,x)\right] \\ &\quad + \left[A(t+\Delta t) - A(t)\right]V(t+\Delta t, x+\Delta t f(t,x)), \end{aligned}$$

and, by using the continuity of the functions, we have

$$d^+W(t,x) \le [A(t)d^+V(t,x) + V(t,x)d^+A(t)].$$

From this inequality and (6.97), we obtain

$$d^+W(t,x) \le g(t, W(t,x))dt, \quad (t,x) \in [R_+ \times R^n],$$

and from (6.99) we note that

$$W(t_0, x_0) \le u_0.$$

The function W satisfies all the hypotheses of Theorem 6.6.1. Therefore, by the application of Theorem 6.6.1 in the context of the definition of W, the estimate (6.98) follows immediately. This completes the proof of the theorem. □

***Remark* 6.6.1.** Assume that V in ($H_{6.6.1}$) satisfies the following more restrictive regularity conditions:
(H_V^*): (a) $V : R_+ \times R^n \to R$ is continuous on $R_+ \times R^n$;
(b) V possesses continuous partial derivatives V_t and V_x on $R_+ \times R^n$.
Then

$$d^+V(t,x) = \left[\frac{\partial}{\partial t}V(t,x) + \frac{\partial}{\partial x}V(t,x)f(t,x)\right]dt. \tag{6.100}$$

Moreover, the right-hand side expression is $LV(t,x)dt$ in (3.40), Section 3.3.

Lemma 6.6.1 ([19, 62, 83, 84]). *Let $A(t)$ be an $n \times n$ matrix function, and let I be an $n \times n$ identity matrix. For each fixed t, let $\|A(t)\|$ be any norm defined on $R^{n\times n}$. For $a > h > 0$, we define*

$$\nu(h) = \frac{\|I + hA(t)\| - 1}{h}. \tag{6.101}$$

Then

(i) $\nu(h)$ *is continuous in* $h > 0$;
(ii) $\nu(h)$ *is nondecreasing in* $h > 0$;
(iii) $\nu(h)$ *is bounded for* $h > 0$;
(iv) $\nu(h)$ *has a right-hand limit at* 0.

Proof. For any fixed $a > h_0 > 0$, and any $a > h > 0$, we consider

$$\begin{aligned}|\nu(h) - \nu(h_0)| &= \left|\frac{1}{h}\left[\|I + hA(t)\| - 1\right] - \frac{1}{h_0}\left[\|I + h_0A(t)\| - 1\right]\right| \\ &\le \frac{1}{hh_0}\left[\left|\left[h_0\|I + hA(t)\| - h\|I + h_0A(t)\|\right]\right| + |h - h_0|\right] \\ &\le \frac{2}{hh_0}|h - h_0| \quad \text{(Observation 1.5.4, norm properties).}\end{aligned}$$

From this inequality the continuity of $\nu(h)$ at h_0 for any h_0, $a > h_0 > 0$, follows immediately. This establishes statement (i).

For $0 < \theta < 1$, $a > h > 0$, we have $\theta h < h$. Now, we consider

$$\begin{aligned}\|I + \theta hA(t)\| &= \|\theta(I + hA(t)) + (1 - \theta)I\| \\ &\le \theta\|I + hA(t)\| + (1 - \theta)\|I\|.\end{aligned}$$

Thus,

$$\frac{\|I + \theta hA(t)\| - 1}{\theta h} \le \frac{\|I + hA(t)\| - 1}{h}. \tag{6.102}$$

This shows that the function ν defined in (6.101) is nondecreasing on $(0, a)$. This proves statement (ii). Moreover $\nu(h)$ is bounded, i.e.

$$|\nu(h)| \le \|A(t)\| \quad \text{[Observation 1.5.4(N4)].} \tag{6.103}$$

This proves statement (iii). Furthermore, from statements (i)–(iii), $\nu(h)$ has a right-hand limit at 0. Thus,

$$\lim_{h \to 0^+} \nu(h) = \mu(A(t)). \tag{6.104}$$

From (6.102) and (6.104), we have

$$\mu(A(t)) = \lim_{h \to 0^+}\left[\frac{\|I + hA(t)\| - 1}{h}\right]. \tag{6.105}$$

This completes the proof of the lemma. □

The above-presented lemma leads to an important concept in the qualitative study of nonlinear and nonstationary dynamic processes [19, 62, 64, 67, 83, 84].

Definition 6.6.1. Let $A(t)$ be an $n \times n$ matrix function defined on R. From Lemma 6.6.1, the limit

$$\lim_{t \to 0+} \frac{1}{h}[\|I + hA(t)\| - 1] \tag{6.106}$$

exists. This limit is called the logarithmic norm of $A(t)$ and is denoted by $\mu(A(t))$.

In the following, we present a problem which shows that the logarithmic norm of a matrix depends on the norm of the matrix.

Problem 6.6.1 ([19, 62, 75, 83, 84]). For an $n \times n$ matrix $A(t)$ function defined on R, we consider the following norms:

(MN_1) $\|A(t)\|_E$ = square root of the largest eigenvalue of $(A^T(t)A(t))$;

(MN_2) $\|A(t)\|_{1R} = \sup \left[\sum_{k=1}^{n} |a_{ik}(t)|\right]$;

(MN_3) $\|A(t)\|_{1C} = \sup \left[\sum_{i=1}^{n} |a_{ik}(t)|\right]$;

(MN_4) $\|A(t)\|_{dC} = \sup \left[\sum_{i=1}^{n} d_k^{-1} d_i |a_{ik}(t)|\right]$;

(MN_5) $\|A(t)\|_Q$ = square root of the largest eigenvalue of $(Q^{-1}A^T(t)QA(t))$ for some positive definite matrix Q.

Then, from the definition of $\mu(A(t))$, it is easy to see that:

(i) (MN_1) implies

$$\mu(A(t)) = \text{the largest eigenvalue of } \frac{1}{2}[A^T(t) + A(t)];$$

(ii) (MN_2) implies

$$\mu(A(t)) = \sup \left[a_{ii}(t) + \sum_{k \neq i}^{n} |a_{ik}(t)|\right];$$

(iii) (MN_3) implies

$$\mu(A(t)) = \sup \left[a_{kk}(t) + \sum_{i \neq k}^{n} |a_{ik}(t)|\right];$$

(iv) (MN_4) implies

$$\mu(A(t)) = \sup \left[a_{kk}(t) + \sum_{i \neq k}^{n} d_k^{-1} d_i |a_{ik}(t)|\right] \text{ for } d_i > 0;$$

(v) (MN_5) implies

$$\mu(A(t)) = \text{the largest eigenvalue of } \frac{1}{2}Q^{-1}[A^T(t)Q + QA(t)].$$

The result presented below provides information about the preservation of regularity properties by the logarithmic norm of a matrix function. For example, if $A(t)$ is a continuous matrix function, then the logarithmic norm $\mu(A(t))$ is also a continuous function.

Lemma 6.6.2 ([19, 62, 75, 83, 84]). *Let $A(t)$ and $B(t)$ be $n \times n$ matrices. The logarithmic norm of a matrix μ possesses the following properties:*

$$\begin{array}{ll} \text{(a) } \mu(\alpha A(t)) = \alpha\mu(A(t)), \quad \textit{for } \alpha \geq 0 & \textit{(positively homogeneous)}; \\ \text{(b) } |\mu(A(t))| \leq \|A(t)\| & \textit{(boundedness)}; \\ \text{(c) } \mu(A(t) + B(t)) \leq \mu(A(t)) + \mu(B(t)) & \textit{(subadditivity)}; \\ \text{(d) } |\mu(A(t)) - \mu(B(t))| \leq \|A(t) - B(t)\| & \textit{(lipschitz condition)}. \end{array}$$

The following result provides an estimate on the real part of any eigenvalue of a given matrix $A(t)$.

Lemma 6.6.3 ([19, 62, 75, 83, 84]). *Let $\sigma(A(t))$ be the spectrum of $n \times n$ matrix function $A(t)$. Then*

$$\operatorname{Re}\lambda(A(t)) \leq \mu(A(t)), \quad \textit{for all } \lambda(A(t)) \in \sigma(A(t)). \tag{6.107}$$

Proof. For any $\lambda(A(t)) \in \sigma(A(t))$, and $c(t)$ is a corresponding characteristic vector with norm 1. Then, for an arbitrary $h > 0$,

$$\begin{aligned} \|(I + hA(t))c(t)\| - \|c(t)\| &= |1 + h\lambda(A(t))|\|c(t)\| - \|c(t)\| \\ &= |1 + h\lambda(A(t))| - 1 \\ &= h\operatorname{Re}\lambda(A(t)) + o(h). \end{aligned} \tag{6.108}$$

On the other hand, from (6.107), we have

$$\begin{aligned} \|(I + hA(t))c(t)\| - \|c(t)\| &\leq \|(I + hA(t))\| - 1 \\ &\leq h\mu(A(t)) + o(h). \end{aligned}$$

This together with (6.108) yields

$$\operatorname{Re}\lambda(A(t)) \leq \mu(A(t)), \quad \text{for all } \lambda(A(t)) \in \sigma(A(t)).$$

This completes the proof of the lemma. □

***Remark* 6.6.2.** Under the assumptions of Lemma 6.6.3, we have

$$d^+c(t) \leq \mu(A(t))\, dt.$$

Illustration 6.6.1. Let us consider the following system of linear time-varying differential equations:

$$dx = A(t)x\,dt, \quad x(t_0) = x_0, \tag{6.109}$$

where $n \times n$ matrix function $A(t)$ is as defined in (4.115). Show that

$$\|x(t)\| \leq \exp\left[\int_{t_0}^{t} \mu(A(s))ds\right] u_0, \quad t \in J, \tag{6.110}$$

where the norm of an n-dimensional vector x depends on the norm of the matrix in the definition of the logarithmic norm.

Solution process. Let $V(t,x) = ||x(t)||$. For $\Delta t > 0$, we have

$$\begin{aligned} V(t+\Delta t, x+\Delta t A(t)x) - V(t,x) &= \|x + \Delta t A(t)x\| - \|x\| \\ &= \|(I + \Delta t A(t))x\| - \|x(t)\|, \end{aligned}$$

and hence

$$\begin{aligned} \frac{V(t+\Delta t, x+\Delta t A(t)x) - V(t,x)}{h} &\leq \frac{\|(I+\Delta t A(t))\|\|x(t)\| - \|x(t)\|}{\Delta t} \\ &\leq \frac{(\|I+\Delta t A(t)\| - 1)\|x(t)\|}{\Delta t}. \end{aligned} \tag{6.111}$$

From Remark 6.6.2, (6.102) and (6.105), we arrive at

$$d^{+}V(t,x) \leq \mu(A(t))V(t,x)dt. \tag{6.112}$$

In this case, the comparison differential equation (6.91) is

$$du = \mu(A(t))u\,dt, \quad u(t_0) = u_0, \tag{6.113}$$

and its maximal solution through (t_0, u_0) is

$$r(t, t_0, u_0) = \exp\left[\int_{t_0}^{t} \mu(A(s))ds\right] u_0.$$

Let $x(t) = x(t, t_0, x_0)$ be any solution to (6.109) satisfying the relation

$$V(t_0, x_0) \leq u_0. \tag{6.114}$$

From (6.112)–(6.114) and the application of comparison Theorem 6.5.1, we have the desired estimate on the solution to (6.109):

$$\|x(t)\| \leq \exp\left[\int_{t_0}^{t} \mu(A(s))ds\right] ||x_0||, \quad t \in J.$$

This completes the solution process of the illustration.

6.6 Exercises

(1) Let $V(t,x) = x^2$, and let $x(t)$ be any solution to a differential equation. Find an estimate for the solution to the following differential equation with initial data (t_0, x_0):

(a) $dx = \left[\cos tx - \frac{x}{1+\sin^2 x}\right] dt$ (b) $dx = \left[2x - \frac{6x}{1+\exp[-x]}\right] dt$
(c) $dx = [2x - x(1 + \exp[1 + \sin x])]\, dt$ (d) $dx = \frac{x}{1+t}\, dt$

(2) Let $A = \mathrm{diag}(d_1, \ldots, d_i, \ldots, d_n)$ be a diagonal matrix, and let $\alpha = \max\{d_1, \ldots, d_i, \ldots, d_n\}$. Show that

$$\|x(t)\| \le \exp[\alpha(t - t_0)]\, \|x_0\|, \quad t \in J,$$

where $x(t)$ is a solution to $dx = Ax\, dt$, $x(t_0) = x_0$.

(3) Prove conclusions (a) and (b) of Corollary 6.6.1.

(4) Provide justification for conclusions (i)–(v) drawn in Problem 6.6.1.

(5) Prove Lemma 6.6.2.

(6) Assume that all the hypotheses of Lemma 6.2.1 are satisfied. Further assume that $f(t,0) \equiv 0$. Then show that

$$f(t,x) = \int_0^1 \left[\frac{\partial}{\partial x} f(t,\theta x)\right] x\, d\theta.$$

Moreover, the system (6.1) can be rewritten as

$$dx = A(t,x)x\, dt, \quad x(t_0) = x_0,$$

where

$$A(t,x) = \int_0^1 \left[\frac{\partial}{\partial x} f(t,\theta x)\right] d\theta.$$

In addition, show that

$$\|x(t)\| \le \exp[\alpha(t - t_0)]\, \|x_0\|, \quad \text{for } t \in J,$$

where $\alpha = \max[\mu(A(t,x))]$ on $J \times D$.

6.7 Variational Comparison Method

In this section, we generalize the comparison theorems presented in Section 6.6. This generalization is based on the ideas of two nonlinear methods, namely the nonlinear variations of parameters and the classical Lyapunov's second method. It is a hybrid of these two methods [70–72].

By employing the concept of the Lyapunov function and differential inequalities of Section 6.6, we present a very general variational comparison theorem. These results connect the solution processes of the dynamic system (6.38) with the maximal solution to the corresponding comparison dynamic equation (6.91) and a solution to an auxiliary dynamic system. A by-product of this is that, several auxiliary comparison results are formulated. These results generalize and extend the existing results in a systematic and unified and hence, this leads to a concept of "Nonlinear Hybrid Method".

Theorem 6.7.1. *Assume that all the hypotheses of Theorems* 6.4.1 *and* 6.5.1 *are satisfied. Further assume that*

(a) $LV(s, y(t, s, x))$ *in* (6.58) *and* $g(t, u)$ *in* (6.70) *satisfy the inequality*

$$LV(s, y(t, s, x)) \leq g(s, Vs, y(t, s, x)), \tag{6.115}$$

where $y(t, s, x)$ *is the solution process of* (6.39) *through* (s, x);

(b) *Let* $r(t) = r(t, t_0, u_0)$ *be the maximal solution to the scalar differential* (6.91) *existing for* $t \geq t_0$;

(c) *Let* $x(s) = x(s, t_0, x_0)$ *be a solution to* (6.38) *existing for* $t \geq t_0$.

Then

$$V(t, x(t, t_0, x_0)) \leq r(t, t_0, u_0), \quad \text{for } t \geq t_0, \tag{6.116}$$

provided that

$$V(t, y(t, t_0, x_0)) \leq u_0. \tag{6.117}$$

Proof. For $t_0 \leq s \leq t$, we imitate the proof of Theorem 6.4.1, and we arrive at (6.59), i.e.

$$\begin{aligned} dV(s, y(t, s, x(s))) &= V_t(s, y(t, s, x(s)))ds \\ &\quad + \frac{\partial}{\partial x} V(s, y(t, s, x(s)))\Phi(t, s, x(s))R(s, x(s))ds \\ &= LV(s, y(t, s, x))ds. \end{aligned} \tag{6.118}$$

From (6.115) and (6.118), we have

$$dV(s, y(t, s, x(s))) \leq LV(s, y(t, s, x))ds, \tag{6.119}$$

provided that $\triangle s > 0$ and $\triangle s = ds$.

Now, we set

$$m(s) = V(s, y(t, s, x(s))), \quad m(t_0) = V(t, y(t, t_0, x_0)), \tag{6.120}$$

and by repeating the argument used in Theorem 6.6.1 the proof of the theorem can be completed. The details are left to the reader. □

Corollary 6.7.1. *Suppose that* $F(t, y) = A(t)y$, *where* $A(t)$ *is an* $n \times n$ *continuous matrix function. From Remark* 4.7.1 (*Observation* 4.3.4), $y(t, t_0, x_0) = \Phi(t, t_0)x_0$, *where* $\Phi(t, t_0)$ *is as defined in* (4.125) (*Remark* 4.7.1). *Further assume that* $g(t, u) \equiv 0$, *as in Corollary* 6.6.1(a). *Then* (6.116) *reduces to*

$$V(t, x(t, t_0, x_0)) \leq V(t, \Phi(t, t_0)x_0), \ \text{for } t \geq t_0.$$

Observation 6.7.1. Under the conditions of Theorem 6.3.1 on (6.39), the conditions of Theorem 6.6.1 on Lyapunov/energy function V in the context of definition analogous to (6.89),

$$\begin{aligned} d^+_{(6.1)}V(s, y(t, s, x)) = \limsup [V(s + \triangle s, y(t, s + \triangle s, x \\ + \triangle s[F(s, x) + R(s, x)])) - V(s, y(t, s, x))]ds, \end{aligned} \quad (6.121)$$

the conclusion of Theorem 6.7.1 remains valid.

Example 6.7.1. Consider a scalar nonlinear autonomous differential equation,

$$dy = -\frac{1}{2}\, y^3 dt, \quad x(t_0) = x_0,$$

and its corresponding perturbed differential equation,

$$dx = -\frac{1}{2}\, x^3 dt + R(t, x), \quad x(t_0) = x_0,$$

where $R(t, x)$ is continuous on $R_+ \times R$ into R, and it satisfies the conditions $R(t, 0) \equiv 0$ and

$$2xR(t, x) \leq \alpha(t)x^2, \quad \text{for } (t, x) \in R_+ \times R,$$

where α is a continuous function defined on R_+ into it-self. Show that

$$|x(t, t_0, x_0)|^2 \leq |y(t, t_0, x_0)|^2 \exp\left[\int_{t_0}^{t} \alpha(s)\, ds\right], \quad \text{for } t \geq t_0.$$

Solution process. From Example 6.2.2, we have

$$\begin{aligned} y(t, t_0, x_0) &= x_0 \left[1 + x_0^2(t - t_0)\right]^{-\frac{1}{2}}, \\ \frac{\partial}{\partial x}\, y(t, t_0, x_0) &= \Phi(t, t_0, x_0) = \left[1 + x_0^2(t - t_0)\right]^{-\frac{3}{2}}, \\ \frac{\partial}{\partial t_0}\, y(t, t_0, x_0) &= -\Phi(t, t_0, x_0)F(t_0, x_0) = \frac{1}{2}\, x_0^3 \left[1 + x_0^2(t - t_0)\right]^{-\frac{3}{2}}. \end{aligned}$$

We pick an energy function:

$$V(t, x) = \frac{1}{2}\, x^2.$$

In this case, we have

$$\begin{aligned} dV(s, y(t, s, x(s))) &= \frac{1}{2}\, d\, |y(t, s, x(s))|^2 \\ &= \frac{1}{2}\, 2y(t, s, x(s)) \left[\frac{\partial}{\partial x_0}\, y(t, s, x(s))dx(s) + \frac{\partial}{\partial t_0}\, y(t, s, x(s))ds\right] \\ &= y(t, s, x(s)) \left[\Phi(t, s, x(s))dx(s) - \Phi(t, s, x(s))F(s, x(s))ds\right] \end{aligned}$$

$$\begin{aligned}
&= y(t,s,x(s))\,[\Phi(t,s,x(s))(F(s,x(s))) \\
&\quad + R(s,x(s))ds - \Phi(t,s,x(s))F(s,x(s))ds] \\
&= y(t,s,x(s))\Phi(t,s,x(s))R(s,x(s))ds \\
&= \frac{x(s)}{[1+x^2(s)(s-t_0)]^{\frac{1}{2}}}\,\frac{1}{[1+x^2(s)(s-t_0)]^{\frac{3}{2}}}R(s,x(s))ds \\
&= \frac{x(s)R(s,x(s))ds}{[1+x^2(s)(s-t_0)]^2} \le \frac{1}{2}\,\frac{\alpha(s)x^2(s)}{[1+x^2(s)(s-t_0)]}ds.
\end{aligned}$$

From this and $0 < \frac{1}{1+x^2(s)(s-t_0)} \le 1$, for $t_0 \le s$ and $0 < ds$, we have

$$dV(s, y(t,s,x(s))) \le \alpha(s)V(s, y(t,s,x(s))ds.$$

Here, the comparison differential equation is

$$du = \alpha(t)u, u_0 = \frac{1}{2}|y(t,t_0,x_0)|^2.$$

Because of its uniqueness of solution (Theorem 2.6.2), its solution is the maximal solution. Thus, its maximal solution is

$$r(t,t_0,u_0) = \exp\left[\int_{t_0}^{t}\alpha(s)\,ds\right]u_0 = |y(t,t_0,x_0)|^2\exp\left[\int_{t_0}^{t}\alpha(s)\,ds\right].$$

Now, using Theorem 6.7.1, we have the desired result. This completes the solution procedure of the example.

Example 6.7.2. We consider Example 6.3.2, i.e. a scalar linear nonautonomous differential equation,

$$dy = \frac{y}{1+t}dt, \quad y(t_0) = x_0,$$

and its corresponding perturbed differential equation,

$$dx = \left[\frac{x}{1+t} + R(t,x)\right]dt, \quad x(t_0) = x_0.$$

We assume that $R(t,x)$ satisfies the condition

$$xR(t,x) \le -\alpha(t)x^2, \quad \text{for } (t,x) \in R_+ \times R,$$

where α is a continuous function on R_+. Show that

$$|x(t,t_0,x_0)|^2 \le |y(t,t_0,x_0)|^2\exp\left[\int_{t_0}^{t}\alpha(s)ds\right], \quad \text{for } t \ge t_0.$$

Solution process. We choose an energy function:

$$V(t,x) = \frac{1}{2}x^2.$$

Following the argument used in Example 6.7.1 and the using expressions in Example 6.3.2,

$$\frac{\partial}{\partial t_0} y(t, t_0, x_0) = -\Phi(t, t_0, x_0)F(t_0, x_0) = -\frac{1+t}{(1+t_0)^2} x_0,$$

$$\Phi(t, t_0, x_0) = \frac{1+t}{1+t_0},$$

we have

$$dV(s, y(t, s, x(s))) \leq \alpha(t)V(s, y(t, s, x(s)))ds.$$

The rest of the argument parallels to Example 6.7.1.

6.7 Exercises

Let $R \in C[J \times R, R]$, and R satisfies the relation:

$$xR(t, x) \leq \alpha(t)x, \quad \text{for } (t, x) \in R_+ \times R$$

for some integrable function α defined on R_+ int R_+.

(1) Let us consider the nonlinear differential equations

$$dx = \left[-\frac{\cos t}{2(\sin t + 2)} x^3 + R(t, x)\right] dt, \quad x(t_0) = x_0,$$

$$dy = -\frac{\cos t}{2(\sin t + 2)} y^3 dt, \quad y(t_0) = x_0.$$

Show that

$$|x(t, t_0, x_0)|^2 \leq |y(t, t_0, x_0)|^2 \exp\left[\int_{t_0}^{t} \alpha(s)\, ds\right], \quad \text{for } t \geq t_0.$$

(2) Given the nonlinear differential equations

$$dx = \left[-\frac{a(t)}{2} x^3 + R(t, x)\right] dt, \quad x(t_0) = x_0,$$

$$dy = -\frac{a(t)}{2} y^3 dt, \quad y(t_0) = x_0,$$

Prove or disprove the following estimate

$$|x(t, t_0, x_0)|^2 \leq |y(t, t_0, x_0)|^2 \exp\left[\int_{t_0}^{t} \alpha(s) ds\right], \quad \text{for } t \geq t_0.$$

(3) Given the nonlinear differential equations

$$dx = \left[-\frac{1}{4}x^5 + R(t,x)\right]dt, \quad x(t_0) = x_0,$$

and

$$dy = -\frac{1}{4}y^5 dt, \quad y(t_0) = x_0,$$

Show that

$$|x(t,t_0,x_0)|^4 \le |y(t,t_0,x_0)|^4 \exp\left[\int_{t_0}^{t} \alpha(s)\,ds\right], \quad \text{for } t \ge t_0.$$

(4) Given the nonlinear differential equations

$$dx = \left[-\frac{1}{2n}x^{2n+1} + R(t,x)\right]dt, \quad x(t_0) = x_0,$$

$$dy = -\frac{1}{2n}y^{2n+1}\,dt, \quad y(t_0) = x_0,$$

where $n > 1$ is an integer.
Prove or disprove the estimate

$$|x(t,t_0,x_0)|^{2n} \le |y(t,t_0,x_0)|^{2n} \exp\left[\int_{t_0}^{t} \alpha(s)\,ds\right], \quad \text{for } t \ge t_0.$$

(5) Given the following nonlinear differential equations

$$dx = \left[-\frac{a(t)}{2n}x^{2n+1} + R(t,x)\right]dt, \quad x(t_0) = x_0,$$

$$dy = -\frac{a(t)}{2n}y^{2n+1}dt, \quad y(t_0) = x_0,$$

where $n > 1$ is an integer.
Prove or disprove the estimate

$$|x(t,t_0,x_0)|^{2n} \le |y(t,t_0,x_0)|^{2n} \exp\left[\int_{t_0}^{t} \alpha(s)\,ds\right], \quad \text{for } t \ge t_0.$$

6.8 Linear Hybrid Systems

A hybrid dynamic system is a pair of interconnected dynamic subsystems operating under both continuous and discrete times [14, 70, 72]. For example, a continuous dynamic process is interrupted by discrete time events. These types of systems provide mathematical models for interconnected dynamic phenomena evolving under different measure chains with state-dependent discrete events.

In this section, we motivate hybrid dynamic phenomena by presenting a dynamic process in medical and biological processes. In addition, we give a few elementary examples to illustrate the role and scope of the mathematical modeling of the dynamic processes that are operating under different timescales.

The section briefly presents a well-known problem, called the "dosage problem" [71]. The common conceptual understanding of this problem can be applied to any problem in the agricultural, biological, chemical, engineering, medical, physical, and social sciences. For example, the term "dose" is not only limited to the measured quantity of a therapeutic chemical/physical agent but also to a portion of an experience or knowledge applied to problem-solving processes in all fields. The main goal of this problem is to determine the correct dose of a substance/knowledge/information/experience and the intradosage time so that the predetermined objective is achieved.

Illustration 6.8.1 (drug prescription problem). Two very important questions to providing the best medical care to patients are:

What is the proper drug dosage?
How frequently should the drug be administered?

We know that the concentration of the drug needs to be within a set of bounds so that it is safe and effective. Also, the frequency of its administration must produce the maximum benefits.

Definition 6.8.1. Let d be the amount of drug concentration in the blood. d is said to be a lower bound/threshold (effective level). If an amount of drug concentration is lower than d in blood, then it is not effective.

Definition 6.8.2. Let D be the amount of drug concentration in the blood. D is said to be an upper bound (safe level). If the drug concentration in the blood is higher than D, then it is harmful.

Let c_0 be an initial single dose of the drug given to the patient, and let c be the drug concentration in plasma at a time t. In addition, let T be the length of the time interval between two consecutive doses. T is called the interdosage time. The goal of this illustration is to prescribe the drug dose with an interdosage time interval so that the drug concentration remains within the prescribed effective and safe bounds (lower and upper).

PDP 1 (drug administration process). There are several ways to administer the drug, and we will consider two of these methods.

(a) Choosing the initial dose c_0 between the upper (D) and the lower (d) bound and the interdosage time T in such a way that there is no drug build up in the plasma. In other words, the difference between the amounts of any two consecutive doses is "zero." In this case, the safe and effective levels are satisfied.

(b) Choosing the initial dose as in (a) considering the interdosage time T as a function of an initially administered amount of the drug dose and the decay rate of the drug concentration, the choice of T is made in such a way that the drug concentration build up due to the difference between the initial drug concentration and the concentration in blood at the end of the immediate past period in each successive period of time remains within the prescribed bounds. In short, the sum of the regular amount of the initial dose and the aggregate residual concentration up to the present dosage time must be within the specified bounds.

In this discussion, we propose to follow the drug administration process outlined in PDP 1(b).

PDP 2 (assumptions). We note that the drug concentration c in plasma at a time t depends on several variables, such as, decay rate, assimilation rate, amount of the dose, dosage interval, body weight, and blood volume.

(i) To reduce the number of variables, it is assumed that the body weight and blood plasma are constant (an average over a specific age group).
(ii) *Homogeneity.* It is assumed that upon the drug dose administration that it is instantaneously distributed in the plasma. In short, the drug concentration is uniformly distributed in plasma.
(iii) *Decay Rate.* The drug elimination from plasma follows Fick's law of diffusion. According to this law, we have

$$dc = -\epsilon c\,dt, \quad c(0) = c_0,$$

where ϵ (coefficient of permeability of the membrane) is a positive constant that depends on many factors, such as the patient in general and Na^+, K, Cl^-, pH value, etc.

PDP 3 (elimination rate). Here, ϵ is the elimination/exercision constant. It is assumed that D and d are as defined in Definitions 6.8.1 and 6.8.2 and are experimentally determined and tested. Here, we choose the initial dose of the drug to be

$$c_0 = D - d.$$

It is measured in milligrams per milliliter per hour (mg/ml/h). From the solution in the procedure in Subsection 2.4.3, we have

$$c(t) = c_0 \exp[-\epsilon t].$$

The right-hand side expression gives the concentration of the drug in the plasma at any time $t \geq 0$.

PDP 4 (assimilation rate). From PDP 2(ii), we recall that when the drug dose with the same amount is administered, then an instantaneous rise in the concentration (like impulse/jerk) takes place. This assumption is not realistic with regard to the drug taken by mouth. However, it is for a drug that is injected directly into the bloodstream. In this case, the discrete dynamic of the drug administration process is described by

$$c(k) = \exp[-\epsilon T]c(k-1) + c_0, c(0) = c_0, \quad \text{for } k \geq 1.$$

Here, the first term on the right-hand side represents the residual of the drug concentration between the kth and $(k-1)$th drug dose administration processes over a time interval of length T for any $k \geq 1$. Thus, the amount of the drug assimilated during the interval of length T is $c(k-1)(1-\exp[-\epsilon T])$.

PDP 5 (drug accumulation under repeated doses). In order to find the drug accumulation after repeated doses, we need to find an expression for $c(k)$ in terms of the initial dose and an arbitrary interdosage time T, i.e. we need to solve the discrete time drug dosage iterative process in PDP 4:

$$\begin{aligned} c(k) &= \exp[-\epsilon kT]c_0 + \exp[-\epsilon(k-1)T]c_0 + \cdots + \exp[-2\epsilon T]c_0 + \exp[-\epsilon T]c_0 + c_0 \\ &= \exp[-\epsilon T][1 + \exp[-\epsilon T] + \exp[-2\epsilon T] + \cdots + \exp[-\epsilon(k-1)T]]c_0 + c_0 \\ &= \exp[-\epsilon T]R(k)c_0 + c_0, \end{aligned}$$

where

$$R(k) = [1 + \exp[-\epsilon T] + \exp[-2\epsilon T] + \cdots + \exp[-\epsilon(k-1)T]] = \frac{1-\exp[-\epsilon kT]}{1-\exp[-\epsilon T]}.$$

From this, we note that $\exp[-\epsilon T]R(k)c_0$ is the aggregate residual of the concentration of the drug due to the first k successively repeated doses over the time interval $[0, kT]$.

PDP 6 (determination of the dosage schedule). From the expression of $c(k)$ in PDP 5, we have

$$\begin{aligned} \lim_{k\to\infty} [c(k)] &= \lim_{k\to\infty} [\exp[-\epsilon T] R(k)c_0 + c_0] \\ &= \lim_{k\to\infty} [\exp[-\epsilon T] R(k)c_0] + \lim_{k\to\infty} [c_0] \\ &= \frac{\exp[-\epsilon T]}{1-\exp[-\epsilon T]} c_0 + c_0. \end{aligned}$$

We note that the above expression is valid for any arbitrary interdosage time $T > 0$. From the choice of c_0 in PDP 3, we choose the interdosage time T so that

$$\begin{aligned} D &= \frac{\exp[-\epsilon T]}{1-\exp[-\epsilon T]} c_0 + c_0 = \frac{\exp[-\epsilon T]}{1-\exp[-\epsilon T]} c_0 + \frac{1-\exp[-\epsilon T]}{1-\exp[-\epsilon T]} c_0 \\ &= \frac{1}{1-\exp[-\epsilon T]} c_0 = \frac{1}{1-\exp[-\epsilon T]}(D-d), \end{aligned}$$

which implies that

$$d = D - (1 - \exp[-\epsilon T])D = D\exp[-\epsilon T],$$

and hence

$$\exp[\epsilon T] = \frac{D}{d}.$$

Thus, the interdosage time (schedule) is determined by

$$T = \frac{1}{\epsilon}\ln\left(\frac{D}{d}\right).$$

PDP 7 (dosage prescription and schedule). From PDP 3 and PDP 6, we summarize that the drug dosage prescription and the interdosage time (the time interval between any two consecutive dosages) are

$$c_0 = (D - d), \quad T = \frac{1}{\epsilon}\ln\left(\frac{D}{d}\right),$$

respectively.

Conclusions

(i) From PDP 3 and PDP 6, we note that

$$D = \frac{\exp[-\epsilon T]}{1 - \exp[-\epsilon T]}c_0 + c_0 = \frac{\exp[-\epsilon T]}{1 - \exp[-\epsilon T]}c_0 + D - d,$$

and hence

$$d = \frac{\exp[-\epsilon T]}{1 - \exp[-\epsilon T]}c_0 = \frac{c_0}{\exp[\epsilon T] - 1}.$$

(ii) From (i), $\frac{d}{c_0}$ is close to 0 if the interdosage time is very large. Therefore, the intermediate aggregate residual concentration at the kth dose is $R(k)$, and hence

$$0 < R(k) < \frac{c_0}{\exp[\epsilon T] - 1}.$$

(iii) If T is chosen in such a way that $\exp[\epsilon T]$ is close to 1 $\exp[\epsilon T] - 1$ is positive. In this case $\frac{d}{c_0}$ is significantly greater than 1. Therefore, the kth dosage concentration $c(k)$ increases as k increases. Thus, from PDP 3, the drug concentration in blood will be very close to c_0 due to the dose. As a result of this and from (i), the drug concentration in blood will be oscillating between D and

$$\frac{\exp[-\epsilon T]}{1 - \exp[-\epsilon T]}c_0 = \frac{c_0}{\exp[\epsilon T] - 1}.$$

Observation 6.8.1.

(i) From Illustration 6.8.1, the mathematical formulation of the drug prescription problem is:

$$\begin{aligned} dc &= -\epsilon c\, dt,\ c(t_k) = c_k,\ t \neq t_k, \quad \text{for } k \geq 0, \\ c_k &= \exp[-\epsilon T]c(k-1) + c_0,\ c(0) = c_0, \quad \text{for } k \geq 1, \end{aligned}$$

where $t_k = kT$, $k \geq 0$.

(ii) From (i), it is clear that the continuous time dynamic process outlined in PDP 2 perturbed by the discrete time event (drug administration) is described by the drug administration process. Thus, the dynamic of the prescribing drug problem is modeled by both discrete and continuous time processes. This type of dynamic process is called a hybrid dynamic process.

(iii) We further note that the hybrid mathematical description in (i) is linear in both discrete and continuous time dynamic processes. Therefore, it is called a linear hybrid system.

6.8.1 *General problem*

In our subsequent discussion, we present a few examples showing the nature of a hybrid dynamic system,

$$\begin{aligned} du &= g(t,u)dt, \quad u(t_k) = u_k, \quad t \neq t_k, \quad k \in I(0,\infty), \\ u_k &= \psi(u_{k-1}(t_k^-, t_{k-1}, u_{k-1}), k), \quad u(t_0) = u_0, \quad k \in I(1,\infty), \end{aligned} \tag{6.122}$$

where we assume $(H_{6.8})$:

(i) $g : R_+ \times R \to R$ are continuous in $[t_k, t_{k+1}) \times R_+$ and for each (t,u), $\lim g(t,u) = g(t_{k+1}^-, \overline{u})$ as $(t,u) \to (t_{k+1}^-, \overline{u})$;

(ii) $\psi : R_+ \times R_+ \to R_+$ is a Borel-measurable function.

From Observation 6.8.1, we further recall that $r(t, t_0, u_0)$ is defined by

$$r(t) = r(t, t_0, u_0) = \begin{cases} r_0(t, t_0, u_0), & 0 \leq t < t_1, \\ r_1(t, t_1, u_1), & t_1 \leq t < t_2, \\ \dots\dots\dots, & \dots\dots\dots, \\ r_{n-1}(t, t_{n-1}, u_{n-1}), & t_{n-1} \leq t < t_n, \\ \dots\dots\dots, & \dots\dots\dots \end{cases}$$

Moreover, the solution process is described by the following iterative process:

$$\begin{aligned} r(t, t_{k-1}, u_{k-1}) &= r_{k-1}(t, t_{k-1}, u_{k-1}), \quad t_{k-1} \leq t < t_k, \\ u_k &= \psi(r_{k-1}(t_k^-, t_{k-1}, u_{k-1}), k), \quad u(t_0) = u_0, \quad k \in I(1,\infty). \end{aligned} \tag{6.123}$$

Observation 6.8.2. From (6.123), we note that for each $k \in I(1,\infty)$, $(r_{k-1}(t), u_k)$ is a solution process of (6.122) on $t_{k-1} \leq t < t_k$, where $r_{k-1}(t) = r(t, t_{k-1}, u_{k-1})$

and $u_k = u(k) = u(k, 0, u_0)$. It is determined by the one-step cyclic (Gauss–Seidel) iterative procedure [77].

We now present a few examples of simple versions of the scalar impulsive dynamic system (6.122) and its explicit solution processes.

Example 6.8.1.

(a) If $g \equiv 0$, $\psi \equiv 0$ in (6.122), then from (6.123) it is easy to show that $r(t, t_0, u_0)$ is defined by $r(t, t_0, u_0) = u_0$.

(b) Let $g \equiv 0$. In addition, let $d : R[\Omega, S] \to R_+$ be a Borel-measurable positive function with $d(\eta_0) = 1$. In this case ψ in (6.122) is $\psi(uk) = d(k)u$, and (6.123) reduces to

$$\begin{aligned} r(t, t_k, u_k) &= u_k, \quad t_k \le t < t_{k+1}, \quad k \in I(0, \infty), \\ u_k &= d(k)u_{k-1}, \quad u(t_0) = u_0, \quad k \in I(1, \infty), \end{aligned} \tag{6.124}$$

which can be solved explicitly, and $r(t, t_0, u_0)$ can be expressed as

$$r(t, t_0, u_0) = \prod_{t_0 < t_k < t} d(k)u_0, \quad \text{for all } t \ge t_0. \tag{6.125}$$

(c) Let $g(t, u) = \lambda(k)u$. d is as defined in (b). In this case ψ in (6.122) is $\psi(u, \eta_k) = d(k)\, u$, and (6.123) reduces to

$$\begin{aligned} r(t, t_k, u_k) &= u_k \exp[\lambda(k)(t - t_k)], \quad t_k \le t < t_{k+1}, \quad k \in I(0, \infty), \\ u_k &= d(k)u_{k-1} \exp\left[\lambda(k-1)(t_k - t_{k-1})\right], \quad u(t_0) = u_0, k \in I(1, \infty), \end{aligned} \tag{6.126}$$

which can be solved explicitly, resulting in

$$\begin{aligned} r(t, t_0, u_0) &= \prod_{t_0 < t_k < t} d(k) \exp\left[\lambda(k)(t - t_k)\right] u_0, \quad \text{for all } t \ge t_0 \\ &= \left(\prod_{k=1}^{n+1} d(k-1) \right) \exp\left[\sum_{k=1}^{n} \lambda(k-1)t_k + \lambda(k)(t - t_n) \right] u_0, \end{aligned} \tag{6.127}$$

for all $t \ge t_0$.

(d) Let $g(t, u) = \lambda(t, k)\, u$, where $\lambda \in C[[t_k, t_{k+1}) \times R_+, R]$, λ is left-continuous at t_{k+1} for $k \in I(1, \infty)$, and d is as defined in (b). In this case ψ in (6.122) is $\psi(u, k) = d(k)u$, and (6.123) reduces to

$$\begin{aligned} r(t, t_k, u_k) &= u_k \exp\left[\int_{t_k}^{t} \lambda(v, k) dv \right], \quad t_k \le t < t_{k+1}, \quad k \in I(0, \infty), \\ u_k &= d(k)u_{k-1} \exp\left[\int_{t_{k-1}}^{t_k} \lambda(v, k-1) dv \right], \quad u(t_0) = u_0, \quad k \in I(1, \infty), \end{aligned} \tag{6.128}$$

which can be solved explicitly, resulting in

$$r(t,t_0,u_0) = \prod_{t_0 \le t_k < t} d(k) \exp\left[\int_{t_k}^{t} \lambda(v,k)dv\right] u_0, \quad \text{for all } t \ge t_0$$

$$= \left(\prod_{k=1}^{n+1} d(k-1)\right) \exp\left[\sum_{k=1}^{n} \int_{t_{k-1}}^{t_k} \lambda(v,k-1)dv + \int_{t_n}^{t} \lambda(v,k)\,dv\right] u_0, \tag{6.129}$$

for all $t \ge t_0$.

6.9 Linear Hereditary Systems

A few elementary examples showing the hereditary effects are presented. They illustrate the role and scope of hereditary dynamic models of real-world problems. In addition, they generate several issues in the modeling of dynamic processes, namely the effects of hereditary perturbations [60, 63, 61, 65, 66, 73, 78].

Differential equations provide mathematical models for dynamic processes in the biological, chemical, medical, physical, and social sciences. They give the description of processes in engineering and sciences. Furthermore, they provide information about the future behavior and its usefulness for the decision-making process.

Example 6.9.1. Let us consider an elementary differential equation,

$$dy = y\,dt, \tag{6.130}$$

where dy stands for the differential of y. Using the method described in Section 2.4, we obtain the fundamental solution to (6.130):

$$y(t) = \exp[t]. \tag{6.131}$$

Observation 6.9.1. From (6.130) and (6.131), we recall that (6.130) is a prototype model of the pure growth of the single species processes. Its state transition process in (6.131) exhibits unlimited growth of the species.

Example 6.9.2. Let us consider an elementary differential equation,

$$dy = -y\,dt, \tag{6.132}$$

where dy is the differential in (6.130). From the elementary technique of finding a solution to (6.130) (Section 2.4), it is clear that

$$y(t) = \exp[-t]. \tag{6.133}$$

Observation 6.9.2. From (6.132) and (6.133), we recall that (6.132) is a prototype model of pure decay of single species processes. Its state transition process in (6.133) exhibits exponential decay of the species.

Example 6.9.3. Let us consider an elementary differential equation,

$$dy^{(1)} = y\,dt, \tag{6.134}$$

where $dy^{(1)}$ stands for

$$dy^{(1)} = \frac{d^2y}{dt^2}\,dt.$$

From the elementary technique of finding a solution to (6.134) in Section 5.3, it is clear that

$$y(t) = \exp[t], \quad y(t) = \exp[-t] \tag{6.135}$$

form a complete set of fundamental solutions to (6.134). From Observation 5.3.2, any solution to (6.134) is given by the set of solutions (6.135).

Observation 6.9.3. We note that (6.134) describes the motion of the particle under repulsive force. Moreover, the solution to the IVPs $y(t_0) = y_0$ and $y^{(1)}(t_0) = y_0^{(1)}$ is

$$y(t) = \frac{1}{2}(y_0 + y_0^{(1)})\exp[(t - t_0)] + \frac{1}{2}(y_0 - y_0^{(1)})\exp[-(t - t_0)].$$

In particular, (a) if $y_0^{(1)} = -y_0$, then

$$y(t) = \frac{1}{2}(y_0 - y_0^{(1)})\exp[-(t - t_0)],$$

and (b) if $y_0^{(1)} = y_0$, then

$$y(t) = \frac{1}{2}(y_0 + y_0^{(1)})\exp[(t - t_0)].$$

6.9.1 *General problem*

The rates of growth of the biological, chemical, compartmental, engineering, physical, and social systems depend on both the present and past history of the state of the system. The time delays are due to the time it takes to respond to the deterministic changes in the external environment of the system. They are finite in nature. These types of dynamic systems can be modeled by systems of functional/delay differential equations of the type

$$dy = \left[A\int_{-\tau}^{0} y(t+s)\,ds\right]dt, \tag{6.136}$$

where y is a state of a dynamic process $\tau > 0$ is the past history which affects the state of the process, and A is an $n \times n$ constant matrix with real entries.

If one imitates the eigenvalue method described in Sections 2.4 and 4.3, one obtains the following characteristic equation:

$$\lambda I = A \int_{-\tau}^{0} \exp[\lambda s] ds. \tag{6.137}$$

This is a transcendental algebraic equation, and it has infinitely many values of λ. As a result of this, it is not possible to find all closed-form solutions to (6.136). In fact, in the following, we present a few simple scalar versions of differential equations.

Example 6.9.4. We consider a very simple scalar delay differential equation

$$dy = y\left(t - \frac{3\pi}{2}\right) dt. \tag{6.138}$$

We note that

$$y(t) = \cos t, \quad y(t) = \sin t \tag{6.139}$$

are solutions to (6.138). However, the corresponding ordinary differential equation (6.130) has the fundamental solution as described in (6.131).

Observation 6.9.4. From (6.138), we note that $\tau = \frac{3\pi}{2}$. The characteristic equation with regard to (6.138) is

$$\lambda = \exp\left[-\lambda \frac{3\pi}{2}\right]. \tag{6.140}$$

We further observe that $\lambda = \pm i$, $i = \sqrt{-1}$ are solutions to the algebraic equation (6.140). From this discussion, (6.139), and Observation 6.9.1, it is clear that the time delay has generated the oscillatory behavior of the solution process [66, 78].

Example 6.9.5. We consider the delay differential equation

$$dy = -y\left(t - \frac{\pi}{2}\right) dt. \tag{6.141}$$

Again, we note that

$$y(t) = \cos t, \quad y(t) = \sin t \tag{6.142}$$

are solutions to (6.141). However, the corresponding ordinary differential equation (6.132) has the fundamental solution as described in (6.133).

Observation 6.9.5. From (6.141), we note that $\tau = \frac{\pi}{2}$. The characteristic equation with regard to (6.141) is

$$\lambda = \exp\left[-\lambda \frac{\pi}{2}\right]. \tag{6.143}$$

We further observe that $\lambda = \pm i$, $i = \sqrt{-1}$ are solutions to the algebraic equation (6.143). From this discussion, (6.142), and Observation 6.9.2, it is clear that the time delay has generated the oscillatory behavior of the solution process [66, 78].

Example 6.9.6. We consider the delay differential equation

$$dy^{(1)} = y(t - \pi)dt. \tag{6.144}$$

Again, we note that

$$y(t) = \cos t, \quad y(t) = \sin t \tag{6.145}$$

are solutions to (6.144). However, the corresponding ordinary differential equation (6.134) has the set of fundamental solutions as described in (6.135).

Observation 6.9.6. From (6.144), we note that $\tau = \pi$. The characteristic equation with regard to (6.144) is

$$\lambda = \exp[-\lambda\pi]. \tag{6.146}$$

We further observe that $\lambda = \pm i$, $i = \sqrt{-1}$ are solutions to the algebraic equation (6.146). From this discussion, (6.145), and Observation 6.9.3, it is clear that the time delay has generated the oscillatory behavior of the solution process [78].

6.9 Exercises

(1) Verify by substitution that each given function/process is a solution to the given differential equation with deviating argument. Moreover, compare the solutions with the solution to the differential equation in the absence of the delay.

  (a) $dy = -y\left(t + \frac{\pi}{2}\right)dt$, $y_1(t) = \cos t$, and $y_2(t) = \sin t$
  (b) $dy^{(1)} = -y(\pi - t)dt$, $y_1(t) = \sin t$, and $y_2(t) = e^t - e^{\pi - t}$
  (c) $dy^{(1)} = y(-t)dt$, $y_1(t) = \sin t$, and $y_2(t) = e^t + e^{-t}$
  (d) $dy^{(1)} = \frac{1}{2}\left[y + y(t - \pi)\right]dt$ and $y_1(t) = 1 - \sin t$

(2) Can you find all possible solutions to the algebraic equation (6.140)? How about (6.143) or (6.146)?
(3) Do the algebraic equations (6.140), (6.143), and (6.146) have a finite number of solutions? Explain.
(4) Is it possible to solve the system of algebraic equations (6.137)?
(5) Compute the solution to the following system of differential equations:

$$dx = Ax\,dt, \quad \text{where } A = \begin{bmatrix} 2 & -1 \\ 1 & 1 \end{bmatrix},$$

and compare the solutions $x_1(t) = e^{\frac{3}{2}t}$ and $x_2(t) = e^{\frac{3}{2}t}$ to

$$dx_1 = 2x_1 - x_2\left(t - \frac{1}{3}\ln 4\right),$$

$$dx_2 = 2x_1\left(t - \frac{1}{3}\ln 4\right) + x_2.$$

6.10 Qualitative Properties of the Solution Process

For the sake of simplicity and to illustrate the basic ideas, a simple mathematical model for the inflation–unemployment process is outlined.

IUP 1. The development of the deterministic mathematical model begins with the basic relationship between inflation and unemployment. This relationship has been described by Phelps [100] and Friedman [26]. It is called the expectation-augmented Phillips curve,

$$p = A - g - \beta \ln U + h\pi, \tag{6.147}$$

where p is the actual inflation rate, g is the rate of increase in productivity, U is the rate of unemployment, π is the expected rate of inflation, and A, β, and h are positive parameters with $h \leq 1$.

IUP 2. The changes in the expected rate of inflation are assumed to be governed by the following differential equations:

$$d\pi = \varphi_{11}(p - \pi)dt, \quad \varphi_{11} > 0, \tag{6.148}$$

where φ_{11} is the speed of adjustment of the expected rate of inflation. From (6.148), we note that when the actual inflation rate exceeds the expected inflation rate, and the expected rate of inflation will increase over time and vice versa.

IUP 3. The level of employment is assumed to be free from a randomly varying monetary policy. This is modeled by the following deterministic differential equation which characterizes the behavior of the rate of employment:

$$dU = -\varphi_{22}(m - p - g)U\,dt, \tag{6.149}$$

where m is the rate of growth of nominal money balances, g is as defined in (6.147), p is as described in (6.147), and φ_{22} is a positive parameter that describes the speed of adjustment of the rate of unemployment.

Observation 6.10.1.

(i) Clearly from (6.147), as the rate of unemployment approaches zero, the actual rate of inflation becomes extremely large, and as the rate of unemployment increases, the actual rate of inflation declines.

(ii) The rate of growth of real money balances is defined to be the difference between the rate of nominal money balances and the actual rate of inflation [], i.e. $m-p$. Therefore, from (6.149), it is clear that when the rate of growth of real money balances exceeds the rate of increase in productivity, the rate of unemployment decreases and vice versa. In the following, using (6.147)–(6.149), we write the deterministic mathematical model in a particular form. This form provides a suitable structure for undertaking the study in a systematic and unified way. From (6.147)–(6.149), we obtain the differential equation

$$dp = [(-\beta\varphi_{22} + h\varphi_{11})p - h\varphi_{11}\pi + \beta\varphi_{22}(m-g)]dt. \tag{6.150}$$

The differential equations (6.148) and (6.150) form the following deterministic system of differential equations:

$$\begin{aligned} d\pi &= \varphi_{11}(p-\pi)dt, \\ dp &= [(-\beta\varphi_{22} + h\varphi_{11})p - h\varphi_{11}\pi + \beta\varphi_{22}(m-g)]dt. \end{aligned} \tag{6.151}$$

The equilibrium state of (6.151) is determined by the system of algebraic equations

$$\begin{aligned} \varphi_{11}(p-\pi) &= 0, \\ (-\beta\varphi_{22} + h\varphi_{11})p - h\varphi_{11}\pi + \beta\varphi_{22}(m-g) &= 0, \end{aligned} \tag{6.152}$$

and its solution is denoted by (p^*, π^*). Moreover, $p^* = \pi^* = m - g$. By setting $x = [x_1, x_2]^T$, where $x_1 = \pi - \pi^*$ and $x_2 = p - p^*$, the system (6.151) can be rewritten as

$$\begin{aligned} dx_1 &= \varphi_{11}(x_2 - x_1)dt, \\ dx_2 &= [(-\beta\varphi_{22} + h\varphi_{11})x_2 - h\varphi_{11}x_1]\,dt, \end{aligned} \tag{6.153}$$

and hence

$$dx = Ax\,dt, \quad x(t_0) = x_0, \tag{6.154}$$

where

$$\begin{bmatrix} -a_{11} & a_{12} \\ a_{21} & -a_{22} \end{bmatrix}, \tag{6.155}$$

with $a_{11} = a_{12} = \varphi_{11}$, $a_{21} = -h\varphi_{11}$, and $a_{22} = \beta\varphi_{22} - h\varphi_{11}$.

The sufficient conditions for the stability of the equilibrium state $(p^*, \pi^*) = (m-g, m-g)$ of the inflation–unemployment mathematical model are presented. We use Lyapunov's second method (Section 6.6). Here, we choose the Lyapunov-like function V, defined by

$$V(x) = d_1|x_1| + d_2|x_2|, \tag{6.156}$$

where d_1 and d_2 are positive real numbers, and compute the derivative of V in (6.89) along the system (6.154), and we obtain

$$d^+_{(6.154)}V(x) \leq [d_1[-a_{11}|x_1| + |a_{12}||x_2|] + d_2[-a_{22}|x_2| + |a_{21}||x_1|]]\,dt.$$

Together with some elementary algebraic rearrangements and simplifications, this gives

$$d^+_{(6.154)}V(x) \leq \mu(A)V(x)dt, \tag{6.157}$$

where $\mu(A)$ is the logarithmic norm (6.105) of the coefficient matrix A in (6.154), defined by

$$\mu(A) = \max_{1\leq j\leq 2}\left\{-a_{jj} + d_j^{-1}\sum_{\substack{i=1\\ i\neq j}}^{2} d_i|a_{ij}|\right\}. \tag{6.158}$$

Theorem 6.10.1. *Assume that the logarithmic norm of matrix A defined in* (6.158) *satisfies the condition*

$$\mu(A) \leq -\nu, \text{ for some positive number } \nu. \tag{6.159}$$

Then the equilibrium state (p^*, π^*) of the deterministic system (6.154) *is exponentially stable. Moreover, the estimate on the rate of decay of the solution process of* (6.154) *is ν.*

Proof. By employing the differential inequalities, Comparison Theorem 6.6.1, and Illustration 6.6.1, we have

$$V(x(t)) \leq V(x_0)\exp[-\nu(t-t_0)], \quad t \geq t_0,$$

which can be rewritten as:

$$d_m\|x(t)\| \leq d_M\|x_0\|\exp[-\nu(t-t_0)], \quad t \geq t_0, \tag{6.160}$$

where d_m and d_M are the minimum and the maximum of $\{d_1, d_2\}$, respectively, and $\|x\| = |x_1| + |x_2|$. It is obvious that the exponential stability of (p^*, π^*) follows from (6.160) in the context of definitions of x_1 and x_2. It is also clear that the estimate on the rate of decay of the solution (6.154) is ν. This completes the proof of the theorem. □

6.11 Linear Stochastic Systems

The development of mathematical models is based on the theoretical and experimental setup, the fundamental laws of science and engineering sciences, and knowledge-based information about dynamic processes. Historically, well-known probabilistic models, namely, the random walk model, Poisson process model, and Brownian motion model, are used as the basis for developing the mathematical models for dynamic processes in science and engineering, thus bringing the existing model-building tool to the classrooms of the 21st century [54]. As a result of this, states of the systems are affected by random environmental perturbations. Hence, the randomness in the mathematical description can occur through [32, 51, 75, 79, 126]: (i) initial and boundary data due to measurement errors, (ii) inherent randomness due to ignorance or uncertainties about systems, and (iii) random forcing processes. For the sake of simplicity, we just present a mathematical problem formulation.

Illustration 6.11.1 (economic dynamics). Let us consider Illustration 2.2.1 [54]. By utilizing all the notations, definitions, and conditions outlined in **Random Walk Model 2.1.1**, we modify the conditions RWM 3 and RWM 4 to incorporate the dependence of the initial state x, and specify the success probability p. The simplicity of these conditions and the specification of p are:

(c*) For $x(t)\triangle x > 0$, it is assumed that the price/value is either increased by $x(t)\triangle x$ (the positive increment in the change of price/value "success") or decreased by $x(t)\triangle x$ (the negative increment in the change of price/value "failure"). We refer to $x(t)\triangle x$ as the microscopic/local experimentally or knowledge-based observed increment per impact on the price/value of an entity at a time t, where E is a collection of economic entities, like an asset, information, product, and service.

(d*) Thus, there is a constant random change Z of magnitude $x(t)\triangle x$ in the price/value of the entity on each subinterval of length τ.

(e) We further assume that

$$p = \frac{1}{2} + \left(\frac{C}{4D}\right) x(t)\triangle x \text{ and } q = \frac{1}{2} - \left(\frac{C}{4D}\right) x(t)\triangle x.$$

By imitating the argument used in **Random Walk Model 2.1.1**, we arrive at

$$x(t + \triangle t) - x(t) \approx Cx(t)\triangle t + \sqrt{2D}\, x(t)\triangle w(t),$$

which implies that

$$dx(t) = Cx(t)\, dt + \sigma x(t)\, dw(t). \tag{6.161}$$

Here, we assume that C and $\sigma = \sqrt{2D}$ are constants. Under the above conditions, $Cx(t)$ is called a measure of the average/expected rate of change of the price, and we refer to C as the specific rate (per capita growth/decay rate) of change of the price/value of the asset/information/product/service at the time t. $\sigma x(t)$ ($\sigma^2 = 2D$) is called the volatility, which measures the standard deviation of the rate of change of the price, and we call σ the random (unpredictable/stochastic/nonanticipated) specific rate of change (per capita growth/decay rate under random fluctuations) of the price/value of the asset/information/product/service at a time over an interval of small length $\triangle t = dt$.

From the elementary technique [54] of finding a solution to the Itô–Doob-type stochastic differential equation (6.161), it is clear that

$$x(t) = \exp\left[\left(C - \frac{1}{2}\sigma^2\right)t + \sigma(w(t))\right] \tag{6.162}$$

is a fundamental solution to (6.161).

Example 6.11.1. Let us consider an elementary Itô–Doob-type stochastic differential equation:

$$dx = 2x\,dw(t). \tag{6.163}$$

From (6.162), we have

$$x(t) = \exp[-2t + 2w(t)] \tag{6.164}$$

as a fundamental solution to (6.163).

Observation 6.11.1. The mean of the stochastic differential equation corresponding to (6.163) is

$$dm = 0\,dt,$$

and its fundamental solution is $m(t) = 1$. One can compare $m(t)$ with $x(t)$ in (6.164).

Example 6.11.2. Let us consider an elementary Itô–Doob-type stochastic differential equation

$$dx = x\,dt + 2x\,dw(t). \tag{6.165}$$

From the elementary technique [54] to finding a solution to (6.165), it is clear that

$$x(t) = \exp[-t + 2w(t)] \tag{6.166}$$

is a fundamental solution to (6.165).

Observation 6.11.2. The mean of the stochastic differential equation corresponding to (6.165) is

$$dm = m\,dt, \tag{6.167}$$

and its fundamental solution is

$$m(t) = \exp\left[t\right]. \tag{6.168}$$

Again, one can compare $m(t)$ with $x(t)$ in (6.166). From the strong law of large numbers, we have

$$\lim_{t\to\infty} \frac{w(t)}{t} = 0,$$

almost certainly and hence $x(t)$ tends to 0 as t tends to infinity.

Example 6.11.3. Let us consider an elementary system of two Itô–Doob-type stochastic differential equations:

$$\begin{bmatrix} dx_1 \\ dx_2 \end{bmatrix} = \begin{bmatrix} -\dfrac{1}{2} & 0 \\ 0 & -\dfrac{1}{2} \end{bmatrix} \begin{bmatrix} x_1 \\ x_2 \end{bmatrix} dt + \begin{bmatrix} 0 & -1 \\ 1 & 0 \end{bmatrix} \begin{bmatrix} x_1 \\ x_2 \end{bmatrix} dw(t). \tag{6.169}$$

The fundamental matrix solution Φ_{s} [54] to (6.169) is given by

$$\Phi_{\mathrm{s}}(t) = \begin{bmatrix} \cos w(t) & -\sin w(t) \\ \sin w(t) & \cos w(t) \end{bmatrix}. \tag{6.170}$$

Observation 6.11.3. The mean of the differential equation corresponding to (6.169) is

$$\begin{bmatrix} dm_1 \\ dm_2 \end{bmatrix} = \begin{bmatrix} -\dfrac{1}{2} & 0 \\ 0 & -\dfrac{1}{2} \end{bmatrix} \begin{bmatrix} m_1 \\ m_2 \end{bmatrix} dt. \tag{6.171}$$

The fundamental matrix solution Φ_{d} to (6.171) is given by

$$\Phi_{\mathrm{d}}(t) = \begin{bmatrix} \exp\left[-\dfrac{1}{2}t\right] & 0 \\ 0 & \exp\left[-\dfrac{1}{2}t\right] \end{bmatrix}. \tag{6.172}$$

We note that: (a) the solution to (6.171) is nonoscillatory; (b) in fact, the trivial solution is exponentially asymptotically stable; (c) furthermore, the equilibrium state $x(t) \equiv 0$ is an attractor. On the other hand, (a) the solution process of (6.169) is oscillatory; (b) moreover, solutions are almost certainly bounded; (c) in fact, the stochastic perturbations have destroyed the exponential asymptotic stability property of the corresponding deterministic system; (d) randomness has generated an orbit.

Observation 6.11.4. For further, detailed study of several examples, we refer to the forthcoming introductory-level book entitled *An Introduction to Differential Equations: Stochastic Modeling, Methods and Analysis* [54].

6.12 Notes and Comments

The content of this chapter provides a short and limited summary of many types of attributes of dynamic processes that can be incorporated in mathematical modeling. This is exhibited by a limited usage of mathematical analysis. Again, our citations are very limited but well-focused. Section 6.2 briefly outlines the fundamentals of model validation of nonlinear, nonstationary dynamic processes. These results are based on class notes of the second author and existing graduate-level books [7, 15, 23, 42, 74, 75, 77, 83, 84]. Section 6.3 contains the most updated version of the method of variation of parameters for nonlinear and nonstationary systems [2, 42, 75, 82]. The generalized method of variation of parameters adapted from Ref. 56 appears in Section 6.4. For its further extensions, see Refs. 76 and 82. The formulation of basic differential inequalities, the Lyapunov function and comparison results of Sections 6.5 and 6.6 are based on Ref. 75,83,84 and 126. The ideas and results regarding the variational comparison method in Section 6.7 stem from Ref. 56. For further extensions, see Refs. 70–72,82. An emerging area of research, namely modeling of hybrid dynamic systems, is introduced in Section 6.8. A brief presentation, of the scope of hereditary effects exhibited in Section 6.9 is adapted from Refs. 61,66 and 78. For further details, see Refs. 60,61,63 and 65. In Section 6.10, an illustration of a qualitative property, namely stability of the equilibrium state of the inflation–unemployment process, is adapted from Ref. 74. For additional related work, see Refs. 26,100–103. Finally, the motivation for the study of stochastic systems [54] is illustrated in Section 6.11.

Bibliography

[1] Ackerman, E. (1962). *Biophysical Science* (Prentice-Hall, Englewood Cliffs).

[2] Alekseev, V. M. (1961). An estimate for the perturbations of the solutions of ordinary differential equations, *Vestnik Moskov. Univ. Ser. I Mat. Meh.* **2**, pp. 28–36.

[3] Arrow, K. J. and Hahn, F. H. (1971). *General Competitive Analysis*, No. 6, Mathematical Economics Texts (Holden-Day, San Francisco).

[4] Auer, J. W. (1991). *Linear Algebra with Applications* (Prentice-Hall Canada, Scarborough, Ontario).

[5] Bailey, N. T. J. (1957). *The Mathematical Theory of Epidemics* (Hafner, New York).

[6] Batschelet, E. (1971). *Introduction to Mathematics for Life Scientists* (Springer-Verlag, Berlin, New York).

[7] Bellman, R. and Cooke, K. L. (1963). *Differential–Difference Equations* (Academic, New York), a report prepared for United States Air Force Project Rand.

[8] Blaxter, K. L., Graham, N. M. and Wainman, F. W. (1956). Some observations on the digestibility of food by sheep and on related problems, *Brit. J. Nutr.* **10**, 2, pp. 69–91.

[9] Bolie, V. W. (1961). Coefficients of normal blood glucose regulation, *J. Appl. Physiol.* **16**, 5, pp. 783–788.

[10] Braun, M. (1993). *Differential Equations and Their Applications: An Introduction to Applied Mathematics*, Vol. 11, Texts in Applied Mathematics, 4th ed. (Springer-Verlag, New York).

[11] Burghes, D. N. and Borrie, M. S. (1981). *Modelling with Differential Equations*, Mathematics & Its Applications (Ellis Horwood, Chichester).

[12] Burghes, D. N. and Downs, A. M. (1975). *Modern Introduction to Classical Mechanics & Control*, Mathematics & Its Applications (Ellis Horwood, Chichester).

[13] Burley, D. M. (1976). Mathematical model of a kidney machine, *Math. Spectr.* **8**, 3, pp. 69–74.

[14] Chandra, J. and Ladde, G. S. (2004). Stability analysis of stochastic hybrid systems, *Int. J. Hybrid Syst.* **4**, pp. 179–198.

[15] Coddington, E. A. and Levinson, N. (1955). *Theory of Ordinary Differential Equations* (McGraw-Hill, New York, Toronto, London).

[16] Cody, M. L. and Diamond, J. M. (eds.) (1975). *Ecology and Evolution of Communities* (Belknap Press of Harvard University Press).

[17] Cope, F. W. (1970). The solid-state physics of electron and ion transport in biology, *Adv. Biol. Med. Phys.* **13**, pp. 1–42.

[18] Cope, F. W. (1976). Derivation of the Weber–Fechner law and the Loewenstein equation as the steady-state response of an Elovich solid state biological system, *Bull. Math. Biol.* **38**, 2, pp. 111–118.

[19] Coppel, W. A. (1965). *Stability and Asymptotic Behavior of Differential Equations* (D. C. Heath, Boston, Massachusetts).

[20] Crow, J. F. and Kimura, M. (1970). *An Introduction to Population Genetics Theory* (Harper and Row, New York).

[21] Defares, J. G. and Sneddon, I. N. (1960). *An Introduction to the Mathematics of Medicine and Biology* (North-Holland, Amsterdam).

[22] Diem, K. (ed.) (1962). *Scientific Tables*, 6th ed. (Geigy Pharmaceuticals).

[23] Driver, R. D. (1977). *Ordinary and Delay Differential Equations*, Vol. 20, Applied Mathematical Sciences (Springer-Verlag, New York).

[24] Edwards, C. H. and Penney, D. E. (2000). *Differential Equations: Computing and Modeling*, 2nd ed. (Prentice-Hall, New Jersey).

[25] Frank, O. (1899). Die Grundform des Arterielle Pulses, *Z. Biol.* **37**, pp. 483–526; a translation appears in *J. Mol. Cell Cardiol.* **22** (1990), 253–277.

[26] Friedman, M. (1968). The role of monetary policy, *Am. Econ. Rev.* **58**, 1, pp. 1–17.

[27] Fulford, G., Forrester, P. and Jones, A. (1997). *Modelling with Differential and Difference Equations*, Vol. 10, Australian Mathematical Society Lecture Series (Cambridge University Press).

[28] Gandolfo, G. (1980). *Economic Dynamics: Methods and Models*, Vol. 16, Advanced Textbooks in Economics, 2nd ed. (North-Holland, Amsterdam).

[29] Gavalas, G. R. (1968). *Nonlinear Differential Equations of Chemically Reacting Systems*, Vol. 17, Springer Tracts in Natural Philosophy (Springer-Verlag, New York).

[30] Gerasimov, Y., Dreving, V., Eremin, E., Kiselev, A., Lebedev, V., Panchenkov, G. and Shlygin, A. (1974a). *Physical Chemistry*, Vol. 1 (Mir, Moscow), transl. from the Russian by G. Leib.

[31] Gerasimov, Y., Dreving, V., Eremin, E., Kiselev, A., Lebedev, V., Panchenkov, G. and Shlygin, A. (1974b). *Physical Chemistry*, Vol. 2 (Mir, Moscow), transl. from the Russian by G. Leib.

[32] Gīhman, Ĭ. Ī. and Skorohod, A. V. (1972). *Stochastic Differential Equations* (Springer-Verlag, New York), transl. from the Russian by K. Wickwire, Ergebnisse der Mathematik und ihrer Grenzgebiete, Band 72.

[33] Giordano, F. R., Fox, W. P., Horton, S. B. and Weir, M. D. (2009). *A First Course in Mathematical Modeling*, 4th ed. (Brooks/Cole, Cengage Learning, Belmont, California).

[34] Goel, N. S. and Richter-Dyn, N. (1974). *Stochastic Models in Biology* [Academic (subsidiary of Harcourt Brace Jovanovich), New York, London].

[35] Goldberg, J. L. and Schwartz, A. J. (1972). *Systems of Ordinary Differential Equations: An Introduction* (Harper & Row, New York).

[36] Goodman, L. A. (1967). On the age-sex composition of the population that would result from given fertility and mortality conditions, *Demography* **4**, 2, pp. 423–441.

[37] Gowen, J. W. (1964). Effects of x-rays of different wavelengths on viruses. In: O. Kempthorne, T. A. Bancroft, J. W. Gowen and J. L. Lush (eds.), *Statistics and Mathematics in Biology* (Hafner, New York), Chap. 39, pp. 495–510.

[38] Gutfreund, H. (1972). *Enzymes: Physical Principles* (Wiley-Interscience, London, New York).

[39] Guyton, A. C. (1971). *Basic Human Physiology: Normal Function and Mechanisms of Disease* (W. B. Saunders, Philadelphia, Pennsylvania).

[40] Halliday, D., Resnick, R. and Walker, J. (2001). *Fundamentals of Physics*, 6th ed. (John Wiley & Sons, New York).

[41] Halmos, P. R. (1960). *Naive Set Theory*, The University Series in Undergraduate Mathematics (D. Van Nostrand, Princeton, New Jersey, Toronto, London, New York).

[42] Hartman, P. (1964). *Ordinary Differential Equations* (John Wiley & Sons, New York).

[43] Hirschfelder, R. and Hirschfelder, J. (1991). *Introduction to Discrete Mathematics* (Brooks/Cole, Pacific Grove, California).

[44] Homans, G. C. (1951). *The Human Group* (Routledge & K. Paul, London).

[45] Huxley, J. S. (1932). *Problems of Relative Growth* (Dial, New York).

[46] Jacquez, J. A. (1985). *Compartmental Analysis in Biology and Medicine*, 2nd ed. (University of Michigan Press, Ann Arbor).

[47] Johnson, F. H., Eyring, H. and Stover, B. J. (1974). *The Theory of Rate Processes in Biology and Medicine* (John Wiley & Sons, New York).

[48] Kendall, D. G. (1949). Stochastic processes and population growth, *J. R. Stat. Soc. Ser. B* **11**, pp. 230–264.

[49] Kimura, M. and Ohta, T. (1971). *Theoretical Aspects of Population Genetics* (Princenton University Press).

[50] Kirby, R. D., Ladde, A. G. and Ladde, G. S. (2010). Energy function method for solving nonlinear differential equations, *Dynam. Syst. Appl.* **19**, 2, pp. 335–352.

[51] Kloeden, P. E. and Platen, E. (1992). *Numerical Solution of Stochastic Differential Equations*, Vol. 23, Applications of Mathematics (Springer-Verlag, Berlin, New York).

[52] Ladde, A. G. (1992). Selecting the best baseball cards for investment, *Math. Comput. Model.* **16**, 10, pp. 135–142.

[53] Ladde, A. G. and Ladde, G. S. (2010). Determinant functions and applications to stochastic differential equations, *Commun. Appl. Anal.* **14**, 3–4, pp. 409–433.

[54] Ladde, A. G. and Ladde, G. S. (2012). *An Introduction to Differential Equations, Vol. II: Stochastic Modeling, Methods and Analysis* (The World Scientific Publishing Company, Singapore).

[55] Ladde, G. S. (1974). *An Introduction to Biomathematics I* (State University of New York at Potsdam), lecture notes.

[56] Ladde, G. S. (1975). Variational comparison theorem and perturbations of nonlinear systems, *Proc. Am. Math. Soc.* **52**, pp. 181–187.

[57] Ladde, G. S. (1976a). Cellular systems. I. Stability of chemical systems, *Math. Biosci.* **29**, 3–4, pp. 309–330.

[58] Ladde, G. S. (1976b). Cellular systems. II. Stability of compartmental systems, *Math. Biosci.* **30**, 1–2, pp. 1–21.

[59] Ladde, G. S. (1976c). Competitive processes and comparison differential systems, *Trans. Am. Math. Soc.* **221**, 2, pp. 391–402.

[60] Ladde, G. S. (1976d). Stability of model ecosystems with time-delay, *J. Theor. Biol.* **61**, 1, pp. 1–13, http://www.sciencedirect.com/science/article/pii/0022519376900990

[61] Ladde, G. S. (1976/77). Competitive processes. I. Stability of hereditary systems, *Nonlin. Anal.* **1**, 6, pp. 607–631.

[62] Ladde, G. S. (1977a). Logarithmic norm and stability of linear systems with random parameters, *Int. J. Syst. Sci.* **8**, 9, pp. 1057–1066.

[63] Ladde, G. S. (1977b). Stability of time-delay compartmental systems. In: X. J. R. Avula (ed.), *Proc. First Int. Conf. Mathematical Modeling* (St. Louis, Missouri, 1977), Vol. V (Univ. Missouri–Rolla), pp. 1613–1622.

[64] Ladde, G. S. (1977c). Stability technique and thought provocative dynamical systems. In: V. Lakshmikantham (ed.), *Nonlinear Systems and Applications Proc. Int. Conf., Univ. Texas, Arlington, Texas, 1976* (Academic, New York), pp. 211–218.

[65] Ladde, G. S. (1978). Time lag versus stability, *IEEE Trans. Autom. Control* **AC-23**, 1, pp. 84–85.

[66] Ladde, G. S. (1979a). Stability and oscillations in single-species processes with past memory, *Int. J. Syst. Sci.* **10**, 6, pp. 621–647, http://dx.doi.org/10.1080/00207727908941607

[67] Ladde, G. S. (1979b). Stability technique and thought provocative dynamical systems. II. In: V. Lakshmikantham (ed.), *Applied Nonlinear Analysis Proc. Third Int. Conf., Univ. Texas, Arlington, Texas, 1978* (Academic, New York), pp. 215–218.

[68] Ladde, G. S. (1981). Competitive processes and comparison differential systems. II, *J. Math. Phys. Sci.* **15**, 5, pp. 435–454.

[69] Ladde, G. S. (2001). Problem solving process, *Bull. Marathwada Mathematical Soc.* **2**, pp. 90–104.

[70] Ladde, G. S. (2005a). Hybrid dynamic inequalities and applications, *Dynam. Syst. Appl.* **14**, 3–4, pp. 481–513.

[71] Ladde, G. S. (2005b). Variational comparison method and stochastic time series analysis. In: R. Arumuganathan and R. Nadarajan (eds.), *Proc. Third Natl. Conf. Mathematical and Computational Models (NCMCM 2005)* (Allied, New Delhi), pp. 16–40.

[72] Ladde, G. S. (2008). Stochastic systems: a class of hybrid systems, *Abstract of Papers Presented to the American Mathematical Society* **29**, 1, p. 173.

[73] Ladde, G. S. (2011a). Competitive cooperative process in biological, engineering, medical, physical and social sciences. In progress.

[74] Ladde, G. S. (2011b). Stochastic stability of inflation–unemployment process. Lecture notes.

[75] Ladde, G. S. and Lakshmikantham, V. (1980). *Random Differential Inequalities*, Vol. 150, Mathematics in Science and Engineering [Academic (Harcourt Brace Jovanovich), New York].

[76] Ladde, G. S., Lakshmikantham, V. and Leela, S. (1976). A technique in perturbation theory, *Rocky Mt. J. Math.* **6**, 1, pp. 133–140.

[77] Ladde, G. S., Lakshmikantham, V. and Vatsala, A. S. (1985). *Monotone Iterative Techniques for Nonlinear Differential Equations*, Vol. 27, Monographs, Advanced Texts and Surveys in Pure and Applied Mathematics [Pitman (Advanced Publishing Program), Boston, Massachusetts].

[78] Ladde, G. S., Lakshmikantham, V. and Zhang, B. G. (1987). *Oscillation Theory of Differential Equations with Deviating Arguments*, Vol. 110, Monographs and Textbooks in Pure and Applied Mathematics (Marcel Dekker, New York).

[79] Ladde, G. S. and Sambandham, M. (2004). *Stochastic Versus Deterministic Systems of Differential Equations*, Vol. 260, *Monographs and Textbooks in Pure and Applied Mathematics* (Marcel Dekker, New York).

[80] Ladde, G. S. and Sathananthan, S. (1992). Stability of Lotka–Volterra model, *Math. Comput. Model.* **16**, 3, pp. 99–107.

[81] Ladde, G. S. and Šiljak, D. D. (1983). Multiparameter singular perturbations of linear systems with multiple time scales, *Automatica — J. IFAC* **19**, 4, pp. 385–394, http://dx.doi.org/10.1016/0005-1098(83)90052-3

[82] Lakshmikantham, V. and Deo, S. G. (1998). *Method of Variation of Parameters for Dynamic Systems*, Vol. 1, Series in Mathematical Analysis and Applications (Gordon and Breach, Amsterdam).

[83] Lakshmikantham, V. and Leela, S. (1969a). *Differential and Integral Inequalities: Theory and Applications. Vol. I: Ordinary Differential Equations*, Vol. 55-I, Mathematics in Science and Engineering (Academic, New York).

[84] Lakshmikantham, V. and Leela, S. (1969b). *Differential and Integral Inequalities: Theory and Applications. Vol. II: Functional, Partial, Abstract, and Complex Differential Equations*, Vol. 55-II, Mathematics in Science and Engineering (Academic, New York).

[85] Larson, E., Roland, Hostetler, R. P. and Edwards, B. H. (1990). *Calculus with Analytic Geometry*, 4th ed. (D. C. Heath, Lexington, Massachusetts), with the assistance of D. E. Heyd.

[86] Latham, J. L. (1969). *Elementary Reaction Kinetics*, 2nd ed. (Butterworths, London).

[87] Leithold, L. (1976). *The Calculus, with Analytic Geometry*, 3rd ed. (Harper & Row, New York).

[88] Levins, R. (1970). Extinction. In: M. Gerstenhaber (ed.), *Some Mathematical Questions in Biology I*, Vol. 2, Lectures on Mathematics in the Life Sciences (American Mathematical Society), pp. 75–108.

[89] Loewenstein, W. R. (1961). Excitation and inactivation in a receptor membrane*, *Ann. N.Y. Acad. Sci.* **94**, 2, pp. 510–534, http://dx.doi.org/10.1111/j.1749-6632.1961.tb35556.x

[90] Lotka, A. J. (1958). *Elements of Mathematical Biology* (Dover, New York), formerly published under the title *Elements of Physical Biology.*

[91] Luce, R. D. (1959). *Individual Choice Behavior: A Theoretical Analysis* (Wiley, New York).

[92] Madison, J. (1976). The Constitution of the United States of America (U.S. Government Printing Office, Washington, D.C.).

[93] Mattson, H. F. (1993). *Discrete Mathematics with Applications* (J. Wiley, New York).

[94] May, R. M. (1973). *Stability and Complexity in Model Ecosystems* (Princeton University Press).

[95] McDonald, D. A. (1960). *Blood Flow in Arteries* (Williams & Wilkins, Baltimore, Maryland).

[96] Murdick, R. G. (ed.) (1971). *Mathematical Models in Marketing* (Intext Educational, Scranton, Pennsylvania).

[97] Needham, J. (1934). Chemical heterogony and the groundplan of animal growth, *Biol. Rev.* **9**, 1, pp. 79–109, http://dx.doi.org/10.1111/j.1469-185X.1934.tb00874.x

[98] Nicosia, F. M. (1966). *Consumer Decision Processes: Marketing and Advertising Implications* (Prentice-Hall, Englewood Cliffs, New Jersey).

[99] Pannetier, G. and Souchay, P. (1967). *Chemical Kinetics* (Elsevier, Amsterdam, New York), trans. by H. D. Gesser and H. H. Emond.

[100] Phelps, E. S. (1967). Phillips curves, expectations of inflation and optimal unemployment over time, *Economica* **34**, 135, pp. 254–281.

[101] Phelps, E. S. (1968). Money-wage dynamics and labor-market equilibrium, *J. Pol. Econ.* **76**, 4, pp. 678–711.

[102] Phillips, A. W. (1958). The relation between unemployment and the rate of change of money wage rates in the United Kingdom, 1861–1957, *Economica* **25**, 100, pp. 283–299, http://dx.doi.org/10.1111/j.1468-0335.1958. tb00003.x

[103] Phillips, R. S. (ed.) (1975). *Funk & Wagnalls New Encyclopedia*, Vol. 7 (Funk & Wagnalls).

[104] Pielou, E. C. (1969). *An Introduction to Mathematical Ecology* (Wiley-Interscience, New York).

[105] Poland, J. (1987). The teaching of mathematics: a modern fairy tale? *Am. Math. Monthly* **94**, 3, pp. 291–295, http://dx.doi.org/10.2307/2323400

[106] Pollard, J. H. (1973). *Mathematical Models for the Growth of Human Populations* (Cambridge University Press).

[107] Rainville, E. D. and Bedient, P. E. (1989). *Elementary Differential Equations*, 7th ed. (Macmillan).

[108] Randall, J. E. (1962). *Elements of Biophysics*, 2nd ed. (Year Book Medical Publishers, Chicago).

[109] Rashevsky, N. (1960). *Mathematical Biophysics: Physico-mathematical Foundations of Biology*, 3rd revised ed. (Dover, New York).

[110] Rashevsky, N. (1961). *Mathematical Principles in Biology and Their Applications* (Thomas, Springfield, Illinois).

[111] Rescigno, A. and Beck, J. S. (1972). Compartments. In: R. Rosen (ed.), *Foundations of Mathematical Biology, Vol. II: Cellular Systems (Dedicated to the Memory of Nicolas Rashevsky)* (Academic, New York), pp. 255–322.

[112] Rescigno, A. and Segre, G. (1966). *Drug and Tracer Kinetics* (Blaisdell, Waltham, Massachusetts), transl. from the Italian by P. Ariotti.

[113] Riggs, D. S. (1963). *The Mathematical Approach to Physiological Problems: A Critical Primer* (Williams & Wilkins, Baltimore, Maryland).

[114] Robinson, J. V. and Ladde, G. S. (1982). Feasibility constraints on the elastic expansion of ecosystem models, *J. Theor. Biol.* **97**, 2, pp. 277–287, http://www.sciencedirect.com/science/article/pii/0022519382901047

[115] Rosen, R. (1967). *Optimality Principles in Biology* (Butterworths, London).

[116] Rosen, R. (1970). *Dynamical System Theory in Biology* (Wiley-Interscience, New York).

[117] Rosen, R. (ed.) (1973). *Foundations of Mathematical Biology, Vol. III: Supercellular Systems (Dedicated to the Memory of Nicolas Rashevsky)* (Academic, New York).

[118] Roughgarden, J. (1979). *Theory of Population Genetics and Evolutionary Ecology: An Introduction* (Macmillan, New York).

[119] Rubinow, S. I. (1975). *Introduction to Mathematical Biology* (Wiley, New York).

[120] Rudin, W. (1976). *Principles of Mathematical Analysis*, 3rd ed., International Series in Pure and Applied Mathematics (McGraw-Hill, New York).

[121] Ryan, F. J. (1964). Analysis of populations of mutating bacteria. In: O. Kempthorne, T. A. Bancroft, J. W. Gowen and J. L. Lush (eds.), *Statistics and Mathematics in Biology* (Hafner, New York), Chap. 15, pp. 217–225.

[122] Saaty, T. L. (1968). *Mathematical Models of Arms Control and Disarmament: Application of Mathematical Structures in Politics*, No. 14, Publications in Operations Research (John Wiley & Sons, New York).

[123] Sapirstein, L. A., Vidt, D. C., Mandel, M. J. and Hanusek, G. (1955). Volumes of distribution and clearances of intravenously injected creatinine in the dog, *Am. J. Physiol.* **181**.

[124] Schwabik, Š. (1992). *Generalized Ordinary Differential Equations*, Vol. 5, Series in Real Analysis (World Scientific, New Jersey).

[125] Segel, I. H. (1975). *Enzyme Kinetics: Behavior and Analysis of Rapid Equilibrium and Steady-State Enzyme Systems* (Wiley, New York).

[126] Šiljak, D. D. (1979). *Large-Scale Dynamic Systems: Stability and Structure*, Vol. 3, North-Holland Series in System Science and Engineering (North-Holland, New York).

[127] Simon, H. A. (1957). *Models of Man: Social and Rational; Mathematical Essays on Rational Human Behavior in a Social Setting* (Wiley, New York).

[128] Simpson, G. G., Roe, A. and Lewontin, R. C. (1960). *Quantitative Zoology* (Harcourt, Brace and Co., New York).

[129] Smith, J. M. (1974). *Models in Ecology* (Cambridge University Press).

[130] Solow, R. M. (1956). A contribution to the theory of economic growth, *Q. J. Econ.* **70**, 1, pp. 65–94.

[131] Solow, R. M. (1987). *Growth Theory: An Exposition* (Oxford University Press, New York).

[132] Stevens, B. (1965). *Chemical Kinetics, for General Students of Chemistry* (Chapman & Hall, London).

[133] Teissier, G. (1960). *Relative Growth*, Vol. 1, Chap. 16 (Academic, New York), pp. 537–560.

[134] Teorell, T. (1937). Kinetics of distribution of substances administered to the body, *Arch. Int. Pharmacodyn. Thér.* **57**, pp. 205–240; article appears in two parts.

[135] van Egmond, A. A. J., Groen, J. J. and Jongkees, L. B. W. (1949). The mechanics of the semicircular canal, *J. Physiol.* **110**, 1–2, pp. 1–17.

[136] Waltman, P. (1974). *Deterministic Threshold Models in the Theory of Epidemics*, Vol. 1, Lecture Notes in Biomathematics (Springer-Verlag, Berlin).

[137] Whittaker, R. H. and Levin, S. A. (eds.) (1976). *Niche Theory and Applications* (John Wiley & Sons).

Index